W9-CEI-538

SCIENCE
1001

153
152
151
150
149
148
147
146
145
144
103
102
101
100
99
98
97
96
95

SCIENCE 1001

ABSOLUTELY EVERYTHING THAT MATTERS IN SCIENCE IN 1001 BITE-SIZED EXPLANATIONS

DR PAUL PARSONS

FIREFLY BOOKS

CONTENTS

INTRODUCTION

SCIENCE ASKS THE FUNDAMENTAL QUESTIONS about why the world is the way it is. And over the course of history the world's greatest scientific minds have provided many of the answers. It's because of science that we now know the Universe began around 13.7 billion years ago in a superheated fireball known as the Big Bang. It's because of science that we know all living things in the world encode their biological blueprint on a chemical molecule known as DNA, which serves as the vehicle by which we pass our traits and characteristics down to our offspring. It's because of science that we know our planet was once ruled by supersize reptiles and that one day 65 million years ago they were wiped out by the impact of a vast comet or asteroid with the Earth. And it's because of science that we have computers in our homes that are more powerful than the fastest university supercomputers of just ten years ago.

Science probably encompasses the biggest areas of human intellectual endeavour. Fitting, then, that it should get billed alongside a suitably big number – 1001. Yet when it comes to dividing up the whole of scientific progress from the last 5,000 or so years into 1001 bite-size nuggets, suddenly it seems woefully small. That's about one entry in this book for every five years – fine for the Dark Ages perhaps, but something of a squeeze for years like 1996 when we had Dolly the sheep (the world's first cloned mammal), claims to have found life on Mars (as fossilized bugs inside a meteorite), and the Deep Blue chess computer won its first game against Garry Kasparov (who was at the time the reigning world chess champion).

In fact, if this book was arranged chronologically I'd probably have run out of space somewhere in the middle of the Renaissance. So instead you'll find the pages within organized by subject. I've taken modern science as it's currently understood and partitioned it into ten major sections: physics; chemistry; biology; the Earth; space; health and medicine; social science;

knowledge, information and computing; applied science; and the future. Each of these categories then splits down again into subsections on key topic areas, and within each you'll find, on average, 12 entries to cover that particular topic. So, physics has subsections on heat, relativity and quantum theory, to name a few. And, for example, the quantum theory subsection has entries on ideas such as Schrödinger's cat, the uncertainty principle, and the 'many worlds' interpretation of quantum theory.

My aim as a writer was to combine the breadth of a reference book – for example, a dictionary of science – with the accessibility and sense of fun that you get from a piece of popular science writing. That was my guiding principle in turning what could have easily been 'Science 100,001' into what you have here. Out went the abstruse and the arcane – topics that the lay reader will neither need nor care about. And what was left I condensed, distilled and clarified into what's hopefully the ultimate balancing act between readability and comprehensiveness.

The entries are written in plain and concise English. Often they are self-contained, but when they aren't there are references in **bold** type to other entries and subsections that either aid understanding or provide further information. If you aren't sure which subsection the entry you are looking for is in, there's a comprehensive index to guide you straight to it. Subsections themselves, meanwhile, are written with as much continuity as possible, so if you're after the complete overview of quantum theory, then that subsection of physics will work as an essay that can be read from start to finish.

Science 1001 is a big book about the biggest subject. I hope you enjoy it.

Paul Parsons

PHYSICS

PHYSICS IS THE MOST FUNDAMENTAL of all the sciences. It governs the behaviour of matter and energy at the most elementary level, from the quarks and smaller subatomic particles that make up the everyday world to the exotic forms of mass and energy that pervaded the universe shortly after its creation in the Big Bang, and which may still lurk out in space today.

Traditionally, physics was divided into disciplines such as mechanics (the science of the movement of physical bodies under the application of forces), as well as heat, light, sound, electricity and magnetism. However, the 20th century brought a revolution in our understanding of physical law with the

MENTUM • NEWTON'S LAWS OF MOTION
FRICTION • DYNAMICS and KINEMATICS
AL DYNAMICS • CENTRIFUGAL FORCE •
IPLE • KEPLERS LAWS • TEMPERATURE
RMAL EXPANSION • CONDUCTION AND
YNAMIC EQUILIBRIUM • ENTROPY • LAWS
NICS • BLACK BODY RADIATION • HEAT
• HOOKE'S LAW • LATENT HEAT • PHASE
E POINT • PLASMA PHYSICS • SURFACE
NCIPLE • VISCOSITY • NEWTONIAN FLUIDS
S • BERNOULLI PRINCIPLE • TURBULENCE
ORY • SOUND WAVES • STANDING WAVES
DOPPLER EFFECT • ELECTRIC CHARGE

discovery of quantum theory (a radical new take on the behaviour of small bodies) and relativity (an equally radical take on the behaviour of objects moving at close to the speed of light). The combination of these disciplines has led to 'quantum field theory', which has brought about a dramatic shift in the way we view the fast-moving subatomic particles that carry the fundamental forces of nature in the universe.

Quantum field theory is now revealing a new unity underpinning physics, where all of the forces of nature are revealing themselves to be just different aspects of the same fundamental entity. Scientists hope this may soon lead them to the holy grail of physics: an all-encompassing 'theory of everything'.

MECHANICS

Speed and acceleration

The rate an object is moving at is given by its speed – just the distance it's travelled divided by the time taken to do so. Accordingly, speed is measured in distance per unit time – e.g., kilometres or miles per hour (km/h or mph) or metres per second (m/s).

Acceleration is the rate at which speed is changing. It's given by the change in speed divided by the time interval. So if it takes a sprinter five seconds to go from standing still to running at 10m/s, then their average acceleration is 2m/s per second – which is usually written as $2m/s^2$.

Inertia and momentum

Inertia is the resistance of a body to move, normally measured by the body's mass. The more inertial mass a body has, the harder it is to move it – that's why pushing a shopping cart is easy but you'll struggle to push a truck. A body's momentum is given by its speed multiplied by its mass, a measure of the impetus a moving body has. And it's why getting hit by a truck hurts a lot more than getting hit by a shopping cart moving at the same speed.

Newton's laws of motion

The 17th-century physicist Sir Isaac Newton came up with three laws that encapsulate the behaviour of moving bodies. The first law sums up inertia, saying that in the absence of any external force, a body will remain at rest or continue in its state of motion in a straight line at constant speed.

What does Newton mean by an 'external force'? He clarifies this in the second law, namely that the external force acting on a body is given by the body's mass multiplied by the acceleration it experiences. Put the numbers in and you can see that you need to apply more force to accelerate a ten-ton truck than to accelerate a 15kg shopping cart by the same amount. Force can be thought of as the rate of change of momentum and is measured in newtons, after Sir Isaac.

Newton's third law says that for every action (that is, every force) there is an equal and opposite reaction (in other words, a force pushing back). It's the reason rockets fly – the rocket exhaust is a jet of hot gas that's forced downwards (action) and this sends the rocket upwards (reaction).

Conservation laws

A key idea in physics is conserved quantities – quantities that cannot change as a physical system evolves. Momentum is a good example. The total momentum before an event – say the collision of two billiard balls – must be the same as the total momentum afterwards. Conservation laws like this enable scientists to predict how

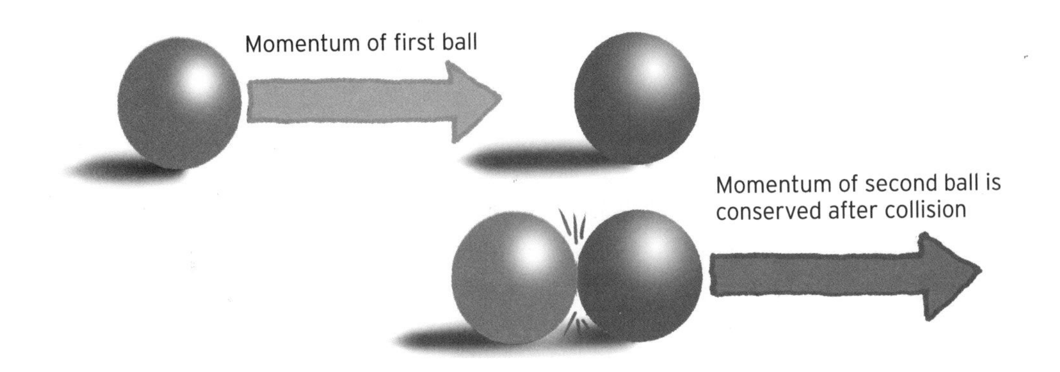

a system will behave. Going back to the billiard balls, let's say one ball comes in, strikes the second and then stops dead. Conservation of momentum means that the second ball must then leave with the exact same momentum the first ball had before the collision. If they're both the same mass it will be travelling at the same speed.

Work and energy

Work and energy are cornerstones of science. To a scientist, work is defined as the force applied to an object multiplied by the distance it moves in response. On the other hand, a system's 'energy' can be thought of as its ability to do work. For example, a truck's engine is able to liberate the chemical energy stored in fuel and then use this energy to do work on the truck and make it move. Energy and work are both measured in joules (J) (after 19th-century physicist James Joule). Roughly speaking, 1J is enough to lift a 100g mass 1 metre above the Earth's surface.

Like momentum, energy obeys a conservation law – it can neither be created nor destroyed. A moving truck has kinetic energy – that is, energy of motion – which has come from the fuel's chemical energy. Likewise, when the truck needs to stop the driver applies the brakes, which get hot, converting the truck's kinetic energy into a different form – heat energy. Energy comes in many different forms – including sound, gravitational, electrical, magnetic and nuclear.

Friction

In an ideal world, all of the energy in a physical system would be available to do useful work. But the real world isn't like that. Friction is the resistive force that tends to slow the motion of physical objects. Two surfaces sliding against each other experience a frictional force as microscopic lumps and bumps on the surfaces rub together. Friction is everywhere; even with the best bearings and lubricants, there will be frictional forces between the moving parts in a car's engine, the gear transmission, wheel axles and so on. And energy must be spent overcoming them.

Friction isn't always a bad thing though – it is responsible for the grip that holds car tyres to the road and it allows you to pick things up in your hands. If there was no friction, objects would just slip between your fingers. And it's why rubbing your hands together on a cold day generates much-appreciated warmth.

Dynamics and kinematics

The mathematics describing motion – namely the equations for an object's position, speed and acceleration at any time, without mention of the forces causing them – is known as 'kinematics'. When the forces causing the motion are included, then the correct term is 'dynamics'. Kinematics and dynamics are the main two branches of classical mechanics – the physics of moving bodies.

Principle of least action

Perhaps the most powerful formulation of dynamics is based on what is known as the principle of least action. The basic idea is that physical systems evolve via the most efficient route possible. Balls don't roll uphill, around a bit, over the top and then back down – they roll straight down.

A physicist using this principle first puts together a mathematical expression that takes stock of all the different kinds of energy in a system. Called the 'action', this formula yields different numerical values depending on which path the system actually follows – and each value can loosely be thought of as a measure of how inefficient that path is. A physicist can select the path for which the numerical value of the action is smallest, and then extract the equations of motion describing it. The principle of least action is used to make dynamics and kinematics tractable in complex areas of theoretical physics, such as relativity and quantum theory.

Rotational dynamics

Like objects moving in a straight line, there are laws governing the motion of rotating objects. Speed is replaced by angular speed (the number of angular degrees moved through per second) and momentum is replaced by its rotational analogue, angular momentum.

Like ordinary momentum, angular momentum increases with angular speed and obeys a **conservation law**. But it also increases with the size of the rotating object. That means that if a rotating object suddenly shrinks, then in order to conserve angular momentum it must spin faster. Ice skaters take advantage of this effect, pulling their arms and legs in tight to make them spin faster. Try it for yourself on a swivel chair.

Centripetal force

People often talk about 'centrifugal force' in relation to spinning objects. However, scientists prefer the term 'centripetal force'. Imagine whirling a weight on a string above your head. Centrifugal force is the outward-pointing force that's trying to snap the string and send the weight flying off in a straight line. But it's the force stopping the weight flying off at a tangent that's actually responsible for making it move in a circle – in this case, the tension in the string. And this is the centripetal force.

That said, centrifugal force most definitely exists – as anyone who's been on a fairground ride will attest. But it's best thought of as a reaction (in the sense of **Newton's laws of motion**) to the more fundamental centripetal force.

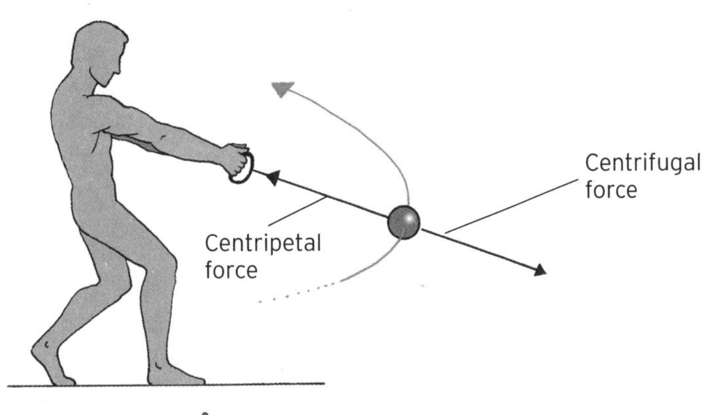

Newtonian gravity

In 1687, Sir Isaac Newton published the first mathematical theory of gravity. Newton's universal theory of gravitation supposed that the gravitational force between two objects increases in direct proportion to both their masses and decreases with the square of the distance between them, linked by a number G – the so-called gravitational constant.

Newton's theory was a remarkable achievement for 17th-century science, a single formula accurately describing phenomena from falling apples to the motion of far-away planets and moons. Newton's law is still an excellent approximation today for calculations involving weak gravitational fields. For stronger fields, however, it has been superseded by Einstein's **general relativity**.

Equivalence principle

In theories of gravity, the equivalence principle states that objects of different mass will fall at the same rate in a gravitational field. The story is that 17th-century scientist Galileo demonstrated this by dropping balls of different weight from the top of the Leaning Tower of Pisa. Astronauts on the Moon conducted their own test by dropping a hammer and a feather – in the absence of air resistance they both fell to the lunar surface in unison. Controlled laboratory experiments have since confirmed the equivalence principle to an accuracy of one part in a trillion.

Newtonian gravity is superficially consistent with the equivalence principle, but Einstein's **general relativity** was the first theory of gravity to encapsulate it completely.

Kepler's laws

The German mathematician and astronomer Johannes Kepler (1571–1630) is credited with discovering the laws governing the motion of **planets** around the **Solar System**. In 1605 Kepler proposed his three laws: the first says that the orbit of every planet is an oval-shaped ellipse with the Sun at one of the foci (an ellipse has two foci, which are analogous to the centre point of a circle); the second law states that a line joining the planet to the Sun sweeps out equal areas in equal times; and the third law is that the square of a planet's orbital period (the time it takes to complete one whole lap of the Sun) is proportional to the long axis of its elliptical orbit raised to the power three.

Amazingly, Kepler made these discoveries before Newtonian gravity had been formulated – even though gravity is the force responsible. This was thanks to the time Kepler had spent working for the Danish astronomer Tycho Brahe. Brahe was renowned for his accurate observations of the positions of the planets – which Kepler used to hone his equations.

HEAT

Temperature and pressure

Thermodynamics is the branch of physics concerned with how energy can be transported and manipulated via heat, and used to carry out useful work. A key property is temperature: heat energy will flow from an area of high temperature to one where the temperature is lower. According to Isaac Newton's law of cooling, the rate of flow is proportional to the temperature difference between the two. So a hot cup of coffee loses heat faster than a lukewarm one.

Pressure is another important variable. A gas heated up inside a container exerts a force on the container walls, but the total force depends on the size of the container. Pressure is just the total force divided by the area of the container walls and is measured in units of newtons per square metre, also known as pascals after 17th-century French mathematician Blaise Pascal.

Kinetic theory

Kinetic theory is a way of explaining the large-scale thermal properties of materials, in particular gases, in terms of the motion of the individual particles – usually **atoms** or **molecules** – they are made up of. Particles inside a gas are all zipping around randomly. The central tenet of kinetic theory is to equate the heat energy of a gas to the sum of the kinetic energies of all these gyrating particles. This means that the hotter a gas gets, the faster, on average, its particles are moving and the harder they are beating against each other and on the walls of the gas's container – so raising the temperature and pressure. The numerical predictions of kinetic theory match exactly with experiments. Kinetic theory also implies that there is a minimum temperature, at which the kinetic energy of the particles reaches zero; this is -273°C (-460°F) – it's impossible for anything to get colder than this.

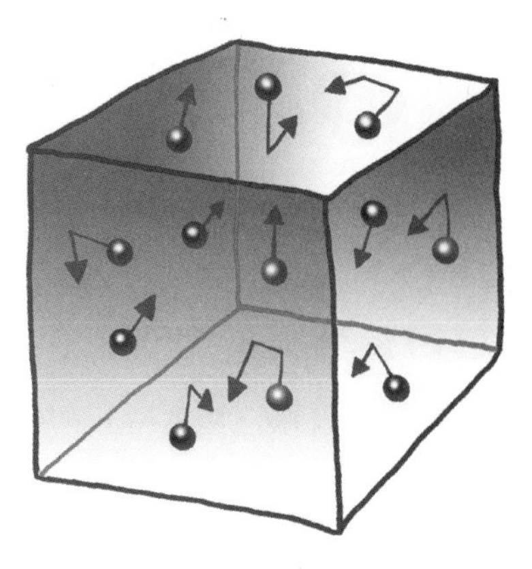

The foundations of kinetic theory were laid by the Swiss mathematician Daniel Bernoulli in 1738 and it was developed further through the 18th and 19th centuries. Scientists were then set on the road towards the even more powerful theory of **statistical mechanics**.

Thermal expansion

Generally speaking, substances tend to increase in volume when they are heated. This property of matter is known as thermal expansion. Kinetic theory explains why it happens – as a substance heats up, collisions between its constituent particles become more violent than when the substance is cooler. Extra recoil from the collisions increases the average distance between the particles, and this is what causes the substance to expand.

Thermal expansion is well understood and can be predicted mathematically; especially useful for engineers designing structures that need to operate in a range of temperatures. For example, bridges incorporate sliding expansion joints so they can endure both the frozen depths of winter and the warmest days of summer without cracking or kinking.

Conduction and convection

Heat energy can take three routes to move from a hot area to a cooler one: conduction, convection and **thermal radiation**. Conduction occurs when hot – and, therefore, fast-moving – particles in a substance collide with those at lower temperature. The collisions transfer kinetic energy to the cooler particles and heat them up, thus spreading heat through the substance.

Convection, on the other hand, can take place only in a liquid or gas. In a hot gas, thermal expansion lowers the gas's density and increases its buoyancy (according to **Archimedes' principle**), and this causes the gas to rise. It's the reason hot-air balloons are able to fly. Meanwhile, cooler gas or liquid sinks for the opposite reason and this can set up the convective cycles that you might see, for example, in a pan of water on a stove.

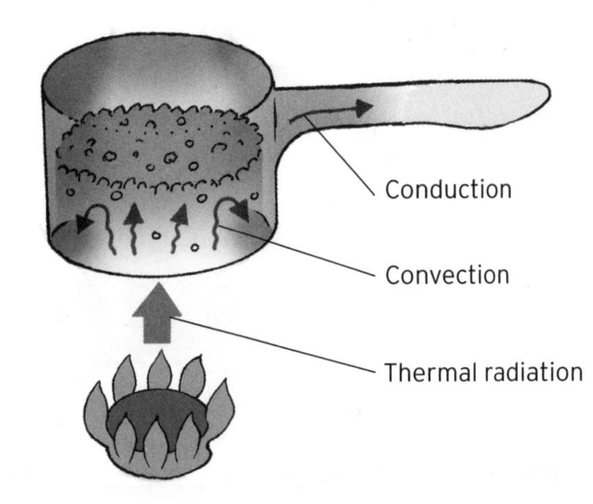

Brownian motion

Particles diffusing in a gas or liquid move with what is known as Brownian motion, after Robert Brown, the physicist who discovered it in 1827. Brown's original observation was of tiny particles in cavities called 'vacuoles' inside pollen grains. Under the microscope, he saw the particles jiggle this way and that as if being buffeted by an unseen force. He later saw the same effect when he examined the motion of dust particles.

Albert Einstein explained Brownian motion, in 1905 saying that the motion was caused by the random kicks the grains received from **atoms** and **molecules** in the air, themselves moving in accordance with kinetic theory. Einstein calculated how far the particles should move with each kick, and his predictions matched with observation. Brown's discovery and Einstein's explanation together became one of the early confirmations of the existence of atoms.

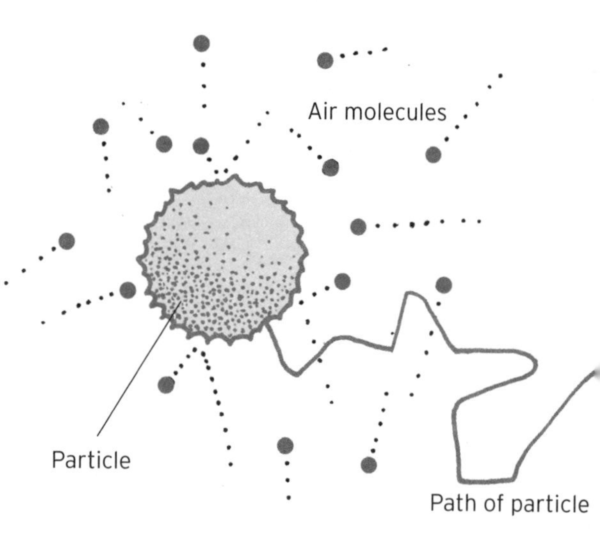

Thermodynamic equilibrium

A system is said to be in thermodynamic equilibrium when its thermodynamic properties, such as temperature, have become constant with time, i.e. when it has ceased to evolve thermally. A simple example is a bucket of ice placed in a warm room – when the ice melts it absorbs heat from the room, raising its temperature and lowering the temperature of the room, until the two converge at one uniform value. This is thermodynamic equilibrium.

Entropy

Entropy is a quantity used in thermodynamics to measure the amount of energy in a system that is available to do useful work. The higher the entropy, the less able the system is to do any work. Work can only be done when a system is out of thermodynamic equilibrium, i.e. thermodynamic equilibrium is a high-entropy state. Imagine a system comprising a heat source and a piston with cold gas inside it. Heat can flow from the source to the piston, expand the cold gas and move the piston arm – so doing work. But if the system is in equilibrium then no heat can flow, and so no work can be done. Another way of thinking of entropy is as the degree of 'disorder' in a system. A tidy desk with everything in neat piles has low entropy; a messy desk with everything strewn evenly across it is in a high-entropy state.

Laws of thermodynamics

Just as **dynamics and kinematics** give mathematical laws governing a mechanical system, so there are four key laws governing the behaviour of a thermodynamic system. The 'zeroth' law (so named because it was formulated after the other three, but considered more fundamental) says that if there are three thermal systems – A, B, and C – and both A and B, and B and C, are in thermodynamic equilibrium, then A and C are also in equilibrium. The first law is a statement of the energy **conservation law**, which says that the change in the total thermodynamic energy of a system is just equal to the heat energy put in minus the work the system does. The second law says that entropy must always increase; in other words, thermodynamic systems inevitably move towards equilibrium and their ability to do useful work thus diminishes. And the third law says that the absolute zero of temperature of -273°C (-460°F), defined by **kinetic theory**, corresponds to a minimum of entropy. Combined with the second law, it means entropy increases with temperature.

Statistical mechanics

An extension of the **kinetic theory** of gases, statistical mechanics takes the properties of individual particles of matter and applies the sophisticated laws of mathematical statistics to draw conclusions about the large-scale, or 'bulk', thermodynamic properties of materials.

For systems of large particles the law of Maxwell-Boltzmann statistical mechanics applies, based on classical kinetic theory. For smaller particles, quantum theory has to be used. These quantum statistical theories are known respectively as Bose-Einstein and Fermi-Dirac statistics, depending on the **quantum spin** of the particles being studied. Statistical mechanics has helped physicists to unravel the internal structure of dead stars, known as **white dwarfs**, and correctly describe the behaviour of thermal radiation given off by hot objects.

Thermal radiation

Stand anywhere near a bonfire and you'll appreciate that hot bodies emit their heat as radiation. In fact, everything with a temperature above absolute zero emits thermal radiation. Calculations using statistical mechanics predict how much energy is emitted at each frequency of the **electromagnetic radiation** spectrum by a hot source. The theory predicts that this is a peaked curve, the wavelength of the peak decreasing as the temperature of the emitter increases.

A poker heated in a fire to hundreds of degrees Celsius/Fahrenheit glows with visible light, usually red or orange and everyday objects emit radiation with a peak in the infrared region of the electromagnetic spectrum – which is why soldiers use infrared goggles to see at night. The difference between thermal radiation and **conduction** or **convection** is that thermal radiation can travel through a vacuum; it is how heat from the Sun crosses the vacuum of space to reach Earth.

Heat capacity

Add heat to a substance and it gets warmer – in scientific parlance, its temperature increases. But the amount of heat energy needed to bring about a temperature rise of, say, 1°C, varies from substance to substance. This is because, according to **kinetic theory**, temperature is a property of the kinetic energy of the atoms or molecules in a substance as they bounce around. But for complex molecules, not all of the energy absorbed gets turned into motion – some, for example, causes vibrations in the internal bonds making up the molecule. Heat capacity is a way of quantifying the amount of heat energy absorbed that gets transferred to kinetic energy in a substance. This will result in a rise of temperature in the substance, measured in joules per degree per kilogram.

MATTER

Solids, liquids and gases

Matter can exist in three major states: solid, liquid and gas, and matter normally moves through this sequence as the temperature increases. Water at less than 0°C (32°F) is a solid (ice), between 0°C (32°F) and 100°C (212°F) a liquid, and above 100°C (212°F) it becomes a gas (steam). This occurs because temperature – or rather, the vigorous motion of the particles in a substance, according to **kinetic theory** – breaks the bonds between **atoms** and **molecules** that keep the substance rigid.

A solid is the most ordered state of matter, its constituent particles held fast in a regimented lattice. Gases occupy the other extreme – in a gas, there is no organization of the atoms or molecules whatsoever and a gas will always expand to fill its container. Liquids sit somewhere between the two – the rigid structure of the solid is gone, but inter-particle forces are still able to hold clumps of atoms or molecules together and maintain some degree of order.

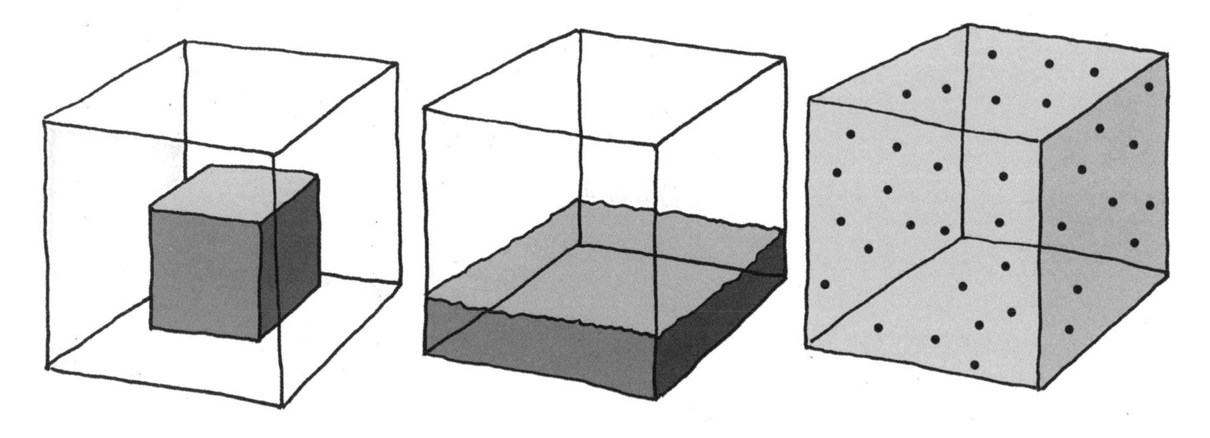

SOLID
Holds shape
Fixed volume

LIQUID
Takes shape of container
Free surface, fixed volume

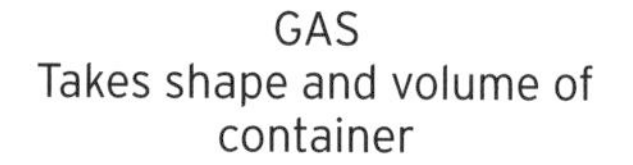

GAS
Takes shape and volume of container

Density

The mass of a substance divided by its volume is known as its density, usually measured in kilograms per cubic metre, though there are other ways of gauging it. One system, called 'specific gravity', compares a substance's density to that of water, which weighs 1,000kg/m^3. Specific gravity (SG) is calculated by dividing the density of the substance by the density of water. So a liquid weighing 1,055kg/m^3 would have an SG of 1.055. SG is used by brewers to assess the amount of sugar that has turned to alcohol in a beverage after fermentation, and also by geologists to determine the density of samples. Density can also be measured in kilograms per litre. Because there are 1,000 litres in a cubic metre, the density of water is 1kg per litre.

Hooke's law

Elasticity is a property of solids that allows them to stretch when subjected to an external force, and then return to their original shape when the force is removed. English physicist Robert Hooke showed in 1678 how the amount that an elastic material stretches by is proportional to the force applied – given by the force multiplied by a number. This number is particular to each material and quantifies how stretchy it is.

However, Hooke's law is only valid up to a point, known as the 'proportional limit'. Stretch a material beyond this and it extends more for each increment of force added. The material still remains elastic though, returning to its original shape when the force is removed. That is, until it reaches the 'elastic limit' point. Then, any extra force causes a permanent deformation. Apply more force again and the material soon breaks.

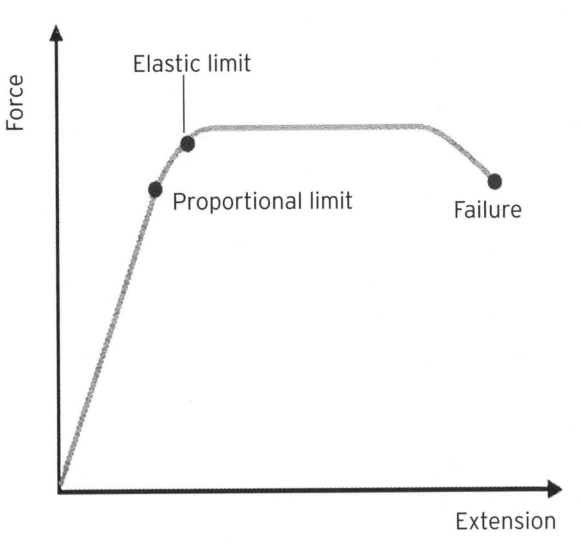

Latent heat

When a substance changes to a more disorganized state as its temperature increases – for example, when a solid melts to become a liquid – it has to absorb extra energy from its surroundings to overcome the attractive forces that are holding it in a cooler, more ordered state. This energy is known as latent heat. To turn water at 95°C (203°F) into steam at 100°C, enough energy must be added to heat the water by 5°C (41°F) (according to its heat capacity) plus the latent heat required to then convert the 100°C liquid into 100°C (212°F) steam. Each substance has a separate latent heat of vaporization (the energy needed to boil liquid into gas) and a latent heat of fusion (the energy needed to melt solid into liquid). The former is normally much larger than the latter.

Phase transitions

The process of a substance changing its physical character is known as a phase transition. Changes of state between solid, liquid and gas are one example. Phase transitions occur when a substance jumps from solid to liquid, from liquid to gas, and from solid directly to gas – a process called 'sublimation'. But phase transitions also describe other transformations of matter, such as **spontaneous symmetry breaking** in **the early Universe** and the conversion of an ordinary metal into a superconductor (see **Superconductivity**).

Phase transitions are classified into two types: first order and second order. First-order transitions happen by the formation of bubbles of the new phase that expand and collide, often violently. Any transition involving **latent heat** happens this way, including the vaporization of liquids – as anyone who has boiled a pan (pot) of water can testify. Second-order phase transitions are much smoother, with the substance gradually evolving from the old phase into the new.

Equation of state

Physicists and chemists characterize any sort of matter by its equation of state – a mathematical formula linking pressure, volume and temperature. One simple equation of state describes what's known as an 'ideal gas' – a model of how gases work, which assumes the particles in the gas have zero volume and exert no forces on one another. It's useful for rudimentary calculations. But for real gases, more accurate formulas are needed, such as the Van der Waals equation (see **Intermolecular forces**) which takes account of **molecule** sizes and the forces acting between them. Astrophysicists use equations of state for modelling everything from planetary atmospheres to the internal structure of **stars** to the behaviour of **the early Universe**.

Triple point

The temperature and pressure at which a substance's three states – solid, liquid and gas – can all coexist together in thermodynamic equilibrium is the triple point. For water, the triple point occurs at a temperature of 0.01°C and a pressure of 611.73 pascals – 0.006 times the standard atmospheric pressure at the Earth's surface. But these numbers are different for every material.

Below the triple-point pressure it is not possible for a substance to exist in its liquid state. Heating a solid at this pressure converts it directly to gas, a **phase transition** known as 'sublimation'. Triple points are used in science as reference points for calibrating thermometers. The triple point of water helps to define the thermodynamic temperature scale used by scientists. The scale is measured in kelvin (K), and zero kelvin = -273°C (-460°F).

Plasma physics

In addition to solids, liquids and gases, there is a fourth state of matter, called plasma. Plasma is a gas that has reached such a high temperature, typically thousands of degrees Celsius/Fahrenheit, that its **atoms** and **molecules** have been torn apart. The process is called ionization and creates a sea of positively charged **atomic nuclei** – or '**ions**' – and negatively charged particles called electrons.

Plasmas play an important part in designing fusion reactors (see **Fission and fusion**) and in astrophysics – cropping up in stars and nebulae in **interstellar space**, and forming the basis for some experimental spacecraft engines. But they are also found in more down-to-earth settings such as when they are formed briefly during lightning strikes or created artificially in plasma television screens to heat phosphor, causing the screen to emit light. When plasma is cooled down, the opposite process to ionization – called 'recombination' – takes place. It takes energy to break the bonds between an atomic nucleus and electrons, and when they recombine this same energy is released as light.

FLUIDS

Surface tension

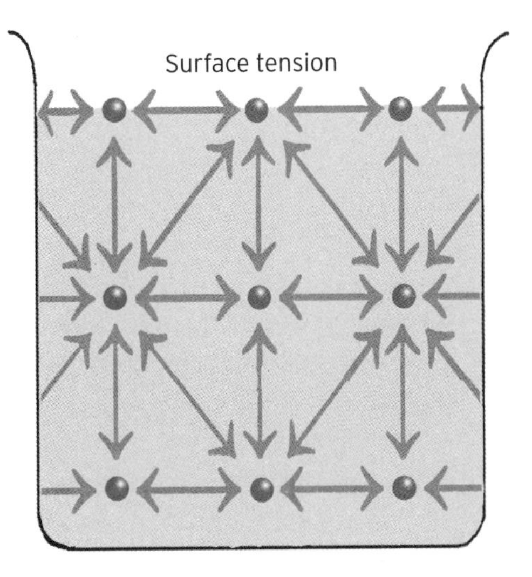

Liquids behave as if they are coated by an elastic skin, an effect called surface tension and it is what shapes liquids into spherical droplets. Surface tension exists because of the forces of attraction between the liquid's **molecules**. Deep inside the liquid, a molecule is surrounded by other molecules and so experiences an equal amount of force pulling it in each direction so the molecules are in equilibrium. But at the surface, molecules only experience forces pulling from below and this creates a net inward force that tries to squeeze the liquid into the smallest volume possible, i.e. a spherical drop.

Surface tension on water is strong enough to support the weight of small objects that would otherwise sink. Insects such as water boatmen take advantage of this effect to walk on the surfaces of ponds and lakes.

Capillary rise

A direct result of surface tension, capillary rise is what causes a liquid in a narrow tube to be drawn upwards. Attractive forces between molecules in the walls of the tube pull the surface of the liquid into a curved shape known as a 'meniscus'. A meniscus forms at any interface between different fluids in a container; the edges of the meniscus curve towards the fluid that feels the least attraction to the container walls. In the case of water and air in a glass tube, the air experiences less attraction towards the glass than the water and so the edges of the meniscus curve upwards. Surface tension then pulls at the edges of the meniscus to haul the water up the tube.

Low-density liquids with high surface tension reach the highest in a tube; the height also increases as the diameter of the tube gets narrower. Capillary rise underpins many everyday phenomena, such as the ability of plants to draw water from the ground and the absorbency of tissue paper.

Archimedes' principle

From corks to ships, anything that floats does so because its average density is less than the density of water – or whatever liquid it is floating on. 'Buoyancy' is the scientific term and the theory underpinning it is known as Archimedes' principle, after the Greek scientist Archimedes of Syracuse who lived in the

3rd century BC. His principle says that an object placed in water experiences an upward force equal to the weight of water that the object has displaced. The heavier the object, the lower it will sit in the water, displacing more liquid until the upthrust force is equal the object's weight and so able to support it. If the upthrust never reaches the weight of the object then it sinks.

Even though modern ships are made of metal, which on its own is heavier than water, the average density of a ship – the metal hull plus all the airspace within – is much lighter, enabling it to float. Submarines are able to dive by taking on board measured quantities of water to control their buoyancy.

Viscosity

Stir your coffee and now stir some treacle (syrup) – the difference is due to a property called viscosity, which can loosely be thought of as the 'gooeyness' of a liquid. Scientists, however, demand a more exacting definition. Imagine two parallel plates a fixed distance apart with fluid between them. Then the viscosity of the fluid is gauged by the resistive force it exerts on the plates as they try to slip past one another.

In this sense, viscosity can be thought of as a kind fluid friction. And, indeed, it is the cause of the forces of fluid dynamics that act to slow down the motion of cars and planes through the air, and of ships through the sea. Viscosity depends on temperature, with most fluids getting less viscous or 'runnier' as they are heated. Water at 10°C (50°F) is over four times as viscous as water at 100°C (212°F).

Newtonian fluids

If the viscosity of a liquid takes a fixed, constant value irrespective of how fast the liquid is flowing then it's called a Newtonian fluid. Examples include water, all known gases, and many industrial lubricants such as motor oil. But some fluids are decidedly non-Newtonian. The commonest are called 'thixotropic' liquids. Their viscosity decreases as the speed increases. Tomato ketchup is a thixotropic liquid, which is why you need to shake it vigorously before getting any out of the bottle. Non-drip paint is another example – it brushes on easily and then becomes viscous again so that it doesn't run.

At the opposite end of the scale are 'dilatant' non-Newtonian fluids. These get more viscous with speed. Cornflour mixed with water is a dilatant liquid. Stir it gently and it remains runny, but stir it vigorously and it instantly thickens up – to the point of becoming virtually rock solid.

Fluid dynamics

Kinematics and dynamics govern the movements of objects acted on by forces, while thermodynamics models systems of heat exchange. Likewise, the laws of fluid dynamics provide a mathematical framework to understand the behaviour of moving fluids. The field can be broken down into hydrodynamics, which describes the flow of liquids, and aerodynamics, for modelling the flow of gases.

Fluid dynamics has manifold real-world applications in engineering, including transport design and hydrodynamic **energy generation**, and is used for understanding aspects of the natural world, such as weather systems (see **Meteorology**) and the locomotion of fish and birds.

Navier–Stokes equations

The Navier-Stokes equations, named after 19th-century physicists Claude-Louis Navier and George Gabriel Stokes, are a set of equations describing the flow of a viscous liquid. They arise from applying **Newton's laws of motion** to the liquid, along with the **conservation laws** of energy, momentum and mass. They are a notoriously difficult set of equations to solve and, as a result, very few exact solutions exist. Instead, most calculations using the equations involve numerical solutions obtained by **scientific computing**.

Bernoulli principle

One of the main applications of aerodynamics is in aviation – explaining the forces that act on an aeroplane during flight. The most important of these is lift – the upward force on a wing caused by the flow of air over it, that keeps the plane in the air. Lift is generated by what's called the Bernoulli principle, first formulated by Swiss mathematician Daniel Bernoulli.

It says that the pressure inside a high-speed stream of air is less then the pressure inside a slower stream. The curved shape of a wing is designed so that the air passing over the top flows faster than the air moving underneath. And this sets up a pressure differential above and below. Just as high-pressure air inside a balloon tries to escape into the lower-pressure surroundings, the high-pressure air beneath the wing wants to move to the low-pressure area above – and this force pushes the wing upwards, generating lift.

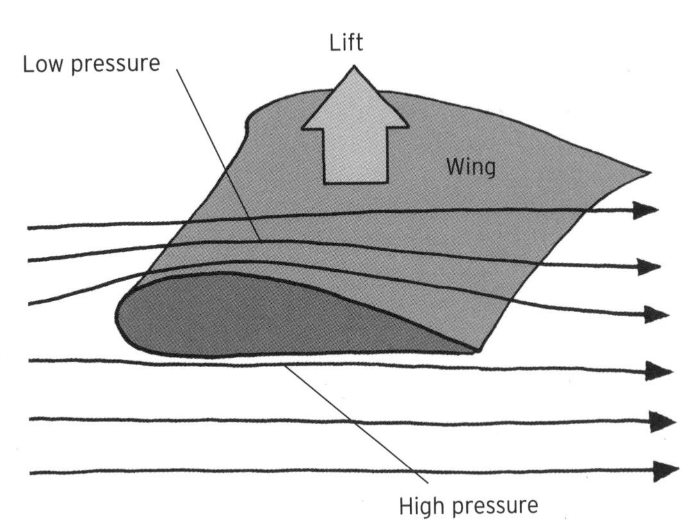

Turbulence

Most of us have experienced turbulence on an aircraft – flying through choppy, restless air that shakes the plane violently. Turbulence crops up in many other areas of engineering where fluid flow plays a part. And yet the behaviour of a fluid as it passes from smooth, well-defined 'laminar' flow to unpredictable, disorganized 'turbulent' flow is poorly understood.

The likelihood that a flow will become turbulent can be measured, given by the so-called Reynolds number – the ratio of the fluid's momentum to its viscosity. Higher Reynolds number flows are prone to becoming turbulent. For example, in a straight uniform pipe, Reynolds values over 3,000 signify the transition to turbulence. Fluid dynamicists believe the secret to comprehending turbulence lies in unpicking the Navier-Stokes equations. So much so, the Clay

Mathematics Institute in Cambridge, Massachusetts, has offered a $1 million prize to anyone who can make 'substantial progress' towards constructing a theory of turbulence from these fiendishly complicated mathematical equations.

Magnus effect

Fans of baseball and soccer know that expert pitchers and kickers can make a ball curve in flight by spinning it. The physics behind this is known as the Magnus effect. When an object moves through a fluid, a layer of the fluid clings to its surface, the 'boundary layer'. When the surface is a spinning ball, the boundary layer creates a swirling vortex around the ball as it moves forward through the air. On one side of the ball, the air in the vortex is moving in the same direction as the air flowing past, speeding that airflow up slightly. Conversely on the opposite side of the ball, the vortex acts to slow the passing airflow down. According to the **Bernoulli principle**, this difference in flow speed sets up a pressure differential. And this creates a force that makes the ball curve towards the side where the flow is fastest.

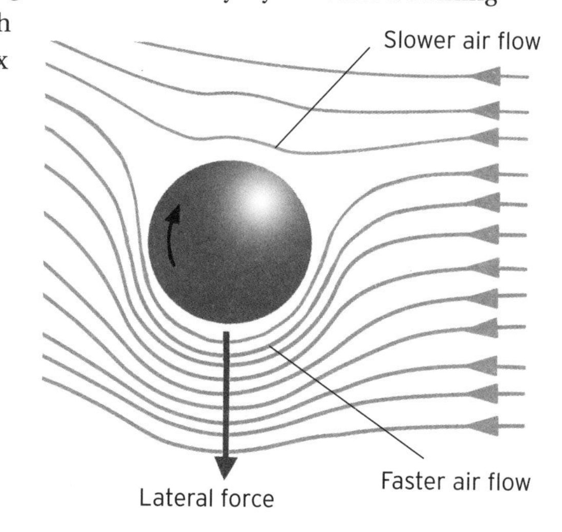

In the 1920s the German engineer Anton Flettner took advantage of the Magnus effect to build a ship that used rotating cylinders as sails – and, later, an aircraft that dispenses with the need for conventional wings.

Shock waves

Every fluid has its own sound speed – the speed at which sound waves move through it. Generally speaking, this increases with the density of the medium. An object travelling through a fluid at faster than its sound speed creates a shock wave – a thin layer of fluid within which there is an abrupt rise in temperature, density and pressure.

An associated affect is a 'sonic boom' – the thunder-like rumble caused by an aircraft travelling faster than the speed of sound in air. It is caused as the shock wave moves outwards, forming a 'shock cone' behind the craft as it flies. The angle of the cone is determined by the aircraft's Mach number – its speed divided by the sound speed in air. A higher Mach number, means the cone is narrower and it takes longer for an observer on the ground to hear the sonic boom after the aircraft has passed over. The fastest jet-powered aircraft ever flown was the Lockheed Martin SR-71, which flew at Mach 3.3.

Bomb blasts and lightning strikes also create supersonic shock waves, as does the cracking of a bullwhip – the 'crack' is actually the sonic boom caused as the end of the whip breaks the sound barrier.

WAVES

Wave theory

Waves are moving disturbances in a medium. They come in two main varieties. In a 'transverse' wave, the disturbance is at right angles to the wave's motion. Ripples on a pond are transverse waves. Throw a rock in the pond and ripples spread out from the impact point. The disturbance is just the height of each ripple above the water's surface. In a 'longitudinal' wave, on the other hand, the disturbance is parallel to the wave's direction of motion. Sound waves fall into this category, as do compression waves on a spring. Squash up a few coils of a stretched spring and let go – the disturbance moves along the spring in the same direction as the spring was compressed.

Physicists assign a wave four main properties. First is wavelength, the physical distance from the peak of one wave to the peak of the next. Second is frequency, the number of waves that pass by a fixed point every second, measured in cycles per second or hertz (Hz). The third quantity is speed – how far the wave travels every second, just given by multiplying its frequency and wavelength together. And finally, the 'amplitude' of the wave is just the size of the disturbance it creates as it passes, e.g. the height of the ripples on the pond.

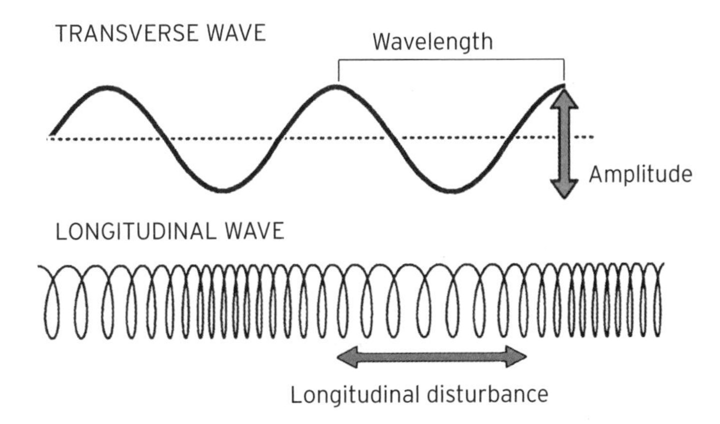

Sound waves

Hit a solid object with a hammer and, assuming there is enough elasticity in the material for it to vibrate, the disturbance will travel through it as a longitudinal wave (see **Wave theory**). If the frequency of the wave falls in the range we can hear – 20 to 20,000Hz – then this is a sound wave. The magnitude of a sound is measured on the decibel (dB) scale, and is a direct reflection of the sound wave's amplitude. The sound of a car ten metres away measures about 30dB, a pneumatic drill at one metre measures 100dB, the threshold of pain is 130dB and the noise of a jet engine at 30 metres clocks an ear-popping 150dB.

A sound wave can travel a huge distance depending on the density of the medium it's moving through. Whale song – travelling through water, which is around a million times denser than air – can be heard thousands of miles away.

Standing waves

Longitudinal and transverse waves (see **Wave theory**) are both examples of travelling waves, which carry energy from one point to another. But some waves go nowhere, being anchored instead to a fixed location; these are known as standing waves, like those on a guitar string. Pluck the string in the centre and it vibrates. Although the two ends are fixed, in between the string is free to vibrate with a shape resembling a half-wavelength transverse wave. Pluck the string a quarter of the way along from either end and it traces out a whole wavelength. A sixth of the way along and the length of string forms a wave of 1.5 wavelengths. And so on. There are an infinite number of these 'modes' for standing waves on a string, each given by setting the length of the string equal to a whole number of half wavelengths. This is an example of a transverse standing wave. Longitudinal standing waves also exist and can be formed, for example, by trapping sound waves in a tube.

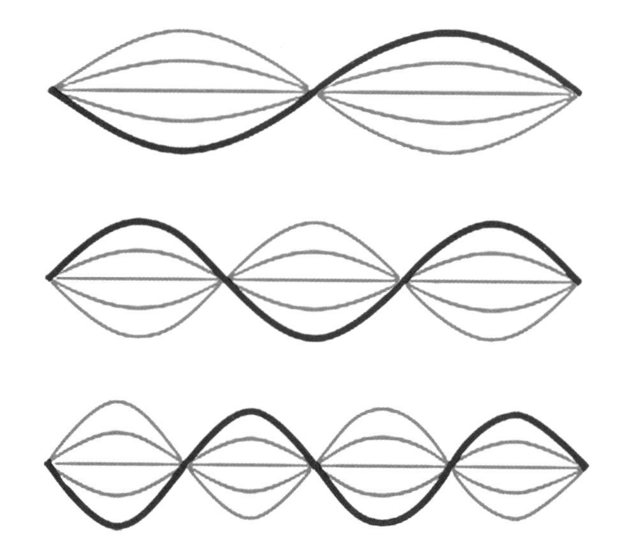

Resonance

Strike a bell and the sound it produces will be a note at its 'natural frequency'. Now attach a sound speaker to the bell and gradually increase the frequency of the sound wave played through it. The bell will vibrate in response to the sound, and the amplitude of the vibrations will steadily increase, reaching a peak when the bell's frequency matches the natural frequency – this is an example of resonance: large-amplitude vibrations caused by relatively small-amplitude inputs.

Resonance is also the reason why a car with its engine idling can sometimes shake violently, experiencing vibrations much larger than when the engine is revving faster. This happens because the engine's idling frequency – the number of revolutions it makes per second while just ticking over – is close to the natural frequency of the car's chassis and body. Engineers sometimes try to limit the effects of resonance using 'dampers' – devices that limit the amplitude of resonant vibrations. This is especially true in the design of tall buildings in earthquake zones.

Simple harmonic motion

A cork bobbing up and down on the surface of the ocean undergoes what is called simple harmonic motion, as its vertical position plotted with time traces out a perfect waveform. Technically, simple harmonic motion isn't wave motion but it is closely related and appears in many branches of physics.

In mechanics, examples include a pendulum swinging back and forth under the action of gravity and a mass bouncing on the end of a spring. The values of **electric current** and **voltage** in certain electronic circuits can also exhibit simple harmonic motion. Meanwhile, in the particle world, 'quantum harmonic oscillators' describe the vibration of some **molecules**.

Doppler effect

Most people have heard the sound of an ambulance siren seem to shift its pitch from high frequency to low frequency as the ambulance passes. This is a manifestation of the Doppler effect, explained using wave theory by Austrian physicist Christian Doppler in 1842. Doppler realized that sound from a source travelling towards you will increase in frequency, or equivalently, decrease in wavelength (because wavelength is just given by the wave's speed divided by its frequency). Wavelength is the distance between successive wave crests. But a moving source catches up with each crest to some extent before the next one is emitted, effectively shortening the wavelength and increasing the frequency. The converse effect happens when the source is moving away, leading to a drop in frequency.

Light waves also experience the Doppler effect. Indeed, travelling fast enough towards red traffic lights will shift their colour to shorter-wavelength green. However, you would need to move at around 18 per cent of the speed of light to do this! The Doppler effect is used by astronomers to calculate the recession or approach velocities of stars. Radar operators also use it for plotting the speed of aircraft.

ELECTRICITY AND MAGNETISM

Electric charge

The fundamental property of electricity is electric charge and is measured in coulombs (C), after the French physicist Charles-Augustin de Coulomb. Many subatomic particles carry electric charge, and it normally appears in discrete chunks equal to whole-number multiples of the charge found on an electron, denoted -e which, in **scientific notation**, takes the value -1.6×10^{-19}C.

Electric charges give rise to electric fields which enable charges to interact with one another from a distance. The fields cause 'opposite' charges, for example +e and -e, to attract one another and 'like' charges, say two electrons each with charge -e, to repel. The magnitude of the interaction between two charges is given by Coulomb's Law, which says that the force they experience increases with the size of the charges and decreases with the distance between them squared. Like energy, momentum and mass, charge obeys a **conservation law** – it's impossible to create or destroy it.

Electric current

A flow of electric charge is known as an electric current, which is measured in amps. One amp corresponds to a rate of flow of charge equal to one coulomb per second. Current is made to flow, for example through a wire, by connecting the wire up to a source of 'electromotive force' (written 'emf'), such as a battery; emf is measured in volts, and is also referred to sometimes as 'electric potential'.

The relationship between electric charge and emf is rather like the relationship between mass and gravitational field. A mass dropped in a gravitational field will fall and the rate it falls at is measured by its speed and determined by the strength of the field. Similarly, an electric charge subjected to an emf will move and the rate at which it moves is measured by the electric current, determined by the strength of the emf.

Resistance

The electric current that flows through a piece of wire connected up to a battery is moderated by the 'resistance' of the wire – the opposition to the flow of charge. Specifically, the current is given by the voltage of the battery divided by the resistance. Resistance is measured in ohms and varies for different materials. Good electrical conductors, such as metals, have low resistance. Current can pass through a conductor because there are charge carriers – usually free electrons – that can be swept along by the voltage of the battery. And metals have plenty of free electrons. On the other hand, poor conductors, such as plastics, contain very few free electrons and therefore have high resistance.

A current must spend energy to overcome electrical resistance. This rate of energy loss is measured in joules per second, or watts, and is given by multiplying together resistance and current squared. A 100W tungsten filament light bulb loses energy at the rate 100J/s because of the resistance of the tungsten and this energy is emitted as light and heat.

Capacitance

A capacitor is a device that is able to store electric charge usually made up of two conducting plates separated by an insulating material known as a dielectric. Examples of dielectrics include air, mylar and ceramic.

Once fully charged up, a capacitor can be discharged – effectively acting as a mini-battery. Large capacitors can hold enough charge to light a torch (flashlight) bulb for a minute or so; other applications include camera flash units, where intense pulses of current are required from a small battery. Capacitance is measured in farads, after the great British electrical engineer Michael Faraday.

Thunderclouds and the ground form a type of natural capacitor – separated by a dielectric layer of air. Under certain circumstances current can arc across the dielectric, discharging with a flash and a bang – thunder and lightning.

Magnetism

Magnets generate magnetic fields around them and are normally dipolar, meaning they have two poles, labelled north and south, with a magnetic field acting between them. Opposite poles attract one another, whereas poles of the same polarity repel.

Magnetic fields also attract so-called ferromagnetic materials, such as iron and cobalt. The atomic properties of these materials make them especially susceptible to magnetic effects; in fact, all materials that form permanent magnets – such as the ones you stick to your fridge – are ferromagnetic. Magnetic fields are measured in units called tesla, and are detected using devices known as magnetometers. Magnetism has found applications in **data storage**, navigation and **medical imaging**.

Induction

It became clear to physicists in the early 19th century that electric current and **magnetism** are somehow connected. Moving a conducting wire through a magnetic field – or, equivalently, subjecting a stationary wire to a varying magnetic field – produced a current in the wire. And likewise, passing a current through a moving wire using a battery generated a magnetic field. These effects became known as induction and the laws governing it were formulated by Michael Faraday and American physicist Joseph Henry. Induction is key to the operation of dynamo generators – which turn rotational motion, produced for example by an internal combustion engine, into electricity – as well as electric motors, which create rotational motion from a current.

The relationships between electric field, magnetic field and motion in each case are given by Fleming's right-hand and left-hand rules, respectively (see diagram), named after British physicist John Ambrose Fleming. The Earth and other planets have magnetic fields, thought to be caused by the dynamo effect of conducting fluids circulating in their cores.

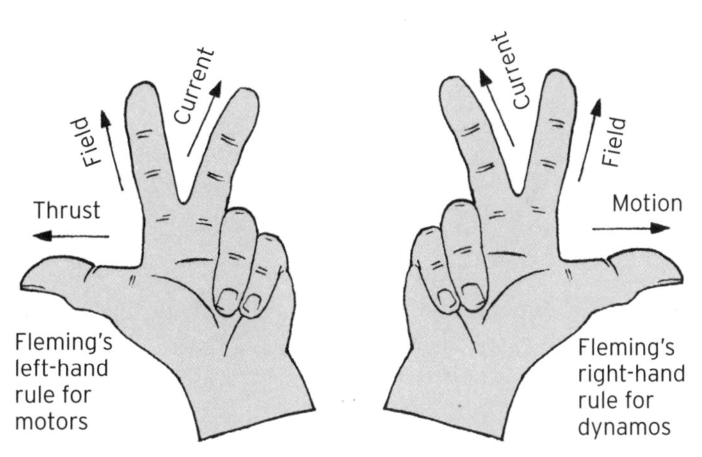

AC/DC

Electric current comes in two forms: AC (alternating current) and DC (direct current). Direct current is constant in time, whereas alternating current, true to its name, varies in time according to **wave theory** with a well-defined frequency and amplitude. Simple battery-operated electrical devices – such as torches (flashlights), cellphones and iPods – all run on DC. However, domestic electricity is AC.

Transformers

Transformers are pieces of equipment for ramping up low-voltage electricity to high-voltage (step-up transformers) or vice versa (step-down). They work through the principle of electrical **induction**. Input voltage comes in along a wire which is coiled around one side of a ring-shaped iron core. The input voltage induces a magnetic field in the core which in turn induces an output voltage in a second wire, wound round the opposite side of the core. The ratio of voltage in to voltage out is just given by the respective number of windings on each side.

Transformers only work for AC voltages because induction requires a varying magnetic field. This is why AC is used almost exclusively for domestic electricity. Household electricity needs to be transmitted over large distances from power stations to consumers and it is most cost-effective to do this with a high voltage and low current, because less heat is lost through electrical resistance. And so overhead power lines carry voltages of hundreds of thousands of volts. However, supplying homes with such high-voltage electricity would be dangerous so power is routed to neighbourhood substations where transformers step it down to a few hundred volts, suitable for home use.

Maxwell's equations

By the early 19th century, it was becoming clear to scientists that electricity and magnetism are just different aspects of the same phenomenon. Scottish physicist James Clerk Maxwell finally formed an overarching theory describing them in 1861, when he published his theory of 'electromagnetism'. It boiled down to four key equations from which all the properties of electric charge and magnetism could be derived, and which encapsulated the connection between them. Maxwell's equations of electromagnetism formed the first example of a **unification theory** – a single scientific framework that brought together the different forces of nature. Others would follow, such as **Kaluza–Klein** theory and **string theory**.

Electromagnetic radiation

The theory of electromagnetism, governed by Maxwell's equations, predicts the existence of waves of electromagnetic energy that travel through space. E-M radiation is made of electric and magnetic fields vibrating, according to wave theory, at right angles to one another and moving through space at the speed of light – 300,000km/s (186,400 miles per second). The radiation is classified across the spectrum by its wavelength. At the long wavelength end, with waves measuring kilometres in length, is long-wavelength radio. Radio occupies a huge swathe of the electromagnetic spectrum, down to wavelengths of around ten centimetres where we enter the microwave region. At around the scale of millimetres this gives way to infrared.

The visible spectrum – the E-M radiation we can actually see – begins at 0.75 microns (that's 0.75 thousandths of a millimetre) and this is red light. The visible spectrum continues up through orange, yellow, green, blue, indigo and violet, finally merging into ultraviolet at around 0.35 microns. White light, which is what we normally encounter, is just a blend of all the colours in the visible spectrum. Ultraviolet continues up to about a millionth of a millimetre, where X-rays – the

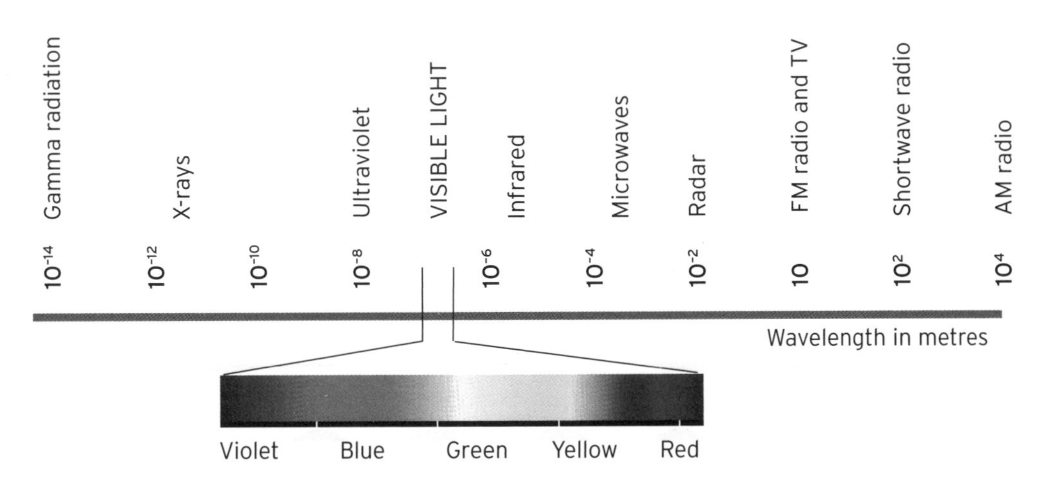

radiation used in **medical imaging** – take over. Gamma rays round off the high-energy end of the spectrum, starting at wavelengths of around a ten-millionth of a millimetre.

Photons

Certain metals emit electrons when light falls on them in a process known as the photoelectric effect, which proves that light can behave like a stream of particles as well as a wave. It was a key observation in the development of **quantum theory**, and it won a Nobel Prize for the young Albert Einstein, who was the first to explain how it actually works.

The photoelectric effect was first observed in 1839 but no one could work out how light waves were capable of kicking electrons out of a solid. Einstein's stroke of genius was to model light as a hail of solid particles called photons – an idea first used by German physicist Max Planck to develop the theory of **thermal radiation**. The photons collide with electrons in the metal like billiard balls, so that when one comes in with sufficient energy, it knocks an electron clean out of the metal. The photoelectric effect is closely related to the 'photovoltaic effect' which produces electricity when light shines on a semiconductor junction – and which underpins the operation of modern solar panels.

Magnetohydrodynamics

The bane of physics and mathematics students worldwide, magnetohydrodynamics (MHD) is a devilishly complex field combining **fluid dynamics** and Maxwell's equations of electromagnetism in an attempt to model the behaviour of electrically conducting fluids in the presence of a magnetic field. MHD has been used to construct generators, where charged fluids flow through a magnetic field to generate a voltage by the same principle of operation as a dynamo. It also forms the basis of an experimental propulsion system for ships and submarines, whereby a current is passed through seawater and a magnetic field then applied to force the seawater backwards like a jet. Not content with the complexity of MHD, some cosmologists studying the **Big Bang** theory have even attempted to merge MHD with Einstein's **general relativity** to model conducting fluids in curved space.

OPTICS

Light waves

Just as sound is a kind of mechanical wave that can be picked up by our ears, so light is a form of **electromagnetic radiation** we can see with our eyes. The wave nature of light was understood long before the development of electromagnetism. Dutch physicist Christian Huygens published a wave theory of light in 1678, much of which still applies today. Later in the 19th century, the theory was developed further by researchers including Englishman Thomas Young and France's Augustin-Jean Fresnel. These scientists took the analogy with sound literally, supposing that – just as sound waves need matter to travel through – so light also needs a medium to carry it. This medium was known as the ether. However, later attempts to detect it drew a blank. Light is now known to be a transverse wave made up from oscillating electric and magnetic fields travelling through space at 300,000 km/s (186,411 miles/s), with no need for a carrier medium. The branch of physics dealing with light and its interaction with matter is known as optics.

Reflection

Two of the most basic optical properties of light are reflection and refraction and these describe what happens to a light wave striking an optically denser surface – for example, a ray of light in air encountering a glass windowpane. Some of the light will be reflected and some refracted – the exact amounts depending on the light's **polarization**. Two laws govern reflection. First, the three lines formed by the incident light ray, the reflected ray and the 'normal' (a line at right angles to the surface) must all lie in the same plane. And, second, the incident and reflected rays must both make the same angle with the normal.

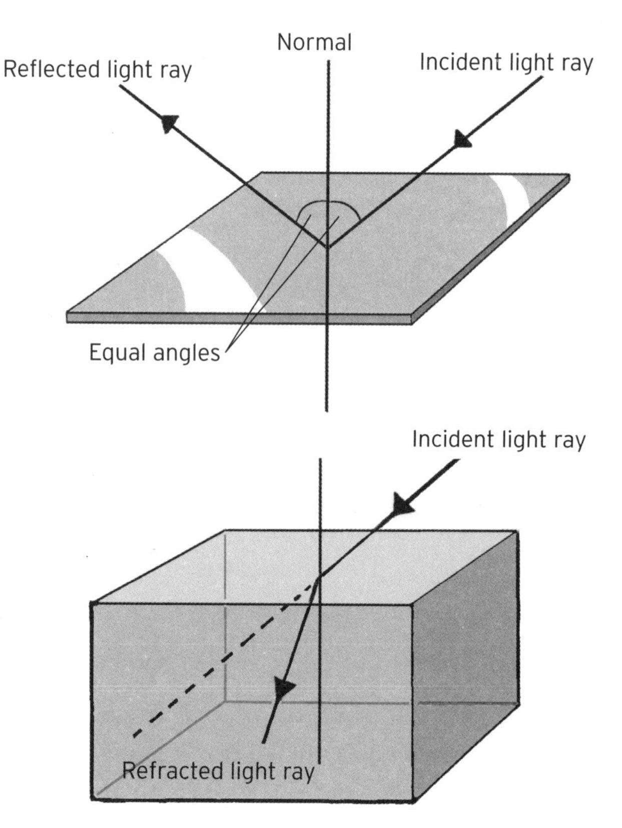

Refraction

Refraction is a little more complex than reflection; it is the bending of a light ray as it moves between media that have different optical densities. Generally speaking, light will bend towards the normal (a line perpendicular to the surface of the glass) when it is moving into a denser medium and away from the normal when moving to one that is less dense. The higher the optical density, the slower the speed of light in the medium; the amount of refractive bending is given by the relative speeds in each medium.

The degree to which a light ray is bent during refraction also depends on its wavelength. In this way, refraction can split a ray of white light into its component colours – a process called 'dispersion'; it is especially pronounced when the light passes through a triangular glass prism.

Magnification

Refraction is the science that explains how the lenses used in telescopes, microscopes and spectacles work – curving the surface of the glass changes the direction of the 'normal' (a line perpendicular to the surface of the glass), and this shifts the direction in which the light gets bent. The degree of bending produced by a lens is engineered so that images of objects passing through appear bigger. This is known as magnification.

Simple refracting **telescopes** use two lenses – an objective lens and an eyepiece. The objective lens gathers light while the eyepiece is responsible for focusing it onto the eye; the magnification of the optical system is determined by the properties of both lenses – and can thus be varied by interchanging eyepieces. Curved mirrors can also be used to magnify light. Whereas a lens concentrates light to a focal point by refraction, mirrors achieve the same effect by reflection. Reflecting telescopes use a curved mirror to gather light and an eyepiece lens to then channel it into the eye.

Diffraction

Refraction is not the only way to bend a beam of light. When light passes through a narrow slit, it spreads out to create a pattern of light and dark bands. The slit needs to be small; shine light through a doorway and you won't observe any diffraction, but when the slit size is comparable to or smaller than the light's wavelength, then the effect becomes pronounced.

The wavelength of light is around a thousandth of a millimetre. Typically slits to diffract light take the form of gratings – pieces of material such as glass with slits made by scoring lines every thousandth of a millimetre or so on their surface. The angle by which the light spreads out after passing through the slit depends on its frequency, meaning that diffraction, like refraction, is another way to disperse light into its constituent colours. All kinds of waves can undergo diffraction, including X-rays, sound and water waves.

Aberration

Systems of lenses or optical mirrors aren't always perfect; defects are known as 'aberrations' and they can occur for various reasons, but the commonest are spherical aberration and chromatic aberration. Spherical aberration occurs when an optical mirror has been ground to the wrong shape so that light falling on different parts of the mirror get reflected to different focal points, forming a blurry image. Normally, an optical mirror is ground so that its cross-section is a deeply concave shape known as a parabola. Spherical aberration occurs when the mirror's shape is more like a sphere, which is the problem that famously plagued the mirror of the Hubble Space Telescope.

Chromatic aberration arises in lens systems because the degree to which a light ray is refracted through glass depends upon its wavelength – or equivalently its colour; multiple images are created, each a different colour and each focused a different distance from the lens. The effect in photographs is to introduce blurry coloured fringes around the edges of objects. The effect can be minimized with 'achromatic' lenses, which use layers of glass with different refractive indices to bring different colours to a focus at more or less the same point.

Polarization

In an ordinary light wave, the light's electric field can point in all possible directions perpendicular to the direction of motion. However, in polarized light this freedom is restricted. The simplest kind is plane-polarized light, where the electric field vibrates in a single fixed direction so that the wave just looks like a vibrating string. Polaroid filters used in cameras and sunglasses are rather like a gate with vertical slats that only allow through waves vibrating in a plane parallel to the slats to produce plane-polarized light. Two overlapping pieces of Polaroid can be rotated relative to one another to vary the amount of light that's allowed to pass through. Indeed, it can be varied right down to zero – equivalent to the slats in each gate being crossed at 90 degrees to each other. Other kinds of polarization exist, such as circular and elliptical, in which the plane of the electric field rotates, tracing out a corkscrew shape as the wave moves forward.

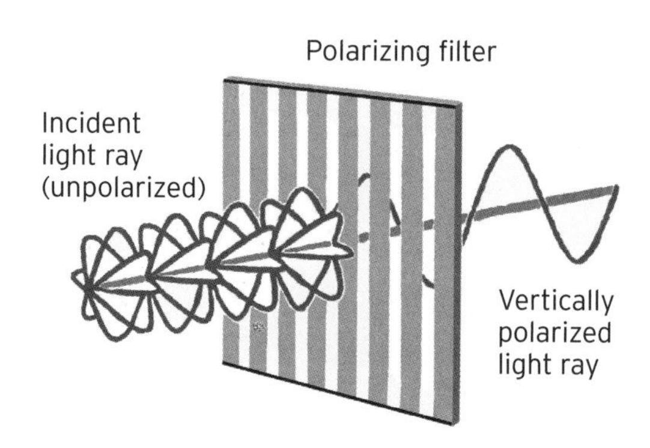

Interference

Throw two stones near to each other in a pond and watch the ripples spread out. Where they collide and overlap a complex pattern of peaks and troughs appears. This phenomenon is called interference. It hinges on the fact that the disturbances cased by waves in a medium add up; where two wave peaks meet, a single large peak forms, equal in height to the two smaller peaks added together. Similarly, two overlapping troughs form a single large trough and these are both instances of 'constructive interference'. However, where a peak and a trough meet the two cancel each other out to some extent, leaving just a small disturbance – or none at all if peak and trough are exactly the same size. This is 'destructive interference'.

Interference affects all kinds of waves – including light. The bright and dark bands produced by **diffraction** are just a manifestation of interference at work, as is the phenomenon of 'beats' – the warbling, pulsating sound produced when two sound waves have almost but not quite the same frequency. Noise-cancelling headphones work on the principle of destructive interference, monitoring external noise and then generating exactly the right sound wave needed to cancel it out.

RELATIVITY

Galilean relativity

Italian physicist and mathematician Galileo was the first to put forward a theory of relativity, in 1632. He stated that the laws of physics are the same for all observers moving at constant speed. So, for example, a scientist studying a swinging pendulum in his laboratory on land sees it make exactly the same movements as someone watching an identical pendulum on a moving ship. It's impossible for them to tell just by looking at the pendulum whether they're moving or at rest.

You can demonstrate Galileo's principle of relativity for yourself next time you're riding on a train. Once the train has accelerated to a constant speed, close your eyes. While the rocking of the train from side to side may give away the fact that you're not stationary, it's impossible to tell without looking out of the window whether you're moving forwards or backwards, or how fast you're going. The Galilean principle set the stage for Einstein's special relativity, nearly 400 years later.

Special relativity

In the late 19th century experiments had begun to reveal niggling disparities between **Maxwell's equations** of electromagnetism and the laws of **mechanics** – which govern the behaviour of moving bodies. The resolution came in 1905 from Albert Einstein, a physicist working at the patent office in Bern, Switzerland. He realized that the problem was in the way mechanics described objects moving at close to the speed of light. In Galilean relativity, an astronaut moving at half the speed of light sees an oncoming light beam move at a relative speed of 1.5 times the speed of light – their velocities add together, just like cars driving head-on towards each other. But Einstein's revolutionary insight was that the speed of light, 300,000km/s (186,411 miles/s), is the same for all observers – it makes no difference how fast you're travelling.

The new laws of motion this led to became known as the special theory of relativity. It not only cleared up the discrepancies between electromagnetism and mechanics but brought new revelations about the nature of space, time and matter – such as the prediction that nothing can travel faster than the speed of light.

Length contraction and time dilation

Einstein's special relativity has some weird consequences. For example, space and time become squeezed and stretched respectively for objects travelling at close to the speed of light. The first effect is known as 'length contraction'. An observer who sees a spaceship fly past them at half the speed of light will measure the length of the spacecraft to be just 85 per cent of what it was when stationary. What's more, clocks on the spacecraft will tick slower. A relativistic effect called 'time dilation' means that it takes roughly 1.15 seconds on the observer's clock for 1 second to pass on the spacecraft. If the spacecraft flew off at the same speed and came back 10 years later, as measured by the astronauts, they'd find the observer had aged by 11.5 years – they return from their journey 1.5 years younger than they should be. Weird as they sound, these effects are real – regularly tested in **particle accelerators**, giant machines that force subatomic particles to near the speed of light.

$E = mc^2$

It's perhaps the most famous equation in the whole of science. $E = mc^2$ essentially states that the energy content of an object, E, is just equal to its mass, m, multiplied by the speed of light, c, squared. The formula, which drops directly out of the mathematics of Einstein's special relativity, has since become the basis for nuclear **fission and fusion** reactions. Physicists found that splitting apart heavy atomic nuclei – or fusing together lighter ones – leads to a net reduction in mass, translating into a colossal release of energy. **Nuclear electricity** plants – and **nuclear weapons** – would later prove the theory.

General relativity

When Einstein's **special relativity** was published, in 1905, the gravitational force – responsible for the orbits of the **planets** – was described by **Newtonian gravity**. In Newton's law, gravity was an instantaneous force – propagating through space infinitely fast. But this was manifestly at odds with Einstein's theory, in which nothing can exceed the speed of light. The problem occupied Einstein for ten years until, in 1915, he published the general theory of relativity – a new theory of gravity that was compatible with the special theory.

Einstein's key insight this time was to identify the gravitational force with curvature of space and time. Empty space, he realized, is like a flat rubber sheet. Roll a marble across the sheet and it goes in a straight line. But now drop a massive object, like a bowling ball, on the sheet and it creates an indentation that curves the marble's path, which is what massive objects do to space to create the effect we observe as gravity. By working out the link between curvature and an object's mass – encapsulated in his all-important 'field equations' – Einstein was finally able to bring gravity in line with relativity.

Bending of star light

One of the first experiments that convinced some scientists to take general relativity seriously took place during a total **solar eclipse** in 1919. Arthur Eddington, an astronomer at Cambridge University, travelled to the island of Principe, Africa, to get a clear view of the eclipse. Here, with the Sun's bright disc obscured by the Moon, he was able to make accurate measurements of the apparent positions of stars whose light had passed near to the Sun.

The deformation of space that general relativity creates doesn't just influence solid objects, but – unlike **Newtonian gravity** – it affects light beams too. This meant that if Einstein was right then the light from distant stars passing near to the Sun should be bent by the Sun's gravity, causing a small shift in the stars' apparent positions on the sky. And this

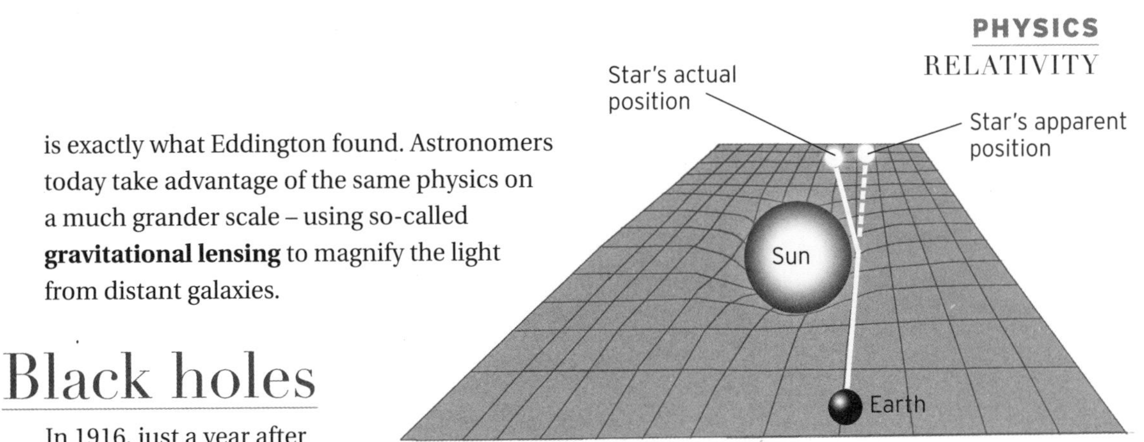

is exactly what Eddington found. Astronomers today take advantage of the same physics on a much grander scale – using so-called **gravitational lensing** to magnify the light from distant galaxies.

Black holes

In 1916, just a year after Einstein published general relativity, German physicist Karl Schwarzschild solved Einstein's complex field equations for a spherical body, thus working out a description of the gravitational field around a star or planet. For weak fields the results boiled down to those of Newtonian gravity. But when the gravitational field was very strong, Schwarzschild noticed something interesting: his equations seemed to be saying that if the radius of the gravitating body was small enough then its gravity would become so strong that nothing, not even light, could escape from it.

Schwarzschild had discovered the mathematical description for what is now known as a black hole. The so-called 'Schwarzschild radius' gives the size that an object of given mass must be squashed down to in order to become a black hole. Even the Earth can become a black hole if it's squashed down small enough – our planet has a Schwarzschild radius of about a centimetre. For most stars it's a few kilometres. The surface delineated by the Schwarzschild radius around a black hole is sometimes known as the hole's 'event horizon'. Black holes can be produced in **supernova** explosions, marking the death of a star. And supermassive black holes – weighing billions of times the mass of the Sun – are thought to lurk at the centres of many **galaxies**.

Gravitational singularities

At the heart of a black hole, where the laws of physics break down, lies a point of infinite **density** known as a gravitational singularity. Gravitational singularities form when the gravitational field of a collapsing object has become so strong that there is no force of nature able to resist it and stop the object collapsing. For example, an ordinary star is supported by its internal gas pressure from **kinetic theory** – the outward force of which balances the inward pull of gravity. Compact objects such as **white dwarfs** and **neutron stars** are held up by different kinds of quantum mechanical pressure but when gravity overcomes these there is nothing that can halt the collapse down to a zero-size point.

A great deal of work on the nature of singularities and how they form was carried out in the 1960s by Oxford University mathematician Roger Penrose and Stephen Hawking at Cambridge. Penrose also conceived the so-called 'cosmic censorship hypothesis', the notion that singularities must always be hidden from view behind an event horizon. Today, most researchers believe gravitational singularities to be a mathematical quirk of classical general relativity, and that quantum effects in a full theory of **quantum gravity** will remove them.

Wormholes

A **black hole** swallows everything that falls within its event horizon but where does this material go? In 1935, Albert Einstein and his Princeton colleague Nathan Rosen argued that it might spew out of counterpart objects known as 'white holes'. Einstein and Rosen supposed that a black hole and a white hole could be connected via a conduit through space and time, later named a 'wormhole'.

Wormholes may offer shortcuts through space, linking regions of the Universe that are otherwise vastly separated, and could even hold the key to **time travel**. However, there is a major hurdle to be overcome if humans are to ever harness wormholes for travel of any kind. Physicists have shown that these spacetime tunnels are inherently unstable. The narrow throat connecting the two mouths tends to pinch shut like a stretched out rubber tube and the only way to hold it open is to use a bizarre kind of material known as exotic matter, which has negative energy. Tiny fractions of a gram have been made in experiments. But to wedge open a wormhole engineers would need a shipment of exotic matter ten times the mass of Jupiter.

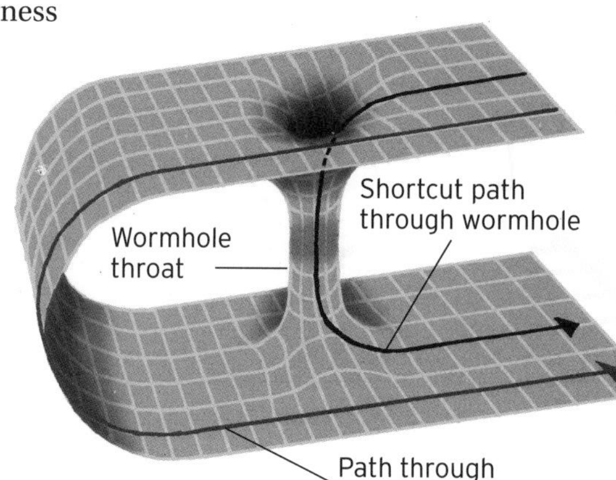

Gravitational waves

One of the few remaining untested predictions of **general relativity** is the idea that strong time-varying gravitational fields – for example, that produced by a pair of black holes careening around each other at half the speed of light – can give off radiation in the form of ripples in the fabric of space and time itself. These 'gravitational waves' travel through space at the speed of light and the passage of a wave would produce tiny fluctuations in the distance between nearby points. Experimental physicists have tried to detect these fluctuations using elaborate arrangements of **lasers** capable of measuring movements as small as 10^{-19}m. But so far their investigations have drawn a blank.

Although gravitational waves haven't yet been detected directly, there is indirect evidence for their existence. Astronomers have observed a pair of binary **neutron stars** – superdense objects formed in supernova explosions – to be coalescing at exactly the rate expected if the system was losing energy by gravitational waves.

Frame dragging

In 1918, two Austrian physicists figured out a consequence of Einstein's general relativity that put it in a league of its own for weirdness. Josef Lense and Hans Thirring calculated that in Einstein's theory, rotating objects drag space around with them, rather like a spoon in treacle (syrup). Any objects occupying the space near to a rotating object therefore get swept around with it. For ordinary planets and stars the effect of this so-called 'frame dragging' is tiny, but near strongly gravitating objects it can be pronounced.

In particular, in a region near to a rotating black hole known as the 'ergosphere', space is swept around so briskly that it is possible to extract useful energy from the rotation. Some scientists have speculated that this could even serve as a power source for an advanced civilization. Speculative evidence for frame dragging was found in 2004 by scientists analysing data from two Earth-orbiting **artificial satellites**. They claimed to have detected the minute frame-dragging effect of our planet.

QUANTUM THEORY

Particles

The smallest components of matter are the so-called subatomic particles, the building blocks of all other materials. The most common types of particles are protons and neutrons: the proton carries an **electric charge** of +e, while the neutron is electrically neutral. They each measure around one Fermi in diameter (10^{-15}m, in **scientific notation**) and weigh about 2×10^{-27}kg. This seems tiny, but protons and neutrons are giants compared with the other common member of the particle world: the electrons, which each weigh just 10^{-30}kg.

Protons and neutrons cluster together in different numbers to form the atomic nuclei of all the known **chemical elements**. Electrons then orbit around these nuclei to form atoms and these in turn bond to each other to make molecules. There are scores more subatomic particles known to exist, from mesons to **neutrinos** to **quarks** and many more which have been theorized but not yet detected. Quantum theory is the branch of physics governing everything that happens in the subatomic world.

Quantization

The key premise of quantum theory is that, at the subatomic level, **matter** is no longer continuous. Instead, it comes in discrete chunks, or quanta. Nature is 'quantized'. One of the first properties that physicists noticed to be behaving in this way was electric charge. In a seminal experiment, the results of which were published in 1910, American physicist Robert Millikan measured the electric charge on droplets of oil, finding them all to be whole-number multiples of a fundamental quantity – the electron charge, -e. Evidence has emerged that other quantities such as energy, momentum – even space and time – are quantized on the smallest scales.

Energy levels

Energy in quantum physics undergoes a process of quantization – that's to say, it's constrained to a discontinuous range of values. For example, an electron orbiting the nucleus of an atom is allowed to occupy only certain 'energy levels' predicted by quantum laws. When an atom absorbs an amount of energy equal to the gap between two energy levels, an electron in the lower level can jump between the two. After a time, it spontaneously drops back down, re-emitting a **photon** of the same energy.

A photon's energy is directly linked to its frequency – the energy is just the frequency times Planck's constant, a number that crops up time and again in quantum theory, equal to 6.6×10^{-34} in units of joule seconds. Because frequency is the same as colour, that means atoms give off light of characteristic colours as their electrons drop between particular energy levels. This can be seen in the real world during **aurorae** – the atmospheric lightshows at the North and South Poles – and in neon signs. Electric fields pump neon gas with energy, which raises the energy levels of its electrons. As they drop back to their original level they give off neon's distinctive red light.

Wave–particle duality

The photoelectric effect proved that waves could behave like solid particles – **photons**. But this wasn't a one-way street; particles too, it soon emerged, could exhibit wave-like properties. For example, electrons can undergo **diffraction** when they pass through the gaps in a crystal lattice. But perhaps the key observation that demonstrated the dual nature of quantum entities as both waves and particles was the so-called double-slit experiment. Light shone through a pair of parallel slits makes an **interference** pattern of many bright and dark bands on a screen, as peaks and troughs in the light waves overlap. But what if the intensity of the light is turned down to the point where single photons pass through the apparatus one at a time? You might suspect each photon to pass though either one slit or the other, and the interference pattern to disappear.

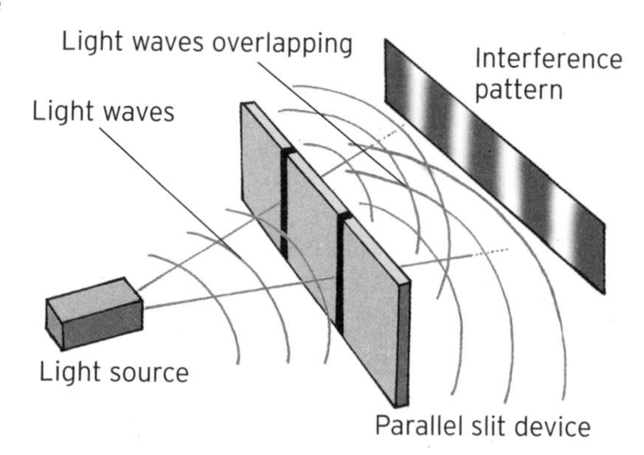

Yet the truth is rather different. The single photon makes a single bright dot on the screen. But repeat the experiment many times and record the position of the bright dot each time, and they all build up to recreate the original interference pattern. Matter really does behave like particles and waves at the same time.

Schrödinger equation

The correspondence between quantum particles and waves was put on a solid footing by Austrian physicist Erwin Schrödinger in 1926. He formulated a wave equation describing a solid particle. The equation doesn't give the particle's definite position in space but rather the probability of finding it at a particular point; it describes a probability wave, the peaks of which correspond to the locations where the particle is most likely to be.

In the case of the double-slit experiment, which was crucial in establishing wave–particle duality, the Schrödinger equation predicts a wave of probability peaks on the screen – these correspond to where each individual photon of light is most likely to end up.

Uncertainty principle

Put forward by German physicist Werner Heisenberg in 1927, the uncertainty principle is perhaps one of the weirdest laws of quantum theory. In essence, the principle says that it's impossible to know both the position and momentum of a particle at the same time. The more accurately you know one, the less accurately you must know the other. Specifically, Heisenberg deduced that the uncertainty in position multiplied by the uncertainty in momentum must be greater than or equal to Planck's constant, a number that arises in the definition of atomic **energy levels**.

Initially, the uncertainty principle was thought to be the result of disturbances introduced by the process of measuring the state of a quantum system. But this view is now outmoded – uncertainty is a fundamental property of quantum particles, whether they are measured or not.

Quantum spin

Quantum particles display many phenomena that are analogous with the large-scale world, but one which is radically different is quantum 'spin'. Whereas spin in the everyday world is a property of motion, like **speed and acceleration**, quantum spin is an intrinsic property of particles – more akin to mass and electric charge. Like other quantum properties, spin is quantized – it normally comes in whole-number multiples of 1/2.

Quantum particles are split into two broad families according to the value of their spin. Those with whole-number spin – 0, 1, 2, etc – are called bosons. While those with half-whole-number spin – 1/2, 3/2, 5/2, etc – are known as fermions. Protons, neutrons and electrons all have spin 1/2, and are thus fermions. Photons, on the other hand, have spin 1, making them bosons. Although the spin of a particle is fixed, it can be flipped between one of two states, dubbed 'up' and 'down' and denoted as +/- respectively. So an electron has spin +1/2 in the up state and -1/2 in the down state.

Quantum numbers

Four so-called quantum numbers describe the behaviour of electrons inside atoms. The first number gives the energy level around the atom's nucleus that the electron occupies, normally assigned the letter 'n' and taking a whole-number value such as 1, 2 or 3. Within each energy level there is scope for the quantum analogue of angular momentum. Normally denoted 'l' this can take any whole-number value between 0 and n-1. So in the n=2 energy level, l can be either 0 or 1. Meanwhile, the magnetic quantum number, 'm_l', can vary between -l and l - so, -1, 0 and 1 are allowed for the case of n=2. Finally, the spin quantum number, 'm_s', records the up or down value of the electron's quantum spin – and so takes the value +1/2 or -1/2.

Quantum numbers all obey **conservation laws** and together specify the 'state' of a quantum system. The four numbers given above are sufficient to specify the state of electrons in an atom; more complex quantum systems require more than these four basic numbers.

Exclusion principle

The Pauli exclusion principle, after Wolfgang Pauli, the physicist who discovered it in 1925, is a theory governing the behaviour of fermions – that is, particles with half-whole-number quantum spin. In its most basic form, the principle says that in an atom no two electrons can have the same values for all four of their quantum numbers. The Pauli exclusion principle, together with the rules of quantum numbers, determine how many electrons can occupy each energy level in an atom. For example, in the lowest energy level, n=1, the only permissible value for quantum numbers l and m_l is zero. The final number, m_s, can be either +1/2 or -1/2. This means there are two distinct states in the n=1 level and therefore it can hold a maximum of two electrons. Similarly, n=2, 3 and 4 can hold 8, 18 and 32 electrons, respectively. The exclusion principle means that bosons and fermions interact with one another very differently, and this affects their large-scale properties, as described by **statistical mechanics**.

Schrödinger's cat

The wave interpretation of quantum theory seems to allow particles to be in two places at once. In the double-slit experiment of **wave-particle duality**, for example, a single particle seems to pass through both slits to create an interference pattern on the screen. Austrian physicist Erwin Schrödinger devised a cunning thought experiment to demonstrate the apparent absurdity of this, which has since become known as Schrödinger's cat.

Schrödinger imagined a locked box, inside which is a cat and a phial (vial) of lethal poison. The poison is rigged to a quantum process – say, the decay of a radioactive atom – so that if the atom decays, the phial (vial) breaks and the cat dies. If the atom remains intact, the cat lives. Schrödinger reasoned that before a measurement is made – i.e. before the box is opened – the atom is governed by a wave equation that describes it as simultaneously decayed and not decayed. And therefore the cat must be simultaneously both dead and alive. Weird.

Copenhagen interpretation

Thought experiments such as Schrödinger's cat forced physicists to think long and hard about what their quantum laws were actually telling them about the physical world. The early view was known as the Copenhagen interpretation of quantum theory, named in honour of quantum physicist Niels Bohr, of the University of Copenhagen in Denmark, who put in much of the work developing it.

Central to the Copenhagen interpretation is a process called 'collapse of the wavefunction'. This says that quantum particles behave like waves – enabling them to be in two places at once – until they are measured, at which point the wave function collapses down and they become solid particles with a well-defined position. In the Schrödinger's cat thought experiment, the wavefunction of the radioactive source – and by association, the cat – only collapse to a well-defined state once the box is opened. Up to that point the cat is both dead and alive – afterwards, it's one or the other.

Many worlds

A modern alternative to the Copenhagen interpretation is what's known as the many worlds interpretation, which was put forward by physicist Hugh Everett in 1957. In essence, this says that every quantum event causes a splitting of our Universe into many parallel universes, in which every possible outcome of the event is realized. While at first sight this seems implausible, many worlds actually offers a more satisfactory interpretation of quantum theory than Copenhagen's collapse of the wavefunction, because it takes the emphasis off the observer.

In the many worlds take on the Schrödinger's cat experiment, our Universe splits into two – one in which the radioactive source decays and one in which it doesn't. In one universe the cat lives while in the other it dies. The equations of quantum theory then give the relative probabilities of us ending up in each universe. Crucially, the cat is never both alive and dead in the same universe and this is down to a feature of the many worlds interpretation called decoherence.

Decoherence

In the Copenhagen interpretation of quantum theory, collapse of the wavefunction determines when a particle stops behaving as a quantum wave and starts behaving like a 'classical' object that can only be in one place at a time. In the many worlds view, the transition from quantum to classical behaviour is referred to as decoherence. Whereas collapse of the wavefunction attributes this transition to the meddling fingers of the observer trying to make a measurement, decoherence puts it down to the inevitable interaction of a delicate quantum system with its environment. It's a little bit like thinking of quantum effects as the balance of forces and air currents needed to create a perfect smoke ring in the air – which are then suddenly swamped by a gust of wind.

In the many worlds view of the Schrödinger's cat experiment, each time the wavefunction of the radioactive source interacts with the particle detector, it forces the wave to decohere into a state where it's either emitted or hasn't emitted a particle of radiation. And it's at exactly this point that the universes corresponding to each possibility peel apart.

Quantum suicide

Will there ever be a way to tell between the many worlds and Copenhagen interpretations of quantum theory? Physicist Max Tegmark, of the Massachusetts Institute of Technology, thinks he's come up with a way: a macabre twist on the Schrödinger's cat thought experiment, called quantum suicide. Tegmark imagines a rifle with an automatic mechanism to pull the trigger. The mechanism incorporates a quantum device so that it will either fire a live round or click harmlessly with 50/50 probability.

To anyone watching, the gun clicks and fires at random. But, says Tegmark, anyone who puts their head in front of the barrel will hear a guaranteed click every time. He says the reason is that in the many worlds interpretation, of which he is a supporter, there are always real universes in which the gun doesn't fire. The experimenter isn't going to be around to see the results in universes where the gun goes off and so must therefore find themselves in a universe in which they've survived. That said, Tegmark adds that he has no plans to try quantum suicide anytime soon.

QUANTUM PHENOMENA

Virtual particles

Empty space isn't really empty. According to the **uncertainty principle** it's a seething mass of subatomic particles popping in and out of existence, which are known as virtual particles. The principle says that the uncertainty in a particle's position multiplied by the uncertainty in its momentum must always be larger than a small number called Planck's constant. But it turns out this is the same as saying that the uncertainty in a particle's energy multiplied by the uncertainty in the time at which you make the observation must also be bigger than a minimum value. Therefore a particle with energy E can spontaneously pop into existence so long as it's gone again in a time t, such that E and t, when multiplied together, satisfy the uncertainty principle. It means you can make short-lived high-energy particles or low-energy particles that stick around for a little longer. Virtual particles are formed in pairs consisting of one matter particle and an **antimatter** partner.

Zero-point energy

The lowest-energy state of a quantum system is known as its ground state, and the **energy level** of the ground state is sometimes called the zero-point energy. It is a phrase that comes from the zero of the thermodynamic temperature scale defined by **kinetic theory**, generally taken to represent the lowest energy a system can possess. And the existence of virtual particles means that the zero-point energy of empty space – also known as the 'vacuum energy' – is actually non-zero. Also the sum of the vacuum energy in all of space – called the cosmological constant, or **dark energy** – is believed to affect the **cosmic expansion** rate.

Casimir effect

A direct consequence of the zero-point energy of space is a phenomenon called the Casimir effect. Two parallel metal plates placed a few millionths of a millimetre apart in a vacuum experience a force pulling them together; the force arises from virtual particles. Quantum theory says that these particles can equally well be thought of as waves. Outside of the plates, waves of all wavelengths can exist, but inside there can only be waves that fit between the plates – **standing waves** for which the distance between the plates is a whole-number of half wavelengths. Now convert back to particle language and this means there are less particles inside than outside, and that creates a pressure difference which forces the plates together.

The Casimir effect was put forward theoretically by Dutch physicist Hendrik Casimir in

1948. It was measured experimentally by Steve Lamoreaux at Los Alamos National Laboratory in 1997.

Quantum entanglement

In 1935, Albert Einstein and two colleagues – Nathan Rosen and Boris Podolsky – put forward a quantum thought experiment that would have lasting consequences. They imagined a quantum process that produces two particles with opposite **quantum spin**. Pion particles can do this – they are known to decay into pairs consisting of an electron and a positron (the electron's **antimatter** counterpart). One decay particle has spin +1/2, the other spin -1/2 but it's impossible to tell which is which until a measurement is made. In fact, the particles themselves don't even decide which is which until a measurement is made that forces their wavefunctions to undergo **decoherence**.

However, when that measurement is made, and one particle's wavefunction is forced to decohere, this instantly fixes the state of the other one too – regardless how far away it may be. Take the particles to opposite sides of the Universe, measure one of them and this instantly determines the state of the other. Einstein famously derided this 'spooky action at a distance', believing it to be unphysical. However, the ability of quantum particles to remain linked in this way is now called quantum entanglement, and it forms the basis of eavesdropper-proof **quantum communication** systems and even **teleportation**.

Bose–Einstein condensate

Fermion particles, those with half-whole-number quantum spin, obey the Pauli **exclusion principle** – that in any given quantum system it's impossible for two or more particles to occupy the same state. For bosons, particles with whole-number spin, it's a different story. Bosons have no qualms about all stacking into the same state. And this means that if a system is cooled very close to the absolute zero of temperature (see **Kinetic theory**), every particle drops into the system's lowest-energy 'ground state'. Such a system is called a Bose–Einstein condensate, after physicists Satyendra Nath Bose and Albert Einstein, who theoretically predicted this state of matter in the 1920s.

Bose–Einstein condensates have since been produced experimentally in laboratories. The wavefunctions of all the particles in a Bose–Einstein condensate overlap to form a single, giant 'quasiparticle' that exhibits quantum effects on the large scale.

Superfluidity

The theory of Bose–Einstein condensates paved the way for the discovery of another amazing property of matter: superfluidity, the ability of supercooled liquids to lose all their **viscosity**. In 1938, it was found that helium-4 (an **isotope** of helium with two extra neutron particles in its nucleus) cooled down to just two degrees above absolute zero forms a Bose–Einstein condensate that also behaves as a superfluid.

Superfluids display strange properties. In addition to zero viscosity, they have zero **entropy** and infinite ability to conduct heat. A superfluid will also form a so-called 'Rollin film' that creeps up the side of its container to spill the fluid over the edge. Weirdest of all, a superfluid in a rotating vessel forms 'quantum vortices', whereby the fluid is only allowed to spin at certain quantized speeds. As the vessel gradually spins faster, the superfluid discontinuously hops up the ladder of allowed rotation rates.

Helium-4 nuclei are bosons, but superfluidity has also been seen in fermion particles such as helium-3. The physics of fermionic superfluidity is closely related to superconductivity.

Superconductivity

Superconductors are materials that when cooled sufficiently assume a state with zero electrical **resistance**. Superconductivity was first observed by Dutch physicist Heike Kamerlingh Onnes in 1911. But a satisfactory theory explaining how it works didn't emerge until the late 1950s.

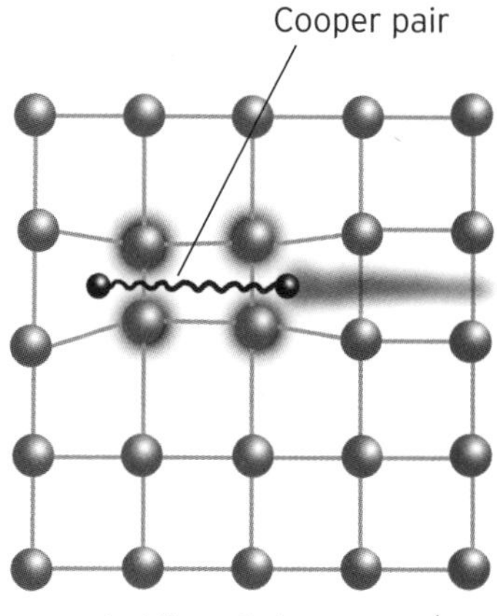

Cooling had long been known to improve electrical conductivity. In an ordinary conductor, negatively charged electrons carry electric current through the material. Heat makes the lattice of atoms in the material vibrate, and collide with the electrons, impeding their motion. Cooling thus lowers the resistance by reducing the vibrations. But in a superconductor, resistance vanishes entirely. Here, electrons cooled sufficiently lock together into so-called 'Cooper pairs' that are able to slip freely through the lattice. Broadly speaking, this works as each electron's negative charge attracts positive atoms, distorting the lattice to create a concentration of positive charge which then pulls the next electron forward, and so keeps the current moving.

Superconductors are now used in the construction of ultra-efficient generators, **particle accelerator** magnets and **medical imaging** machines. But all these devices require cooling; the holy grail now is to develop superconductors that can operate at room temperature.

PARTICLE PHYSICS

Dirac equation

Schrödinger's equation provided a good description of quantum particles that were slow moving. But it was no good for particles travelling at close to the speed of light, which Einstein had shown required his theory of **special relativity**. In 1927, British theoretical physicist Paul Dirac constructed a version of the Schrödinger equation that incorporated relativity to describe fast-moving charged electron particles. It became known as the Dirac equation.

The equation appears deceptively simple and it can be solved exactly for electrons in the hydrogen atom, but solving it in other cases requires approximation techniques or computers.

The Dirac equation was an outrider heralding the start of a new era for quantum physics. Not only did it predict the existence of antimatter, but – incorporating quantum descriptions of both electrons and the electromagnetic field – it offered the first quantum field theory.

Antimatter

When scientists first studied the Dirac equation in detail, they found that it doesn't just describe the behaviour of electrons. Two solutions actually drop out of the mathematics – one describing the electron, and another describing an identical particle but with opposite electric charge. Electrons carry negative charge, so the new particle was positively charged – and for this reason, it was named the 'positron'.

Experimental physicists duly found the first positrons in 1932, just four years after the Dirac equation was published. The positron is now recognized as the first example of antimatter. An antimatter particle has the same mass but opposite electrical charge to its matter partner. When a particle and its antiparticle meet the result is annihilation, with the mass of both being converted into a photon of electromagnetic radiation, with energy given by Einstein's formula $E = mc^2$.

Quantum field theory

electromagnetic force, described by **Maxwell's equations**, turns out to be one of four fundamental forces of nature that operate in our Universe. It exhibits 'action at a distance' – electrons, and any other charged particles, can 'feel' the electric charge of particles nearby. This electromagnetic interaction is mediated across space by photons of electromagnetic radiation. Photons are the quanta of the electromagnetic field. And the physics describing them is known as a quantum field theory (QFT).

QFT began in 1927 with the **Dirac equation**, which took account of the field around single electrons. The theory was developed further in the 1940s by physicists, including Richard Feynman, and refined into what is now called quantum electrodynamics, a quantum field theory providing a full description of the interaction between matter and **electromagnetic radiation**. Next, the challenge was to build quantum descriptions of the other fields in nature: gravity, and the strong and weak nuclear interactions.

Feynman diagrams

There can be no doubting that quantum field theory is extremely complicated, and obtaining exact solutions to the mathematical equations that it throws up is not always possible. In this case, physicists often turn to approximation techniques in order to obtain useful numerical results. And Feynman diagrams – named after their creator, Nobel prize-winning physicist Richard Feynman – are one such method. Each diagram (examples are shown on the right) represents a particular interaction of particles. Feynman came up with rules for converting the form of a diagram into relatively simple mathematics describing the process it shows. Making a

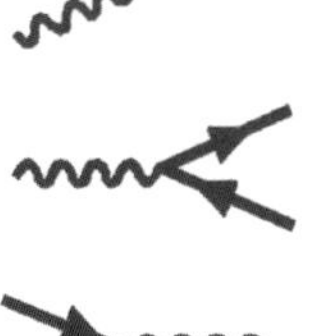

quantum calculation then involves drawing Feynman diagrams for all the relevant interactions, converting them into mathematical terms and adding all the terms up.

Renormalization

Even using Feynman diagrams doesn't always make the mathematics of quantum field theory tractable. Sometimes the numbers that drop out of calculations are gobbledegook – predicting that the real-world values of physical quantities rise to become infinite. These mathematical catastrophes are known as 'divergences'.

Renormalization is a mathematical scheme for getting rid of unwanted divergences. When it was first used, in quantum electrodynamics, the reaction to it was mixed, with many physicists regarding it as the mathematical equivalent of getting a square peg into a round hole with the help of a very large hammer. Nobel laureate Richard Feynman called it a 'dippy process' and 'hocus-pocus'. However, the simple fact is that it gives the right answers. More recently, theorists have warmed to renormalization. They've found that it can be interpreted physically, as a new phenomenon that causes the parameters of a theory to 'run' – varying with the energy of the particles involved.

Weak force

The weak force is one of two fundamental interactions of nature that operate between particles inside the nuclei of atoms; and so named because it's some 100 billion times weaker than the force of electromagnetism described by **Maxwell's equations** and quantum electrodynamics. Whereas electromagnetism is moderated by photons of the electromagnetic field, the carriers of the weak force are known simply as W and Z. Both are massive particles of **quantum spin** -1. W carries the same electric charge as the electron – but either positive or negative – and the Z particle is uncharged. Both W and Z have been detected experimentally in **particle accelerators**.

The weak force is responsible for the **radioactive decay** of atomic nuclei and is described at the quantum level by the unified **electroweak theory**, formulated in the late 1960s. It is extremely short-ranged and doesn't exist outside atomic nuclei. For this reason, the weak force is a purely quantum phenomenon and, unlike electromagnetism, has no analogous large-scale, or 'classical', counterpart theory.

Quarks

Protons and neutrons, the particles that make up atomic nuclei, are themselves made of smaller units, called quarks. Quarks come in three main 'flavours', labelled u, d and s – for 'up', 'down' and 'strange'. Protons and neutrons are each made up of three of these quarks – a proton is two u quarks and a d quark; a neutron is two d quarks and a u. In addition to u, d and s, there are three further quark flavours, dubbed c, t and b – for 'charm', 'top' and 'bottom'. Quarks are fermions, with quantum spin 1/2, and electric charge equal to a whole-number multiple of e/3. The six quark flavours can be transmuted between each other by the weak force and all have been detected experimentally.

The quark model was put forward independently in 1964 by physicists Murray Gell-Mann and George Zweig. Gell-Mann coined the name from a line in James Joyce's novel *Finnegan's Wake*: 'Three quarks for Muster Mark!'

Strong force

The strong force – so called because it is 100 times stronger than the electromagnetic force – is what binds quarks together to make protons and neutrons and, in turn, binds those particles to form atomic nuclei. It is mediated by a quantum field of massless, spin-1 particles called gluons, and the **quantum field theory** governing their behaviour is known as quantum chromodynamics (QCD).

The 'chromo' prefix comes from the fact that quarks possess their own kind of charge, called 'colour'. It's a bit like electric charge, but more complicated in that it comes in three types – red, green and blue – and each can be positive or negative (denoted, for example, as red and anti-red). QCD colour has nothing to do with colour in the real world – it's just an arbitrary name for an abstract quantum concept. As with the weak force, the strong interaction only operates at very short, quantum-scale ranges, and so has no 'classical' analogue.

Particle families

Particle physics is all about grouping together particles according to their properties and many groupings, or 'families' of subatomic particles exist. Fermions and bosons, the grouping of particles according to their **quantum spin**, provide one example. Hadrons and leptons are another: hadrons are particles that feel the **strong force**, leptons are particles which feel every force except the strong force. Hadrons include quarks and protons. Leptons include electrons, along with muons and tauons.

Baryons, meanwhile, are a subset of hadrons and are particles that contain three quarks. This family includes protons and neutrons and makes up the bulk of what we might call 'normal matter' in the Universe. Another group of subatomic particles is the mesons, which are quark doublets – that is, pairs of quarks – and they include kaons and pions. Some experimental physicists claim to have found evidence for new particle families. Called 'tetraquarks' and 'pentaquarks', each member is a cluster of four and five quarks, respectively.

Neutrinos

Neutrinos are members of the lepton particle family; they have zero electric charge, negligible mass and quantum spin 1/2, making them fermions. Wolfgang Pauli first put forward their existence in 1930 to satisfy various **conservation laws** in **radioactive decay**. The first neutrino was detected in 1956.

Neutrinos are notoriously hard to observe because they interact poorly with all other forms of matter. Indeed, billions of them are streaming through your body from space every second as you read this. Modern-day neutrino detectors use enormous tanks of water; of the many billion neutrinos that pass through the water, occasionally one will collide with a water molecule and emit a measurable flash of light called 'Cerenkov radiation'.

Stars give off a neutrino burst prior to going **supernova**, and the detection of such a shower is a prompt for astronomers to turn their telescopes on the source. Neutrinos are also seen pouring from the Sun on a daily basis, as by-products of the **nuclear reactions** that power it. Studies of solar neutrinos reveal that the particles come in three types – labelled 'electron', 'mu' and 'tau' after the members of the lepton particle family that emit them. These types can 'oscillate' – spontaneously transmuting between one another.

Standard model

The standard model is our overarching theory of subatomic particle physics. Scientists have assembled it by pulling together what they know about **quantum field theory** descriptions of the four forces of nature, and the existing classifications of particle families.

Broadly, the standard model breaks particles down into three major groups. There are six leptons – electrons, muons and tauons, plus a neutrino particle for each. Then the six hadrons, namely the quarks – these 12 particles all have **quantum spin** 1/2 and are thus fermions. Finally, there are four particles that propagate the interactions between hadrons and leptons – electromagnetic photons, the W and Z particles of the weak force and the gluons that mediate the strong force (these all have **quantum spin** 1, and so are bosons).

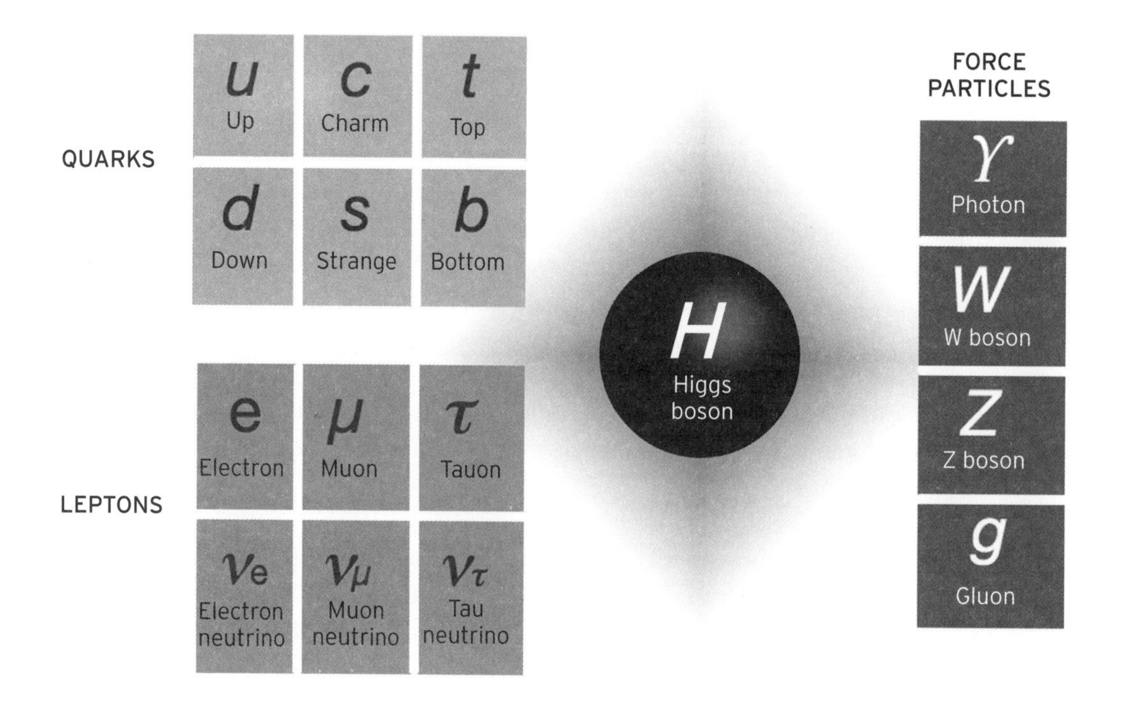

Higgs boson

The Higgs mechanism, proposed in 1964 by physicist Peter Higgs, explains why particles in the Universe have mass; an essential part of the standard model, it is yet to be detected experimentally. Higgs proposed that a field of particles, which have become known as 'Higgs bosons', permeates all of space. As a massless particle travels through space it distorts the field, causing Higgs particles to cluster around it, so it becomes massive itself. Particles attract Higgs bosons in different numbers, explaining why different particles have different masses – and, indeed, why some particles remain massless. The Higgs mechanism plays a fundamental role in the **electroweak theory**, the description of the **weak force**. The W and Z particles in this theory are massive, and this can only be explained with the help of the Higgs. The race is now on for particle accelerators to detect the Higgs boson experimentally.

Particle accelerator

The bulk of the experimental work in particle physics is done using giant machines called particle accelerators, which consist of circular underground tunnels, around which 'beams' of subatomic particles are fired – their paths curved into a circle by powerful magnets.

By carefully controlling the degree of magnetism, two opposing beams can be accelerated to near the speed of light and then collided head-on. The idea is that by smashing particles together like this they can be cracked open to reveal their inner workings. By studying the fragments that fly out from the collisions scientists occasionally spot new particles of matter. This is how researchers using the Large Hadron Collider (LHC) – the world's biggest and most powerful particle accelerator – hope to catch a fleeting glimpse of the elusive Higgs boson. The LHC is situated at the European particle physics lab CERN, on the French–Swiss border. It measures 27km (almost 17 miles) in circumference.

Quantum gravity

There is, of course, a final force in the description of the four fundamental interactions in nature: gravity. Einstein's **general relativity** provides an excellent description of gravity in the large-scale, or 'classical' world, yet attempts to construct a gravity theory that's consistent with the principles of quantum theory have run into trouble.

Whereas the electromagnetic force is mediated by the humble **photon**, the gravitational interaction requires a more complex carrier particle. Called the graviton, it's predicted to be massless and have a **quantum spin** of 2. Quantum field theories describing the interaction of spin-2 particles can be constructed, but infinities crop up in them which cannot be removed by the usual method of **renormalization**.

String theory and **M-theory** offer promising lines of research that could lead to a working theory of quantum gravity. As does an idea called 'loop quantum gravity', which supposes that the curved space of general relativity is quantized into a chainmail-like network of loops and links.

Hawking radiation

In 1974, Cambridge University physicist Stephen Hawking argued that **black holes** might not be so black after all. He showed how taking account of quantum effects in the physics of these merciless consumers of matter could allow them to give a little back. In lieu of a full quantum gravity theory, Hawking built a standard **quantum field theory** of particles on the curved space described by general relativity – an approximation known as 'semi-classical gravity'. When he did this for the space just outside the black hole's outer surface – the event horizon – he found something interesting.

Like anywhere else, pairs of virtual particles are created here, but Hawking calculated that every now and again one particle in the pair gets dragged over the horizon by the black hole's gravity, while the other escapes away. The particle that falls in has negative energy relative to the one that escapes and the net effect is a steady radiation of particles while the black hole's mass diminishes over time – as if the hole is evaporating. The process has become known as Hawking radiation.

NUCLEAR PHYSICS

Atomic nucleus

The nucleus is the central core of an atom, where its proton and neutron particles reside. Existence of the nucleus was revealed experimentally in 1909 when New Zealand physicist Ernest Rutherford and colleagues directed positively charged alpha particles (produced by **radioactive decay**) at a thin sheet of gold foil. Most of the particles passed straight through, but a very small number got deflected through large angles. Rutherford interpreted this to mean that now and again the alpha particles were running into tiny concentrations of electric charge within atoms – the nuclei. This meant that atomic nuclei must be tiny compared with the rest of the atom; indeed, if an atom could be scaled up to the size of a football stadium then the nucleus would measure about the size of a pea. Nuclear physicists have since learned how to exploit the energy locked away in the atomic nucleus – for both good and bad.

Nuclear shell model

The internal structure of the atomic nucleus is governed by the so-called nuclear shell model. Like the theory of electron energy levels, this predicts that particles in the nucleus – called nucleons – inhabit their own energy levels that fill up gradually as the nucleon number increases.

Like electron levels, the number of nucleons that each shell can hold obeys a sequence: 2, 6, 12, 8, 22, 32. But whereas atomic energy levels hold just electrons, nuclear energy levels are home to both protons and neutrons, and each have their own set of levels. This sequence leads to what are sometimes called the 'magic numbers' for atomic nuclei – the number of neutrons or protons for which a nucleus fills up its shells exactly. They are given by summing terms in the previous sequence to give: 2, 8, 20, 28, 50, 82. A nucleus containing a magic number of neutrons or protons turns out to be especially stable against radioactive decay. The nuclear shell model accurately predicts other properties such as the total **quantum spin** of atomic nuclei.

Nuclear reactions

When two or more atomic nuclei come together – or a single nucleus is bombarded with particles or photons – then a nuclear reaction may take place. Specifically, a reaction is said to have occurred when the nuclei you get out are different from the ones that went in. That's because nuclear reactions have the power to turn one **chemical element** into another. The chemical type of a nucleus is determined by its **atomic number**, or equivalently the number of protons it has; for example, nitrogen has seven protons; carbon has six. A simple nuclear reaction might involve firing a neutron at a nitrogen nucleus, the neutron gets absorbed and a proton is spat out in return – nitrogen is thus turned into carbon.

Some nuclear reactions generate energy, while others soak it up. The amount of energy liberated or absorbed is given by Einstein's famous formula $\boldsymbol{E = mc^2}$, where m is the difference between the mass that went in to the reaction and the mass that has come out.

Nuclear binding energy

The binding energy of an atomic nucleus is the energy required to overcome the **strong force** and tear it apart into its constituent particles. You might expect the binding energy per nucleon – that is, the total binding energy of a nucleus divided by the number of particles it contains – to be roughly constant for all elements. But it's not; it's small for light elements, rises to a peak at the chemical element iron and then decreases again.

Any nuclear reaction that increases the binding energy per nucleon will release energy. Think of it this way: binding energy is the energy that has to be put in to divide a nucleus. Therefore, assembling that nucleus requires the opposite – that the net increase in binding energy be released.

This rationale is the basis for **fission and fusion**. Breaking up heavy atomic nuclei or fusing together light ones generates energy – and a large amount of it, at that – which can be harnessed.

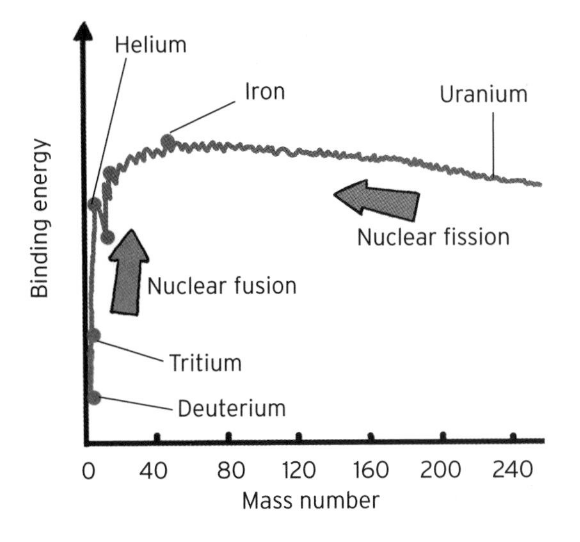

Fission and fusion

Some nuclear reactions give off huge amounts of energy – these are known as fission and fusion. In fission reactions, heavy weight atomic nuclei are split into two smaller ones resulting in the release of energy; fusion, on the other hand, involves joining together lighter nuclei. The positively charged nuclei must overcome the electromagnetic repulsion between each other (due to their protons) so that the short-ranged **strong force** can bind them together. This is done by heating the nuclei to high temperature – so that their collision energy, according to **kinetic theory**, is extremely high; the dependency on temperature is why fusion is sometimes known as a thermonuclear reaction.

In addition, a phenomenon called 'quantum tunnelling' comes into play, whereby the **uncertainty principle** temporarily gives each colliding nucleus a higher energy than it actually has. Energy released from a **nuclear reaction** is typically millions of times the energy available from burning chemical fuels such as petrol.

Chain reactions

Nuclear reactors and atomic weapons rely on the fact that as soon as one or two atomic nuclei have undergone fission or fusion, the process becomes self-sustaining – a chain reaction. Fission is started by bombarding the nuclei of heavy elements such as uranium or plutonium with neutron particles. Absorbing a neutron makes the heavy nucleus unstable; it splits apart, releasing energy. Also released is a shower of neutrons which then repeat the process with other heavy nuclei – this is a nuclear chain reaction. Loosely, fusion can also be regarded as a chain reaction; once the temperature has risen sufficiently for one or two nuclei to fuse, the energy released keeps the temperature high enough for the process to continue.

Radioactive decay

Some atomic nuclei undergo nuclear reactions without the need to be bombarded by other nuclei or subatomic particles; these nuclei spontaneously emit particles in a process called radioactive decay. The particles emitted are known as 'nuclear radiation' and come in three varieties. 'Alpha particles' are essentially the same as the nuclei of helium – bunches of two protons and two neutrons. 'Beta particles' are high-speed electrons. Atomic nuclei have no electrons to eject but these particles can still be given off when a nucleus converts one of its neutrons into a proton, plus an electron, plus a neutrino, a process driven by the **weak force**.

Especially volatile nuclei can give off the third type of nuclear radiation, 'gamma rays' – high-energy electromagnetic **photons**, which can arise as a by-product in the emission of alpha and beta radiation. They can also be emitted if particles in the nucleus have been boosted to higher shells, according to the **nuclear shell model**, and then drop back down – emitting a photon as they go, much like an electron changing energy levels. The degree of radioactivity of a nucleus is given by its 'half life' – the time taken for half the nuclei in a sample to decay.

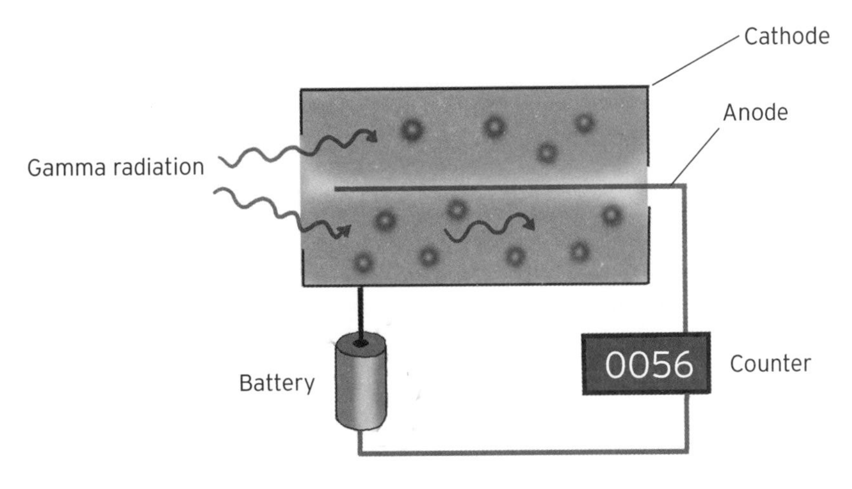

Geiger tube

Radioactive decay – or rather, the radiation it gives off – can be detected using a device called a Geiger tube. It consists of a handheld tube containing a gas that becomes briefly conductive when a particle of radiation passes through it. The tube has a central metal spike raised to a high voltage relative to the outer metal casing. A particle of radiation passing through knocks an electron from an atom in the gas to make that atom electrically charged. The high voltage then accelerates the charged atom, causing it to collide into more atoms in the gas. Electrons are knocked from these atoms by the collisions, making them charged and leading to a 'cascade' effect that culminates with a pulse of electricity flowing through the gas. The pulse can be registered on an electrical counter hooked up to the tube – or sent to a loudspeaker, to give an audible 'click'.

UNIFICATION THEORIES

Kaluza–Klein theory

Einstein's **general relativity** pulled space and time together into a single four-dimensional entity known as 'spacetime'. The crux of his theory was that spacetime curvature could then explain the force of gravity. In the 1920s, mathematicians Theodor Kaluza and Oskar Klein added **Maxwell's equations** of electromagnetism into the mix, positing the existence of a fifth dimension and used curvature of the resulting five-dimensional spacetime to construct a unified theory describing both gravitational and electromagnetic forces. We don't see the **extra dimension**, they argued, because it's 'compactified' – curled up so tightly that it's hidden from view.

To make the theory work, Kaluza and Klein also had to introduce a new field of particles pervading space into their equations. Back in the 1920s physicists didn't believe such particles fields existed; today, however, new particles are routinely predicted – and in some cases are detected experimentally. Kaluza and Klein's model is acknowledged to have formed the basis for modern ideas in the unification of the forces of nature, such as **string theory** and **M-theory**.

Extra dimensions

We're all used to inhabiting a world that has three dimensions of space and one of time, but some theories in physics suggest that there may be more to it – that there exist extra dimensions hidden from view. Extra dimensions crop up in theories that seek to unify the forces of nature. Most common are extra dimensions of space, although **M-theory** has hinted at the existence of extra time dimensions too.

The reason we don't see extra dimensions is that they have been 'compactified'. Take a sheet of paper and roll it up into a tube. As the radius of the tube gets smaller, the 2D sheet of paper begins to look increasingly like a 1D line, which is what happens when a dimension is compactified. A particle trying to move in the extra dimension soon arrives back where it started. In 2007, physicists at the University of Wisconsin-Madison suggested that studies of the **microwave background** radiation could be one way to find out how many dimensions our Universe really has.

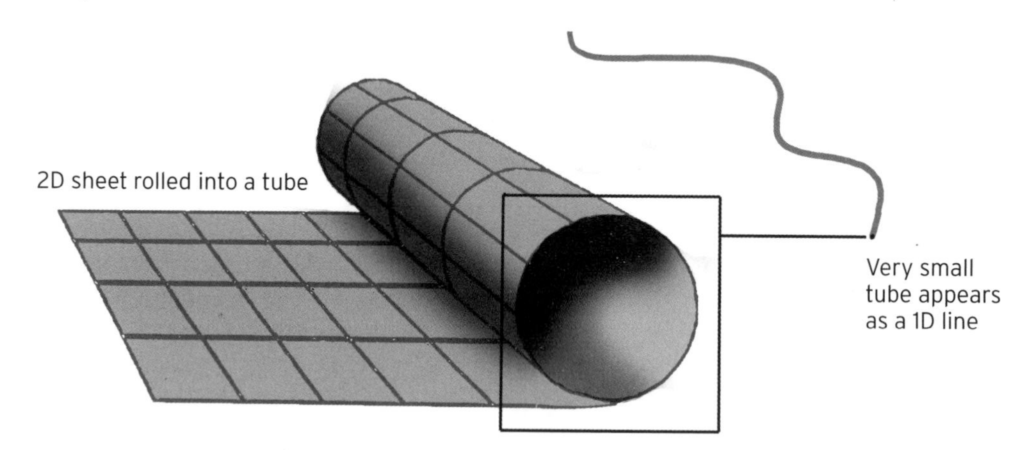

Electroweak theory

One of the defining breakthroughs of 20th-century physics was the discovery in 1968 that **Maxwell's equations** of electromagnetism and the **weak force** – the force responsible for **radioactive decay** within the nucleus of an atom – are just facets of the same fundamental phenomenon: the electroweak force. At first glance, the two seem incompatible; the weak force is just that – extremely weak. And while electromagnetism has an extremely long-range – electromagnetic waves, such as light and radio, can cross the entire Universe – the weak force is confined to the atomic nucleus.

Nevertheless, physicists Abdus Salam, Sheldon Glashow and Steven Weinberg showed how these disparate entities could be unified at high energies, such as during the first moments of the Universe after the **Big Bang**. Their theory, which incorporates spontaneous symmetry breaking, predicted two new subatomic particles – called W and Z. And these were promptly detected at the European particle accelerator lab CERN in 1973. The electroweak theory was such a breakthrough in physics that its originators were awarded the 1979 Nobel Prize.

Spontaneous symmetry breaking

This is a phenomenon in unified theories of particle physics, where forces brought together by a unified theory at high energy split apart from one another as the energy gets lower. For example, the electroweak force existed in the Universe for a few fractions of a second after its creation in the hot, high-energy **Big Bang**. Once the Universe cooled sufficiently, it underwent a phase transition and decoupled into electromagnetism and the weak force – which are what we see today.

The word 'symmetry' here refers not to simple mirror images, but to complex self-similarities in the mathematics describing each theory. During spontaneous symmetry breaking, the overarching symmetry of the unified theory breaks down into the individual symmetries of the component theories. The high-energy symmetric state is rather like a pencil balanced on end. But once the pencil topples in a particular direction then the symmetry is broken.

Grand unified theory

Just as the electroweak theory successfully bonded electromagnetism with the weak force, physicists have been searching for a counterpart theory that can bring together the electroweak interaction and the **strong force** – the force that binds together the particles in atomic nuclei. Such a scheme is known as a grand unified theory (GUT).

However, scientists are yet to agree on the form that the correct GUT should take. One reason is lack of experimental data. As with the electroweak theory, GUT models involve spontaneous symmetry breaking, meaning that the electroweak and strong forces were joined together during the high-energy **early Universe** before splitting apart as it cooled down. But the GUT unification energy is thought to be around 100 million million times higher for the electroweak force, and particle accelerators aren't yet anywhere near powerful enough to probe this range.

Supersymmetry

Theoretical physics is all about symmetries – things you can do to a theory that leave its predictions unchanged. A simple example is time symmetry. Drop a stone from your bedroom window on Tuesday, and the same equations of motion will govern its descent when you repeat the experiment on Wednesday. Far deeper and more subtle symmetries are used to characterize the complicated laws of the subatomic particle world; one example is supersymmetry. Elementary particles of matter come in one of two distinct families – called bosons and fermions. The difference between them hinges on an abstract quantity known as **quantum spin**.

Under supersymmetry, every boson particle is given a fermion counterpart, and vice versa. All of these so-called 'superpartners' would have existed in the first split second after the **Big Bang**, before a spontaneous symmetry breaking event skewed the Universe to leave just the particles we see today. Theoretical physicists find that supersymmetry, sometimes written SUSY, helps remove some of the unrealistic 'divergences' – infinite values for physical quantities such as particle masses – that are thrown up by grand unified theories and theories of everything, and which can't be dealt with by **renormalization**.

Mirror matter

Like supersymmetry partners, mirror matter is another hypothetical family of subatomic particles forecast to exist by fundamental symmetries of nature.

Mirror matter concerns a symmetry known as 'parity' – quite literally, mirror reflection symmetry. Our hands respect parity symmetry; for every left hand, there is a mirror-image right hand. Similarly, electromagnetism, the strong force and gravity all respect parity too – for every left-handed particle in these theories there is a right-handed counterpart. However, the weak force seems to violate it.

Mirror matter is an attempt to redress the balance by hypothesizing that for every particle in the theory of weak interactions there is a mirror partner. If these mirror particles exist they must only interact with normal matter through gravity. Conveniently, this makes them virtually invisible – which is why mirror matter hasn't been found yet. Although some scientists believe it has, speculating that mirror particles could account for the Universe's **dark matter**.

Theory of everything

The ultimate unified theory wraps up all four of nature's forces – electromagnetism, the strong and weak nuclear forces, and gravity – into a single coherent mathematical framework. Theoretical models such as this are known as 'theories of everything' and finding the one that describes our Universe is widely regarded to be the holy grail of modern physics. A common misconception is that the theory of everything will allow scientists to calculate the state of every single particle of matter that exists at any time,

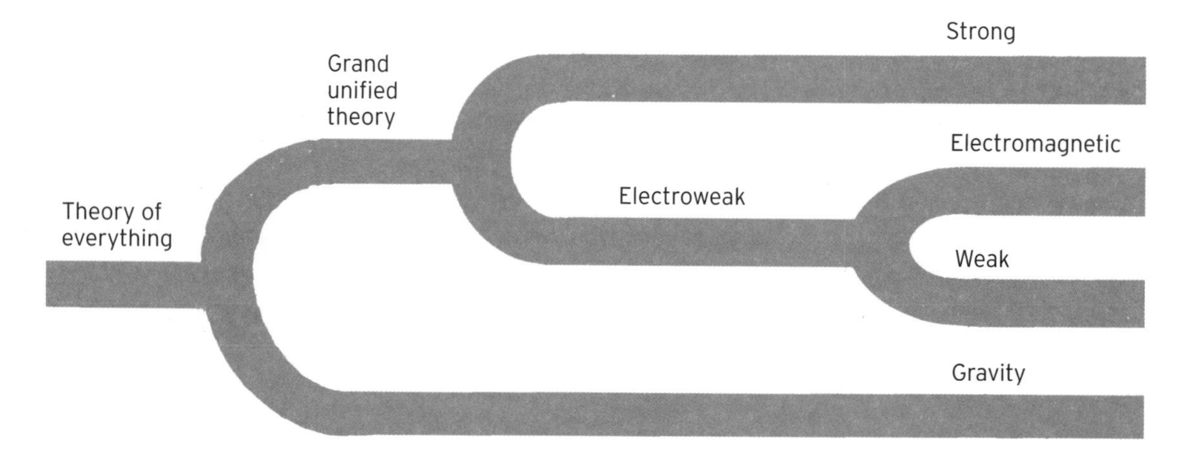

thus rigidly mapping out the entire future of the Universe. It won't – any more than the theory of electromagnetism will tell you what song is currently playing on the radio.

The principal difficulty in constructing a theory of everything is marrying up **quantum theory**, which describes electromagnetism and the nuclear forces, with Einstein's general relativity, which describes gravity. Attempts to quantize general relativity – to make 'quantum gravity' – often yield unphysical results. However, progress is being made with candidate models such as string theory and its big brother, M-theory.

String theory

Developed in the early 1970s, string theory is a new way of looking at particle physics – by doing away with the particles. The real particles that are observed in nature – such as electrons and protons – have finite size, and this led some physicists to criticize the point-like nature assigned to them in calculations. String theory is an attempt to address this by replacing point particles with entities that have some degree of extent in space – in this case, one-dimensional vibrating 'strings' of energy. The particles we know can then be thought of rather like **standing waves**, or 'notes' being played on the string.

String theory is one of several possible schemes to unify **general relativity** with **quantum theory** and thus create a theory of everything. Like unified models that have gone before, such as Kaluza–Klein theory, string theories require the existence of extra dimensions – most commonly, a total of 10 or 26 spacetime dimensions are needed. Some string theories also incorporate supersymmetry – leading to what are known as 'superstrings'.

M-theory

In 1995, the many versions of string theory that exist – with no obvious way to choose between them – led scientists to put forward a new umbrella theory, of which each possible string theory is just a special case. The new model was named M-theory. Rather than modelling particles as one-dimensional 'strings', M-theory now treated them as two-dimensional 'membranes'. The strings are still there – they're just 1D slices through these 2D membranes, with the particular orientation of the slice determining the particular variant of string theory you're looking at. For this to work, space has to have one extra dimension on top of the many dimensions string theory already demands. So, for example, all of the ten-dimensional string theories are encapsulated by a single eleven-dimensional M-theory. No one seems sure what the 'M' in M-theory actually stands for; 'matrix' and 'membrane' are common suggestions, as are 'mother' and 'master'.

CHEMISTRY

THE STUDY OF THE CHEMICAL ELEMENTS, the reactions that take place between them, and the compounds they form makes up a strand of science known as chemistry. Chemistry derives directly from physics – the laws of physics predict the behaviour of particles called electrons that orbit around the nuclei of atoms. And the physical properties of different substances, and the way they react with other substances, are all determined by their unique electron structure.

The first chemists are thought to have lived in ancient Egypt, around the year 2000 BC. They perfected chemical techniques for extracting medicines from plants, producing soap and tanning leather. In the West, the first chemists were 'alchemists', who believed that it was possible to turn base

ATOMIC MASS • CHEMICAL ELEMENTS
ES•SUBSTANCES•MIXTURES•SOLUTIONS
CTROLYTE • COMPOUND • IONIC BONDS
FORCES • METALLIC BONDS • BINARY
ES•CHEMICALFORMULAS•MOLECULAR
Y • STRUCTURAL FORMULAS • ISOMERS
MICAL REACTIONS • REACTION ENERGY
• COMBINATION AND DECOMPOSITION
ION • BUNSEN BURNER • CHEMICAL
ILIBRIUM•FREERADICALS•ANALYTICAL
LYSIS • TITRATION • EBULLIOMETRY
TALLOGRAPHY • CHROMATOGRAPHY
• LAB ON A CHIP • CHEMOMETRICS
CHEMISTRY

metals into gold (in fact, this is possible – but only through exotic nuclear reactions, not table-top chemistry).

Chemistry as a science began with the work of 17th-century philosopher Robert Boyle, who was the first to describe mathematically the behaviour of gases. His work was followed by that of French scientist Antoine Lavoisier in the late 18th century, who threw out the pseudoscientific ramblings of the alchemists and ushered in a new chemistry based on rigorous scientific principles.

Later came the theory of atoms and the discovery of how atomic properties define the chemical elements. And so modern chemistry was born.

ATOMS

Atomic numbers

Chemists and physicists classify the **electric charge** of an atomic **nucleus** by a quantity known as the atomic number, denoted by the letter Z. It's equal to the number of positively charged proton particles inside the nucleus. But the number of protons is also equal to the number of smaller, negatively charged electrons buzzing around the nucleus – and this is useful for working out the chemical properties of a particular atom.

Electron shells

Electron particles orbit around the nucleus of their atoms in a series of concentric shells, which correspond to the **energy levels** of the electrons. An atom's atomic number tells chemists how many electrons the atom has. The **exclusion principle** of **quantum theory** prevents these from all packing into the same shell. Instead they gradually fill up the shells that are available; the maximum number of electrons in each shell, from the inside out, follows the sequence 2, 8, 18, 32 and so on. If the shells are labelled by the letter n, which takes the values 1, 2, 3 and so on, then the maximum in each shell is just $2n^2$.

The outer shell of an atom's electrons is the one responsible for **chemical reactions** and is known as the 'valence shell'. The fewer electrons there are in this shell the more reactive a **substance** is. Sodium has an atomic number of 11 – in other words, it has 11 electrons. Its $n = 1$ shell is full, as is $n = 2$, while the remaining electron sits alone in the $n = 3$ shell – which is oxygen's valence shell. This makes sodium extremely reactive.

Atomic mass

Adding up the total number of protons and neutrons gives the atomic mass of an atom, denoted by the letter A. Protons and neutrons both weigh approximately the same amount: 1.67×10^{-27}kg (in **scientific notation**).

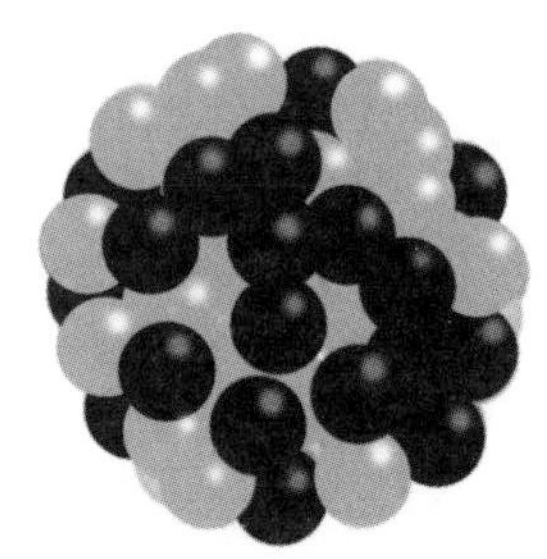

Chemical symbol, X
Mass number, A
Atomic number, Z

For example, oxygen has an atomic mass of 16 – meaning each atom weighs 2.67×10^{-26}kg. The mass of oxygen's eight electrons aren't included in this figure because they are some 2,000 times lighter than neutrons and protons and so make only a tiny contribution to the total mass of the atom. The number of neutrons in an atom's nucleus, denoted by N, is just given by the formula A–Z, the atomic mass number minus the atomic number. Oxygen therefore has eight neutrons nestled in amongst its eight protons.

Chemical elements

Naturally occurring **substances** are known as chemical elements. An element can be broken down into individual atoms. Contrast that with a chemical **compound**, of which the smallest building blocks are molecules made up of two or more elements joined together. Each chemical element is specified by its unique atomic number, and denoted in **chemical formulas** by a one- or two-letter abbreviation of its name. For example, hydrogen is denoted by the letter H and has atomic number 1; oxygen, written O, has atomic number 8; and iron, Fe, has atomic number 26.

Chemical elements are displayed graphically on a chart known as the periodic table (see below). The table is divided into vertical columns known as groups – the elements in each group have a similar number of electrons in the valence electron shell, giving them similar chemical properties. These similarities in the outer electron configuration crop up over and again as the atomic number increases – each repetition corresponds to a new horizontal row or 'period' in the table. This is why the upper half of the periodic table has a strange shape.

Period \ Group	1	2	3	4	5	6	7	8	9	10	11	12	13	14	15	16	17	18
1	1 H																	2 He
2	3 Li	4 Be											5 B	6 C	7 N	8 O	9 F	10 Ne
3	11 Na	12 Mg											13 Al	14 Si	15 P	16 S	17 Cl	18 Ar
4	19 K	20 Ca	21 Sc	22 Ti	23 V	24 Cr	25 Mn	26 Fe	27 Co	28 Ni	29 Cu	30 Zn	31 Ga	32 Ge	33 As	34 Se	35 Br	36 Kr
5	37 Rb	38 Sr	39 Y	40 Zr	41 Nb	42 Mo	43 Tc	44 Ru	45 Rh	46 Pd	47 Ag	48 Cd	49 In	50 Sn	51 Sb	52 Te	53 I	54 Xe
6	55 Cs	56 Ba		72 Hf	73 Ta	74 W	75 Re	76 Os	77 Ir	78 Pt	79 Au	80 Hg	81 Tl	82 Pb	83 Bi	84 Po	85 At	86 Rn
7	87 Fr	88 Ra		104 Rf	105 Db	106 Sg	107 Bh	108 Hs	109 Mt	110 Ds	111 Rg	112 Uub	113 Uut	114 Uuq	115 Uup	116 Uuh	117 Uus	118 Uuo
				57 La	58 Ce	59 Pr	60 Nd	61 Pm	62 Sm	63 Eu	64 Gd	65 Tb	66 Dy	67 Ho	68 Er	69 Tm	70 Yb	71 Lu
				89 Ac	90 Th	91 Pa	92 U	93 Np	94 Pu	95 Am	96 Cm	97 Bk	98 Cf	99 Es	100 Fm	101 Md	102 No	103 Lr

Isotopes

An isotope of a chemical element has the same atomic number as atoms of that element – and so the same number of protons and electrons – but a different number of neutrons. As a result, isotopes have different atomic mass numbers. The atomic mass number is often used to classify different isotopes of the same chemical element. So for example, whereas carbon-12 is the normal atomic form of carbon, with 6 protons and 6 neutrons in the nucleus, other isotopes such as carbon-13 and carbon-14 also exist.

Because isotopes result from changes to an atom at the nuclear level, their chemical properties – determined by their electron structure – are usually the same as those of the standard atom. The exception is deuterium – hydrogen with one extra neutron in its nucleus. Deuterium is twice as heavy as ordinary hydrogen and this slows down its **chemical reaction** rate considerably.

Ions

Usually an **atomic nucleus** has the same number of positively charged protons inside as it does negative electrons buzzing around it. And so the net **electric charge** of the atom is zero. Ions, however, are atoms that have lost or gained electrons to give then an overall charge.

Ions are split into two types. A 'cation' has lost some of its electrons so that the nucleus gives the atom positive electric charge. An 'anion', on the other hand, has gained some electrons to acquire a net negative charge. Cations are formed when an atom absorbs enough energy to knock an electron out of the atom. Anions are made by adding electrons to an atom's valence **electron shell**. Atoms with a full valence shell are especially stable. Like a ball rolling down a hill, an atom will always try to move into the most stable state possible. This means that if the atom is one electron short of having a full valence shell, then it's likely to capture any electrons that pass by – forming an anion.

Transuranium elements

Uranium is a naturally occurring chemical element with an **atomic number** of 92. Uranium plus most of the elements with smaller atomic numbers occur naturally on Earth. All of the elements with higher atomic numbers exhibit **radioactive decay**, which has caused all natural deposits of them to decay away. For this reason, the so-called transuranium elements must be created artificially in controlled **nuclear reactions**.

Uranium itself has applications as a fuel for reactors in **nuclear electricity** plants. A by-product of these reactors is plutonium, a transuranium element with atomic number 94, which is used in **nuclear weapons**.

Elements with higher atomic numbers are increasingly costly to manufacture – a gram of Californium (atomic number 98) costs $10 million – and they have few practical applications. Nevertheless, in the interests of science, a total of 20 transuranium elements have now been produced in labs around the world.

Moles

In chemistry, a mole is a measure of the amount of a substance. It is defined as the number of atoms in 12 grams of ordinary atomic carbon (**atomic mass** 12), and has the value 6.022×10^{23} (in **scientific notation**). This number is sometimes called the 'Avogadro constant', named in 1909 after the 19th-century Italian scientist Amedeo Avogadro, who first put forward the concept. Any mass of a substance equal in grams to its atomic mass will then contain a mole of atoms. But the mole isn't confined to atoms. It's possible to specify a mole of ions or electrons – or molecules, once you known the molecule's **molecular mass**.

CHEMICAL SYNTHESIS

Substances

It's easy to use the term glibly, but a chemical substance is the name given by chemists to **matter** that can be identified by its chemical make-up – in terms of the fundamental **chemical elements**. So the chemical element oxygen is a simple example of a substance. As is carbon dioxide, given by the chemical symbol CO_2, meaning each molecule is made of one atom of carbon and two atoms of oxygen. A chemical substance is still the same substance regardless what state it's in – **solid, liquid or gas**. So water is still water whether it's in the form of ice or steam. **Photons** of light are an example of something that's not a chemical substance, because photons aren't made from chemical elements.

Mixtures

When two chemical substances are mixed together but not bonded chemically to form a **compound**, the result is known as a mixture. Air is an example of a mixture; it is made of the gases oxygen, nitrogen, argon and carbon dioxide mixed together, but there's no such thing as a molecule of air – the atoms of the different gases do not bond together.

Mixtures can be either homogeneous or heterogeneous. In a heterogeneous mixture, the components are in different phases (**solid, liquid or gas**) and so easily separated – water and ice is an example. Homogenous mixtures have all their components in the same phase and so are harder to separate.

Mixtures comes in three different types. A 'solution' is a homogeneous mixture where the different substances are evenly distributed, e.g. stirring salt into water. A '**colloid**' is a heterogeneous mixture where microscopic particles of one substance remain – an example is milk, a mixture of tiny fat globules in water. Finally, a 'suspension' is a heterogeneous mixture with larger solid or liquid particles in a liquid or gas. Muddy water is an example. Left to stand, particles in a suspension will separate out from the medium they're suspended in – for example, muddy water separates into clean water with a layer of sediment at the bottom.

Solutions

A solution is a homogeneous mixture of two chemical substances. Even if the components are in different phases when they are combined, they quickly adjust to the same phase to form an even mixture. The process of forming a solution is known as dissolving. In a two-component solution the larger component is called the 'solvent' while the smaller component is known as the 'solute'. Dissolving sugar granules into water is an example – the solute is the sugar while the water is the solvent. Solutions such as this, with water-based solvents, are known as 'aqueous'.

Chemists refer to the ability of one substance to dissolve in another as solubility. Two substances that will not dissolve – say, water and oil – are called 'immiscible'. The strength of a solution is called its concentration, expressed as **moles** of solute per litre of solvent. This volume-based measure is sometimes known as the 'amount concentration' of the solution. Concentration can also be given by mass – the 'molal concentration' of a solution is the number of moles of solute per kilogram of solvent.

Diffusion and effusion

The natural mixing of the components of a **solution** is a process known as diffusion; random motions of molecules – as described by **kinetic theory** – makes them intermingle until they have become a homogeneous **mixture**. Diffusion occurs in both gases and liquids. Gases can also undergo a related process called 'effusion', whereby they gradually escape through microscopic holes in the container holding them. Although tiny, the holes are bigger than the molecules of the gas.

The rate of effusion is inversely proportional to the square root of the gas's molecular mass in **atomic mass** units. So in a mixture of equal volumes of hydrogen (atomic mass 1) and helium (atomic mass 4) the hydrogen will escape twice as fast as the helium. Effusion means that carbonated drinks stored for long periods in plastic bottles will eventually lose their fizz through tiny pores in the plastic.

Osmosis

Take two solutions of different concentrations and separate them by a semipermeable membrane – a thin barrier with holes small enough for molecules of the solvent to pass through, but which are too big for the solute molecules. Solvent will flow from the weaker solution to the stronger one; until the two concentrations are the same; this process is known as osmosis. The solutions on either side of the membrane fall into one of three categories: the stronger solution is known as 'hypertonic', the weaker solution is 'hypotonic', and when both have the same concentration they are 'isotonic'.

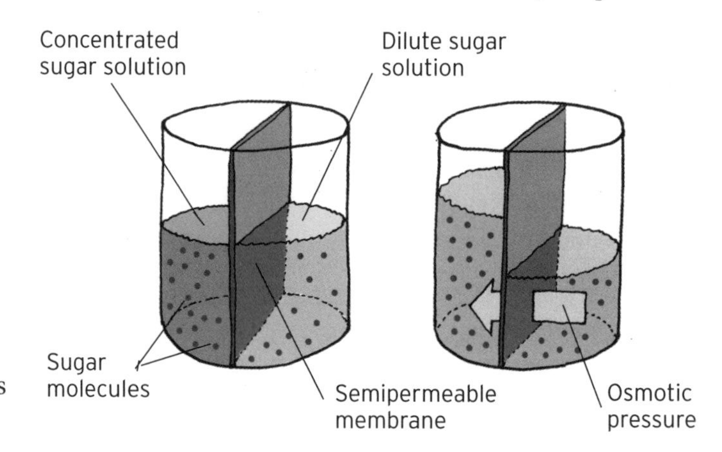

The flow of solvent across the membrane creates a pressure, known as osmotic pressure – and applying the same pressure to the hypertonic side of the membrane is sufficient to halt the process. Osmosis is essential to the functioning of living **cells**, allowing them to draw in liquid through their outer membranes.

Electrolytes

An electrolyte is a substance, usually a liquid, that conducts electricity because of the presence of positive or negative **ions**. Electrolytes are sometimes solutions, made by dissolving salts in a solvent such as water. A common example is table salt (sodium chloride), which when added to water breaks apart into positively charged sodium ions and negatively charged chlorine ions.

Electrolytes can be classified as either strong or weak, according to their ion concentration. Living organisms make use of electrolytes for transmitting the electrical signals that comprise nerve impulses (see **Neurobiology**), used for returning sensations to the brain and for making muscles work. Electrolytes are also essential in the operation of **batteries**.

Compounds

Mixtures are formed when the atoms of two chemical substances intermingle with one another. But when the atoms of the components join together to form a new substance, the result is known as a compound. Unlike a mixture, the components of a compound cannot be separated by physical processes such as filtration.

Compounds containing water are known as 'hydrous', and those that do not are 'anhydrous'. Despite absorbing water, hydrous compounds can still exist in solid form. Take, for instance, cobalt (II) chloride – the (II) just indicates that the cobalt is missing two electrons and is thus doubly ionized. This solid compound, formula $CoCl_2$, is anhydrous. But when it absorbs water it becomes cobalt (II) chloride hexahydrate $CoCl_2 \cdot 6(H_2O)$ – still solid but now hydrous. The fundamental particles of a compound – **molecules** – are specified in terms of their **chemical formula**. They are made by the formation of chemical bonds. And these can take two main forms: ionic bonds and **covalent bonds**.

Ionic bonds

An ionic bond joins two oppositely charged chemical ions by electrostatic attraction – the force that pulls together opposite **electric charges**. Ionic bonds usually form between a metal and a non-metal. Metals typically have a small number of electrons in their outer **electron shell** – known as the valence shell – while non-metals have a valence shell that is almost full (see **Metallic bonds**). Having either an empty or a full valence shell places an atom in a more stable configuration. Nature tends to favour stable configurations – that's why pencils prefer to be on their sides rather than balanced on end – and that means there's a strong chance that an electron will hop from the valence shell of the metal to the valence shell of the non-metal, making both more stable (see **Ions**). Because this also makes both atoms oppositely charged, electrical forces then hold them together as a molecule.

Sometimes more than two atoms are involved. For example, magnesium has two electrons in its valence shell. Chlorine, however, has just one space remaining. But if two chlorine atoms combine with each atom of magnesium, then magnesium can give an electron to both the chlorine atoms – placing all three in a more stable state. The negatively charged chlorine atoms then stick to the positively charged magnesium to form an ionic bond, turning the three atoms into a compound: magnesium chloride, $MgCl_2$.

Covalent bonds

As well as **ionic bonds**, another way for atoms to join together to form compounds is known as covalent bonding. Here, atoms share electrons in their valence electron shells. Two hydrogen atoms can bond together to form molecular hydrogen, or H_2 – a molecule known as an **allotrope**. Each hydrogen atom alone has one electron in its valence shell. But because this shell can hold a maximum of two electrons – and because the full shell is the most stable configuration – then pairs of hydrogen atoms can come together, sharing their electrons to put them both in a state where their valence shells are full.

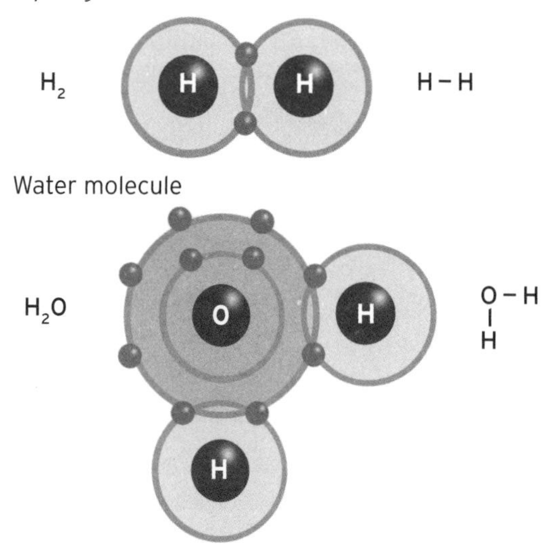

Water (H_2O) is made by covalently bonding oxygen to two hydrogen atoms. Oxygen has two spaces in its valence shell, which are filled by sharing the hydrogen electrons. In return, hydrogen atoms borrow two outer electrons from the oxygen.

Covalent bonds can be either polar or non-polar, depending on how the electrons are shared between the different atoms. The bonds attaching hydrogen to oxygen in water are polar – because the oxygen makes greater use of hydrogen's electrons than vice versa. Conversely, hydrogen–hydrogen bonds are non-polar because the electrons are shared evenly. Pairs of complex atoms can also share several electrons to form very strong double or even triple bonds.

Intermolecular forces

Forces don't just exist between the atoms that make up a molecule but also between molecules themselves. Intermolecular forces are responsible for holding molecules together. And this explains how materials shift between solid, liquid and gaseous states, depending on the temperature. There are two principle types of intermolecular force. Van der Waals forces are due to complex electrical interactions between molecules. They complicate the theoretical description of gases, meaning that corrective terms must be added to the 'ideal gas' theory (see **Equation of state**).

The other type are hydrogen bonds, which form when hydrogen within a molecule tries to bond with other nearby atoms and molecules. Hydrogen bonds are present in water ice and are also the mechanism by which **DNA** bases join together to form the base pairs that encode the genetic sequence of living organisms. Hydrogen bonds are stronger than Van der Waals forces; however, all intermolecular forces are weaker than the covalent bonds and **ionic bonds** that hold together the molecules themselves. A less important type are dipole–dipole interactions, due to the electrical attraction and repulsion of molecules with pronounced **chemical polarity**.

Metallic bonds

Metals are chemical elements that occupy the left-hand side of the periodic table. Technically, they are defined as all the elements to the left of a diagonal line joining the elements boron (B) and polonium (Po). Elements to the right of the line are non-metals. Those on the line are **semiconductors**.

Metallic bonds are the forces that hold a metal – or an **alloy** – in a rigid **crystal** lattice. It's why metals are so strong. Atoms of a metal lose the electrons from their valence **electron shell**, and these form a sea of particles that slosh between the positively charged **atomic nuclei**. This sea of negatively charged particles is free to move through the solid and is what gives a metal its high conductivity, both of **electric current** and heat. But the electrons perform another function. The strong force of attraction between their opposite electric charges and the positive charge of the nuclei hold the atoms firmly in place, making the metal difficult to stretch, bend or break. And this force is referred to as a metallic bond.

Binary compounds

A binary compound is a compound that contains two component chemical elements. Ordinary table salt is an example, made by joining an atom of sodium (Na) with an atom of chlorine (Cl) to form sodium chloride, NaCl. The name of the resulting compound is determined by a number of factors. For a covalently bonded compound, the name is given by combining the names of the two elements, changing the ending of the second element to 'ide' and adding a prefix to each depending on the number of atoms of each species needed (1 = mono, 2 = di, 3 = tri, and so on) – but drop the 'mono' if there's only one atom of the first element. Under this scheme, CO_2 is carbon dioxide, SF_6 is sulphur hexafluoride, and H_2O is dihydrogen monoxide – better known as water.

When a compound is formed by **ionic bonds**, its name is given by taking the name of the positive ion (the cation) followed by the negative ion (the anion) and changing the ending of the anion's name to 'ide' – hence, sodium chloride. Polyatomic molecules formed from ionic bonds use more complex naming conventions involving endings such as 'ite' or 'ate'. For example, sulphate is a compound made from sulphur and four oxygen atoms. Add two atoms of hydrogen and you get hydrogen sulphate, H_2SO_4, which is sulphuric acid.

Organic compounds

The majority of compounds containing the chemical element carbon are called organic. Humans and all other forms of life on Earth are based on carbon. Accordingly, some of the carbon compounds classified as organic form the building blocks for the chemistry of life – an area known as **biochemistry**. **Carbohydrates**, **DNA**, **proteins** and other nutrients are all examples of organic compounds.

However, others have little connection to life processes at all. For example, a subset of organic compounds are the hydrocarbons – made entirely by combining carbon and hydrogen in different ways. They include natural gases such as methane and propane, as well as **petrol** and other flammable liquid fuels.

MOLECULES

Molecules

Whereas an atom is the smallest indivisible unit of a chemical element, a molecule is the smallest unit of a chemical compound. Molecules range in their complexity from the very simple, like salt – made from joining an atom of sodium with an atom of chlorine – to the extremely complex, such as the **DNA** molecule that life is based on, each molecule of which is built up from millions of individual atoms.

Atoms are held together to make molecules by **covalent bonds**. Compounds formed by **ionic bonds** are sometimes not regarded as molecules in the strictest sense, as they are just ions pulled together by the attraction of their opposite **electric charges**.

Chemical formulas

The molecules making up any given compound can be specified by a chemical formula, indicating the chemical elements that each one is made from – and in what proportions. The formula represents each element by its chemical symbol with a subscript to indicate how many atoms of that element go into each molecule. For example, water is a chemical compound made from one atom of oxygen and two atoms of hydrogen, and is represented by the chemical formula H_2O.

Brackets can sometimes be used to show when one molecule has bonded to another to form a new compound. For example, each molecule of cobalt chloride, $CoCl_2$, can absorb six water molecules to become $Co(H_2O)_6Cl_2$. Ions are sometimes marked with a right-hand superscript denoting their electric charge, such as Cl^- or Cu^{2+}. Meanwhile, **isotopes** are shown by a left-hand superscript indicating their **atomic mass**. So, heavy water, which is water made from an isotope of hydrogen that has an extra neutron in its nucleus, can be written ^2H_2O.

Molecular mass

The molecular mass of a compound is the mass of one molecule measured in **atomic mass** units. It's given simply by adding up the masses of all the molecule's constituent atoms – taking care to add or subtract units if any of them are **isotopes**. A water molecule, for example, consists of two hydrogen atoms (each atomic mass 1) and an oxygen atom (atomic mass 16) – and so has a molecular weight of 18 atomic mass units. That also means that 18 grams of water contains one **mole** of water molecules.

Allotropes

Some chemical elements can exist in different forms, and each form has different properties. These are known as allotropes. Oxygen, for instance, has three common allotropes: ordinary atomic oxygen (O), dioxygen (O_2) and ozone (O_3). The latter two are molecules made by bonding oxygen to other oxygen atoms. Although ordinary oxygen is harmless, dioxygen can be harmful at high pressures, while ozone is toxic. But the word allotrope can have another meaning too, referring to different arrangements of the atoms or molecules within a solid. For instance, the most common forms of carbon are diamond – in which the atoms are arranged in a rigid crystal lattice – and graphite, where the atoms are bonded in hexagons laid over one another in layers.

The formation of allotropes has nothing to do with changes in phase, the transitions between solid, liquid and gas. However, the change from one allotrope to another can be triggered by changes in temperature and other environmental conditions. Iron, for example, changes from a body-centred cubic crystal lattice to an allotrope with a face-centred cubic structure when it is heated above 906°C (1,663°F) – see **crystal**.

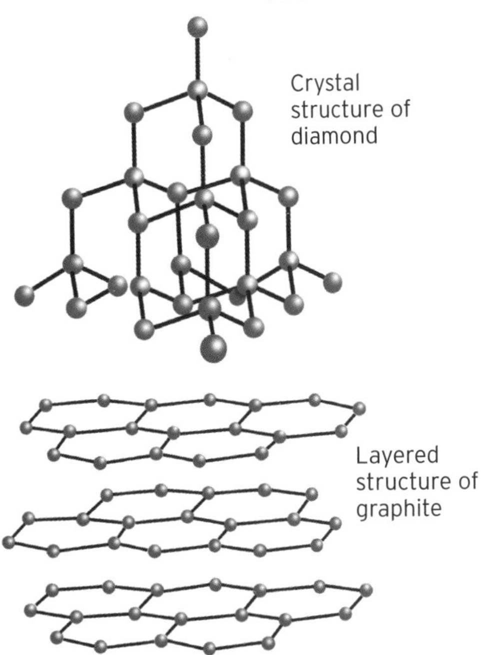

Crystal structure of diamond

Layered structure of graphite

Molecular geometry

Chemical formulas show what a molecule of any given compound is made from. But what does the molecule actually look like? How are the atoms within it actually laid out? This is the province of molecular geometry. A molecule's structure can be determined accurately from theoretical considerations, by applying **quantum chemistry** to the structure of the molecule and crunching out the results in a computer.

Of the many different molecular geometries, there are five principle types. 'Linear' molecules are the simplest, with the atoms all lying on a straight line. 'Trigonal planar' molecules form a flat triangular shape. 'Tetrahedral' molecules have a three-dimensional shape that resembles a four-sided, pyramid-like solid. 'Trigonal-bipyramidal' molecules resemble two tetrahedral molecules stuck back to back. And 'octahedral' molecules trace a shape resembling an eight-sided solid. Experimentalists can measure the molecular geometry of a compound by bouncing **electromagnetic radiation** or subatomic **particles** off of its molecules.

Linear

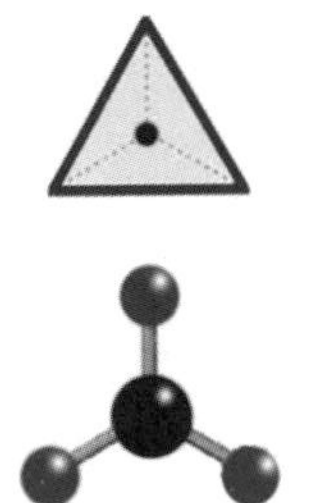

Trigonal planar

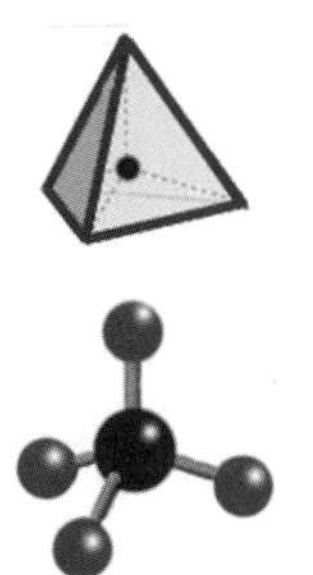

Tetrahedral

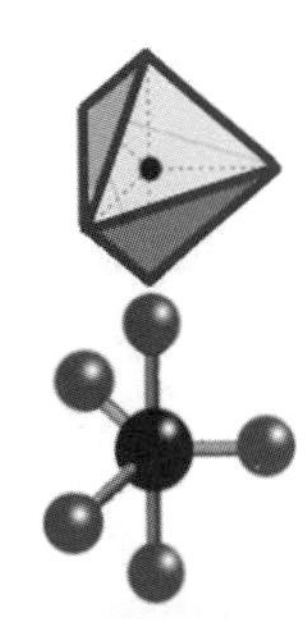

Trigonal-bipyramidal

Octahedral

Structural formulas

Structural formulas are a simple way for chemists to write down the **molecular geometry** of a chemical compound. It works by breaking the **chemical formula** for a molecule down into groups that represent how its component atoms fit together. So, for example, ethanol has the chemical formula C_2H_6O. However, its structural formula is CH_3–CH_2–OH. In other words, it looks like CH_3 (a 'methyl' group carbon) bonded to CH_2 (a 'methylene' group carbon) bonded to OH (hydroxyl group oxygen).

The structural formula can also be drawn out graphically. In the case of ethanol, it looks like the first diagram below. This kind of representation is sometimes called a 'Lewis structure'.

Lewis structure of ethanol

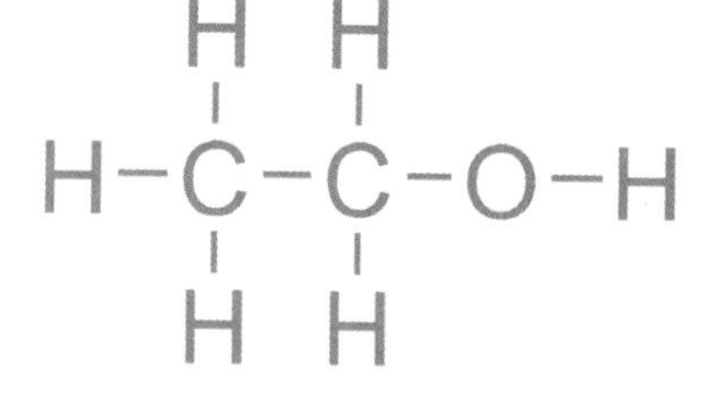

Drawing structural formulas for 3D molecules is harder. One method is known as a Natta projection, and uses a triangular-shaped bond line to show atoms sticking up out of the plane of the paper and a stripy or dotted bond line to show atoms that are behind the paper. A Natta projection for methane (CH_4) is shown below.

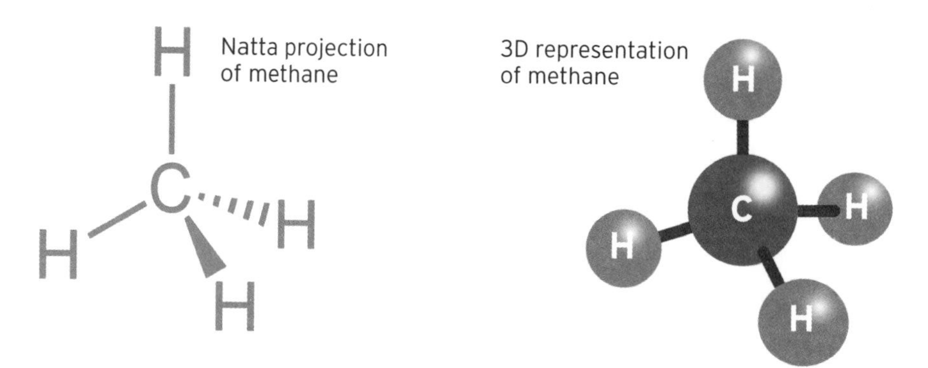

Isomers

Isomers are compounds that have the same **chemical formula** but different **molecular geometry** and, therefore, a different structural formula. Ethanol and dimethyl ether (DME) have the identical chemical formula C_2H_6O. Yet whereas ethanol has the structural formula CH_3–CH_2–OH, DME is very different, looking like CH_3–O–CH_3.

Chemical properties can vary radically between isomers of the same compound. Different isomers arise through different chemical reaction processes, which make the constituent atoms and molecules bond together in different ways.

Chirality

Take a look at your hands – they're mirror images of one another. There's no way you can physically position both hands so that they look identical. Some molecules have this property too – scientists call it chirality.

Two molecules which are mirror images of one another – and so have opposite chirality – are called optical isomers. These have an interesting property in that they rotate the plane of **polarization** of light that passes through them. These substances are also known as optically active. Those that rotate light in a clockwise direction (as seen if you're looking towards the light) are given the label '+'. Their optical isomers, which rotate light counter-clockwise, are labelled '–'. Many biological molecules can exhibit chirality, including **amino acids** – the basic units from which **proteins** are made.

Chemical polarity

The distribution of **electric charge** within a molecule, owing to the way the atoms are arranged, can give parts of the molecule positive charge while others are more negative. This asymmetry is called chemical polarity.

Perhaps the most common polar molecule is water. Its bent structure places an excess of negative charge around the oxygen atom and an excess of positive charge on the side where the two atoms of hydrogen are attached. In the diagram below, the positively charged end of water is shaded pale grey, and the negatively charged end is dark grey.

Because water is polar, other polar molecules tend to dissolve easily in water to form a solution. Polar molecules are also able to bond with other polar molecules through dipole–dipole **intermolecular forces** – with the positively charged end of one molecule attracting the negatively charged end of another, and vice versa.

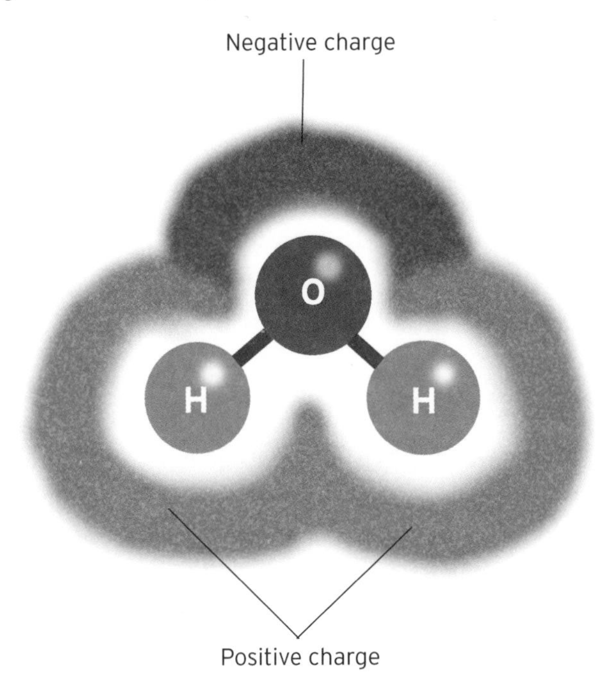

CHEMICAL CHANGE

Chemical reactions

Different chemical elements and compounds bring about changes to one another to produce new compounds, a process known as a chemical reaction. The simplest are combination and decomposition reactions, where chemicals come together or break apart to form new compounds. But chemical reactions can take a variety of forms, including **combustion** and the neutralization of **acids and bases**.

The substances put into the reaction are the 'reactants', while those that come out are known as the 'products'. Sometimes the reactants only need to be put together in order to react. At other times they may need some sort of stimulus, such as stirring, or heating – as is the case for the chemical reaction between petrol and oxygen in a motor car engine.

Some reactions even require a little chemical help, from a 'catalyst', which is a substance that speeds the reaction along, but emerges at the end unchanged. For example, the catalytic converters in cars pass exhaust fumes through a filter impregnated with a platinum catalyst. The precious metal encourages poisonous carbon monoxide in the exhaust to bond with oxygen to form carbon dioxide which is then released, leaving the platinum unaltered.

Reaction energy

Chemical bonds store energy and so breaking or creating them, respectively, releases or absorbs energy. Reactions that absorb energy are called endothermic, while those that give it off are called exothermic. As suggested by the suffix 'thermic', the energy manifests itself as heat. **Photosynthesis** is an example of an endothermic reaction. Here, plants absorb heat from the Sun and use it to react carbon dioxide with water to make glucose and oxygen. An example of an exothermic reaction is **combustion** – when a material combines with heat and oxygen to give off even more heat. The energy in or out of a reaction can be gauged by **calorimetry**.

A more dramatic exothermic reaction is the reaction produced by burning a powdered mixture of aluminium and iron oxide (rust). The powder is very hard to light, but once you do the reactants rapidly heat to produce a blob of molten iron and aluminium oxide at 2,500°C (4,532°F). This 'thermite' reaction is used in military incendiary devices and for welding.

Chemical equations

Chemists write down the reactants and products involved in a chemical reaction as a chemical equation. Equations are written using an arrow with reactants to the left and products to the right. So, for example, if compounds A and B combine to make a third compound, C, the reaction is written $A + B \rightarrow C$. If a reaction is reversible, meaning it can proceed in the opposite direction as well, then a double arrow is used, $A + B \rightleftharpoons C$.

Stoichiometry

The relative proportions of reactants and products in a chemical reaction are determined by a branch of chemistry called stoichiometry. Sometimes a reaction will involve more than one molecule of one reactant. For example, the equation $Al + O_2 \rightarrow Al_2O_3$ describes the formation of aluminium oxide. Except this equation is unbalanced – one atom of aluminium and two atoms of oxygen go in but two atoms of aluminium and three atoms of oxygen come out. The balanced form of the equation is $4Al + 3O_2 \rightarrow 2Al_2O_3$. The numbers in front of each element indicate how many molecules of each element are involved in the reaction. Now equal amounts of each element are present on both sides, and the equation is said to be balanced. The numbers are called 'stoichiometric coefficients'. A 'stoichiometric compound' is one formed from reactants combined in whole-number proportions.

Stoichiometry determines the optimum air–fuel mixtures for the internal combustion engines that power motor cars, ensuring there is just the right amount of oxygen to burn all the fuel. For petrol, the optimum air–fuel proportions are 14.7:1.

Combination and decomposition

The simplest kind of chemical reaction is the combination reaction, also known as synthesis. Here, the reactants are two or more chemical elements or compounds that join together by forming chemical bonds, to create a single product.

Decomposition is the reverse process, where a single reactant breaks down into two or more products. Often, this happens through the action of some kind of stimulus, such as heating or the passage of an **electric current**. For example, passing a current through water leads to a kind of decomposition known as electrolysis, where the current breaks water molecules apart into hydrogen and oxygen. The chemical equation for the reaction is $2H_2O \rightarrow 2H_2 + O_2$.

Redox

A chemical reaction that adds and/or removes electrons from the reactant atoms or molecules is known as a redox reaction, short for reduction-oxidation. Specifically, oxidation is a loss of electrons, while reduction refers to a gain of electrons. The two reactions occur together, resulting in the transfer of electrons from one element or compound to another.

For example, a redox reaction occurs between dihydrogen (H_2) and difluorine (F_2) to form hydrogen fluoride, with the chemical formula HF. The chemical equation for this reaction is $H_2 + F_2 \rightarrow 2HF$. This can be broken down into the oxidation of hydrogen ($H_2 \rightarrow 2H^+ + 2e^-$) and the reduction of fluorine ($F_2 + 2e^- \rightarrow 2F^-$) culminating with the formation of hydrogen fluoride ($2F^- + 2H^+ \rightarrow 2HF$).

Oxidation is so named because it was originally thought that only oxygen was capable of bringing about this sort of chemical change. Similarly, the word 'reduction' referred to the resulting depletion of oxygen as it reacted. Now there are other known oxidants, such as fluorine and chlorine – as well as reducers, such as hydrogen. The rusting of iron is a redox process, whereby electrons are transferred from iron to oxygen before the two bond together to form a flaky red compound called iron oxide.

Acids and bases

An acid is a compound that contains hydrogen so that when dissolved in water the hydrogen is released as positively charged ions, H^+. These hydrogen ions can bond to other substances and have a corrosive effect on them. Acids conduct electricity and have a sharp, sour taste.

On the other hand, a base – sometimes known as an alkali – is the opposite of an acid. It soaks up H^+ ions, thus reducing the acidity of a solution. Negatively charged hydroxide ions, OH^-, are a simple example of a base. They react with hydrogen ions to form ordinary water, chemical equation $H^+ + OH^- \rightarrow H_2O$. The combination of an acid and a base in this way is called a neutralization reaction. More complex acidic solutions usually produce some kind of salt in addition to water when they are neutralized. For instance, when hydrochloric acid (HCl) is neutralized by the base sodium hydroxide (NaOH), it forms water and table salt (NaCl).

Combustion

Combustion is an example of a chemical reaction where the **reaction energy** is exothermic. Heat given off helps the reaction to continue, or even speed up. For some materials, rapid combustion can lead to an explosion. Normally a combustion reaction is a form of oxidation (see **Redox**), combining the combustible fuel with oxygen to create reaction products and heat. When butane gas (C_4H_{10}) burns, it goes through the combustion reaction $2C_4H_{10} + 13O_2 \rightarrow 8CO_2 + 10H_2O$.

This is smoke-free burning. Some compounds produce other solid and liquid residues when they burn, leading to the formation of ash, soot and clouds of smoke – made of gases along with solid and liquid particles. For example, aluminium powder will ignite in a Bunsen burner flame, according to the chemical equation $4Al + 3O_2 \rightarrow 2Al_2O_3$ to form aluminium oxide particles.

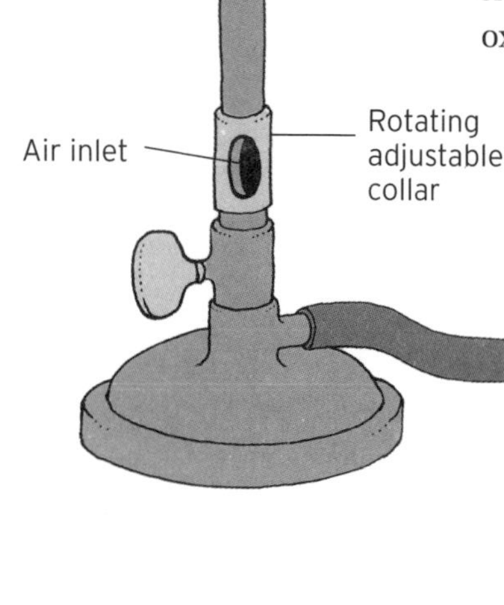

Bunsen burner

A Bunsen burner is an essential piece of laboratory equipment, widely used for heating mixtures to stimulate chemical reactions – and for other tasks such as sterilization of equipment. It was invented in 1852 by German chemist Robert Bunsen. His university was constructing a new building with piped-in gas, and Bunsen decided to make the most of it by designing a gas-powered burner that delivered a hot, clean, controllable flame suitable for scientific work.

The Bunsen burner consists of a vertical metal tube. Gas is fed in at the base and then rises to the top where it undergoes combustion. Bunsen's stroke of genius was to add an air inlet at the foot of the tube, adjustable by a rotating collar, allowing the gas to suck in air before it is burnt. With oxygen from the air mixed in, the gas burns very efficiently – giving an intense, clean blue flame with no soot to contaminate equipment. By contrast, closing the air intake puts the burner into safety mode – for use in between experiments. Without the extra oxygen, the flame is yellow and much cooler, and leaves a sooty deposit on any glassware placed in it.

Chemical precipitation

A chemical reaction that causes solid particles to form in a liquid, and then sink to the bottom, is known as precipitation. When two solutions are mixed together and a chemical reaction occurs between the solutes that produces an insoluble compound, particles of the new compound drop out of the liquid. Some metathesis reactions have this effect. The new compound is called a precipitate and, once it has settled out, the layer of liquid above it is known as the supernate.

Precipitation can also occur when a solution is mixed with a liquid in which the solute has low solubility. Such liquids are known as antisolvents. Chemical precipitation should not be confused with the process of meteorological precipitation, which is another name for rainfall.

Metathesis

Sometimes a chemical reaction is more about the bonds holding the reactants together than it is about the chemicals themselves. In a metathesis reaction, the bonds between two reactant compounds swap over to create new products. An example is the reaction given by the chemical equation $NaCl + AgNO_3 \rightarrow AgCl + NaNO_3$. Here, sodium chloride and silver nitrate swap bonds to form silver chloride and sodium nitrate.

Metathesis reactions involve reactant molecules that are joined together by **ionic bonds** – where pairs of **ions** are held together by the attraction of opposite **electric charges**. The reactions happen when the ions from which two reactants are made become more stable by breaking their ionic bonds and re-bonding with different ions from another reactant. Metathesis reactions crop up in chemical precipitation and the neutralization of acids and bases.

Chemical equilibrium

When a reversible chemical reaction – a reaction indicated by a double arrow in its chemical equation – proceeds in both directions at the same rate, and the quantities of reactants and products are both constant with time, then the reaction is said to be in chemical equilibrium.

If for any reason equilibrium of a chemical reaction is upset, then the reaction rate will increase in the direction needed to counteract the change. For instance, let's say the reaction $A + B \rightleftharpoons C$ has reached equilibrium. Now if someone adds a quantity of compound C, upsetting the balance, the reaction rate from right to left will increase, converting C into A + B more quickly than it is converted back. The process continues until equilibrium is re-established; it is known as Le Chatelier's principle.

Free radicals

Electrons orbiting in atoms and molecules tend to pair up with other electrons that have opposite **quantum spin**. Free radicals are a group of extremely reactive atoms and molecules that have an unpaired electron in their outer valence **electron shell**. It's the tendency of this electron to bind with electrons in other chemical elements and compounds that makes free radicals so reactive. Indeed, they contribute to some destructive chemical reactions, including **combustion**, food spoilage and depletion of the **ozone layer** in the Earth's atmosphere. Examples include dioxide (O_2) and hydroxyl ions (OH^-).

In biology, free radicals are cited as one cause of the process of ageing, and are implicated in the onset of **dementia, cancer** and many other diseases. Action of free radicals on the body can be combated by 'antioxidants', chemicals that soak up free radicals and block the oxidation reactions in the body (see **Redox**) that produce them. Examples of antioxidants are vitamin E, beta-carotene and the polyphenols found in wine and chocolate.

CHEMICAL ANALYSIS

Analytical chemistry

Faced with an unknown chemical compound or solution, how do chemists go about finding out what's in it? This is where chemistry meets experimental science – it's called analytical chemistry. Broadly speaking, analytical chemistry is a two-step process. The first stage is 'qualitative' analysis. This involves applying general chemical tests to ascertain what the mystery substance is made of. What does it look like? What colour is it? Is it an **acid or base**? Does it undergo **combustion** when heated, and if so, with what colour does it burn? All these tests – and others – allow chemists to build up a picture of which chemical elements and compounds are present.

The next stage is to determine the relative abundance of these elements and compounds – and this is where the second step, 'quantitative' analysis, comes in. Its aim is to apply specific chemical tests – such as titration, gravimetry and **spectrometry** – to determine the exact proportions of elements present. A substance measured in such a procedure is referred to as an analyte.

pH indicators

Indicators are chemicals used to test for the presence of **acids and bases** in a solution. The standard measure of acidity or basicity is the solution's pH, which stands for 'potential of hydrogen'. It's measured on a logarithmic scale defined as $-\log_{10}c$, where $-\log_{10}c$ is the opposite of raising 10 to the power c (so for any variable x, $\log_{10}(10^x) = x$). In this case, the number *c* is the amount of hydrogen ions (H^+) in a 10cm cube of the solution, measured in **moles**. Water has neutral pH of 7. A pH level of less than 7 indicates an acid, while a pH of more than 7 indicates a base.

A solution's pH level can be measured electrically or by using a chemical indicator, a solution that changes colour in response to the presence of acids or bases – acid turns the indicator red, while bases turn it blue. Indicators are a cocktail of many chemicals – although the juice obtained by steeping a chopped-up red cabbage in boiling water for 20 minutes is a good substitute.

Gravimetric analysis

A quantitative technique in analytical chemistry, gravimetric analysis is used to determine the mass of solute in a solution or suspension (see **Mixtures**). It works by physically or chemically extracting the solute as a solid and then weighing it. In the case of a suspension or **colloid**, the extraction might be done by filtering.

For solutions, **chemical precipitation** reactions are often used to bring the analyte out of the solvent. This might involve adding another compound to make a new solvent in which the analyte is not soluble. Once solid particles are collected, they are washed to remove impurities, dried by heating, and finally weighed on a high-precision balance. Gravimetric analysis is a simple yet effective procedure that doesn't require expensive equipment.

Titration

Titration is a way of determining the concentration of a chemical element or compound in a solution, by adding a reactive chemical and noting how much of the chemical must be added before the reaction is complete.

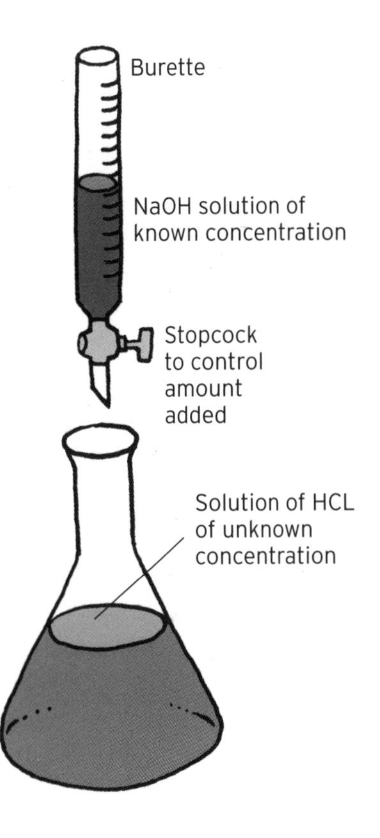

For example, the chemical reaction describing the neutralization of hydrochloric acid and sodium hydroxide into water and salt is given by the chemical equation $HCl + NaOH \rightarrow H_2O + NaCl$. Let's say we have a beaker of HCl solution of unknown concentration. A chemist could determine the concentration by dripping in an NaOH solution of known concentration and noting when the reaction has reached its endpoint. From the reaction equation, the amounts – that is the number of molecules – of HCl and NaOH must be equal at this point. The number of molecules of HCl in the beaker is then given by measuring the volume of NaOH delivered and multiplying by its concentration (see **Solutions**).

So if the concentration of NaOH is 0.5 **moles** per litre and 0.025 litres of it must be dripped into the

HCl then the number of HCl molecules in the beaker is 0.025 × 0.5 = 0.0125 moles. If the volume of the HCl solution is 0.05 litres then its concentration can be calculated as 0.0125/0.05 = 0.25 moles/litre.

During titration, the reactant is delivered from a burette – a long glass tube with a stopcock at the bottom and graduations up the side to make volume measurements easy. The tricky part of titration is measuring when the reaction has reached its endpoint. This is usually done by adding an indicator to the analyte (see **Analytical chemistry**) – such as a pH indicator – that changes colour when the reaction is complete.

Ebulliometry

Measuring the boiling point of a solution in an effort to determine the **molecular mass** of the solute is called ebulliometry. Dissolving any substance in a solvent will raise its boiling point – that is, the temperature at which the solvent turns into vapour. Chemists call this phenomenon boiling-point elevation.

When the temperature of a liquid is raised to its boiling point, its atoms or molecules escape from the liquid's surface as a gas (see **Solids, liquids and gases**). But solute particles get in the way of the escaping gas, knocking it back into the liquid. Energy of the particles has to be given extra 'oomph' to barge past the solute and escape, and this is done by increasing the temperature (see **Kinetic theory**). The resulting boiling-point increase tells chemists how many solvent particles are present. The final step is to measure the mass difference between the solution and the same volume of pure solvent which, combined with number of solute particles, gives each particle's weight.

Spectrometry

Spectrometry is done using a device called a spectroscope, which measures the chemical composition of a sample by measuring the brightness of the light it gives off at different wavelengths. Compounds and chemical elements each absorb light at specific wavelengths, corresponding to the energy of the gap between their various **electron** shells. Similarly, when a sample is heated it can become incandescent, giving off radiation at a wavelength that corresponds to the energy gap between two of its electron shell levels.

When the intensity of the light from a sample is plotted against its wavelength, this absorption and emission shows up as dips and peaks, respectively. And particular patterns of dips and peaks correspond uniquely to particular elements and compounds.

Distillation

One way to separate out liquid components of a mixture is distillation, a process relying on the fact that different liquids have different boiling points. Once a chemist knows what liquids are present, they can heat the mixture to the respective boiling points of those liquids, collect the vapour produced and then re-condense it into liquid form by cooling. Distillation is used to produce spirit beverage drinks such as whiskey. This is possible because alcohol boils at 78.5°C (173.3°F). Meanwhile, **chemical engineering** plants use a form of distillation to refine crude oil.

Crystallography

Determining the interatomic and intermolecular structure of a solid is an area of analytical chemistry known as crystallography. Atoms and molecules can be bolted together in different ways (see **Allotropes**), giving the resulting material quite different properties. The simplest structures have a cubic form, with atoms or molecules forming the nodes of a three-dimensional lattice of horizontal and vertical lines. There are more complex variations, such as body-centred cubic and face-centred cubic (see **Crystals**), as well as many more.

Crystallographers probe this structure by firing short-wavelength radiation through the sample – such as X-rays, or subatomic **particles** such as neutrons and electrons (which have an effective wavelength owing to **wave–particle duality**). The regular spacing of the atoms and molecules in the solid then acts like a **diffraction** grating, so that the emerging radiation, or particles, cast a diffraction pattern of bright and dark spots that reveal the form of the structure within.

Chromatography

Chromatography is a technique for separating out different solutes from a solution.

It works by passing the solution through a tall column packed with solid particles. Different chemicals cling to the particles to varying degrees – a process known as 'adsorption'. As the solution is poured into the column at the top and then flushed down with more solvent, chemicals that are strongly adsorbed by the particles take a long time to trickle down the column. Conversely, those which experience little adsorption will move down rapidly and reach the bottom first. This way, the different chemical components in the solution can be separated and their composition analyzed using, say, spectrometry.

The name chromatography comes from an experiment conducted by botanist Mikhail Tsvet, who invented the technique in 1906. He passed plant pigments down a chromatography column, finding them to separate into different coloured bands on their way down. In addition to analysis, chromatography can be used to separate or purify compounds before they are used in further chemical processes. A variant called gas chromatography can be used to separate the components of gaseous mixtures.

Calorimetry

Calorimetry, a branch of analytical chemistry, determines the amount of heat produced or absorbed during a chemical reaction. The energy is measured using a device called a calorimeter; the most common variety used in laboratories is the so-called 'bomb' calorimeter consisting of a sealed chamber (the 'bomb') in which the reaction takes place. The bomb is immersed in an insulated vessel of water, and a remote system enables the experimenter to initiate the reaction inside. Energy produced/absorbed then heats/cools the water, producing a rise/fall in temperature that can be measured using accurate thermometers. The well-known **heat capacity** of water then enables chemists to calculate the total energy difference.

Electrochemical analysis

A **battery** works using two electrodes made of different metals, separated by a conducting liquid called an **electrolyte**. Chemical reactions between the electrolyte and the electrode metals cause an **electric current** to flow. Electrochemical analysis works by replacing the electrolyte with a mystery substance – and uses the resulting electrical behaviour to draw conclusions about the substance's properties. Electrochemical analysis consists of tests such as measuring the voltage across the two electrodes, measuring the electric current flowing through the analyte (see **Analytical chemistry**) or applying a current to the analyte to induce a redox reaction.

Lab on a chip

The miniaturization of devices, brought about by fields such as **molecular engineering** and **microchips**, has enabled apparatus that are just a few square centimetres in size to perform chemical tests. This technology, known as 'lab on a chip', can analyze the smallest volumes of fluids, sometimes less than a picolitre (see **Scale prefixes**). These tiny labs can perform tasks such as determining the composition of a sample, and run tests on human body fluids – such as blood, saliva and urine – looking for the biomarkers that warn of illnesses including influenza and cancer.

The microscopic scale of these devices means that gone are the days when carrying out chemical tests required bulky apparatus in a large, immobile laboratory. A laptop computer connected to a few handheld pieces of kit can perform literally thousands of tests – in tandem if need be. Lab on a chip devices are of use to paramedics, field scientists operating in remote locations and for planetary probes sent to explore distant worlds.

Chemometrics

It's easy for a human observer to spot a link between one or two data variables in a chemistry experiment, say the temperature and concentration of a solution, and the rate at which a chemical reaction takes place – but when there are hundreds, or maybe even thousands of variables, human brains simply aren't up to the task. And that's where the computer-based discipline of chemometrics comes in – rather like **datamining** applied to libraries of chemical data.

The idea is to use computers to search for correlations between the data and the observed properties of chemicals and their reactions, giving chemists additional insights when confronted by similar patterns in newly gathered data. Chemometrics researchers at the University of Bristol, in England, have used the technique to figure out how the likelihood that someone is involved in drug crime is linked to the degree of drug residue found on the banknotes (dollar bills) in their pocket. The findings of their research have already been used in court.

PHYSICAL CHEMISTRY

Thermochemistry

How does heat affect chemical reactions? The interplay of temperature with chemistry is a field known as thermochemistry. Heating a substance can break the bonds between molecules (see **Intermolecular forces**), leading to phase transitions between solid, liquid and gas – that is, melting and boiling. Extreme heating can also break the bonds within molecules, turning molecules into individual atoms – water heated to above 3,000°C (5,430°F) begins to 'dissociate' into hydrogen and oxygen.

Thermochemistry is also used to calculate properties such as the **latent heat, heat capacity** and the temperature at which substances undergo **combustion**, all from chemical considerations. Scientists working in the field of thermochemistry use **calorimetry** for experimental work.

Photochemistry

Paint, ink and dye all fade after prolonged exposure to sunlight. Such fading is an example of photochemistry, the interactions of atoms and molecules with **electromagnetic radiation**. In the same way that applying heat can activate and accelerate chemical reactions, so absorbing **photons** of light can raise electrons in reactants to higher **electron shells**, increasing their reactivity. Energetic photons of light can also break chemical bonds, splitting molecules into their components. But it's not just visible light – infrared, ultraviolet, X-rays and radio waves can also influence chemical processes.

Photochemistry describes the process of **photosynthesis** – by which plants extract energy from sunlight – and photography. A photographic film is made by coating plastic with silver halide, a compound that undergoes a chemical reaction when exposed to light, converting some of it into atoms and ions of silver. In this way, the light imprints a faint image on the film which is then amplified when the film is developed. The analytical technique of **spectrometry** is also a consequence of photochemistry.

Quantum chemistry

Chemical reactions are caused by the movement and interaction of electrons between atoms and molecules, a process governed by the laws of **quantum theory**. Using quantum theory to make chemical predictions is a field called quantum chemistry. Quantum laws describe the electrons around the nucleus of an atom as wave-like entities, with the peaks of the waves giving the locations where the electrons are most likely to be found.

The theory predicts the **energy levels** and **electron shells** occupied by electrons. And this in turn leads to the prediction of physical quantities such as how reactive an atom is, how much energy it takes to remove electrons to create an ion, and even simple concepts like how big an atom or molecule is. A hydrogen atom has a diameter of about 0.1nm (where nm is nanometres – see **scale prefixes**), an atom of oxygen is 1.2nm across, and when one atom of hydrogen and two of oxygen combine the result is a water molecule, 0.278nm in size. Quantum chemistry also underpins the group and period structure of the periodic table of the chemical elements.

Electrochemistry

Electrochemistry is a discipline that has a lot in common with **electrochemical analysis**. It deals with the chemical reactions of **electrolyte** solutions with conducting **metals**. Simple electrical **battery** cells are an example of electrochemistry, operating through **redox** reactions. The electrodes of the battery are made from different metals – zinc and copper are common choices. Zinc reacts with an acidic electrolyte through oxidation, causing the metal to lose electrons and become positively charged. Conversely, the copper reacts with the electrolyte through reduction, causing it to gain electrons and become negatively charged. An **electric current** will then flow through any conductor connected across these positive and negative electrodes.

Electroplating is an electrochemical process to plate metals. It works using a similar arrangement but applying an external voltage source across the two metal electrodes. Using as electrodes a piece of silver and an iron spoon, suspended in a silver nitrate electrolyte, would result in the iron acquiring a thin plating of silver. Silver ions are formed through oxidation which are then electrically conducted through the electrolyte and deposited on the iron by reduction – creating a silver-plated spoon. Galvanization – plating metals with zinc to prevent them corroding – is another application of this process.

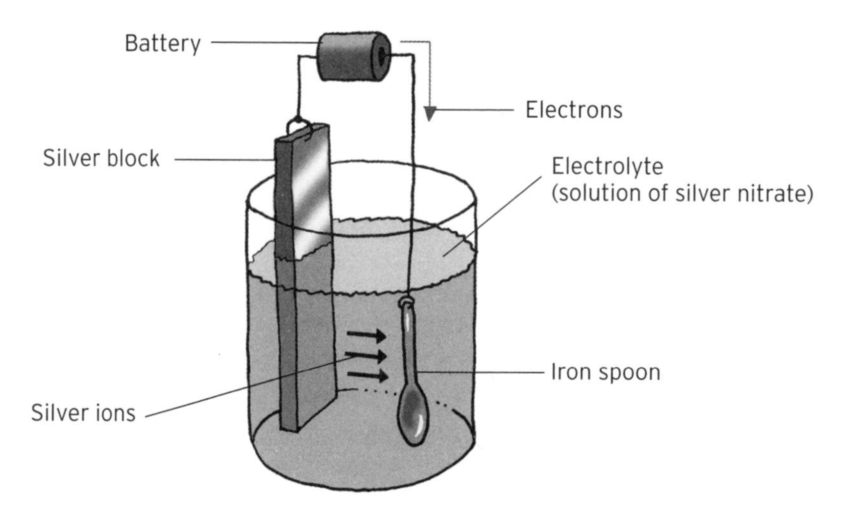

Sonochemistry

The interaction of **sound waves** with chemical systems is called sonochemistry. Technically speaking, it's not the sound waves themselves that interact with chemical elements and compounds, but rather the energy released during the collapse of bubbles seeded in a solution by the passage of sound, a process known as 'cavitation'. It happens when a sound wave briefly lowers the pressure of a small volume of liquid enough to vaporize it, forming bubbles of gas. The bubbles then collapse supersonically, triggering shock waves that heat and compress the liquid. And it is these effects that can influence chemical reactions. Ultrasound is the principle source of sound waves for sonochemistry, delivering high-intensity sound that can induce considerable cavitation in solutions. The frequency of the sound is several tens of kilohertz, kHz, just above the highest frequency a young human ear can detect – approximately 20kHz.

One phenomenon that's baffling sonochemists is 'sonoluminescence'. Here, collapsing cavitation bubbles in a liquid give off faint flashes of light. Some kind of interaction is taking place between the bubbles and the electron structure of molecules in the liquid to create the light – but so far, no one is quite sure what.

MATERIALS CHEMISTRY

Colloids

A mixture of a liquid with solid particles which are bigger than individual molecules yet small enough not to sink is known as a colloid. Unlike a solution, passing a colloid through a filter will separate the liquid from the particles. Common colloids are blood and ink. The liquid part of the colloid is known as the 'continuous medium' while the solid particles are called the 'dispersed medium'.

There exist other kinds of colloids where the continuous and dispersed media aren't liquid-solid. For example, an 'emulsion' is a colloid where both media are liquids – such as milk (fat droplets in water) and some paints. Aerosols are another – these colloids are made of solid particles dispersed in a gas. And foam is a colloid made of gas dispersed into a continuous liquid. In fact, the only kind of colloid that doesn't exist is gas-gas, as all gases can be mixed freely with one another.

Auxetics

Stretch a lump of solid material and you would expect it to become thinner around the middle as you tried to pull it apart. Not so with auxetic solids – these solids actually get fatter as you stretch them out. The secret to making them is to engineer the bonds between molecules so that when the structure is stretched it also expands in a direction perpendicular to the stretching. Imagine each line on the diagram, right, is a length of wood and at each corner there is

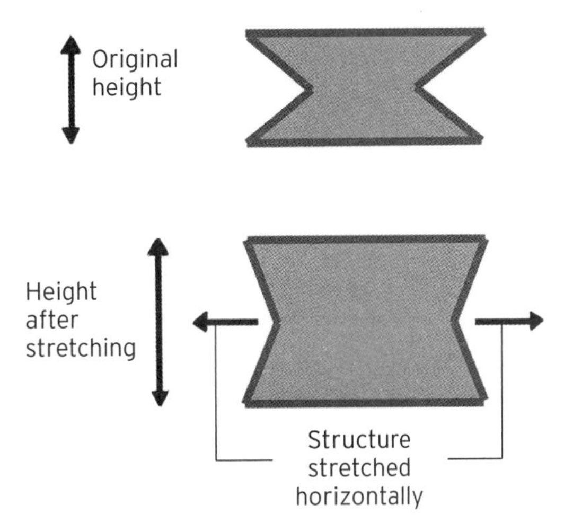

a hinge. If you pull this structure apart horizontally, then the horizontal lengths of wood at the top and bottom move vertically outwards. Now replace the pieces of wood with intermolecular bonds, and the hinges with molecules, and you start to see how an auxetic material might look on the smallest scales.

The field of auxetics is still in its infancy, but there are already minerals and manmade **macromolecules** known that exhibit auxetic behaviour. Future applications for these amazing materials could include body armour, pipes that can be unblocked simply by pulling on them and building materials that self-heal cracks.

Surfactants

Wash your windows and the water's **surface tension** causes it to contract into beads on the glass, leaving some areas dry. A surfactant is a chemical that lowers the surface tension of water, and other liquids, allowing them to spread out and form an even film over the glass. That's why surfactants are used in detergents, where the reduction of surface tension enhances the ability of the water to wet surfaces that it comes into contact with.

A surfactant – short for surface active agent – is made up of elongated molecules, one end of which is attracted to water (hydrophilic) while the other end is repelled by it (hydrophobic). When a surfactant is added to water, the molecules form a layer on the surface with their hydrophilic ends pointing into the water and the hydrophobic ends pointing outwards. With no water at the surface to create surface tension, the liquid is now able to spread out and coat the glass of your windows evenly. Common surfactants used in detergents include sodium laureth sulphate and cocamidopropyl betaine.

Alloys

An alloy is a mixture of a metal and another chemical element or compound, formulated to give desirable engineering properties. If an alloy is made from two metals (see **Metallic bonds** and **Ionic bonds**) both components are usually melted into a molten state, and then mixed to form a solution, which is then solidified. Brass, an alloy of copper and zinc, is made this way. Some alloys are made from a metal and a non-metal. Steel, for example, is made by mixing iron (metal) with a precisely controlled amount of carbon (non-metal).

Whereas pure metals have well-defined melting points, this isn't the case with alloys. They usually have a range of temperatures for which they begin to soften, as the lower melting-point components become liquid, only becoming completely molten when the temperature rises higher. Alloys are tailored to their intended purpose, with different elements and compounds added to give just the right properties such as tensile strength, corrosion resistance and elasticity.

Crystals

When the atoms or molecules of a solid bond together to form a regular, rigid lattice the resulting material is called a crystal. Diamond and quartz are examples, although the atoms and molecules of most metals also form a crystal lattice, the formation of which depends on conditions such as temperature and pressure. This can result in different crystal structures, known as **allotropes**, being formed from identical chemical elements or compounds. On the other hand, solids that retain the same crystalline structure when they go through a chemical reaction are known as 'allomers'.

Particles can be held in a crystal lattice by all kinds of **intermolecular forces**, including the attraction of opposite **electric charges** and hydrogen bonds. Arrangement of atoms or molecules within crystals can take various forms (see diagram). As particles are added to a crystal, they can amplify this underlying structure, giving some chunks of crystal a striking angular or polygonal shape. Studying the crystal structure of solids is a field known as **crystallography**.

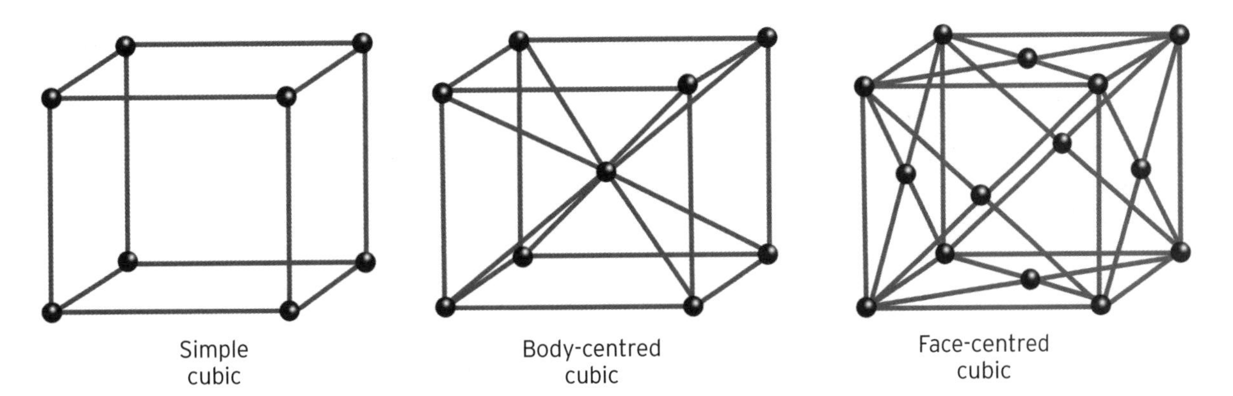

Amorphous solids

At the opposite end of the scale from crystals, which have a well-ordered atomic and molecular structure, are the amorphous solids, which have a highly disordered structure. Common examples are glass, amber and many plastics. Technically speaking, it's incorrect to say amorphous solids are totally lacking in structure. For example in glass, silicon dioxide (SiO_2), individual molecules bond to their neighbours according to well-defined rules – the silicon in one molecule will bond only to an oxygen part of other molecules.

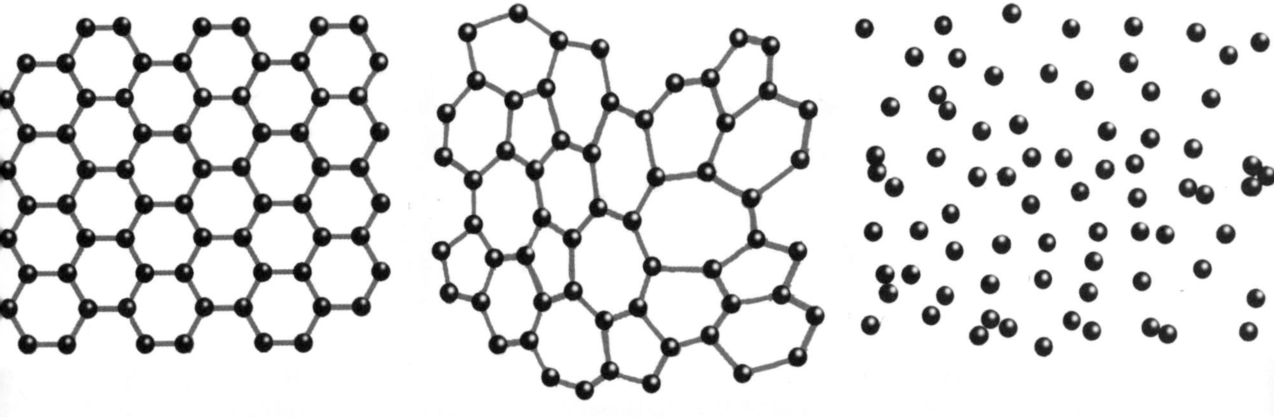

But in glass there is none of the structure across large groups of molecules that's present in crystals, and this is typical of all amorphous solids. Amorphous materials are also sometimes called 'vitreous'.

Pyrophoric solids

Pyrophoric solids are materials that will ignite spontaneously upon contact with air.

All materials have what's called an autoignition temperature, above which they will undergo **combustion**. Most materials commonly encountered have a high autoignition temperature, meaning that a flame or other intense heat source must be applied to start them burning. Pyrophoric solids, however, have an autoignition temperature that is room temperature or below.

The autoignition temperature of a material is determined by its readiness to undergo oxidation reactions (see **Redox**), combining with an oxidant (normally oxygen in the air) to release heat through combustion. Pyrophoric materials are extremely hazardous to handle and they are usually stored in tightly sealed containers that have had all the oxygen inside replaced with an inert gas such as helium. Examples of pyrophoric solids include plutonium and phosphorus (used in military incendiary bombs and tracer bullets). Some gases and liquids are pyrophoric too, such as hydrazine rocket fuel.

Macromolecules

Large molecules, each containing hundreds or even thousands of atoms, are called macromolecules and are built up from smaller units called 'monomers'. Each monomer is a molecule in its own right but, unlike **crystals** – which are joined by **intermolecular forces** – macromolecules join their molecular components together by means of **covalent bonds**. A macromolecule is made from these repeating building blocks 'bolted' together over and over again.

Because macromolecules are built from many monomers, they are sometimes known as 'polymers'. Polyethylene, a polymer used to make disposable plastic shopping bags, is made from a large number of monomer molecules called ethylene (C_2H_4), bolted together to form long polymer chains. Indeed, all plastics are polymers. Macromolecules also play an important role in biochemistry. **Proteins**, **carbohydrates**, **lipids** and nucleic acids such as **DNA** and **RNA** are all polymers.

Plastics

Plastics are synthetic materials made from hydrocarbon macromolecules. The name 'polymer', an alternative term for macromolecules, has become largely synonymous with plastic, however not all polymers are plastics. Plastics come in two principle varieties. 'Thermoplastic' materials re-soften each time heat is applied, allowing them to be remoulded and reused many times. However, 'thermosetting' plastics can only be heated and softened once, after which they will remain solid when reheated.

Bakelite was an early thermosetting plastic developed in the first decade of the 20th century, and used to make casings for electrical goods. It is made by mixing phenol (C_6H_5OH)

and formaldehyde (CH_2O). Most plastics are made of molecules built up from the same basic chemical elements – hydrogen, carbon, oxygen, nitrogen, chlorine and sulphur. Common plastics you may encounter are polyvinyl chloride (PVC), used in plumbing; polyamides, used to make nylon stockings; and polystyrene, used for plastic cutlery and, in its expanded form, foam packaging.

Combinatorial chemistry

Combinatorial chemistry is all about making tiny tweaks to the structure of a complex molecule to see how its properties change. This is done using computer-controlled robotic lab systems to synthesize and assess huge numbers of molecules en masse.

Components of large molecules can be put together in thousands of different ways and combinatorial chemistry allows this huge parameter space to be rapidly explored. It has proved especially fruitful in the hunt for new pharmaceutical drugs, where combinatorial chemistry techniques now turn out 100,000 new compounds every year. The properties of these compounds are added to a computerized database that can be queried whenever researchers are looking for a new chemical with specific properties. One drug developed this way and now approved for clinical use is Sorafenib, used to treat **cancers** of the liver and kidneys.

Computational chemistry

Computational chemistry is the application of computers to chemistry. **Chemometrics** is one branch – along with its big brother 'chemoinformatics', an area using **datamining** techniques to trawl through the vast libraries and databases created through the course of combinatorial chemistry research. Say, for instance, chemists are looking for a compound with specific properties but it isn't in their database. How can the properties of the closest compound they have be tweaked to make it fit the bill? Chemoinformatics researchers can help by applying algorithms to find cases in the database where the properties of other compounds have been shifted in the same way – and thus advise experimentalists on new approaches they can try.

Other computational chemists are involved in 'molecular modelling', using computer-based theoretical calculations to unlock the properties of molecules. Many applications of molecular modelling lie in **biochemistry** research, building the theoretical understanding of molecules including **proteins** and **DNA**.

BIOLOGY

THE THIRD OF THE BIG THREE SCIENCES (after physics and chemistry), biology is about applying the principles of chemistry to explain the science of living things. It covers everything from the operation of the smallest biological units – cells – through to microorganisms, the structure and behaviour of plants and animals, as well as explaining how these organisms arose and evolved in the first place.

The history of biology begins with the ancient Greeks, who were inveterate studiers of fauna and flora. The real theoretical understanding of biology, however, didn't begin until much later, with the invention of the microscope in the late 16th century; this enabled the discovery of cells and later became instrumental in fathoming their structure and how they work together inside organisms.

RBOHYDRATES • LIPIDS • NUTRIENTS •
ERGETICS • RECEPTORS • HORMONES •
RYOTES • PROKARYOTES • CELL NUCLEUS
• RIBOSOMES • PLASMIDS • AUTOPHAGY •
AR DIFFERENTIATION • MICROORGANISM
IMAL • MICROBES • MICRO-PLANTS AND
UCLEOTIDE • DNA • DOUBLE HELIX • RNA
GOSITY • GENETIC DOMINANCE • GENETIC
S • GENE • SEQUENCE • HUMAN GENOME
MAINS • KINGDOMS • PHYLA • CLASSES •
HISTORY • ANIMALS • MORPHOLOGY
MENTAL BIOLOGY • ETHOLOGY
PHOTOSYNTHESIS •

The big biological breakthrough of the 20th century was genetics. This was the discovery that all of the information determining the structure of our bodies – such as how many fingers we have and how our organs work – as well as our identifying features such as hair and eye colour, is all encoded on a molecule called DNA that lies at the centre of every single one of our cells.

Now genetics is promising further exciting discoveries. While the 20th century was the golden age of physics, some are predicting that the 21st century will turn out to be the golden age of biology.

BIOCHEMISTRY

Amino acids

Biochemistry is the study of the chemical make-up of living organisms and the interactions between the chemicals that are responsible for sustaining life. Some of the fundamental chemical building blocks of life are a group of **organic compounds** known as amino acids.

Amino acids join together into short polymer chain molecules (see **Macromolecules**) known as 'peptides', or longer 'polypeptides', which consist of several hundred amino acid molecules. A common kind of polypeptide playing a major role in biology are protein molecules. Amino acids also contribute in other areas such as **metabolism** and **neurobiology**; consequently, amino acids are an essential part of any animal's diet. They also have applications in technology, including **chemical engineering**.

Proteins

Proteins are polypeptide molecules made up of long chains of amino acids bonded together. They are built from basic amino acids ingested by an organism and assembled according to the instructions encoded within the organism's **DNA**. The resulting different types of protein create the different types of tissue that serve a multitude of different roles within the organism (see **Cellular differentiation**).

A protein's properties can be quantified by a number of factors that describe its structure. The sequence of amino acids making up the protein is known as its 'primary structure', and the resulting molecule can then coil around itself like a telephone wire – the form of this coiling is called its 'secondary structure'. The overall 3D shape of the resulting coil is the 'tertiary structure' and the process by which the molecule assumes its 3D structure is known as 'protein folding'. Interactions between a number of protein molecules can form a final structural type called the 'quaternary structure'. Study of proteins and their role in living cells is a field of science called 'proteomics'.

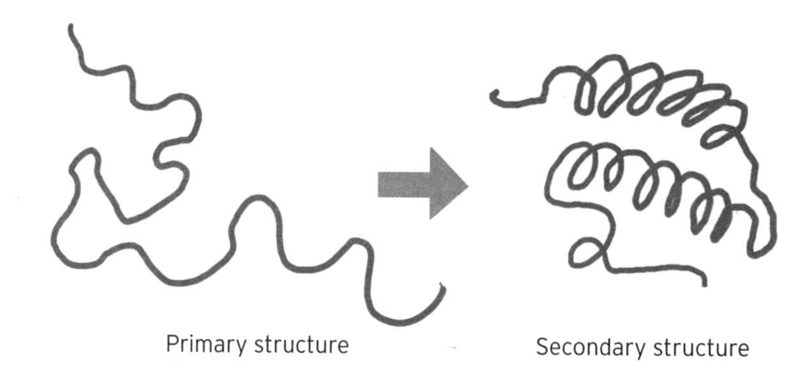

Enzymes

An enzyme is a special kind of protein that acts as a catalyst to speed the rate of a biological chemical reaction. Like other proteins in a living organism, enzymes are created according to the instructions within DNA; this enables DNA to make different kinds of **cell**, serving different biochemical purposes through their different enzyme concentrations. Enzymes are important in a great many biological processes such as the digestion of **nutrients**, muscle action (see **Musculoskeletal system**) and in propagating chemical signals through cells.

Changes in acidity (see **Acids and bases**) and temperature (see **Heat**) also influence the activity of particular enzymes in a process called 'enzyme inhibition'. Enzymes are named by adding the suffix 'ase' to the substance which they act on – for example, lactose is broken down by the enzyme lactase.

Enzymes aren't exclusive to natural biology. They are used in technological applications such as biological washing powders, where they help to increase the rate of chemical reactions involved in the breakdown of stains caused by biological matter such as fat (see **Lipids**), blood and **plant pigments**.

Carbohydrates

Carbohydrates are **organic compounds** made of carbon bonded to molecules of water; they include sugars and starches. Organisms use them primarily as a source of energy, though carbohydrates also play a role in the operation of **cells** and are important in the structure of plants and the hard shells of some invertebrate animals (see **Vertebrates**). Carbohydrates often carry the suffix 'ose'. For example, the simplest carbohydrates are sugars, such as sucrose, fructose and glucose. These sugars are bonded together to form larger molecules, called 'polysaccharides', to form 'complex carbohydrates' such as starch and glycogen, which is used by living creatures for energy storage.

Despite the popularity of low-carb diets, most experts in **food science** recommend that adult humans get most of their dietary energy from complex carbohydrates, by eating foodstuffs such as pasta, potatoes and bread.

Lipids

Another group of **organic compounds** important for energy production and storage in organisms are fats. And these are part of a broader group of molecules known as lipids, which also include waxes, steroids such as cholesterol, and certain kinds of vitamins (see **Nutrients**) that can form solutions in fat. Like carbohydrates, lipids are a component of cell membranes and important for energy production and storage – and for chemical signalling processes that assist in the functioning of biological cells.

Fat-based lipids are an essential nutrient in the human diet, needed to dissolve the vitamin groups that are only soluble in fat – namely, vitamins A, D, E and K. Meanwhile, certain fatty acids, such as omega-3, are believed to have beneficial properties of their own in staving off disease. However, excessive intake of saturated fats (chiefly animal fat) and so-called 'trans fats' (such as partially hydrogenated vegetable oils, which have had hydrogen added to increase their shelf life) have been shown to lead to an elevated risk of heart disease.

Nutrients

Nutrients are organic chemicals essential to the survival of biological organisms; they provide organisms with energy, material with which to repair damaged tissue, as well as chemicals needed for life processes.

Animals, including human beings, require three main groups of nutrients – **proteins**, **carbohydrates** and fats (see **Lipids**). These are supplemented by various vitamins – chemicals that promote healthy bones, skin, vision and nervous system – as well as minerals such as iron, which is essential for making the red blood cells needed for transporting oxygen around the body. Proteins, carbohydrates and fats are examples of 'macronutrients' – chemicals needed in large quantities – while vitamins and minerals are 'micronutrients', only needed in small amounts.

Metabolism

An organism's metabolism is the set of chemical processes by which it both extracts energy from the nutrients it consumes and then uses this energy for basic life functions and for growing new cells. Metabolism is governed by enzymes; the particular set of enzymes an organism possesses determines which are the most effective chemical routes – known as 'metabolic pathways' – to produce energy and put it to use. A metabolic pathway consists of a series of **chemical reactions**, each one sped along by a particular enzyme catalyst. In this way, the particular suite of enzymes at an organism's disposal determines which foods are nutritious to it and which are poisonous.

The metabolic pathways that break down molecules – for example, to digest food – are referred to as 'catabolism', while the pathways that manufacture new molecules – such as **proteins** and other components of cells – are known as 'anabolism'. The metabolism requires its own source of energy to run, which it spends at a pace known as the 'basal metabolic rate' (BMR). Someone with a fast BMR – sometimes known simply as a 'fast metabolism' – can eat more food without gaining weight than someone who has a slow BMR.

Chemosynthesis

Whereas animals derive their energy and nutrients from the digestion of **organic compounds**, and plants make energy through **photosynthesis**, some **microorganisms** have a third method at their disposal – chemosynthesis. These microorganisms – known as

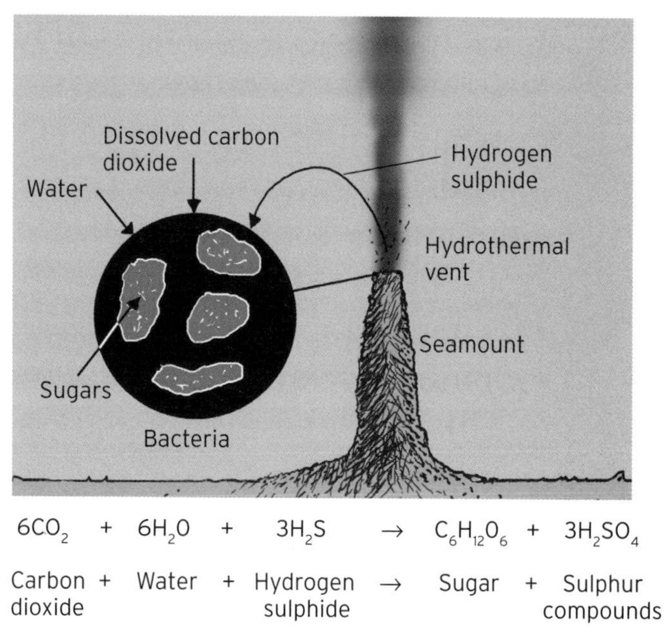

$$6CO_2 + 6H_2O + 3H_2S \rightarrow C_6H_{12}O_6 + 3H_2SO_4$$

Carbon dioxide + Water + Hydrogen sulphide → Sugar + Sulphur compounds 'chemoautotrophs' – are able to ingest inorganic chemical compounds and convert them to organic compounds through oxidation chemical reactions (see **Redox**).

Chemoautotrophs inhabit environments where there is little natural sunlight or organic matter to feed on, such as in the dark depths of the ocean. Here, **hydrothermal vents** in the sea floor can provide a source of inorganic chemicals, such as hydrogen sulphide, to sustain chemoautotrophic bacteria.

Bioenergetics

Study of the energy flowing through living organisms is a field called bioenergetics, which involves calculating the energy budget for the organism, and balancing the energy intake – from food and/or sunlight – against the energy expenditure in terms of metabolism, growth, waste and **heat** losses.

Organisms store energy through the chemical bonds (see **Ionic bonds**) between their constituent chemicals. When weakly bonded **atoms** and **molecules** are rearranged into more strongly bonded **chemical compounds**, the extra energy in these bonds can be released and used at a later time by breaking the bonds. In the body the main molecule serving this purpose is adenosine triphosphate (ATP).

Receptors

Similarly to the radio receivers of biochemistry, receptors are molecules that sit on the outside of cells and bond to other chemicals. The other chemical acts like a messenger, altering the chemical properties of the receptor and in turn triggering a response within the cell. Receptors are proteins found in the outer 'plasma membrane' of the cell and each receptor is tuned to respond to a specific kind of messenger chemical. These might be chemicals produced by other parts of the organism, such as hormones or neurotransmitters (see **Neurobiology**), or they may be engineered chemicals designed to have a beneficial effect on the cell (see **Pharmacology**) – or even a detrimental one (see **Toxicology**). The chemicals that bond to a receptor are sometimes known as 'ligands'; they change the 3D structure of the receptor protein and that is what triggers the cell's response to the ligand's arrival. The response is normally a secondary chemical process within the cell.

Hormones

Hormones are chemical messengers that carry signals from one part of an organism to another, and regulate everything from your mood to telling you when you're hungry. They work by binding to receptors within cells; for example, when receptors in liver or muscle cells bond to the hormone insulin they stimulate the cells to absorb the carbohydrate glucose from the bloodstream and store its energy as the molecule glycogen.

Hormones can either be transmitted via the blood (known as 'endocrine' hormones) or through a dedicated network of ducts – called 'exocrine' hormones. Some hormones also carry messages within cells, and these are called 'intracrine' hormones. Glands made up of specialized groups of cells produce hormones. Examples of these glands include the thyroid gland, which produces thyroxine – which controls metabolism – and the adrenal gland, responsible for producing stress hormones such as epinephrine, which increase the body's performance at times of increased demand. All large organisms, such as plants and animals have a hormone system.

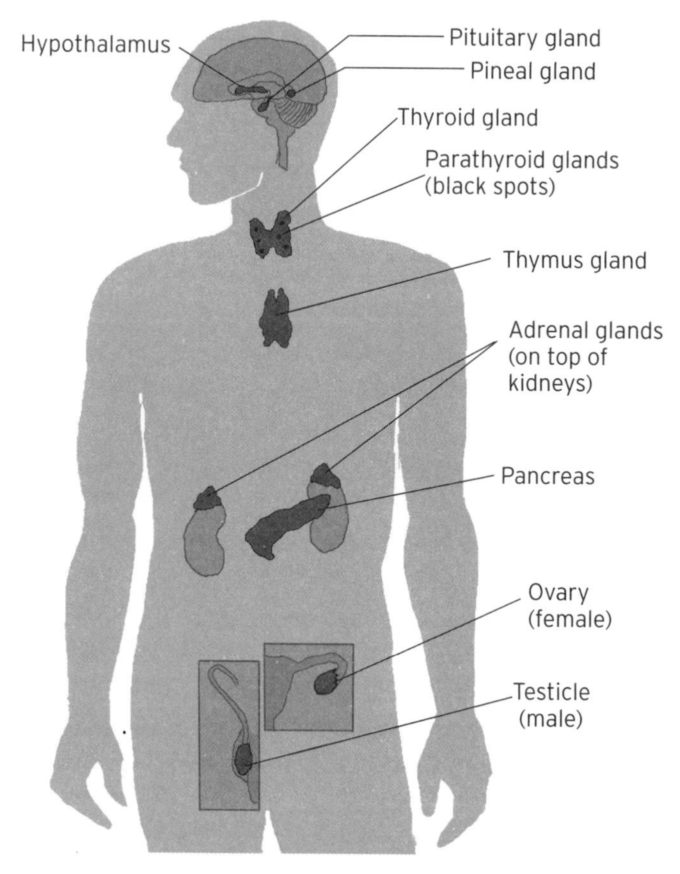

Homeostasis

Maintaining the processes essential for life to exist requires careful regulation of the conditions inside an organism, to ensure that parameters such as temperature and chemical composition, remain within acceptable bounds. Regulating these parameters is a process known as homeostasis.

Perspiration and panting to regulate temperature are examples of homeostasis at work. Similarly, chemical processes keep the chemical balance of cells in check; for example, 'osmoregulation' controls the balance of fluids and salts in animal cells through **osmosis**. Meanwhile, chemical waste products are filtered from the blood by the kidneys and excreted as urine. Other mechanisms moderate **blood** glucose, food intake, acidity in cells and the operation of the **immune system**. Nearly all biochemical homeostatic processes are controlled by hormones.

Respiration

Respiration is the name given to a set of **chemical reactions** that occur within cells to turn biochemical energy from nutrients in food into the molecule ATP (see **Bioenergetics**). ATP can then be shipped around the organism and tapped into whenever energy is required at a later time.

Respiration takes one of two forms. 'Aerobic respiration' involves oxygen – here, nutrients such as amino acids from protein, as well as carbohydrates and lipids, undergo oxidation reactions (see **Redox**) to form ATP. When energy is needed ATP reacts with water, breaking chemical bonds to form adenosine diphosphate (ADP) and releasing the bond energy. 'Anaerobic respiration', on the other hand, generates ATP without the input of oxygen, a process also known as 'fermentation'. One type occurs in human muscles when there is insufficient oxygen in the blood – for example, during extremely strenuous exercise such as sprinting. Again, blood sugars are converted into ATP, but this time there's a by-product – lactic acid, the acid that produces the familiar burning sensation in the over-worked muscle.

CELL BIOLOGY

Cells

The basic unit of all living organisms is the cell and large animals and plants contain trillions (10^{14} – see **scientific notation**) of them. These organisms are called 'multicellular', however, there are life forms – such as bacteria – that are composed of just one, and these are called 'unicellular'. Cells are made of protein material and are the fundamental machines of biology, where all the processes responsible for sustaining life take place: energy production, tissue growth, homeostasis and **hormone** production. Each kind of cell has a particular function determined by the cocktail of **enzymes** within it, which influence the rate of particular biochemical reactions. Cells come in two principle types: **prokaryotes** and eukaryotes, and individual components of cells are called 'organelles'. The cell theory of biology was first put forward in 1839 by German biologists Matthias Jakob Schleiden and Theodor Schwann. However the term 'cell' is much older – first used by the 17th-century English scientist Robert Hooke.

Eukaryotes

Eukaryotes are the most common of the two main types of cell, and form the constituents of almost all multicellular plants and animals. However, some single-cell organisms are also eukaryotes, known as **protists**. Sometimes the term eukaryote is also applied to the complete organism as well as its component cells. Eukaryotic cells measure from around a few microns (thousandths of a millimetre) up to about a millimetre in size and are more complex and therefore larger than the other main cell type, the **prokaryotes**.

A typical animal eukaryote cell has an outer sheath, known as the plasma membrane. The body of the cell within is referred to collectively as the 'cytoplasm', and consists of a watery liquid, called 'cytosol', immersed in which are the various organelles, each enclosed in their own membranes. Each organelle carries out a specific function within the cell. **Mitochondria** generate most of the cell's energy, in the form of ATP; enzymes are stored in cavities known as vacuoles, vesicles and lysosomes; while **ribosomes** are the protein factories of cells, manufacturing new tissue according to the blueprint stored in

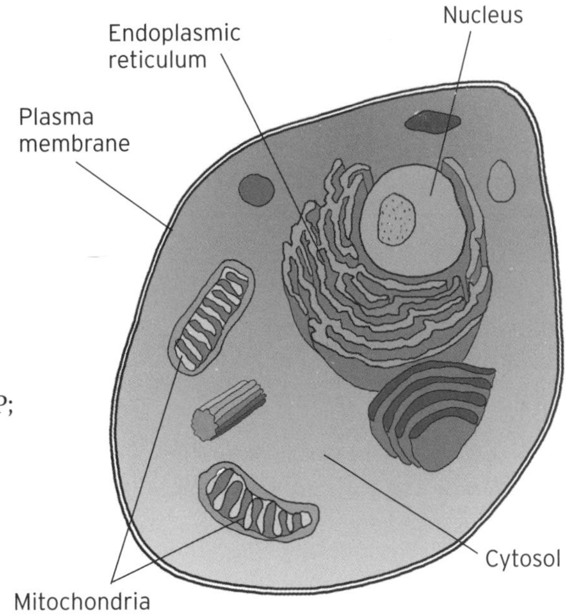

the organism's **DNA**. The DNA itself resides within the cell nucleus. Just outside the nucleus is a porous structure called the 'endoplasmic reticulum' which handles the transport and folding of newly made proteins. And supporting the whole cell is a scaffold-like structure known as the 'cytoskeleton'.

Prokaryotes

Prokaryotes are the simpler of the two major kinds of cell. The principle difference distinguishing them from their cellular cousins, the **eukaryotes**, is the total lack of a cell nucleus – instead, their genetic material is in a free-floating bundle called the 'nucleoid' at the centre of the cell. Whereas eukaryotes can be multicellular (see **Cells**), prokaryotes are always unicellular – the chief examples being bacteria and archaea (see **Prokaryote microbes**).

Prokaryotes have fewer organelles in their cytoplasm (see **Eukaryotes**). Like eukaryotes, they have ribosomes and a plasma membrane; they also have an outer casing beyond the membrane, called the 'cell wall'. Most bacterial prokaryotes are also home to **plasmids** – lengths of DNA not connected to the nucleoid. Some prokaryotes also have a tail, called a 'flagellum', which they use for propulsion. Their simplicity means prokaryotic cells tend to be much smaller than eukaryotes – indeed, they can be as small as a tenth of a micron across (0.0001 of a millimetre).

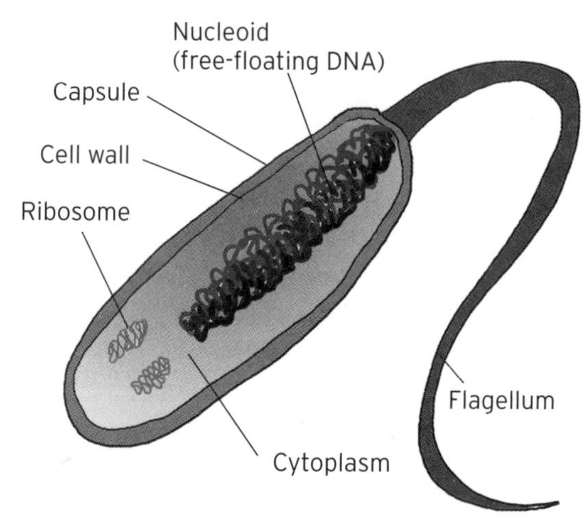

Cell nucleus

A principle purpose of the nucleus is to control the other functions of the cell. The nucleus is the centre of a **eukaryote** cell, and is home to the chromosomes of DNA that carry the organism's genetic code. Surrounding the nucleus is a double-walled membrane called the 'nuclear envelope', which is perforated by tiny pores through which small protein molecules and **RNA** can cross, serving as chemical messengers between the nucleus and the rest of the cell. However, it cannot be crossed by larger protein or DNA molecules.

Whereas the body of a cell is supported by the cytoskeleton structure, the nucleus is built around a fibrous framework called the 'nuclear lamina'; the contents of the nucleus are sometimes referred to collectively as 'nucleoplasm'. A class of **microorganisms** called protozoa have two nuclei – one to control cell division and another to regulate the cell's other functions.

Chromosomes

Chromosomes are lengths of **DNA** inside the cell nucleus of **eukaryote** cells that carry the genetic information (see **Genes**) to build identical copies of the cell through cell division, and to manufacture proteins through **gene expression**. Humans have a total of 22 distinct chromosomes, called 'homologous chromosomes'; however, each cell of the body – known as the 'somatic cells' – contains two copies of each chromosome, arranged in pairs. These 44 chromosomes are supplemented by one pair of sex chromosomes, determining whether the organism is male or female (see **Reproductive biology**), giving a total number in the nucleus of 46. For other species this figure can be quite different – for example, cats have 38 chromosomes while maize has just 20.

Each chromosome is of a different length and carries distinct genes – in humans, the gene for brown eyes is on chromosome 15, while chromosome 2 is thought to play a big part in determining intelligence. Some diseases are caused by chromosomal abnormalities; for example, people with Down's syndrome have an extra copy of chromosome 21.

Ploidy

The number of copies of each chromosome found inside a cell nucleus is determined by the ploidy of the cell. In humans, the 'somatic cells' comprising the body are 'diploid', meaning that there are two copies of each chromosome present inside every cell. **Gametes**, on the other hand – male sperm cells and female egg cells (see **Reproductive biology**) – are 'haploid', which means each cell contains only one copy of each chromosome. When a sperm and an egg combine at conception, a chromosome from each parent will make up each chromosome pair in the resulting zygote cell.

Other types of organism may have more than two copies of each chromosome in their cell nuclei – a situation known as 'polyploidy'. Durum wheat, for example, is 'tetraploid' – every somatic cell has four copies of each chromosome in the nucleus.

Mitochondria

Mitochondria are another kind of organelle found within eukaryote cells. They are responsible for generating most of the cell's energy from nutrients through the formation of the molecule adenosine triphosphate (ATP) – see **bioenergetics**. They also manufacture ribosome RNA, messenger RNA, and various proteins, as well as controlling the metabolism inside the cell. Mitochondria measure between a half and a few microns (thousandths of a millimetre) across, and are about ten microns long. A cell may contain between one and thousands of mitochondria depending on its type (see **Cellular differentiation**).

Like the cell nucleus, mitochondria also contain DNA – arranged as a single chromosome joined at both ends into a loop. Mitochondrial DNA (mtDNA) makes up less than 1 per cent of the total DNA in cells. It is only inherited from an organism's mother, and is passed on virtually unchanged from generation to generation. This has led to the idea of 'Mitochondrial Eve', the common female ancestor of all humans from whom all of our mtDNA is directly descended. It's currently believed she lived in Africa around 200,000 years ago (see **Out of Africa**).

Ribosomes

Ribosomes are factories where proteins are made from amino acid building blocks. These spherical components of cells, measuring just 20 nanometres across (see **Scale prefixes**) and made of **RNA** and protein, take so-called messenger RNA (mRNA) and use the information it stores to manufacture long polypeptide protein molecules. Messenger RNA is produced in the cell nucleus, where it takes a copy of the genetic information on the DNA of chromosomes and encodes it in the sequence of **nucleotides** making up its own structure, a process called 'transcription'. The mRNA then travels through the pores in the nuclear envelope surrounding the nucleus to carry the blueprint of the proteins to a ribosome site. There, the ribosome moves along the mRNA chain, reading off the sequence of information and bolting together amino acids accordingly; this manufacturing process, whereby proteins are made from mRNA, is called 'translation'. Some **antibiotic** drugs work by selectively knocking out the ribosomes inside bacteria (see **Prokaryote microbes**), destroying their ability to function.

Plasmids

These are circular pieces of DNA most commonly found within prokaryote cells, especially bacteria (see **Prokaryote microbes**). Thousands of them can exist in a single cell. Plasmid DNA typically stores a few thousand base pairs (see **Nucleotides**) of genetic information. Its genetic code stores instructions for functions to be carried out by the cell, often defensive processes such as building resistance to toxins, breaking down potentially harmful chemical compounds or manufacturing proteins that attack other organisms.

Plasmids are used in **genetic modification** as a way to insert modified genetic code into an organism's cells, and as a method for mass-producing particular proteins – by inserting the genetic code for the protein into the plasmid DNA and inserting the plasmid into a bacteria cell, the cell can be fooled into replicating many copies of the new protein.

Autophagy

This is a cannibalistic process that cells sometimes undergo, whereby they literally eat themselves. Autophagy plays a vital role in the overall health of the host organism. The cells sacrifice some of their non-essential parts to provide nutrients to fuel cell components needed for essential processes. Cells may do this when their usual source of nutrients is in short supply, or to dispose of damaged organelles (see **Eukaryotes**), or even to rid themselves of bacterial infections (see **Prokaryote microbes**). A double membrane forms around the part of the cell to be devoured, forming a packet known as an 'autophagosome', which merges with another cellular component called a lysosome, infusing enzymes through the double membrane wall to digest the contents.

Cell division

The process by which biological cells make copies of themselves is called cell division and proceeds in different ways depending on the type of cell – **eukaryote** or **prokaryote**. Eukaryotic cells divide through a two-step process of 'mitosis' followed by 'cytokinesis'. Mitosis starts with DNA replication inside the cell nucleus so that the double-helix structure of each DNA molecule unzips along its length. Then the nucleotide bases (see **DNA**) along each half join with new bases to form two copies of the original strand. Mitosis is followed by cytokinesis, where the cell nucleus divides in half to form two nuclei, with one copy

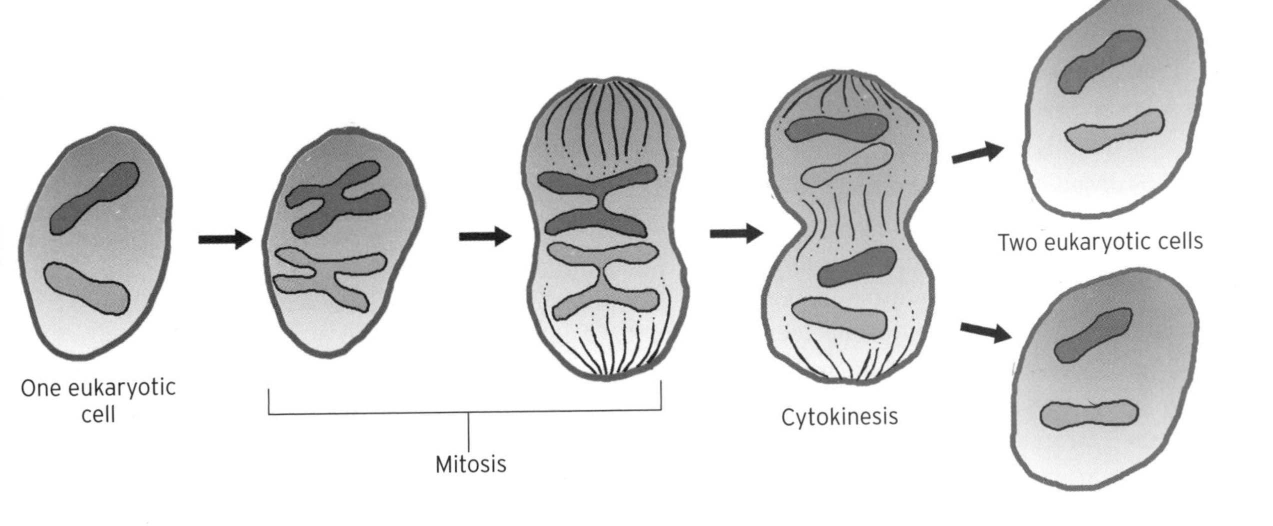

of each DNA strand going to each one. Finally the rest of the cell's cytoplasm splits to create two eukaryote cells where before there was only one.

The equivalent process in prokaryote cells, which have no nuclei, is called 'binary fission'. In this case the long stringy mass of DNA at the centre of the cell unzips and replicates in a manner similar to mitosis to form two copies. The cell then begins to expand, separating the new DNA strands and stretching the cell's plasma membrane until it breaks in half to spawn two new prokaryote cells. Colonies of cells grow by these processes – as do entire organisms, developing from that first fertilized cell through to adulthood (see **Reproductive biology**).

Gametes

Organisms that reproduce sexually (see **Reproductive biology**) have germ cells known as gametes. In animals, including humans, the male gametes are called 'sperm' and the female gametes are 'eggs' or 'ova'. Whereas ordinary body cells in a human are 'diploid' (see **Ploidy**) – that is, they have two copies of each chromosome, one from each parent – gametes have only one, they are 'haploid'. At the moment of conception, the gametes from both parents fuse to form the first cell of

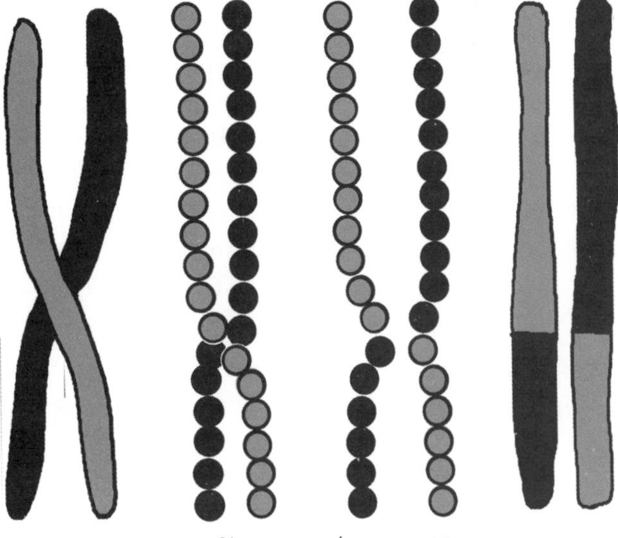
Chromsomal crossover

the offspring – called a 'zygote', which has two copies of each chromosome, one taken from each parent's gamete.

Gametes are made in a process called 'meiosis' which works rather like cell division, except that rather than making two cells it makes four – each one with just a single copy of each chromosome. During meiosis, a secondary process called 'chromosomal crossover' takes place; this effectively shuffles the DNA between each of the two chromosomes in the cell that is dividing, so that the chromosomes in the resulting gametes aren't direct copies.

Tissues

Organisms are made up of a multitude of different cell types, each performing a specific function. In humans there are around 210 different cell types accounting for the properties of different body components. A large ensemble of cells of one type is known as tissue. Different tissue types can then combine to form the body's **internal organs**, such as liver, heart and brain, as well as blood, bone, skin and **immune system**.

In animals, there are four major classes of tissue: 'nervous tissue' pipes electrical impulses to and from the brain, carrying sensory information and muscle signals; 'muscle tissue' is able to contract in response to nerve impulses, enabling animals to move; 'connective tissue' is responsible for holding other tissue types together and includes bone and the cartilage in joints; lastly, 'epithelial tissue' makes up the skin as well as protective coverings for internal organs and ducts.

Plant tissues are simpler, with three principal components: 'epidermis tissue' which forms the plant's outer covering, 'vascular tissue' which is responsible for circulating plant nutrients, and 'ground tissue' which manufactures and stores energy from **photosynthesis**.

Cellular differentiation

In addition to the 'somatic cells' that make up an organism's tissue types, the body manufactures other cell types. 'Germ cells', or **gametes**, are used in **reproductive biology** for passing the parents' genes to their offspring. A third type are the 'stem cells', which can grow into any of a wide range of more specialized cells in the body. The process, where stem cells change into other types of somatic cell, is called cellular differentiation; it first takes place in the embryo stage of an organism's development when so-called embryonic stem cells first differentiate. Adults possess a supply of stem cells too, which can differentiate on demand into the cells needed to repair tissue damage.

Some creatures, such as freshwater hydras, can even differentiate one type of somatic cell into another – first converting the cell back into a stem cell and then re-differentiating it into the type needed to patch up injuries. This is how a hydra that's cut in two can grow into two new hydras.

MICROBIOLOGY

Microorganisms

Microbiology is the study of organisms made of a single cell, or a small cluster of cells still too small to be seen with the naked eye. These life forms are known as microorganisms, or sometimes 'microbes', and can be observed with the help of a **microscope**. Microorganisms can be made from either **prokaryote** or **eukaryote** cells, and they can live in all parts of the Earth – from the depths of the ocean to high up in the atmosphere. Most of the species of life on Earth are microbes. Prokaryote microbes are divided into two groups – bacteria and archaea. Eukaryote microbes exhibit greater variety, being split into a number of categories: '**protists**' are single-cell eukaryote microbes; there are also animal microbes, as well as **micro-plants and fungi**.

Microorganisms confer both hazards and benefits to larger creatures. Some bacteria species, such as anthrax and E. coli, can cause harmful infections. However, microbes are essential to the healthy operation of animal **gastrointestinal systems** and for the well-being of the environment as a whole, breaking down and recycling waste organic material.

Prokaryote microbes

Microorganisms made from prokaryote cells occupy two principal domains: bacteria and archaea, and both of these **domains** are made up of microbes that are single-celled. Bacteria are able to replicate (see **Cell division**) in just minutes. They are classified according to their shape, or 'morphology'; spherical bacteria are known as 'cocci', while elongated, rod-shaped bacteria are called 'bacilli'. These base names then take a prefix depending on how the cells in a bacterial colony group together: species that tend to pair up with another cell take the prefix 'diplo'; those that form long chains take 'strepto'; those that cluster in triangular groups have the prefix 'staphylo'. These are the main types, but others exist too.

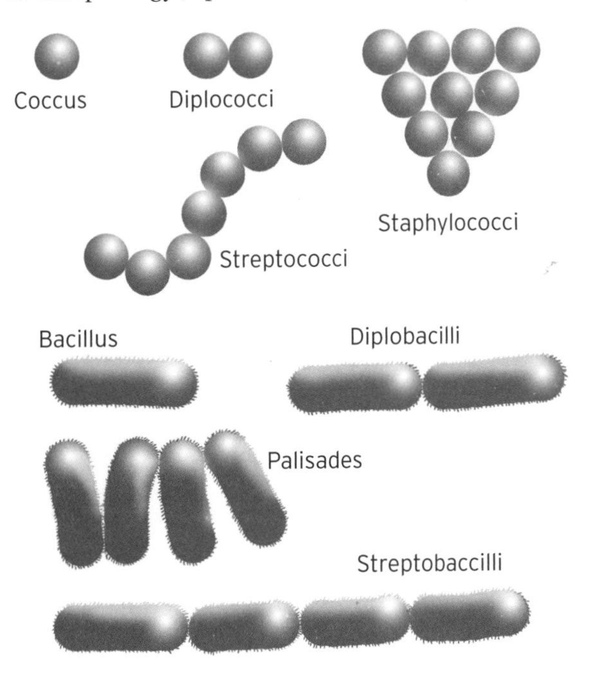

Archaea, while being of similar size and shape, have a subtly different chemical make-up to bacteria; for years, until 1990, their differences were not recognized and the two were believed to be part of the same **kingdom**. Archaea are believed to have been the earliest life forms on the Earth.

Protists

Protists are a **kingdom** of eukaryotic microorganisms and form an umbrella term for single-celled **eukaryotes**. In some alternative classification schemes, there are now a number of kingdoms classifying what used to fall under the banner of protists. The 'chromalveolata' kingdom includes dinoflagellates, which can form so-called red tides in the oceans and are toxic to marine life. This kingdom also includes the microbes responsible for the disease malaria. Many other parasites are included in the kingdom 'excavata'; while the kingdom 'rhizaria' are highly mobile, using primitive feet, called 'pseudopods', to move around. Various species of algae are in the 'archaeplastida' kingdom – many algaes, surprisingly, are not classed as plants. There are thought to be as many as 40 distinct phyla (see **Phylum**) of protists. Some biologists also use the term 'protozoa', which refers to the subset of protists that feed on organic compounds.

Animal microbes

Multicellular animal life forms that are too small to be visible to the unaided eye are known as animal microbes. They include many members of the 'arthropod' **phylum**, such as dust mites – from the arachnid (spider) **class** – which thrive in human homes and are implicated in some allergic conditions, including asthma. Spider mites are another micro-animal in the arachnid class, measuring up to a millimetre in size.

Other animal microbes come from the arthropod phylum's 'crustacean' subphylum, a group of marine animals – which includes crabs and lobsters – and is home to millimetre-size microorganisms such as 'cladocera', better known as water fleas. It also includes 'copepods', a species that makes up some of the mass of animal microorganisms in the ocean known as 'zooplankton'. Further examples of animal microbes are tiny worms from the 'nematode' phylum, and 'rotifers' – tubular marine animals that measure up to half a millimetre in length.

Micro-plants and fungi

Eukaryote microorganisms that aren't animal microbes can occupy the kingdoms of either plants or fungi. Tiny microorganism species of plant are known as plant microbes. Chlorophyta, for example, is a 'division' (see **Phyla**) of the plant kingdom and includes microbes that live in water as algae. Phytoplankton in the oceans are also plant microorganisms.

Like plants, fungi are a common form of organism on land – growing in the wild, for example, as mushrooms. However, fungal microbes also exist. These include yeast, single-celled microbes used in brewing and baking, and the wide variety of moulds – multicelled microorganisms that can often be found growing on food that has passed its expiry date. The earliest **fossils** from this kingdom hail from the Proterozoic aeon, some 1.4 billion years ago.

Symbiosis between species of fungal microbes and plant microbes of the chlorophyta division leads to lichen – moss-like growths that can be found on trees and stones.

Chemotaxis

The mechanism that bacteria and other microorganisms use to guide their movement in response to their chemical environment is known as chemotaxis. Organisms are able to sense concentrations of particular chemicals and then either move towards them – for example, if the chemical serves as a nutrient for that particular microorganism – or away from it, in the case of a toxin.

Using **receptors** tuned to detect sources of both food and poison, organisms sense their chemical environment; their receptors tell the organism whether it's moving towards or away from the chemical source. For propulsion and steerage, prokaryote bacteria cells use a flagellum – a long, flailing tail. Eukaryote microbes are thought to use different propulsion methods, such as hair-like 'cilia' on their bodies, or by growing foot-like protusions – called 'pseudopods'. Other forms of 'taxis' also exist, enabling microorganisms to direct their motion according to stimuli including light levels ('phototaxis'), heat ('thermotaxis') or electric fields ('galvanotaxis').

Viruses

Viruses are smaller than bacteria and other microorganisms, typically no bigger than a few hundred nanometres in diameter (see **Scale prefixes**). Not strictly life forms in their own right, they consist of just a strand of genetic material (DNA or RNA) surrounded by an outer protein jacket and are usually too small to be seen in an ordinary **microscope**, requiring a **scanning electron microscope** instead.

Viruses replicate by invading cells and hijacking their protein-copying machinery, using it to build new viruses. The new viruses then burst forth, destroying the cell in the process; each new virus can then go on to infect a new cell and the process repeats. Their destructive nature means viruses are often the cause of life-threatening illnesses, such as **influenza**, rabies, hepatitis and **AIDS**. Viral diseases can be treated or prevented using **vaccines and antivirals**. Not all viruses are bad though. So-called 'bacteriophage' viruses can selectively target harmful bacteria cells, and viruses are also used in **gene therapy**.

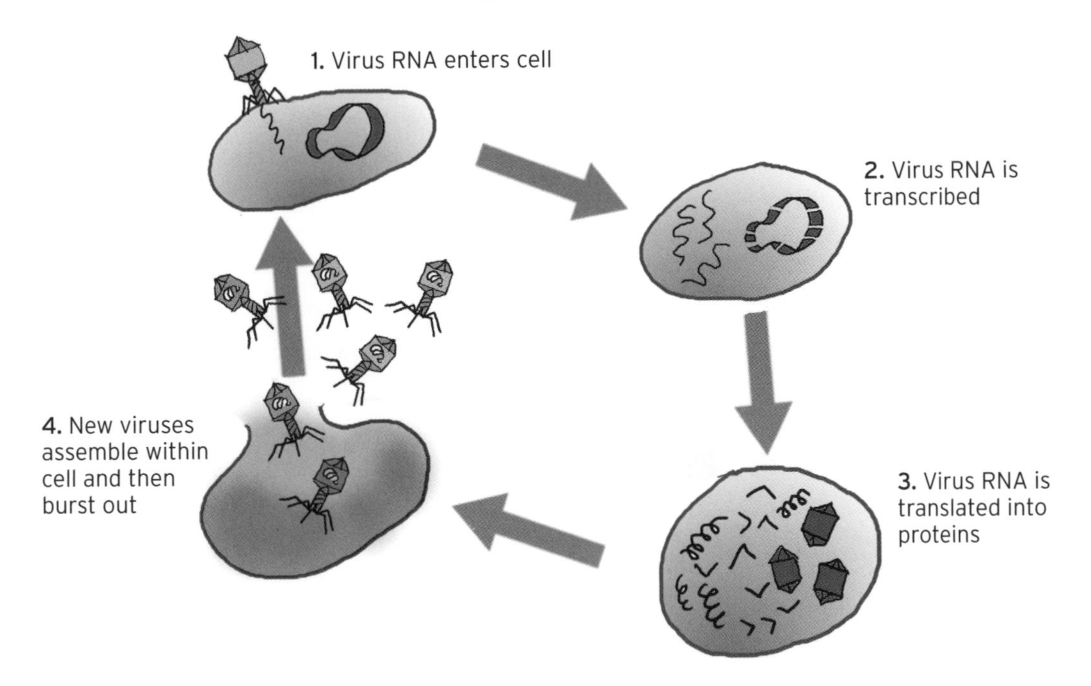

Nanobes

Discovered in 1996, nanobes may well be the smallest form of life known – measuring just 20 nanometres across (see **Scale prefixes**), one tenth the size of the smallest microorganism. Nanobes were first found by Australian scientist Philippa Uwins, of the University of Queensland, in samples of rocks laid down during the early- to mid-**Mesozoic era**. The finger-like tendrils resemble the microscopic structure of some micro-fungi (see **Micro-plants and fungi**) and Uwins claims they represent a new form of life. However, the claim has attracted controversy with some researchers arguing the structures are not life forms at all, but simply crystalline growths in the rock. In 2001, Uwins announced the results of new research which seems to reveal the presence of DNA within the nanobes.

MOLECULAR BIOLOGY

Nucleotides

Molecular biology is about applying the principles of **biochemistry** to explain genetics (see **Genes**) – how the blueprint of an organism is stored inside its cell nuclei, and also how this information is passed down to offspring during reproduction, known as **heredity**.

Cells store and process information on molecules called ribonucleic acid (RNA) and deoxyribonucleic acid (DNA). These are polymers (see **Macromolecules**) made of long chains of sometimes hundreds of millions of smaller molecules, known as nucleotides. Each nucleotide is made of sugars, either ribose (found in RNA) or deoxyribose (found in DNA), bonded to various phosphate molecules – compounds of hydrogen, oxygen and phosphor. But there's an all-important third ingredient – nucleotide molecules also contain chemicals called 'bases', which are organic compounds of hydrogen, nitrogen and carbon. There are five nucleotide bases relevant for molecular biology: adenine (A), guanine (G), thymine (T), cytosine (C) and uracil (U). It is the bonds that form between these bases, making so-called 'base pairs', that build up the characteristic double-helix structure of DNA, storing the information that underpins life.

DNA

DNA, short for deoxyribonucleic acid, is a polymer **macromolecule** made of a long chain of nucleotides each built from deoxyribose sugar, phosphates and one of the four nucleotide bases adenine (A), guanine (G), thymine (T) and cytosine (C). In large eukaryote organisms, such as humans, the DNA is arranged in the cell nucleus in lengths known as chromosomes. They can be extremely long – the longest human chromosome is a chain of over 200 million nucleotides.

The sequence of the nucleotide bases along a DNA molecule's length, for example CTTCGTA . . . ., encodes all of the information about the organism's make-up, rather like bits of **binary data**. Each group of three bases in the sequence makes up the equivalent of a byte of genetic information, called a 'codon'. Each organism's DNA is unique – a fact that's turned out to be useful in paternity testing and **forensics**.

Double helix

Chromosomes aren't made of a single strand of DNA but two, intertwined around each other in a double-helix structure. Nucleotide bases on one strand bond to bases on its opposite number to hold the helix together. Each type of base can only bond to one other type – G bonds exclusively to C, and A to T. This means that the two strands are not identical, yet either one specifies the organism's genetic code uniquely, a crucial fact in DNA replication – a key stage in **cell division**. Here, the helix unzips into two separate strands and the bases on each strand re-bond one by one to new counterpart bases to form two complete new double helices. The double helix structure of DNA was discovered by molecular biologists Francis Crick and James Watson in 1953.

RNA

RNA, or ribonucleic acid, is the molecular cousin of DNA. It too is a long-chain **macromolecule** of nucleotides, but RNA is based on ribose sugar, rather than deoxyribose. Like DNA, each nucleotide incorporates one of the bases adenine (A), cytosine (C) and guanine (G) – but also uracil (U), rather than DNA's thymine (T). Where DNA exists as a double helix, RNA is usually single stranded.

With the exception of viruses, RNA is not the principle carrier of genetic information but serves a host of secondary purposes. So-called 'messenger RNA' (mRNA) is used for the production of proteins within cells, carrying genetic information from the nucleus to ribosome sites. Here, 'transfer RNA' (tRNA) guides amino acids together in the sequence dictated by the mRNA chain. The ribosome itself, meanwhile, is made from 'ribosomal RNA' (rRNA).

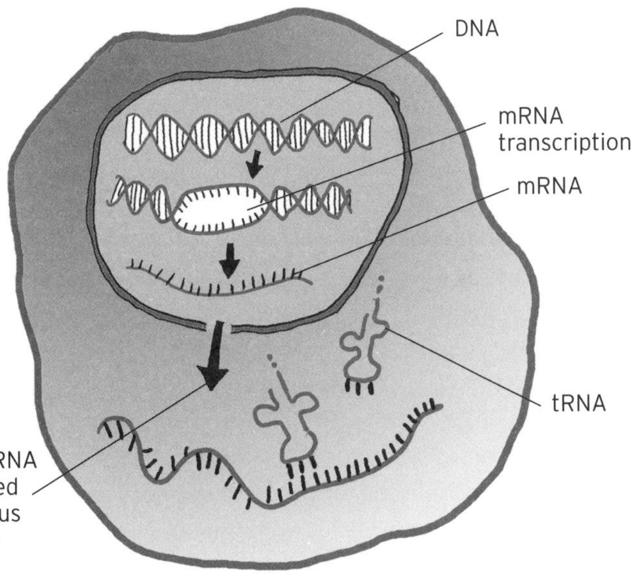

Genes

A gene is a section of the DNA sequence in the cell nucleus of an organism containing the information needed to make a particular protein used in the organism's body. Each codon (see **DNA**) specifies a particular amino acid in the chain making up the protein. In this way, genes are nuggets of data that record everything about the organism – from the colour of its skin, through the layout and functions of its internal organs right down to the workings of individual cells. But they don't just define the organism – the genes of two parents merge to compose the genetic make-up of their offspring. Our children inherit characteristics from us – such as facial features, intelligence and blood type – and our genes are responsible for this (see **Heredity**).

In between the genes are stretches of so-called 'non-coding DNA'. Some of this is thought to play a role in regulating gene expression. However, the function of the rest is largely unknown, leading to it sometimes being called 'junk DNA'. DNA was first observed in 1869 by Swiss biologist Johannes Miescher, and its **double-helix** structure was discovered in 1953. But it was a team led by scientist Oswald Avery working in New York in 1944 who first realized that DNA was the carrier molecule for genes.

Gene expression

Using the information stored as genes on strands of DNA to manufacture proteins in an organism is a process known as gene expression. It is carried out in one of the cell's ribosomes. Every cell in an organism contains a copy of its entire genetic code, but any particular cell only uses a small part of this code. For example, skin cells don't need to know how the pancreas works. Cells regulate which portions of the genetic code they express by a process called 'DNA methylation', where a methyl compound – chemical formula CH_3 – is added to cytosine **nucleotides** in genes that aren't needed to prevent them from being expressed.

Alleles

Alleles are different variants of the same gene. Take two copies of the same chromosome of DNA. Now look at a sequence of **nucleotide** bases at exactly the same point on each chromosome corresponding to a particular gene – if they are different then the two chromosomes are said to possess different alleles of that gene. An example would be human blood groups, the gene for which is located on chromosome 9. Different sequences of nucleotides along the stretch of chromosome 9 where this gene is located correspond to different alleles, which manifest themselves as the different human blood groups – A, B and O.

Zygosity

When a sperm cell and an egg cell combine during sexual reproduction, the single copies of each chromosome found within the cell's nuclei come together to give the pairs of chromosomes that reside in the resulting zygote cell (see **Gametes**). Zygosity is about comparing the alleles of different genes found on the chromosomes in each pair. The combination of alleles from the two chromosomes is known as the 'genotype' of the organism, while the physical proteins – that is, the features or 'traits' of the organism – that this translates into (see **Gene expression**) are referred to as the 'phenotype'.

When the two copies of a chromosome have the same allele of a gene for a particular trait, the genotype is called 'homozygous', and when they are different it's 'heterozygous'. If one allele is absent they are called 'hemizygous', while if both are missing the genotype is 'nullizygous'.

Genetic dominance

When alleles of a gene on two chromosomes in a pair are heterozygous (see **Zygosity**), it's not clear which one gets expressed (see **Gene expression**), which is where genetic dominance comes in. An example is given by the genotypes that produce different blood groups in humans, determined by a gene on chromosome 9. The alleles for blood groups can take on one of three forms – A, B or O. When allele A is present on both copies of chromosome 9 (a homozygous genotype) or when it has the heterozygous combination of A and O, the resulting phenotype is blood group A. Here, A is called 'dominant' while O is 'recessive'. B is dominant over O in the same way. Only when both chromosomes have the OO homozygous genotype is the resulting blood group O. A special case arises for the AB genotype which leads to the phenotype blood AB – alleles A and B are then said to be 'co-dominant'. Another case is blue and brown eyes in humans – the allele for brown eyes is dominant while blue is recessive.

ABO genotype in the offspring	Alleles inherited from the mother: A	B	O
Alleles inherited from the father: A	A	AB	A
B	AB	B	B
O	A	B	O

Genetic mutations

Changes introduced to the genes stored on the DNA sequence of an organism are known as genetic mutations, and the shifts in the organism's genotypes (see **Zygosity**) that they produce can cause shifts in the corresponding phenotype – the organism's physical traits. Sometimes these changes can be beneficial; for example, **evolution** by **natural selection** is driven by natural mutations introduced by the shuffling of an organism's DNA by 'chromosomal crossover' when it forms its reproductive gamete cells.

Other mutations are less beneficial. Those introduced to DNA by certain chemicals and by ionizing radiation such as that from **radioactive decay** can lead to cancer (see **Radiobiology**). Further mutations can be caused by renegade stretches of DNA known as 'transposons', but are sometimes given the nickname 'jumping genes' after their ability to hop to different locations in the DNA sequence of a cell. Transposons are known to cause serious diseases including cancer and the blood-clotting disorder haemophilia.

Recombinant DNA

Genetic mutations are a way in which the DNA code of an organism can become altered in an uncontrolled way, either by natural causes or by the effects of pollutants in the form of chemicals or radiation. But there's another more deliberate way to tamper with an organism's genes, and that's recombinant DNA. It works by synthesizing DNA sequences artificially, in a laboratory, and then splicing them into the organism's existing genetic sequence to bring about physical changes to the organism once the new DNA is expressed (see **Gene expression**).

The new DNA can be inserted into the organism using various techniques. Viruses work by injecting their genetic material into a host organism, and so replacing the viral genetic code with the recombinant DNA and infecting the host with the virus is one way to do it; bacterial **plasmids** can also be used in much the same way. Techniques for making recombinant DNA were first developed in the early 1970s, and now form the basis for **genetic modification** technologies such as **GM food** and **synthetic biology**.

Genomics

The entire genetic code of an organism – the sequence of A, G, T and C nucleotides you get when you lay all of its DNA chromosomes end to end, and the map of how this sequence divides into the chunks that make up the genes – is referred to as the organism's 'genome'. The scientific study of the genomes of different organisms is called genomics, the goal of which is to be able to analyze the complete gene sequence of any given organism in order to predict its physical traits.

Genomics breaks down into three areas: 'structural genomics' uses chemical processes and amounts of computing power to try to map the genome of organisms; 'functional genomics' is about **gene expression** – how exactly genes translate into traits – and how this relationship changes under different conditions; finally, 'comparative genomics' tries to draw parallels between the genomes of different **species** in a bid to try to gain a better understanding of one species by analyzing the genome of another.

The first organism to have its genome sequenced was a bacteriophage virus in 1977. Genomics really began to take off in the 1980s, and in 2001 the first draft of the genome of human beings was completed – the Human Genome Project.

Gene sequence

A key tool in genomics is gene sequencing – using chemical methods to read out the sequence of nucleotide bases making up a strand of DNA. There are two principal methods. The Maxam–Gilbert method involves chopping up the DNA strand using enzymes and then treating the pieces with various chemicals that each react with only one of the four DNA nucleotide bases in order to deduce the sequence.

The Sanger method is the other technique, whereby a new strand of DNA is synthesized from the test sample. Synthesis of the new strand can be stopped at any point by adding a chemical that reacts with one of the four nucleotide bases. Observing which chemical stops the synthesis reveals the next base, and by repeating the process the entire DNA sequence can be obtained. As it can be applied to RNA as well as DNA, and can be fully automated, the Sanger method is generally considered superior.

Human Genome Project

Beginning in 1988 the Human Genome Project was an international effort to map the human genome, cataloguing base pairs that comprise each gene in the DNA sequence of human beings. The genome of human beings is a huge 3 billion base pairs long (see **Nucleotides**); a typical human being, however, only has about 24,000 useful genes. Costing a total of $3 billion dollars, the rough draft of the sequence was completed in 2001, and the full sequence in 2003.

Researchers are now trying to gain further insight into the functionality of all the genes that make up the human genome – a project that will bring considerable health benefits through **genetic medicine**. Other projects aim to map variations in the genome between different ethnic groups, and to map the DNA sequences of individuals – in 2007 American biologist Craig Venter published his entire DNA sequence, becoming the first person in history to do so.

BIOLOGICAL TAXONOMY

Biological classification

There are currently more than 1.5 million species of animals, plants and microorganisms known to science. Biologists split up this teeming diversity of life into categories depending on each organism's shape and appearance (see **Morphology**) and its **anatomy** using a scheme referred to as 'biological taxonomy'. Life on Earth is organized into a hierarchical structure by a modern classification system using eight 'taxonomic ranks', or taxa (singular 'taxon'). The top rank is known as a **domain**, each of which is divided into several **kingdoms**, which then split into a number of **phyla**. These in turn break down into **class**, then **order**, **family**, **genus** and, finally, **species**.

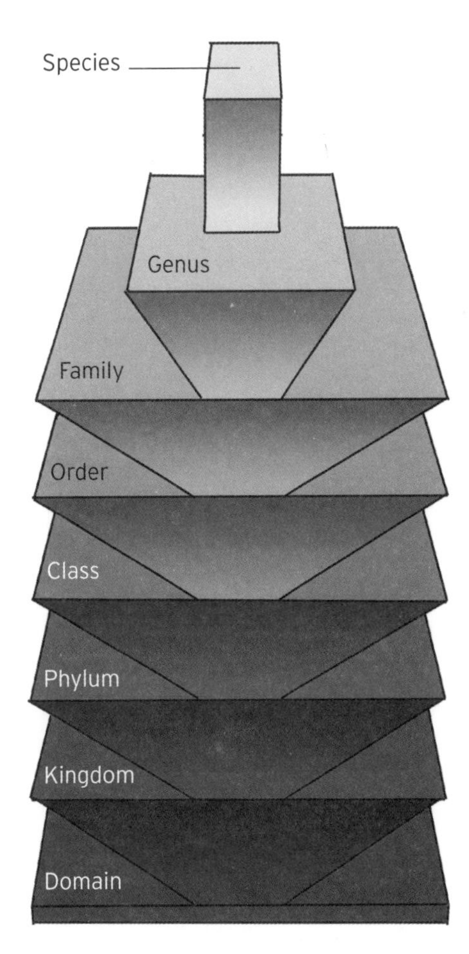

Carl Linnaeus, the Swedish botanist and zoologist, laid down the foundations of this system in the early 18th century. He also introduced the convention of two-part biological naming, known as 'binomial nomenclature', by which modern organisms are named. It uses two words – the first being the genus of the organism, which is capitalized, followed by the species name in lower case – both words italicized. For example, modern humans – *Homo sapiens* – are of the genus *Homo*, species suffix *sapiens*. A number of intermediate classes have also been added to the system, such as 'subspecies' and 'infraclass'.

Domains

Domains are the highest taxonomic rank of organisms in the natural world. In the modern classification scheme there are three of them, based on the genetic make-up of the organism's cells – Archaea and Bacteria, both comprised of **prokaryote** cells, and Eukarya, which are organisms made up of **eukaryote** cells.

Kingdoms

The next group down from domains in biological classification is kingdoms. In Carl Linnaeus's initial scheme of classification there were just two kingdoms: animals and plants. However, this view has evolved gradually and there are now a total of six kingdoms of life recognized to science. The domain of Archaea is home to the kingdom of Archaebacteria, which derive their nutrients from **chemosynthesis**; the domain of Bacteria also holds just one kingdom: Eubacteria, which feed via many processes other than chemosynthesis. The remaining four kingdoms all fall within the Eukarya domain – they are: Animals, Plants, Protists and finally Fungi – which are distinguished from plants by the fact that they don't **photosynthesize**.

Phyla

The taxon of phylum lies below kingdom and above class in the current biological classification scheme. In the animal kingdom there are nearly 40 phyla. These include Mollusca (marine molluscs), Nematoda (nematode worms) and Chordata – the phylum that contains birds, reptiles, amphibians and mammals (including human beings).

In **botany** – the study of plants and fungi – phyla are instead known as 'divisions'. Divisions of the plant kingdom include Angiosperms (flowering plants), Pteridophyta (ferns) and Bryophyta (moss), and many others. The fungus kingdom, meanwhile, is split into six divisions: Basidiomycota, Ascomycota, Aycophycophyta, Zygomycota, Deuteromycota and Glomeromycota – all different types of mushroom, classified according to differences in their reproductive organs. There are phyla of microorganisms too. The Archaebacteria kingdom has five phyla: Crenarchaeota, Euryarchaeota, Korarchaeota, Nanoarchaeota and Thaumarchaeota, and there are many more phyla in the kingdom of Eubacteria.

Classes

Class is the name given to the groups of life in the natural world directly below phylum. This is where the number of groups in our scheme of biological classification starts to get really large. And yet the names of classes become slightly more familiar too – including reptiles (Reptilia), mammals (Mammalia), birds (Aves) and amphibians (Amphibia).

Orders

The taxon in biological classification that lies below class and groups together similar biological families is 'order'. Examples include Carnivora, which are the meat-eating order of the mammal class, rodents (Rodentia), which are small members of the mammal class, and Rosales, which covers roses and orchids in the plant kingdom. Humans are in the order Primata – primates, which includes monkeys and apes.

Families

Family is the level of taxonomic classification one down from order. Family names obey a strict syntax, with animal names ending in 'idae' – for example, Felidae (cats) and Crocodylidae (crocodiles) – and botanical names, i.e. plants and fungi, ending in either 'aceae' or just 'ae' – such as, Aceraceae (maples) and Bambuseae (bamboo). Humans occupy the family Hominidae, also known as the 'great apes', and which also includes chimps, gorillas and orangutans. Pigeons occupy the family Columbidae, which is now the sole occupant of the order Columbiformes, following the extinction of its other family, Raphidae – including the dodo.

Genus

The category of biological classification lying between family and species is called genus.

In the binomial naming of species the first of the two words used is always the genus of the organism, and it is always capitalized. For example, domestic cats carry the name *Felis catus* – where Felis is the genus. Whereas other taxa in the scheme of classification of life on Earth are determined by rigorous biological considerations, partitions between genera are determined in a relatively arbitrary way – by looking for natural gaps between obvious groupings of species. Some researchers have suggested a definition of genus with a little more scientific merit – in that it should constitute a group of organisms that can interbreed to form hybrids. Lions and tigers, for instance, can breed to create 'ligers'. Though at present this scheme is only a proposal.

Common names of organisms often bear a similarity to the name of the genus – such as the genera Acacia (acacia trees) and Elephas (elephants). This is especially true when the distinction between individual species of a genus is very subtle.

Species

Species is the final and most specific taxon in the biological classification of life on Earth. It is commonly regarded as a group of organisms whose biology is similar enough to enable them to breed with one another to produce fertile offspring. Species that are similar but not identical though which are able to interbreed to produce hybrids (most of which are sterile) are still usually members of the same genus.

Even so, biologists argue over the exact definition of species and how to assign species names to newly discovered organisms. This is known as the 'species problem' and there are numerous proposed solutions, from organizing species according to their genetic heritage to their physical form, or **morphology**, to classifying them according to the particular niche they occupy in their environment.

The number of species on the planet – microorganisms, as well as plants, animals and fungi – could run to hundreds of millions. There are 1.25 million species of animals alone, hundreds of thousands of plants and millions more bacteria. Over time, organisms of the same species in different geographical locations adapt to their different surroundings and evolve through **natural selection** to ultimately create new species (see **Speciation**).

Natural history

Taken together, the gamut of life on the Earth's surface – from domains down to species – forms an area of science known as natural history. Dating from the time of the ancient Greek scientists in the 4th century BC, early definitions of the term also include non-living aspects of the Earth, such as **geology**, though nowadays it is normally reserved for the study of animals, plants and fungi.

Natural history breaks down broadly into two disciples: botany (the study of plants and fungi) and zoology (the study of animals, including, birds, insects, reptiles and amphibians) – usually with the emphasis on observational fieldwork rather than on hard scientific study. For this reason, some scientists regard natural history as a branch of popular science – the province of television documentaries and magazines – rather than of true scientific endeavour.

ZOOLOGY

Animals

Animals are the members of the biological kingdom Animalia. They include birds; cold-blooded reptiles; amphibians, which can live in water as well and on land; insects with their hard exoskeletons; and mammals, which give birth to live young.

Animals are multicellular **eukaryote** organisms that procreate by sexual reproduction – fusing gamete cells from two parents of opposite sex to create offspring. Having no capacity for **chemosynthesis** or **photosynthesis**, they are unable to manufacture their own food and so must ingest other organisms in order to acquire nutrients. The scientific term for this is 'heterotrophy'. For this reason, most animals are mobile, enabling them to seek out food and this has also necessitated the development of a nervous system (see **Neurobiology**), enabling them to recognize food sources and respond to other stimuli from their environment, and to coordinate their movements. The first animals appeared on Earth during the 'Cambrian explosion' of life, during the **Paleozoic era**.

Morphology

Studying the shape or form of an organism – both the internal and external features – is a biological discipline known as morphology, a science used in the classification of organisms according to **biological taxonomy**. Morphology breaks down into two main areas: 'eidonomy', taking stock of the external appearance of an organism, and 'anatomy', which looks at the structure of a creature's internal organs.

Variations in the form of different species are quantified by a field called 'morphometrics' that involves making detailed measurements. It's a more scientific approach to morphology than comparing sketches and verbal descriptions, allowing hard and fast comparisons of different morphological types through the application of rigorous mathematical analysis and computer techniques. Usually, morphology is supplemented by DNA analysis when trying to identify an unknown species.

Vertebrates

Put simply, vertebrates are animals that have backbones. In the scheme of biological taxonomy, they form a so-called 'subphylum' – called 'vertebrata' – placed just below the **phylum** of Chordata. Backboned animals began with early creatures that had a so-called notochord, a rod-like bone running along their length, which evolved, with the notochord dividing up into a number of smaller jointed bones – the vertebrae – to form a flexible 'spinal column'. This bony linkage also carries the spinal cord – a bundle of nervous tissue that is the central information highway for nerve impulses to travel between the brain and the rest of the body (see **Neurobiology**).

Most of the classes of large animal life forms in the world today are vertebrates: mammals, birds, reptiles, amphibians and fish. Vertebrates arose soon after the first animals during Earth's **Paleozoic era**. Invertebrates – animals without backbones – also exist today; the most prominent group is the Arthropod phylum, which holds among its number insects, spiders and crustaceans. Other invertebrates include worms, corals, anemones and jellyfish.

Reproductive biology

Animals reproduce sexually, which means that a gamete cell from a male and a female parent animal combine in the mother to form a 'zygote' – the first cell of their offspring. The zygote then grows by **cell division** and **cellular differentiation** to eventually form the baby animal.

The sex of the child is determined by the combination of 'sex chromosomes' in the two gamete cells; in a human gamete there are 22 chromosomes that carry forward the parent's genes, together with a sex chromosome, which is labelled either X or Y. Female egg cells can carry only the X type; male sperm can carry either X or Y. When the two come together to form an XX combination a female child is created; and when the combination is XY the result is a male.

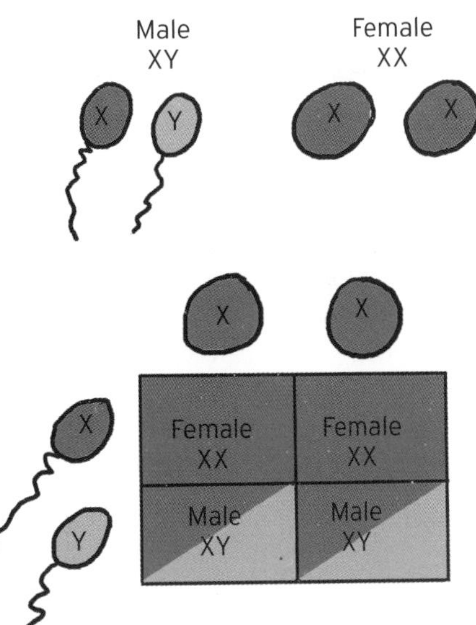

In other organisms, such as birds, the situation is reversed, with variations in sex chromosomes in the female parent determining the sex of the child. Other species have different, and often more complex, sex-determination systems – the platypus, for example, has no less than ten sex chromosomes.

Developmental biology

Once the first cell of an organism has been produced from a zygote cell (see **Gametes**) its subsequent growth is governed by a field known as developmental biology. First, the zygote undergoes cell division and grows into a blob of cells known as a 'blastocyst' – in humans this can be made of up to 100 cells. The blastocyst, which forms a few days after conception, then grows on to become an embryo. Here, chemical processes regulate the expression of genes into proteins across the body of the growing embryo to create the first basic structures. The embryo grows inside the mother animal, gaining cells and transforming these via **cellular differentiation** into blood, bones, nerves and internal organs.

In mammals the stage after embryo formation is called the 'fetus', which is brought to full term (nine months, in the case of humans) and delivered live. Other **classes**, such as birds and reptiles, deliver their young as eggs which later hatch. Some offspring go through an additional 'metamorphosis' stage after birth, during which their appearance changes radically – the pupation of larvae into butterflies is an example.

Ethology

The field of ethology is concerned with the behaviour of animals – in particular, instinctive behaviours such as acquiring food and protecting their young. A mother graylag goose, for example, will automatically try to roll an egg it sees back to those in its clutch – even if the egg isn't actually one of its own. And some animals will offer themselves as a distraction to divert predators from their offspring (see **Altruism**). Other instinctive behaviours are learned. Russian psychologist Ivan Pavlov famously trained dogs to salivate on the sound of a bell ringing, by ringing the bell each day just before feeding time (see **Behaviourism**).

Ethology draws upon aspects of **evolution**, **neurobiology** and environmental science to explain the behaviour of animals in the natural world. It is closely related to **sociobiology**, which attempts to ascribe not just survival instincts but also animal social behaviour to biological processes.

Branches of zoology

The huge number of animal species in the world means zoology is a massive subject. To make this field of science more manageable, zoologists divide their studies up into a number of sub-disciplines. Ethology, **ecology**, and **evolution** are all fields that contribute. But most zoologists who choose to specialize do so according to considerations of biological taxonomy – in other words, they concentrate on particular groups of animal species. For example: mammologists study mammals; ornithologists study birds; herpetologists amphibians; entomologists insects; arachnologists spiders; ichthyologists fish; and helminthologists worms. One final group of zoologists is concerned with animals that aren't even alive – palaeozoologists study animal remains recovered from **fossils**.

Cryptozoology

The study of animals not recognized by science is called cryptozoology, which sounds something of a contradiction and, indeed, many serious scientists believe cryptozoology to be nothing more than **pseudoscience**. Cryptozoologists base their studies on eye-witness reports, investigating the claims of people who believe they have made sightings of 'cryptids' – beasts ranging from the Loch Ness monster to the Yeti (Bigfoot) to the Orang Pendek, a diminutive primate believed by some to inhabit the forests of Sumatra. Critics argue, however, that the methods employed by cryptozoologists are unscientific.

Nevertheless, the field was lent some support in 2003 when archaeologists working on the island of Flores in Indonesia uncovered skeletal remains of a new race of hominids just a metre tall, and dating from as recently as 12,000 years ago – proving, argued cryptozoologists, that there are unusual species that have evaded the eye of science. Botany also has scope for its own mythic species – through the field of 'cryptobotany'.

BOTANY

Plants

Botany is the study of plants and fungi. Plants, members of the kingdom Plantae, are multicellular **eukaryote** life forms that produce nutrients from sunlight through **photosynthesis** – which is what gives them their green hue. They include organisms such as flowering plants, trees, bushes, grasses and mosses. Because plants have no need to pursue prey or actively seek out other sources of food, they are immobile organisms with no nervous system and are slow to react to stimuli from their immediate environment. Their cells have a tough outer coating of cellulose – known to diners as 'dietary fibre'.

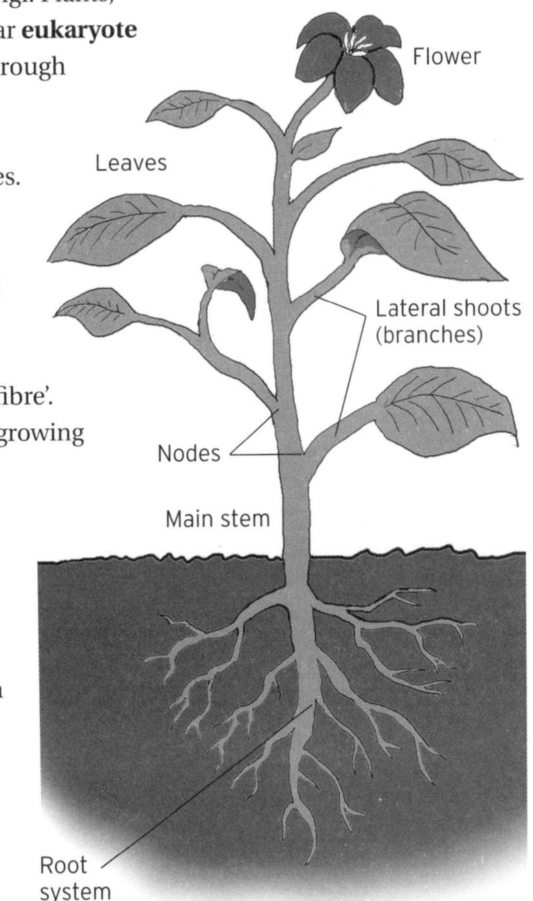

Plants are usually composed of a main stem growing out of the ground, where it is anchored by a root system that grows through the soil to suck up water and chemical plant nutrients. Lateral shoots, also known as branches, grow outwards from points on the stem called nodes; the lateral shoots are adorned with leaves, which collect sunlight for photosynthesis. Plants of the division (see **Phyla**) Angiospermae, of which there are more species than any other, also produce flowers. The earliest plant remains are fossilized green algae, dating from the Cambrian period of the **Paleozoic era**.

Fungi

Fungi are members of the kingdom of eukaryote organisms containing moulds, yeasts and mushrooms and can be either single-celled or multicellular. Unlike plants, fungi lack chlorophyll in their tissue and so do not photosynthesize – instead, they get their energy and nutrients as parasites (see **Parasitism**), by growing on other organisms. This is why mould grows on spoiled **food**. The fungus kingdom is believed to be home to some 1.5 million species and the scientific study of these organisms is known as 'mycology'.

Fungal reproductive biology works by producing spores that then grow into new fungi; these can be produced both sexually or asexually, depending on the particular species of fungus. A single mushroom can release billions of spores at a time. Fungi have proven extremely useful to humans; we can eat them as they are, or use their **microbiology** properties to make bread – along with beer and wine to wash it all down. Blue cheese is made by infusing cheese with strains of mould. Perhaps most important of all is their role in **antibiotic** drugs – which began with the discovery of penicillin mould.

Photosynthesis

Plants produce energy from sunlight, water and atmospheric carbon dioxide by photosynthesis. It occurs in green plants, algae and some bacteria species (see **Prokaryote microbes**). Photosynthesis is, arguably, the most important chemical reaction on Earth and its by-product is oxygen – planet Earth's plant life is literally what enables animals and other aerobic organisms to breathe. Photosynthesizing plants also form the basis of the food chain, and even serve as fuel for heating and cooking through wood burning.

The reaction takes place within parts of plant cells known as 'chloroplasts'. Here light-absorbing **plant pigment** – the green-coloured chlorophyll – uses the energy from the Sun to power the combination of water drawn up through the plant's roots with CO_2 taken in through pores in the leaves called 'stomata'. The chemical equation for the process is: $CO_2 + H_2O + \text{sunlight} \rightarrow CH_2O + O_2$, where CH_2O is energy-producing **carbohydrate**.

Transpiration

The pores in the leaves of plants, known as 'stomata', through which they take in carbon dioxide for photosynthesis, also act rather like pores in the skin of an animal, allowing water vapour to escape in a process known as transpiration. It serves to keep the plant cool, and also encourages the root system to take up new water from the ground bringing with it minerals and plant nutrients.

Roots take in liquid from the ground via **osmosis**, which is then circulated around the plant in the 'xylem' – porous tissue through which watery sap can flow. Xylem tissue can also be quite rigid, and this is what forms wood in bigger plants. A second type of vascular tissue, known as 'phloem', is responsible for transporting the carbohydrates manufactured in photosynthesis from the leaves to the rest of the plant.

Plant nutrients

Plants require a different set of nutrients from animals; as with animals, these are divided into macronutrients – those which are needed in large amounts – and micronutrients, effectively the 'vitamins' a plant needs. Plant macronutrients include nitrogen, phosphorus, potassium, carbon, hydrogen and oxygen. Carbon is essential for building the structure of the plant and is extracted from carbon dioxide in the atmosphere through photosynthesis. Oxygen and hydrogen are used in the formation of carbohydrate during photosynthesis; phosphorus is involved in energy transport. Potassium helps to regulate the opening and closing of the 'stomata' through which CO_2 goes in for photosynthesis and water vapour leaves via transpiration. Nitrogen is used in protein building – although the air is 78 per cent nitrogen, plants must extract it from the ground through their roots. Plant micronutrients include minerals such as zinc – needed for **gene expression** – and chlorine, which is used to drive **osmosis** in the plant's roots.

Carnivorous plants

Not content sucking in carbon dioxide through their leaves and nitrogen through their roots, some plants instead have a taste for flesh and these are known as the carnivorous plants. Probably the best known is the Venus flytrap, which boasts a jaw-like trap that can spring shut on hapless insects, allowing the plant to feed on the organic matter in their bodies. But there are many other species. The 'sundews' use sticky flypaper-like traps to snag unsuspecting insects. Others still, such as the pitcher plant, brandish long funnels, down which insects fall and are unable to escape.

Carnivorous plants are thought to have evolved independently six times; however, it takes special circumstances for such an adaptation to be useful to a plant. Trapping insects and then digesting them takes energy and resources that could be used to find plant nutrients in the usual way. Unless it can guarantee a steady influx of prey, a carnivorous plant could find its energy budget in the red.

Seeds

Plants reproduce by scattering seeds, tiny plant embryos, from which new plants can grow. Seeds are produced by flowering plants, also known as 'angiosperms', and non-flowering varieties that scatter seeds from cones or directly from a central trunk, known as 'gymnosperms'. These plants reproduce sexually, scattering pollen from the male 'anther'. This is then received by the 'stigma', the female reproductive organ, of another plant of the same species to fertilize an egg that then develops into a seed.

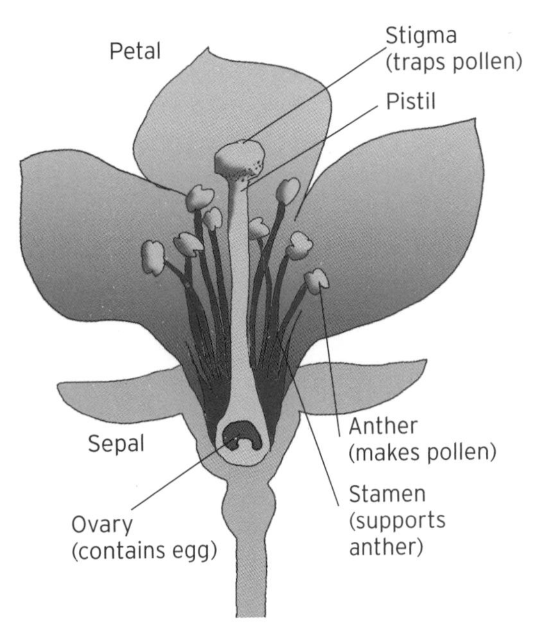

Other plants reproduce using spores, which are scattered from the plant like seeds. Plants that do this include mosses and liverworts, as well as ferns and algae. Fungi reproduce this way too. Spores are just a single cell containing the genetic blueprint of the organism. Seeds, on the other hand, contain not one but a bundle of germ cells, together with a supply of plant nutrients and usually a hard protective shell, maximizing their chances of survival. This is why the evolution of seeds was such a revolution in plant biology. Seeding plants first appeared in the Devonian period, during the **Paleozoic era** of Earth's prehistory.

Plant pigments

Plants get their spectrum of colours from a range of chemical pigments. Primary among these is chlorophyll, the green pigment that colours the stems and leaves of plants and is responsible for **photosynthesis** by which they generate their own energy. But other pigments exist too. Carrots and some related root vegetables get their distinctive orange colour from 'carotene'. Other related compounds – known as 'carotenoids' – are 'lutein', a yellow colouring found in kale and peppers, and 'lycopene' which is what gives tomatoes their red hue. Meanwhile, various 'anthocyanins' are responsible for the colorations in the petals of flowering plants. And 'betalains' give beetroot its redness. Plant pigments can be extracted and used to make dyes.

Phytochemistry

Plants are cooking pots of chemical activity and the study of these chemicals and the reactions between them is a field known as phytochemistry. Certain chemicals in plants afford them protection against insects and diseases, and help in pollination (see **Seeds**). For example, flying insects called thrips feed on the pollen produced by the male part – the 'cones' – of the cycad plant. In response, the cones give off a toxic odour, driving out the pollen-covered insects. At the same time, the female parts waft an enticing scent to draw the fleeing bugs in – so pollinating the plant.

Plants also have a suite of **hormone** chemicals at their disposal to carry messages from one part of the organism to another. These hormones flow through the plant's vascular channels to transmit the cues that trigger flowering, shedding of leaves and ripening of fruit. Some of the chemicals cooked up inside plants are toxic; others can have therapeutic effects and are used in the field of phytopharmacology.

Phytopharmacology

Phytopharmacology is the use of plant chemicals as **medication**. Plants have been used as a source of herbal medicine for thousands of years; however, this is often regarded as something of a **pseudoscience**. The aim of phytopharmacology is to put plant-based medicines through the same rigorous selection procedures and clinical trials as other drugs to fully understand their benefits and side effects before approving them for use in patients.

Notable examples of effective clinical drugs extracted from plants are digoxin, which is used to treat heart conditions (made from the foxglove plant); quinine, which is an anti-inflammatory drug used to treat malaria and other ailments (extracted from the bark of the cinchona tree); and the anti-clotting agent aspirin, which also has applications as an **analgesic** and anti-inflammatory (and was originally derived from the willow tree). Recently concern has arisen over 'biopiracy' – where pharmaceutical companies plunder the natural fauna and folk medicines of developing countries for new drugs, while offering little in the way of remuneration.

Horticulture

The many applications of plants and their chemicals, through phytochemistry, and for food, have led to the scientific field of horticulture – the study of how to cultivate plants for human use. It includes the study of soil quality, fertilizers to provide extra plant nutrients, pesticide chemicals to control harmful insect species, the treatment of plant diseases, and selective plant breeding to improve the quality of specimens – both naturally and through deliberate **genetic modification**.

Horticulture is normally concerned with developing techniques for cultivating plants and applying them on small scales, whereas **agriculture** deals with their larger-scale application. Some amateur horticulturalists cultivate their home gardens for food, growing fruit and vegetables, which has become popular with environmentalists as a way of reducing the 'food miles' on produce. Modern horticultural techniques include **hydroponics** and tissue culture – growing clones of strong plant specimens from slivers of cells.

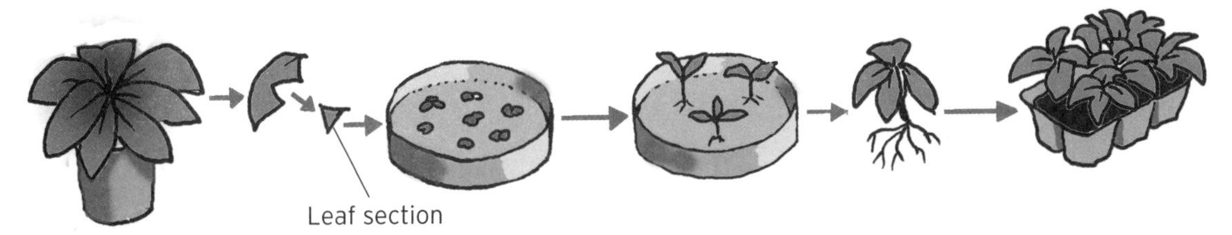

How tissue culture works

Plant behaviour

It's easy to think of plants as sessile, inert life forms that simply sit in one place, soaking up sunlight and water. Yet plants are capable of some complex behaviours. For example, time-lapse photography of plants growing reveals them moving over time towards light sources – to maximize the amount of energy they generate by photosynthesis, a motion referred to as 'tropism'. Root systems also undergo a kind of tropism too, growing preferentially towards nutrient sources.

Another kind of plant behaviour is known as 'nastic movement', where the motion is rapid and – unlike tropism – bears no relation to the direction in which the stimulus causing the movement has come from. The snapping shut of a Venus flytrap (see **Carnivorous plants**) is a good example of nastic movement. Plants achieve these movements by shifting fluids within their foliage – taking fluid from one side of a stem and transferring it to the other causes the stem to bend. Venus flytraps achieve their rapid motion using changes in acidity to bring about a split-second reduction in the size of the cells holding the trap open, causing it to snap shut. Plant behaviour is closely linked to the field of **plant intelligence**.

Plant intelligence

Plant behaviour – the ability of plants to adapt and respond to stimuli from their immediate environment – has led some botanists to suggest that plants possess a primitive form of intelligence. Several pieces of evidence have drawn them to this conclusion; for example, acacia trees are able to sense when a herbivore is chewing on their leaves and give off bitter-tasting tannin in response. Other acacia trees seem able to smell the tannin of nearby trees and start producing their own long before the hungry herd arrives. Even the Venus flytrap (see **Carnivorous plants**) is smarter than you might think. Its trap is triggered by tiny hairs on its surface, but each hair must be touched twice in a short space of time for the trap to spring – to prevent it from being accidentally triggered by, say, raindrops. In other words, the plant must remember which hairs have been touched recently – it has a primitive memory. Plants have no brain or nervous system (see **Neurobiology**). Instead, the researchers believe these basic cognitive abilities are arising through chemical interactions in their **hormone** systems.

ECOLOGY

Environment

Ecology is the study of the interaction between organisms and their natural environment. An organism's environment is defined as the chemistry, physics and biology that make up the surroundings in which it lives. For an ocean fish, the environment is the water around it, the seabed and the other life forms it must share this space with. In recent years ecology has focused on the interaction of human beings with their global, planetary environment – humans can influence, and be influenced by, pretty much any part of the Earth. At the other end of the scale, the environment of a microorganism might solely consist of the innards of another organism that it inhabits.

Ecology and the environment is also an important consideration in **evolution**. Darwinian natural selection causes an organism to adapt to its environment over progressive generations. Change the environment and the organism ultimately changes too.

Carbon cycle

Animals give out carbon in the form of carbon dioxide as they breathe. Natural **wildfires** and deforestation take carbon from trees and transfer it to the atmosphere. On the other hand, plants take in carbon to create carbohydrate through photosynthesis, while other processes, such as the death of organisms, both on land and in the sea, return carbon to the ground. These competing factors work together to create the 'carbon cycle' – a constant circulation of carbon from the ground, through living plants and animals, to the atmosphere and back to the ground again.

Ordinarily, the amount of carbon in the cycle would remain roughly constant. But human beings are dredging up carbon in the form of fossil fuels – such as coal and oil – from underground. This is steadily adding extra carbon to the cycle responsible for **climate change** and the **greenhouse effect**.

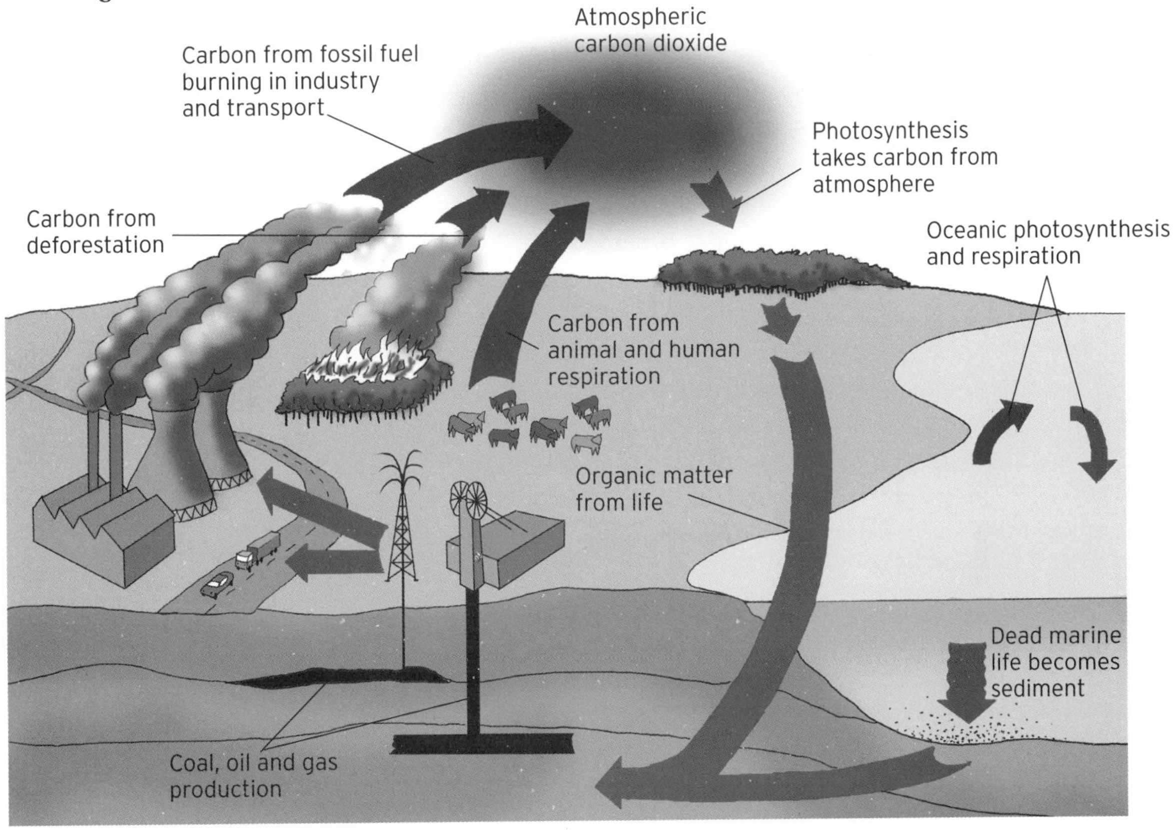

Biomass

The total mass of living organisms within a particular ecological niche at a given time is referred to as biomass. On the Earth the total biomass is believed to be as much as 2,000 billion tonnes. Of these, crop plantations make up approximately 2 billion tonnes, domesticated animals 700 million tonnes, and human beings somewhere around 400 million tonnes. Most of the Earth's biomass – around 1,600 billion tonnes – is locked up in the planet's forests.

Different **terrain** types generate biomass at different rates. **Wetlands** are the fastest producers, cranking out around 2.5kg per square metre every year, closely followed by rainforests, reefs and estuaries at the **coast**. The principle driver for biomass growth is photosynthesis – developing plant mass and thus providing food for other organisms. Biomass is often used as a measure of the food stocks available to particular organisms at each stage in the food chain.

Biodiversity

Biodiversity is the range and richness of different species found in a particular ecological niche. It's the product of **evolution** filling the planet with different species – up to 100 million of them, according to an estimate by the National Science Foundation. Biodiversity is important because it regulates the chemistry and biological interactions of the natural world. It keeps the environment in balance – if one species dies out, say because of disease, then high biodiversity makes it likely there will be a similar species that can step into the breach.

Cropland has perhaps the lowest biodiversity on Earth – especially 'monoculture', when a single crop species is planted over a wide area. On the other hand, forests and jungles hold the greatest biodiversity. And there has been much concern over the impact that the deforestation of these regions for agricultural land will have on their biodiversity, and on the planet as a whole. Biodiversity is closely related to the idea of 'genetic diversity' (see **Conservation genetics**).

Biogeography

The interaction of biodiversity with the different and ever-varying forms of Earth's **terrain** – or 'biomes' as they're also known – is a field known as biogeography. It deals with charting the distribution of species across the face of the Earth and how the distribution evolves with time due to factors such as **ice ages** and **continental drift**.

Biogeography is especially useful when applied to island communities, where geographical isolation has produced species that can be wildly different from those on mainland locations. This phenomenon can apply to **islands** but also to organisms in geographical areas isolated by landforms such as **deserts** or **mountains**. Biogeographical study began in the latter years of the 19th century with the work of British biologist and geographer Alfred Russel Wallace.

Biological interaction

Environments occupied by living species are delicately balanced, with the life cycles of all the species in an ecosystem interlinked, interacting with one another through processes such as competition and predation. A number of different interactions can occur between organisms. 'Amensalism' is the name given to behaviour of one species conferring no advantage to it but inhibiting the development of another – for example, the roots of some walnut trees exude chemicals that are toxic to the roots of other trees. 'Commensalism' is the opposite, where one organism benefits from another at no cost to the other; barnacles, for example, attach themselves to large creatures, giving them a place to live at no cost to the host. 'Competition' is interaction between two species that works to the detriment of both; while 'mutualism' is a mutually beneficial relationship between two species such as symbiosis. 'Neutralism' leaves both species unchanged. And, finally, both predation and parasitism involve one species benefiting at the expense of another.

Symbiosis

When two species work closely together to their mutual benefit, the relationship is known as symbiosis, an extreme form of mutualism (see **Biological interaction**) whereby the two species interact closely. An example is found in the convoluta worm, which lives in shallow ocean waters and gives a home to photosynthetic algae. The algae live under the worm's skin, generating carbohydrates from sunlight, which the worm feeds on; in return, the algae get a safe haven to live in, away from predators. This is an example of 'endosymbiosis' – when one species lives inside the other. When a species lives on the surface of the other, the relationship is known as 'ectosymbiosis', an example of which would be the small 'cleaner fish' that attach themselves to the skin of larger species, a relationship providing the cleaner fish with a source of food and also helping keep the host free from dead scales and parasites.

Parasitism

Parasitism is a form of interaction between two species where one species feeds off a host to the host's detriment. An example is Plasmodia, a **genus** of **protist**, which is transmitted by the bite of the anopheline mosquito and causes the deadly illness malaria. Perhaps the strangest is the cordyceps fungus, a species infecting the brains of ants, that compels the ant to climb to the top of a tall plant stem, at which point the ant's head then explodes, scattering new fungal spores (see **Seed**) over a wide area.

As with symbiosis, biologists draw a distinction between parasites that live within their host (so-called 'endoparasitism') and those that live on the surface ('ectoparasitism'). And there are a further two classifications: 'obligate parasites' can only exist via parasitism of living organisms; whereas 'facultative parasites' can also survive by soaking up dead organic material. Parasitic species often exhibit 'coevolution', evolving through **natural selection** in lockstep with their host species (see **Red Queen hypothesis**).

Conservation

Using our understanding of the Earth's natural environment to try to preserve it is an area of science known as conservation. It includes the careful use of natural resources, preserving **biodiversity**, reducing air pollution and making efforts to protect endangered species and the ecology.

There are a great many threats to our environment such as deforestation, overfishing and **wetland** draining. Many of these threats are brought about through industrial and economic considerations. As a result, much effort on the part of conservationists goes into convincing governments to pass legislation to protect species and habitats, even if that means curtailing economic growth in the short term. In the long term, having a strong economy will rely on us also having a stable and thriving environment.

Conservation biologists keep tabs on whether a species is endangered by rating its

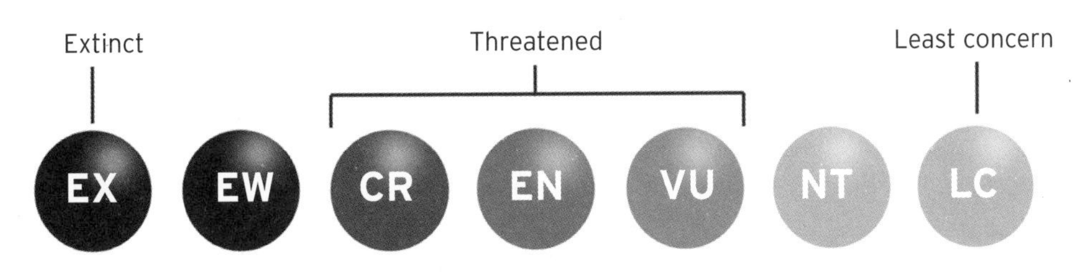

EX = extinct; EW = extinct in the wild; CR = critically endangered; EN = endangered; VU = vulnerable; NT = near threatened; LC = least concern

'conservation status' on a scale that runs from 'extinct' (EX) through 'endangered' (EN) and 'vulnerable' (VU) to 'least concern' (LC) at the opposite end. In 2004, an international team of researchers, writing in the journal *Science*, estimated that up to 50 per cent of the species alive today will be extinct by 2060 – prompting suggestions that the Earth is in the grip of a sixth **mass extinction**.

EVOLUTION

Natural selection

The British ship HMS *Beagle* reached the Galapagos Islands in September 1835; onboard was a young naturalist called Charles Darwin. As the *Beagle* sailed from island to island, Darwin gathered samples of the **species** that he found. And very soon a picture started to emerge – on each island, Darwin found similar species but with subtle differences. He began to theorize that identical species had acquired these small differences as they adapted to the slightly different conditions on each island. According to his theory random mutations (see **Genetic mutation**) would be introduced to each new generation of a species. Any mutations that proved to be beneficial would increase that individual's likelihood of living long enough to reproduce and so pass those characteristics on to its own offspring. This tendency was called 'survival of the fittest', and the theory as a whole became known as natural selection. All that was needed now was a mechanism of heredity, by which characteristics could be handed down from one generation to the next.

Heredity

In the late 19th century, at around the same time that Charles Darwin was developing the theory of natural selection, Gregor Mendel – a monk at the Augustinian Abbey of St Thomas in Brno, in what is now the Czech Republic – found something interesting. Mendel discovered how we pass characteristics on to our children.

Between 1856 and 1863, he grew and cross-pollinated some 29,000 pea plants in the monastery's garden. In one experiment, he crossed a smooth, yellow variety of pea with one that was green and wrinkly, expecting a new, slightly wrinkly strain of pea, with a colour somewhere between green and yellow. But Mendel was in for a surprise. The peas in the new generation were either just as smooth or just as wrinkly as their parents, and either green or yellow – there was no middle ground. More surprising still, in some cases the traits had become jumbled up, making some peas yellow and wrinkly while others were green and smooth. It was as if traits were being passed from one generation of pea to the next in indivisible, discrete chunks. Long before the discovery of DNA, Mendel had discovered genes. His work wouldn't be noticed by the scientific community until the 20th century, but when it was it gave natural selection the mechanism its supporters had been looking for.

Peppered moth

One of the most striking examples of natural selection in action is the evolution of the peppered moth. The species originally had light-coloured wings with dappled markings that helped camouflage it from predators in its natural habitat, where it would settle on light-coloured lichens growing on trees. However, during the Industrial Revolution in mid-to late-19th century England, rising levels of **air pollution** killed many species of lichen, and left the bark of the trees blackened with soot. The light-coloured peppered moth then became easy prey for birds. As a result, any natural mutations (see **Genetic mutation**) that made the wings of any particular moth darker increased that individual's chances of survival. The environment changed, so natural selection switched from favouring moths with light wings to those with dark wings. Before long all the peppered moths living in pollution-stained England had inky-black wings.

Original peppered moth markings

Post-industrial peppered moth markings

Red Queen hypothesis

The Red Queen hypothesis is the name given by biologists to the evolutionary 'arms race' that takes place between two species 'coevolving' – that is, whose characteristics each evolve in response to the characteristics of the other. For example, in the case of a predator feeding off a prey species, the prey might evolve new markings to camouflage it better in its natural habitat, making it harder for the predator species to spot. This will simply spur the predators to evolve better eyesight through natural selection – those animals with random genetic mutations making their eyes more able to spot the new markings will be more likely to survive and pass their genes on, and will eventually come to dominate the predator population.

The name 'Red Queen' comes from Lewis Carroll's *Through the Looking Glass*, in which the Red Queen muses that Alice must run as fast as she can to stay in the same place – like two competing species who must evolve as fast as they can to maintain the same level of relative Darwinian fitness. Scientists have argued that the Red Queen hypothesis also explains why so many organisms reproduce sexually (see **Reproductive biology**) – because the substantial reshuffling of the genes from one generation to the next in sexual reproduction maximizes the rate at which evolution takes place.

Speciation

Evolution has driven different populations of the same species to develop traits that make them better adapted to their home climate. For example, humans with dark skin are better able to exist in hot climates than those with white skin. But when the changes brought about by natural selection become so great that a population can no longer interbreed with other members of its species then a new species is said to have been created, a process is known as speciation.

There are two major types of speciation – 'allopatric' and 'sympatric'. In allopatric speciation, species diverge because of geographic differences that isolate one or more populations in different environments to which they then adapt differently. This is the mechanism by which the subtly different species observed by Charles Darwin on the Galapagos Islands emerged. On the other hand, in sympatric speciation new species are created from populations inhabiting the same region. Here, it is behavioural differences that emerge between populations that leads to the creation of new species. For example, one population might discover a new source of food and become adapted for seeking out that food source instead of the food eaten by the rest of its species.

Convergent evolution

Evolution seems to favour specific solutions to the problems thrown up by the natural world. Particular traits – such as flight and eyesight – have emerged multiple times among completely unrelated groups of organisms. This phenomenon is known as convergent evolution.

Wings have emerged independently in birds, insects and bats. Meanwhile, eyes are even more prolific, evolving on no fewer than 40 separate occasions. Other examples include the echolocation used both by bats (order Chiroptera), as well as whales and dolphins (order Cetacea); while leaves have evolved across the plant kingdom. Convergent evolution can be contrasted with 'divergent evolution', when initially similar species evolve apart, and 'parallel evolution', when similar traits emerge in species that are different but have a common ancestor.

Out of Africa

Modern humans – the species *Homo sapiens* – are believed to have evolved in Africa between 100,000 and 200,000 years ago from a species of great ape. From here, they migrated outwards and around the world, where their large brains gave them the wit

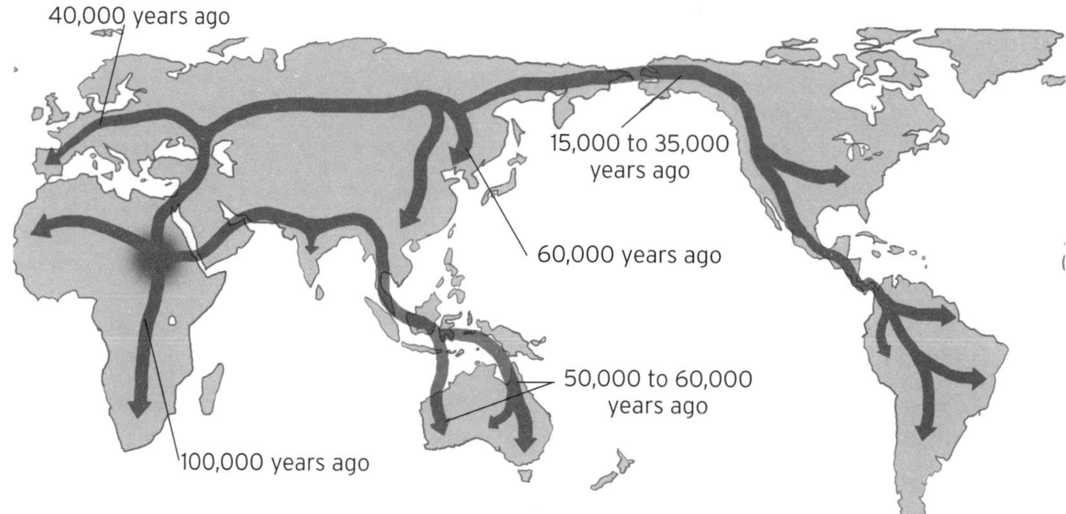

to outperform and ultimately replace more primitive hominid species, such as Neanderthals and *Homo erectus*. This theory for the evolution of our species is known as 'out of Africa'. The theory was first suggested by the father of evolution, Charles Darwin, in his book *The Descent of Man*, and since been confirmed by the field of 'archaeogenetics', which compares DNA patterns found in ancient human remains with those of modern peoples. In particular, archaeogenetics looks at mitochondrial DNA (see **mitochondria**), which is passed from generation to generation without change in females, and the Y chromosome which is passed intact to males (see **Reproductive biology**). The mtDNA and Y chromosomes of modern humans around the world have been found to match those of Africa's ancient population of *Homo sapiens*.

Missing link

Archaeology, and in particular **palaeontology**, have allowed the evolution of species to be traced by unearthing **fossil** remains of organisms from different time periods from **Earth**'s past. But when a sequence of fossils showing the evolution of a particular species seems to be missing a step, the gap is sometimes referred to as a missing link. The term missing link is used predominantly in popular reporting. Scientists tend to prefer the alternative term 'transitional fossil' – a fossil of a species that forms a transitional step in the evolutionary chain. An example is the fossilized remains of archaeopteryx – a dinosaur with feathers, which provides strong evidence to support the theory that modern birds evolved from dinosaurs. More recently, in 2009, scientists uncovered the fossilized remains of a 47-million-year-old species of primate. Dubbed 'Ida', the primate provides a key link between humans and the rest of the animal kingdom.

Lazarus taxon

Taxa (see **Biological taxonomy**) that seem to disappear from the fossil record and then reappear again at a later time are known as Lazarus taxons – because they have apparently come back from the dead, just like Lazarus in the story from the New Testament. Unlike their biblical namesake, however, Lazarus taxons don't miraculously rise from the grave. Rather, their existence is down to the patchy nature of the fossil record. For fossilization to occur requires very specific conditions, meaning that only a proportion of species that have ever lived leave behind a permanent record – and even fewer of these fossils are actually found by **palaeontologists**. A prime example of a Lazarus taxon is the coelacanth – a fish that was thought to have gone extinct 80 million years ago, until, that is, a living specimen was found off the coast of South Africa in 1938.

Punctuated equilibrium

The conventional view of **natural selection**, as put forward by Charles Darwin, says that evolution is a steady, continuous process. But a controversial slant on this was proposed in 1972 by American biologists Niles Eldredge and Stephen Jay Gould. Called punctuated equilibrium, their theory suggested that evolution takes place in fits and starts with rapid flurries of development – the 'punctuations', each lasting no more than 100,000 years (a heartbeat in geological terms) – interspersed with long periods of relative stability, the 'equilibria'. The evidence for punctuated equilibrium is mixed. The lineages of some species, as revealed through the **fossil** record, seem to fit the theory well – for example, ammonite molluscs. Others, however, seem to have undergone a more gradual development.

Sociobiology

Sociobiology is the contention that we inherit from our parents not just our physical features, but our temperament, mental outlook and behavioural traits as well. The theory hinges on the idea that our behaviour is shaped in the same way that our physical characteristics are shaped by natural selection – with the behaviours that bring the greatest rewards being the ones most likely to spread through the population.

While few scientists dispute the application of sociobiology to the animal world, its application to human behaviour is controversial. Critics claim culture is the driving force behind human behaviour, yet proponents argue that the theory can explain the proliferation of certain human social behaviours, including criminality. Harvard University biologist E.O. Wilson coined the term 'sociobiology' and has done much of the pioneering work in the field.

Lamarckism

Do bodybuilders have muscular babies? That was essentially the thesis of a theory of evolution that pre-dated Darwin's, put forward in 1809 by the French biologist Jean-Baptiste Lamarck. The idea was that organisms passed on to their young not just their innate characteristics, but also those they had acquired in life. So professors have intelligent children, bodybuilders do indeed have muscular babies, and – as Lamarck actually theorized – giraffes acquired their long necks from the accumulated stretching of many generations of animals all trying to reach the leaves at the very tops of the trees. As the evidence accumulated for natural selection and Darwin's theory gained in popularity, little evidence emerged to support Lamarckism and the idea fell from favour. Recently, it has enjoyed a mild resurgence of interest in the new field of **epigenetics**.

EVOLUTIONARY GENETICS

Selfish gene

The Selfish Gene is the title of a 1976 book written by the eminent British biologist Richard Dawkins, and has since become a metaphor for how evolution by Darwinian **natural selection** is driven by a competition for dominance between our genes. Natural selection favours genes that produce behaviours and physical traits that propagate those genes most effectively. In this way, organisms effectively function as survival vehicles for their genes. As well as explaining the overtly selfish nature of evolution – for example, driving predators to develop better hunting skills – the theory also explained more altruistic behaviours, such as organisms sacrificing themselves so that their relatives may live, a phenomenon called inclusive fitness.

Inclusive fitness

The British mathematical biologist John Haldane once remarked: 'Would I lay down my life to save my brother? No, but I would to save two brothers or eight cousins.' His point is that you share half your genes with your brother (or your sister) and to ensure that two brothers survive confers the same evolutionary advantage in terms of passing your genes on (see **Selfish gene**) as ensuring that you yourself survive. Similarly, you share one eighth of your genes with your cousin and so ensuring that eight cousins live at your expense does the same job. The idea is known as inclusive fitness, and it explains why many species exhibit altruism towards members of their own family – behaviour which, at first glance, might seem to be at odds with the selfish gene interpretation of natural selection. A prairie dog, for example, will sound the alarm if it spots an approaching predator. The whistle-blower warns its family members of the threat – yet places its own life in danger by drawing attention to itself.

Altruism

Natural selection, and in particular its genetic interpretation through the selfish gene, might seem incompatible with the idea of organisms helping each other out; but, in fact, the natural world is replete with seemingly random acts of kindness. Inclusive fitness is one such example – where an organism places itself in danger to help members of its own family. But there's another form of Darwinian charity that drives the behaviour of creatures, called 'reciprocal altruism'. It occurs when organisms help others who are in need, in the hope that one day, when they themselves are in need, others will help them back. This behaviour is observed in ant colonies. Ants have extra stomachs to store reserve food to feed other members of the colony who are hungry, and any ant who fails to help a comrade is starved by other members of the colony when it needs food itself.

Evolution also drives us humans to this behaviour – even though we may not realize it. That warm glow we feel inside whenever we give money to charity or help someone in need is our genes' way of rewarding us, not just for being nice, but for doing something that actually promotes our own survival. It's the same reward we feel whenever we indulge in other essential activities – like eating or reproducing. The theory of reciprocal altruism was developed by Harvard University biologist Robert Trivers in 1971.

Population genetics

Population genetics looks at the occurrence of different **alleles** of particular genes (for example, 'brown' and 'blue' are different alleles of the gene for eye colour) within a population of organisms of the same species, to determine how this occurrence changes with time.

A key statistic in population genetics is the 'allele frequency' – the total number of alleles of a particular gene, expressed as a fraction of the total number of those genes in the population.

For example, in a population of N people, if a number X of them have blue eyes and Y of them have brown eyes, then the allele frequencies for blue and brown eyes are X/N and Y/N, respectively. The set of alleles of the gene, in this case blue and brown, is called the 'gene pool' – and the bigger the gene pool, the bigger that gene's 'genetic diversity' (see **Conservation genetics**). Evolution manifests itself as variations in X and Y with time. Population genetics is affected by four main factors: **genetic mutations**, **natural selection**, gene flow and genetic drift.

Gene flow

A principle contributor to **population genetics** is a phenomenon known as gene flow – the movement of alleles of particular genes through a population due to the physical migration of organisms within it. Organisms with greater mobility experience greater gene flow within their populations. Gene flow leads to higher genetic diversity (see **Conservation genetics**), giving the population greater stability, but decreases the odds of new species arising through **speciation**. Geographical features that impede migration – such as mountains and oceans – can inhibit gene flow, creating isolated gene pools of the kind discovered by Charles Darwin on the Galapagos Islands.

In the development of **genetically modified organisms**, scientists try to actively inhibit gene flow to prevent engineered genes from jumping to wild populations – **genetic pollution** – and this is achieved using techniques such as **terminator genes**.

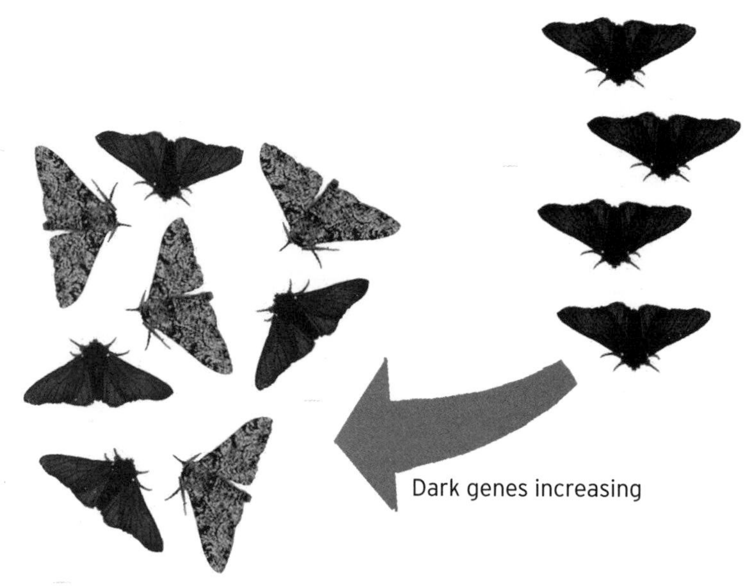

Genetic drift

Genetic drift is an effect in population genetics that causes the frequency of different alleles of genes in the population to vary at random. It is caused by statistical fluctuations in factors such as whether particular organisms live long enough to reproduce, and – in those who do get to reproduce – whether alleles make it through the random shuffling process of chromosomal crossover (see **Gametes**).

There is much debate over the importance of genetic drift compared with natural selection in driving the evolution of species. Generally speaking, genetic drift becomes significant in small populations of organisms, where the small statistical sample size makes it less likely that the random variations in allele frequency will average out. In larger populations the effect is small.

Nature vs nurture

Which plays the bigger role in determining the traits and characteristics of an organism – the genes it inherits from its parents or the acquired traits it learns and accumulates through environmental influences and its life experiences? This is known as the 'nature vs nurture' debate. The key to answering this question lies in what are called 'twin studies' – where scientists look at the traits exhibited by pairs of identical twins who have had different life experiences. In theory, any differences between them should all be down to nurture rather than nature.

Most twin studies conducted to date show neither nature nor nurture to be the sole determinant – it's a bit of both. And while some traits are obviously influenced almost entirely by nature – such as eye colour – others are clearly more open to influence, such as muscular strength, phobias and even our sense of humour. Nature vs nurture lies at the heart of the debate over whether **sociobiology** can be applied to the behaviour of human beings.

Evo devo

Evolutionary developmental biology, or evo devo for short, examines how factors such as genes and evolution have influenced **developmental biology** – the subset of **reproductive biology** that deals with development from a single fertilized cell to an adult organism.

One of the most important concepts is how a group of genes – called 'Hox genes' – shape the early development of embryos. They do this by carefully regulating **gene expression** in different ways across a developing embryo, to turn the small cluster of cells into a fetus. Experiments to alter the structure of the Hox genes in fruit flies have caused new generations to grow extra pairs of wings or to have limbs in the wrong places. The sequence of DNA making up the Hox genes is an evolutionary antique – having remained more or less the same in organisms for hundreds of millions of years. So much so that replacing the Hox genes in flies with those from mice – which both come from the same common ancestor – produces normal insects.

Epigenetics

Epigenetics is a term used by biologists to refer to instances where traits and characteristics are observed in organisms that can't seem to be accounted for in the genes written into their DNA. Epigenetic phenomena are thought to be caused by changes in the biological machinery responsible for **gene expression**. Some biologists have hailed epigenetics as 'the new **Lamarckism**', claiming it offers a mechanism by which organisms can pass on to their offspring non-congenital traits, which they have acquired during their lifetimes.

ORIGIN OF LIFE

Abiogenesis

How did life first get going on Earth? Scientific investigation of how biological processes can arise from non-living chemicals is called abiogenesis. The first **fossils** of living organisms date to around 3.5 billion years ago – around a billion years after the Earth formed. They were stromatolites, structures formed by the deposition of **sedimentary rock** and microorganisms. However, the very first life forms on Earth, are believed to have been unicellular **prokaryote microbes**. Conditions on the young planet were harsh and avoiding the pummelling being delivered by **asteroids** and **comets** means life couldn't have begun any earlier than 3.7 to 4 billion years ago on the surface, and perhaps 4 to 4.2 billion years ago on the **ocean** floor.

The Miller–Urey experiment and others like it have shown how basic biochemicals might have emerged. And theories such as RNA world, **iron–sulphur world** and **clay theory** offer possible mechanisms by which these chemicals turned into living organisms. But to date, scientists don't know for sure which, if any, of these theories are correct.

Miller–Urey experiment

In 1953, Stanley Miller and Harold Urey at the University of Chicago carried out a famous experiment to find out whether conditions on the young Earth could have brought about the chemicals needed for life. They cooked up water, methane, hydrogen and ammonia – chemicals thought to exist on the early Earth – and fired an **electric current** through the mixture to simulate the young planet's stormy conditions. After a week's continuous running, the experiment had generated various **organic compounds**, including amino acids – the building blocks of proteins. A number of follow-up experiments have been carried out, in which scientists have succeeded in producing not just amino acids but also nucleotide bases – the fundamental units that make up DNA and RNA molecules.

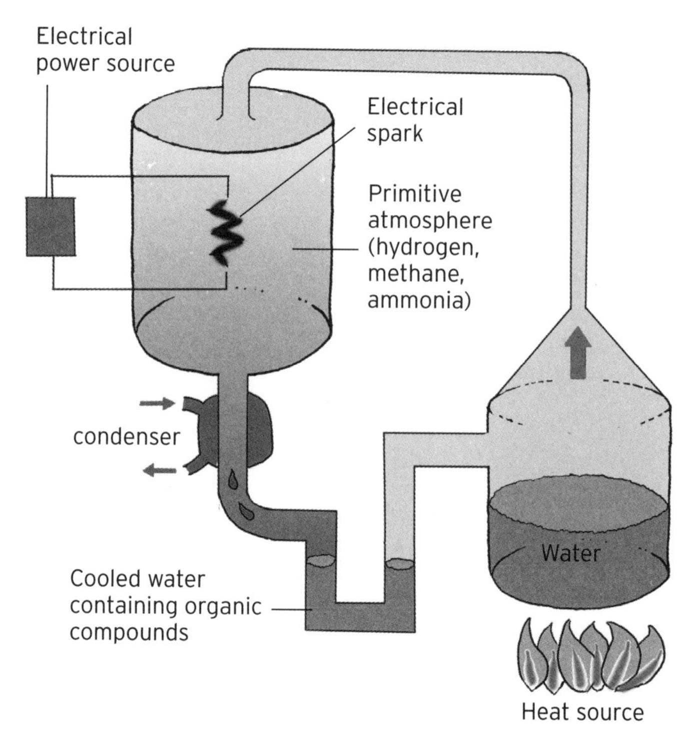

In 2008, scientists from the USA and Mexico repeated the Miller–Urey experiment, but with a twist – they tweaked the set-up of their apparatus to recreate the conditions near a volcanic eruption. This variant of the experiment produced more organic chemicals than any other – suggesting that volcanoes (see **Volcanoes**), while deadly to life today, may have played a central role in its origin.

RNA world

RNA world is a theory for how life might have begun on Earth, in which the organisms based on DNA and proteins that inhabit the planet today were preceded by an era during which life was based on DNA's molecular cousin RNA. Organisms in the modern world use the molecule DNA to store the genetic information (genes) needed to build proteins, some of which form the enzymes needed to replicate these genes and so create new generations of organisms. But this has led to a chicken-and-egg scenario: namely, which came first – protein or DNA? It seems the two would have to come into existence simultaneously, which seems unlikely. In the RNA world theory, first put forward by American biologist Carl Woese in 1968 and developed further by Walter Gilbert in 1986, both information storage and replication are handled by RNA. The idea was suggested after it was discovered that RNA can have similar catalytic properties (see **Chemical reactions**) to enzymes. Many biologists believe an RNA world existed – though the jury is out on whether it represents the first life on Earth.

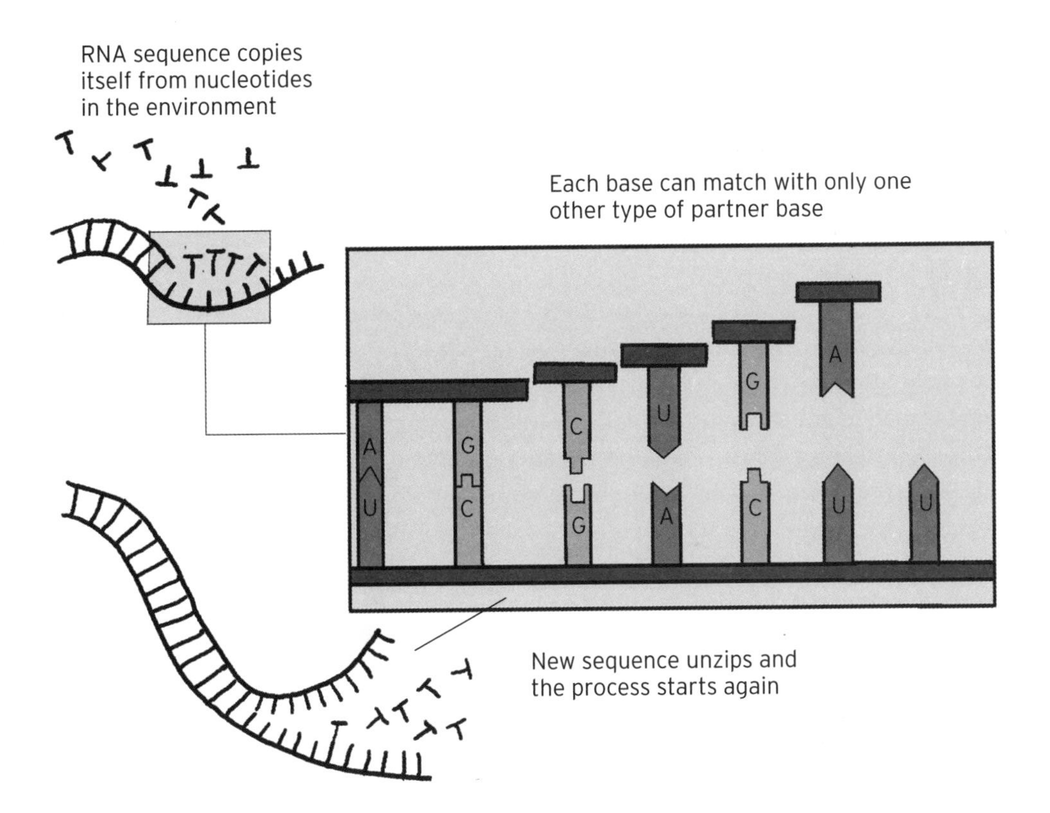

Iron–sulphur world

Could life on Earth have begun inside iron-sulphide rocks? A theory known as iron–sulphur world holds that early life forms didn't rely on DNA or RNA but instead were first propagated through inorganic chemical compounds around **hydrothermal vents** in the ocean floor. Iron–sulphur world was first proposed by German chemist Günter Wächtershäuser in the late 1980s and early 1990s. Iron and sulphur are able to set up chemical cycles that resemble **metabolism** processes in modern organisms. The theory is especially interesting because it predicts the transition from inorganic life to organic life, made of carbon-based protein, of the type that exists today. Iron and sulphur together make acetic acid which, when it combines with carbon and ammonia, forms amino acids that in turn hook up to make proteins – a process that Wächtershäuser and colleagues demonstrated experimentally in 1997.

Deep hot biosphere

What if life didn't get started at the Earth's surface, or even at the bottom of the ocean, but many kilometres underground? This is the theory put forward by British scientist Thomas Gold in the 1970s, and which he later named the 'deep hot biosphere'. According to Gold the first life forms would have been archaea (see **Prokaryote microbes**). It's an appealing theory because underground offers one of the safest environments for organisms to shelter from the turmoil – **asteroid** bombardments and volcanic eruptions – that wracked the Earth during its formative years. Today, it's known that bacteria and other microorganisms thrive underground, down to a depth of at least five kilometres.

But Gold had another surprise up his sleeve. If his theory is correct then underground bacteria, feeding on methane gas expelled from the Earth's mantle, are what actually generate underground oil reserves – not the compressed remains of organisms from the planet's surface, as in the standard theory (see **Coal** and **Petrol**). If he's right then oil is being continually produced, and concern over **peak oil** may be misplaced.

Clay theory

Proposed by British biologist Graham Cairns-Smith in 1968, clay theory is an innovative idea for the origin of life on Earth and says that the first life forms took shape on the surface of ancient lumps of clay. The theory holds that crystals of clay have chemical properties that enable them to self-replicate without complex molecules like DNA or RNA. What's more, clay crystals are able to adapt to their environment by natural selection – with certain forms of clay being preferentially selected for by particular environments. For example, sticky clays tend to form riverbed silts, and riverbeds silted with sticky clays attract more sticky clays. Similar selection processes exist on the surfaces of the clays, enabling them to selectively snag particular kinds of molecule, and this, believes Cairns-Smith, may be what ultimately transferred the lineage of clay life to a new molecule: DNA.

Extraterrestrial origins

Some scientists think life on Earth might have originated in space – that organic material descending to our planet from the outer reaches of space billions of years ago may have been what seeded life here in the first place. In that case, we are all aliens.

In 2009, NASA scientists announced that they had detected amino acids in material gathered from Comet Wild 2 by the Stardust spacecraft during a fly-by in 2006. The water of Earth's oceans is believed to have been deposited here by **comets**; now it seems plausible that the chemical foundations for life may have come with it.

Some researchers even believe that microorganisms themselves may have come from space, through **panspermia**. In 1996, NASA scientists announced that they had found what looked like fossilized bugs in a meteorite from Mars (see **Meteors**). While this claim is now largely discredited, other scientists have shown how microbes could survive an arduous journey through space – even travelling between star systems – locked away inside tiny grains of dust.

BIOPHYSICS

Mathematical biology

Biology doesn't come across as a discipline heavy in mathematical content – certainly not as much so as physics. However, biologists are increasingly developing numerical representations of their subject, allowing them to apply powerful techniques from **pure mathematics** and **applied mathematics**, leading to dramatic new insights. The power of mathematics lies in its ability to help humans visualize and analyze concepts that are beyond the range of our senses and intuition. Were it not for mathematics, we would have little chance of making sense of the great pillars of theoretical physics – **relativity** and **quantum theory**. Similarly, biologists are bringing this discipline to bear on the big problems of biology – such as gene structure and gene expression, neurobiology and cell biology, and to help them unpick the myriad processes that contribute to the state of the environment.

Placing biological science on a firm mathematical footing has also enabled researchers to call on the problem-solving power of computers which are already bringing new results, and indeed whole new fields of science – such as **systems biology**. Application of computers to biology is sometimes known as 'bioinformatics'.

Biomechanics

Mechanics is the branch of **physics** that deals with the behaviour of objects in response to forces applied to them. Likewise, biomechanics is concerned with the effect of physical forces on biological systems. It is applied on all scales throughout organisms – from calculating the dynamics of fluids and organelles within cells to the stresses on bones (see **Musculoskeletal system**), applying **Navier–Stokes equations** to model blood flow or using aerodynamics (see **Fluid dynamics**) to understand how birds fly.

A major application has been within sport science, where the use of 3D video capture, together with computer analysis, has enabled sports coaches to dramatically improve the performance of athletes – for example, figuring out how to tweak a golfer's swing to yield maximum distance. Biomechanics calculations are typically harder and more complex than their counterparts in pure physics owing to the imprecise shapes and many degrees of freedom in biological systems.

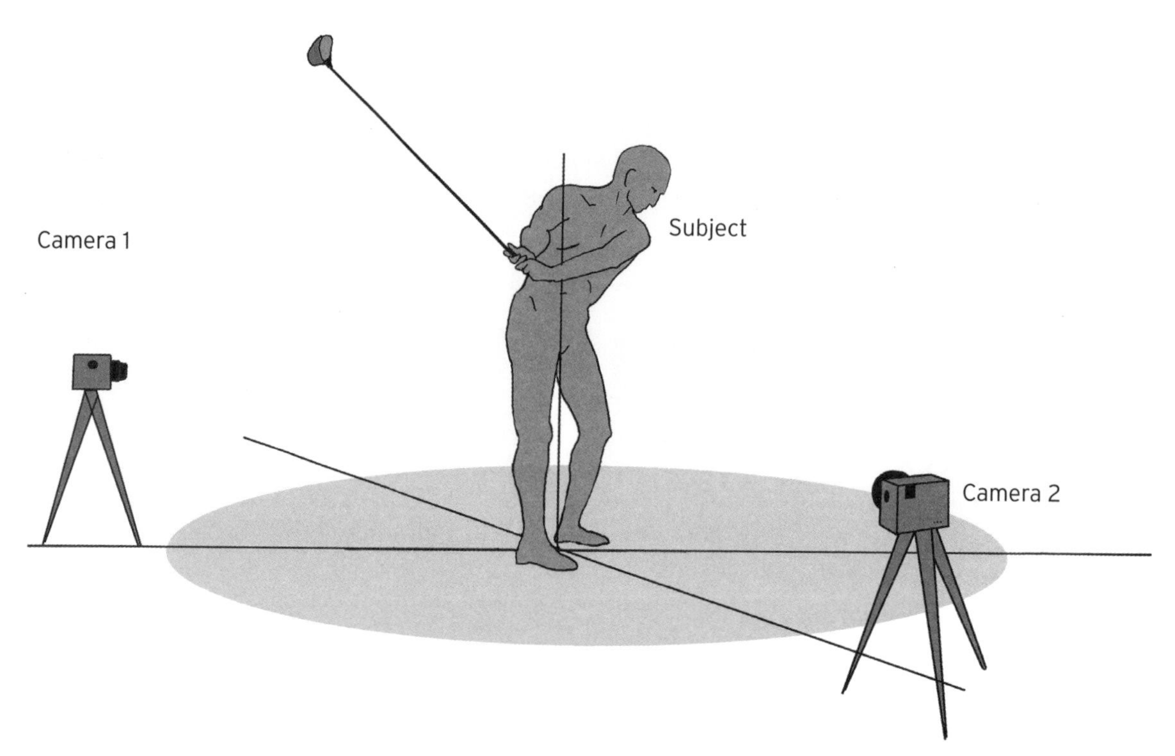

Bioengineering

Whereas biomechanics applies principles from the physics of moving bodies to solve problems in biology, bioengineering does the same by drawing upon a whole raft of techniques and technologies from engineering. Examples include the development of artificial hearts, livers, bones, eyes and a host of other such **prosthetic** and **cybernetic implants**. It also encompasses the development of medical machinery and drug delivery systems, genetic modification, synthetic biology and **biomimetics**.

But bioengineering is more than that. It's about applying the 'engineering method' to design and build solutions to problems in biological science. Even the term 'bioengineering' itself has been reverse-engineered to describe the application of biology to engineering – for example, when **civil engineers** might use biological materials in the construction of a building, such as concrete reinforced with plant fibre.

Radiobiology

Radiobiology is concerned with the impact of radiation on biological tissue. That includes ionizing radiation – such as alpha, beta and gamma (see **Radioactive decay**) – as well as thermal radiation (see **Black body radiation**) and various wavelengths of **electromagnetic radiation**. Applications of radiobiology include assessing the threat posed to astronauts from **cosmic rays**, as well as particles and waves from the Sun. It has also been used to investigate the claimed link between cellphone use and incidences of brain tumours (see **Cancer**) – though so far the results remain inconclusive.

As well as dealing with the potential hazards of radiation, radiobiology also deals with the benefits. Radio-imaging techniques are used to diagnose diseases and monitor their development; these include X-rays, as well as technologies such as SPECT scans (see **Tomography**), where mildly radioactive substances are injected into the body to act as a tracer of blood flow. In cancer treatment, radiation therapy is sometimes used to kill malignant cells by bombarding them directly with gamma rays.

Agrophysics

Agrophysics is the application of **physics** to agronomy – the science of growing crops and other plants for human use. Agrophysicists construct mathematical theories of agronomy and test them through rigorous measurement and experiment – with the aim of ultimately improving crop yield.

There are two major areas of interest – the first is using physical considerations to improve the delivery of air, water and plant nutrients, through techniques such as irrigation, **hydroponics** (where plants are grown in liquid) and aeroponics – where they grow in a nutrient-rich mist. The second area is the supply of light and heat using, for instance, greenhouses and artificial illumination. Scientists are using agrophysics today to come up with methods of sustainable agriculture that maximize production while minimizing the impact upon the environment.

Systems biology

Up until recently, biology was a divided discipline. There were **cell biologists**, **molecular biologists**, **biochemists**, and many others – and never the twain shall meet. Systems biology is a holistic approach to biological science that looks at each biological organism as a whole unit – taking care to understand not just the processes inside it but also the interactions between each process, which can be equally important.

A prime example is the function of the **immune system**. No single mechanism is responsible for the immune response of organisms to infection – rather, it's an emergent response that comes from genes, proteins and other biological pathways all acting in concert. Systems biology draws heavily on computer power to run simulations of how different biological systems within an organism function and interact. The first such simulation was a crude mathematical model of the heart developed by British biologist Denis Noble, at Oxford University in the 1960s. More recently, he and his colleagues have extended this work to create a sophisticated 'virtual heart' running inside a supercomputer (see **Parallel computing**).

THE EARTH

OUR HOME WORLD, PLANET EARTH, came into being as the incipient fragments of the Solar System coalesced around the young Sun 4.5 billion years ago. Human beings arose only in the last couple of hundred thousand years. If the time since the planet formed was represented as a 24-hour day, then our species has been around for about the last three seconds.

But humans haven't always realized this much. As recently as the turn of the 20th century, scientists – notably the British physicist William Thomson (aka Lord Kelvin) – believed that the Earth was just 100 million years old. Kelvin based his claim on cooling times, but had ignored the complex physics of the Earth's interior and a new factor that was yet to

E AND LONGITUDE • DAY AND YEAR • TIME
NS • CORIOLIS EFFECT • PRECESSION •
• PROTEROZOIC AEON • PALEOZOIC ERA
Y PERIOD • TOPOGRAPHY • CONTINENTS
RIVERS • VALLEYS • LAKES • WETLANDS
CAVES • STALAGMITES AND STALACTITES
ST • MANTLE • CORE • GEOMAGNETISM •
OCEAN LAYERS • SEAS • TIDES • TIDAL
HES • HYDROTHERMAL VENTS • MAGMA
ENT FAULTS • CONVERGENT FAULTS •
NENTAL DRIFT • SUPERCONTINENTS •
SLIDES • LIMNIC ERUPTION • FLOODS

be discovered – radioactivity, an effect that was generating extra heat in the planet's interior.

Today, the Earth remains a complex dynamical system that scientists still cannot claim to fully understand. The big unknown is climate change. The Earth's climate is an extraordinarily sensitive system, the behaviour of which is determined by many factors – not just man-made pollutants, but cloud cover, volcanic activity, ocean currents and more.

While we know global warming is taking place and that it's probably our fault, quite how bad it's going to get or what (if anything) we can do about it will only become clear in the years and decades to come.

EARTH SCIENCE

Curved Earth

Our planet, the Earth, is a spinning ball of rock, metal, gas and liquid nearly 13,000km (over 3000 miles) in diameter with a circumference of just over 40,000km (nearly 25,000 miles). Earth gets its spherical shape due to gravity. This is the force that makes rivers flow downhill, and it has the same effect on the solid material that the Earth is made from – tending to smooth out the planet's surface until everything is the same distance from the centre. The only shape that allows this is a sphere. Gravity has to a large extent evened out the lumps and bumps on Earth's surface to make it smoother than a billiard ball. Yet Earth isn't quite spherical; the planet is rotating roughly once every 24 hours creating an outward force that makes the planet slightly fatter at the equator than around the poles, by about 43km (27 miles) (see **Centripetal force**). Scientists call this flattened shape an 'oblate' spheroid.

Equator

The equator is an imaginary line dividing the Earth into two equal-size half spheres, known as the Northern and Southern Hemispheres. It forms a plane at right angles to the planet's axis of rotation. Two further imaginary lines circle the Earth parallel to the equator, called the Tropic of Cancer and the Tropic of Capricorn, lying 23.5° north and south of the equator, respectively. The warm region in between these lines is known as the 'tropics' (see **Seasons**).

Outside of these limits lie the North Temperate Zone and the South Temperate Zone, in which temperatures are much cooler. These extend towards the Arctic and Antarctic Circles – the Earth's chilly polar zones (beginning at 66.5° north and south of the equator), where temperatures really plummet.

Poles

Imagine travelling further and further north or south on the Earth's surface, and drawing circles parallel to the equator on the planet's surface as you go. As you move further from the equator the circles become smaller and smaller, eventually shrinking to two points at the top and bottom of the planet, and these are known as the poles.

In fact, Earth has two distinct sets of poles. The 'geographical poles' are defined by the planet's rotation, such that if you could skewer the planet on a giant spindle – upon which it turns through its night/day cycle – then the spindle would protrude from the north and south geographical poles. But Earth also has a pair of 'magnetic poles', like the poles of a bar magnet and are caused by electrical currents in the Earth's core (see **Geomagnetism**). The magnetic poles wander from year to year; in 2005, the North Magnetic Pole was about 7° from geographic north, while the South Magnetic Pole was almost 30° astray.

Latitude and longitude

Anyone who has played 'battleships' knows how a coordinate system works. A two-dimensional space is divided up into a grid and numbers along each side of the grid form pairs of coordinates that give each square of the grid a unique reference. This is the basis of longitude and latitude coordinates; the only difference is that rather than a flat two-dimensional surface, the Earth is a sphere, and so each coordinate is measured as an angle around the sphere. So, for example, opposite points on its surface are 180° apart, while moving 360° around the sphere brings you back to where you started.

The first coordinate is latitude, measured along a line parallel to the Earth's equator, which is taken to have latitude 0°. The North and South Poles then have latitudes +90° and –90° respectively. Longitude, on the other hand, is given by the distance around the equator. The zero point is known as the 'prime meridian', a line connecting the North and South Poles and passing through the Royal Observatory in Greenwich, England. Adding a suffix 'E' or 'W' to the longitude coordinate indicates whether the distance is measured east or west of the prime meridian. Coordinates on the planet's surface are usually stated as the latitude followed by the longitude. New York, for example, is at 41°N, 73°W.

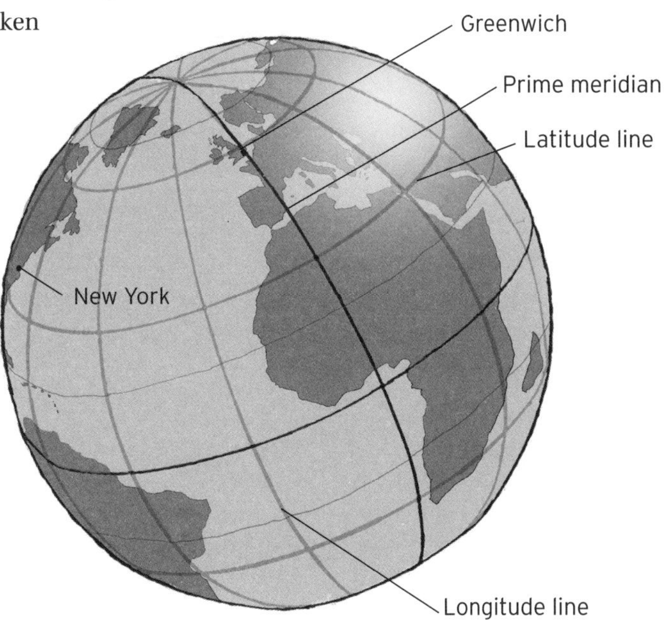

Day and year

The length of the day and year on Earth are determined by the planet's rotation and orbit around the Sun. As Earth rotates, the Sun appears to cross the sky from east to west. One day is defined as the time taken for the Sun to return to the same point overhead after one complete rotation. Earth turns once on its axis every 23 hours 56 minutes 4 seconds. But in this time the planet has also moved around the Sun slightly due to its orbit, meaning Earth must rotate a little more to bring the Sun back to where it was the previous day, and the extra time this takes brings the length of the day up to 24 hours.

The Earth's orbit carries it around the Sun once every 365.25 days, and this is where our definition of the year comes from. Ordinary calendar years are just 365 days long, but we make up the quarter-day shortfall by adding an extra day to the calendar every fourth year – these are the leap years. Sometimes tiny wobbles in the planet's orbit must be corrected for by the addition of 'leap seconds'.

Time zones

The rotation of the Earth means that it's always daytime somewhere, while on the opposite side of the planet it is night (see **Day and year**). That is why we have created time zones, meaning you have to adjust your watch when you fly to faraway locations. The time zones are usually fixed by longitude – so the time at the prime meridian (see **Latitude and longitude**) is considered the zero point. This time is known as Greenwich Mean Time (GMT) or, more internationally, as Coordinated Universal Time (UTC). Some countries are so big they have several time zones within their borders; for instance, in the United States where there is a four-hour time difference between the east and west coasts.

International date line

Having time zones around the world means that not only can the time of day change from country to country, but also the day itself. After all, a new day must start somewhere – and this point is known as the International Date Line (IDL). The IDL is a line of constant longitude, passing through both the North and South Poles, and lies 180° around the planet from the prime meridian (see **Latitude and longitude**). The time at the IDL is 12 hours different from that at the prime meridian. So, for example, if at the prime meridian it's 3 p.m. on December 10, then the time at the IDL is 3 a.m. A fraction to the east of the IDL it's 3 a.m. the same day, December 10, but to the west of the line it's 3 a.m. the next day – December 11. If you're in a plane flying over the IDL the time instantly jumps by 24 hours as you cross it, forcing you to adjust not only your watch but your calendar too.

Seasons

As the Earth travels around the Sun, there are cycles of warmth and cold known as seasons, caused by the tilt of the planet relative to its orbital plane around the Sun. The tilt is 23.5°, which means that for part of the year the planet's northern hemisphere leans towards the Sun – so that the Sun's light and heat is concentrated on a smaller area than when the Earth is at the opposite point of its orbit half a year later. The first position corresponds to summer in the northern hemisphere, and the latter is winter. While the northern hemisphere tips towards the Sun, the southern hemisphere tips away – which is why those south of the equator enjoy summer in December and winter in June.

In the north and south temperate zones (see **Equator**) there are four seasons. However, within the tropics the climate is constantly warm, and the small variations in temperature

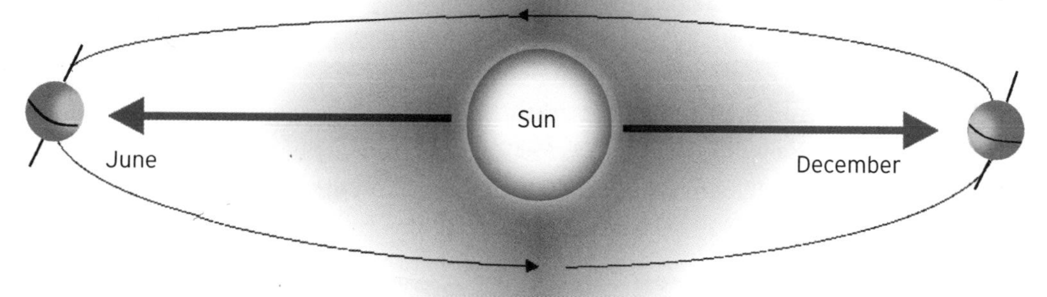

that result manifest themselves as varying levels of rainfall, broadly dividing the year into two seasons: wet and dry. Also, the high latitudes of the Arctic and Antarctic Circles means Earth's tilt plunges these regions into constant darkness for six months during winter, followed by six months of constant daylight, with no sunset.

Coriolis effect

The Coriolis effect is a phenomenon caused by the Earth's rotation tending to make air in the northern hemisphere swirl in an anticlockwise direction and air in the southern hemisphere swirl clockwise. Each day the Earth makes one complete revolution – 360° – but because the equator is the widest part of the planet, the surface there rotates fastest – 1,670km/h (1,038mph). By comparison, at the latitude of New York (41°N) the planet's surface is moving at just 1,260km/h (782mph). And it is this difference in speed that's behind the Coriolis effect, creating swirling motions in the atmosphere which are responsible for weather systems including **cyclones**. Often it is said that the direction in which bath water swirls as it goes down the drain is also caused by the Coriolis effect. However, this is not the case – as swishing the water in the opposite direction with your finger will quickly confirm.

Precession

A child's spinning top wobbles from side to side as it spins, a phenomenon known as precession. In fact, all spinning objects can exhibit precession – and the Earth is no exception. Our planet's rotation axis wobbles about its central position, returning to its starting point every 26,000 years and this changes the night sky; the rotation of the planet makes the celestial sphere appear to wheel overhead. At present, the Earth's axis of rotation points towards the star Polaris – known as the Pole Star – but 5,000 years ago it pointed towards the star Thuban, while in AD 14,000 the role of the Pole Star will fall to the brilliant star Vega in the constellation Lyra.

Precession is the biggest of several effects that shift the rotation axis of the Earth. Others are 'nutation', a small wobble of the axis about its precession path, and 'polar motion' – a tiny and unpredictable variation caused by factors such as ocean currents and winds.

AGES OF EARTH

Deep time

Our planet's history spans a colossal 4.5 billion (4,500,000,000) years. Just as we break up each year into months, weeks and days, scientists break Earth's past into more manageable units. The biggest are called 'super aeons' spanning several billion years; there has only been one super aeon in the history of our planet – called the Precambrian. Next down are the 'aeons' lasting between 500 million and 2 billion years. Aeons are subdivided into 'eras', each a few hundred million years long. And these are split down further into 'periods', each lasting several tens of millions of years, and 'epochs', which are usually 10 to 20 million years long. The smallest unit of deep time is the 'age', which lasts typically a few million years or less.

Whereas our everyday units of time have fixed duration, the precise lengths of units of deep time are determined from **stratigraphy** of the **fossil** record. Each chapter in Earth's past corresponds to a layer of rock in the record, which can be identified and dated using **archaeology**. The difference in date between the top and bottom of the layer fixes the duration of that particular aeon, era, period or epoch.

Hadean aeon

The first aeon in Earth's history was the Hadean aeon, stretching from 4.5 to 3.8 billion years ago, a period of time that saw the formation of the **solar system** and planet Earth, and the creation of the Moon in a colossal impact event known as the Big Splash. Earth's oldest minerals and rock formations were forged during this time, and the earliest life forms may have been created then as well.

A major event at the end of the Hadean aeon was the Late Heavy Bombardment, during which a hail of **cosmic impacts** peppered the inner solar system. Much of the cratering seen on the Moon today is the result of this ancient cosmic blizzard. The Late Heavy Bombardment may have been caused by the migration of one of the giant planets – such as Neptune – outwards through the solar system, disrupting smaller bodies as it went.

Archean aeon

Following immediately from the Hadean aeon, the Archean aeon stretched from 3.8 billion years to 2.6 billion years ago (bya). Plant life able to carry out **photosynthesis** is thought to have emerged at this time, when the continents were also created and **tectonic** activity is thought to have begun too. However, the layout of the continents was very different from the world map of today – with most of the Earth's land mass concentrated into a single **supercontinent**, which would later break up through **continental drift**. Surface water was plentiful. The Archean aeon is split into four eras – the Eoarchean (3.8–3.6bya), Paleoarchean (3.6–3.2bya), Mesoarchean (3.2–2.8bya) and Neoarchean (2.8–2.6bya). From the Archean aeon's Paleoarchean era come the oldest known fossilized remains of life.

Proterozoic aeon

Longest of all aeons, the Proterozoic, during which the first multicellular microorganisms emerged, extends from 2.6 billion years to 0.57 billion years ago (bya) and is divided into three eras. During the Paleoproterozoic era (2.6–1.6bya), Earth's early atmosphere became oxygenated through the action of bacteria performing photosynthesis. There was also the appearance of the first eukaryote life forms. Following this was the Mesoproterozoic era (1.6–1.1bya), bringing the evolution of sexual reproduction. Finally the aeon ended with the Neoproterozoic era (1.1–0.57bya), when an extreme ice age is thought to have taken place. Together the Hadean, Archean and Proterozoic aeons together make up the Precambrian super aeon.

Paleozoic era

After the Proterozoic aeon came the Phanerozoic aeon, beginning 0.57 billion years ago (bya) and extending right up to the present day. The Paleozoic was the first era of this aeon, lasting from 570 million years ago (mya) to 248mya. It is divided into six periods – the Cambrian (570–470mya), Ordovician (470–438mya), Silurian (438–408mya), Devonian (408–360mya), Carboniferous (360–285mya) and Permian (285–248mya).

Rocks from the Cambrian period show a marked increase in the number of **fossils** as a growth spurt of life took place in the oceans – known as the 'Cambrian explosion' – bringing the appearance of hard-shelled marine animals and continuing into the Ordovician. The Silurian period saw life begin to move from the seas on to the land – plants at first and later, in the Devonian period, the first land animals with backboned fish thriving in the oceans. In the Carboniferous period, much of Earth's land mass became covered with lush forest giving a home to the land's new inhabitants. Insects emerged during the Permian period, growing in abundance due to the presence of oxygen being churned out by Earth's forests through photosynthesis. A **mass extinction** at the end of the Permian period killed off many of these insect species, paving the way for the rise of reptiles and amphibians.

Mesozoic era

The second era in the Phanerozoic aeon, the Mesozoic spanned from 248 million years ago (mya) to 65mya. It's the time when dinosaurs ruled the Earth and for this reason it's often regarded as the 'age of reptiles'. It splits down into three periods beginning with the Triassic period (248–213mya), during which most of the Earth's land mass was again concentrated into a giant **supercontinent**. Amphibians and reptiles were common on the land; meanwhile, giant creatures evolved in the seas.

The heyday of the dinosaurs was the Jurassic period (213–144mya). Plant life burgeoned at this time too – with ferns, conifers and dense jungle embracing the planet. Animals first took to the skies as well, with the emergence of pterosaurs (flying dinosaurs) and the first feathered birds. The era ended with the Cretaceous period (144–65mya). Many of the iconic dinosaur species – including Tyrannosaurus rex and the nimble Velociraptor – lived during this time. The **cosmic impact** of a comet or asteroid on the Earth ended the period and wiped out the dinosaurs.

Cenozoic era

We are now in the Cenozoic era, which spans the period from the time of the dinosaur extinction 65 million years ago (mya) up to the present day. After the demise of the dinosaurs, the mammals inherited the Earth as large warm-blooded creatures developed from small animals like rodents to fill the evolutionary niches the dinosaurs had left vacant. Birds evolved into their present form.

The Cenozoic era is divided into three periods – the Paleogene (65–23mya), Neogene (23–2.6) and **Quaternary** (2.6mya to present day). At some point during the Neogene, most likely between 5–7mya, it is thought the common ancestor of both chimpanzees and modern humans existed. Under some naming schemes the Paleogene and Neogene are joined together into one period, called the Tertiary.

Quaternary period

Earth's current period is the Quaternary, the final period of the Cenozoic era; it began 2.6 million years ago (mya). At the start of the Quaternary the continents were more or less as they are now. Later, four **ice ages** gripped the Earth as the polar caps expanded, encasing large areas of the planet in ice; it was around 10,000 years ago that the most recent ice age ended.

The Quaternary is broken down into two epochs – the Pleistocene (stretching from 2.6mya to 11,700 years ago) and the Holocene (from 11,700 years ago to the present day). Perhaps most significantly, the Pleistocene saw the emergence of modern humans. Our species, *Homo sapiens*, first appeared in Africa between 100,000 and 200,000 years ago and quickly spread from there to colonize the world (see **Out of Africa**). During the Holocene epoch humans became skilled in agriculture, leading in turn to the development of civilization.

TERRAIN

Topography

The science underpinning the 'lay of the land' is known as topography and aims to chart the three-dimensional shape of surface terrain. A topographic map of an area, also known as a 'relief map', shows the elevation using a system of 'contour lines'. Think of these as a series of equally spaced horizontal slices through the landscape, the outlines of which are then projected down onto a flat sheet of paper (see diagram). Numbers can indicate the height of each contour and the closer together the contour lines, the steeper the terrain is.

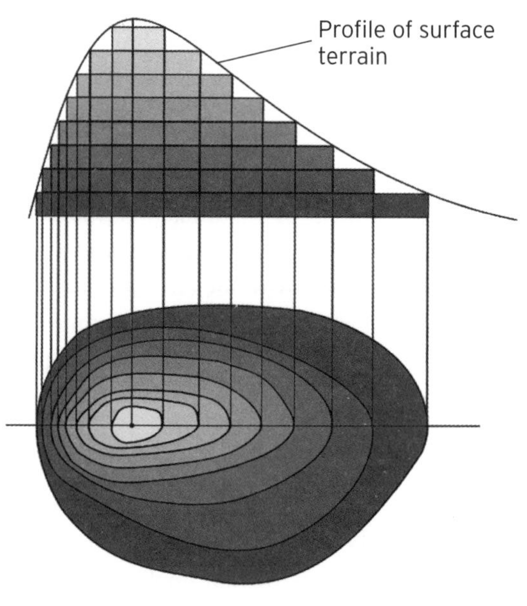

To make accurate relief maps, surveyors must make detailed measurements of the land, which is achieved through a host of instruments used on-site for measuring the tilt of the land and the elevation angles of surface features. Aerial photography is also employed, as is data gathered from space. In February 2000, the space shuttle Endeavour used radar to map the topography of the Earth in unprecedented detail.

Continents

The continents are the major land masses of the Earth and occupy around 29 per cent of the planet's surface. Liquid water oceans fill the rest of the area. There are seven continents: Asia, Africa, Antarctica, Australia, Europe, North America and South America; almost all are either separated by expanses of ocean or have been in the past, because of **continental drift**. The exceptions are Europe and Asia, which have always been joined, and for this reason they are classified in some schemes as a single continent: Eurasia. Antarctica is a continent because there is dry land beneath the frost; the Arctic is not because it's just an ice sheet afloat upon the sea.

The height of ocean water means that the edges of the continents are actually submerged, boundaries known as 'continental shelves', and beyond them the seabed drops precipitously. Continental shelves vary in width from a few to thousands of kilometres.

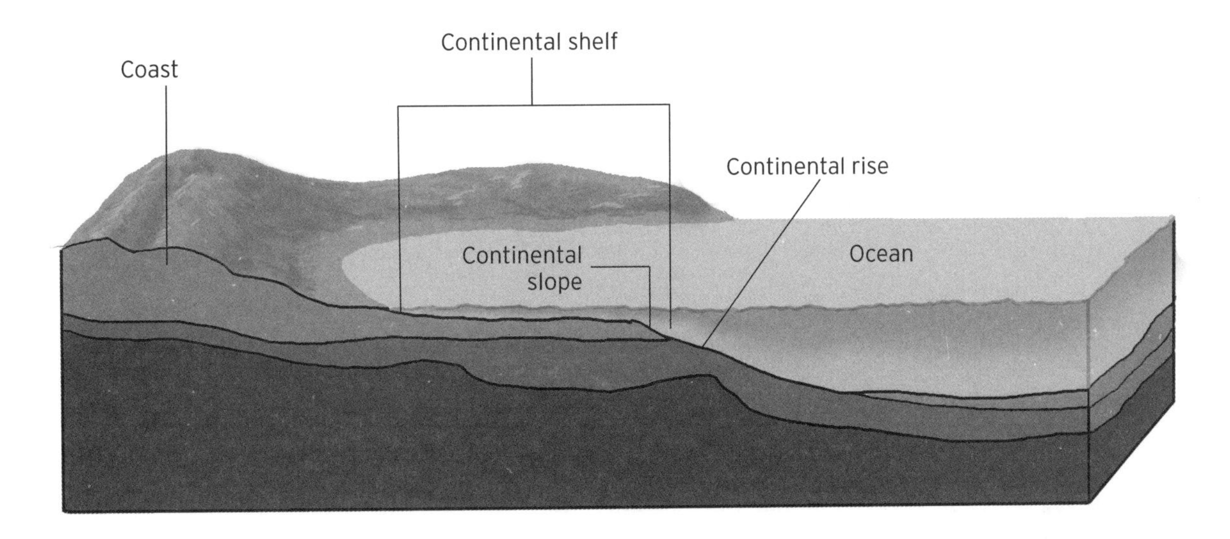

Islands

Islands are bodies of land separated from the major continents by water and they come in three principle types. Continental islands, such as the United Kingdom, are part of their nearby continental land mass – Europe, in this case – because they are situated upon that continent's shelf. Other examples include Tasmania, Sumatra and Greenland (which is part of continental North America). By contrast, oceanic islands are not connected to any continental land mass but are normally the result of undersea **volcanoes** or **tectonic plate** activity pushing up the sea floor. Macquarie Island in the Pacific was formed tectonically; the Hawaiian islands are volcanic. A group of islands formed tectonically is known as an archipelago. A third type are the atolls, which are grounded on coral reefs, the accumulated skeletons of small marine creatures formed around volcanic islands that have since subsided. The Maldives, in the Indian Ocean, are atolls.

Glaciers

Vast masses of fresh-water ice that cover the landscape are known as glaciers; they form in low-temperature areas where the rate of deposition of ice and snow has exceeded the rate of melting and erosion for very many years. Every continent except Australia has glaciers and stunning examples can be found in New Zealand, Alaska and Tibet. Glaciers make up the world's largest reservoir of fresh water and this is why there is such concern over the possibility of them thawing as a consequence of **climate change**, causing catastrophic **sea-level rise**.

When glaciers merge to cover a large area of terrain the result is a featureless landscape known as an 'ice sheet'. Ice sheets exist in Greenland and Antarctica, though during an **ice age** they can cover a substantial fraction of the Earth's surface. Glaciers move across the land as the weight of overlying ice creates pressure, literally making the glacier ooze outwards and spread. The ice can spread in this way by up to tens of metres per day, leaving scars on the landscape as it scrapes by, and serving as evidence of past glaciations. Where a glacier meets the coast it can spread out across the sea to form an 'ice shelf'. Portions that break away from the shelf to float freely are known as 'icebergs'.

Sea ice

In contrast to icebergs, which break away from a coastal **glacier**, sea ice forms when ocean water itself freezes. The salt content of the oceans lowers the freezing point of the water from 0°C to -1.8°C. Vast crusts of sea ice cover the ocean inside the Arctic circle, forming the Earth's northern polar cap. Much of this is a solid mass; however, at lower latitudes, where the temperature begins to rise, these giant ice fields start to break up into smaller chunks called 'ice floes'. Sea ice on Earth is a seasonal phenomenon, with ocean ice coverage growing in winter and then receding again during summer months.

Mountains

Mountains are tall outcrops of rock that cover 24 per cent of the Earth's land mass. They can be formed in several ways. Regions with active plate tectonics churn up the landscape to create rugged terrain. Similarly, the collision of **tectonic plates** can cause the formation of mountain ranges as the ground jackknifes upward in the collision. The Himalayan mountain range was formed this way, when the Indo-Australian plate crashed into Eurasia 70 million years ago. Other mountains may be **volcanoes**, formed by the gradual build-up of erupted lava.

The summits of mountains are cold because of their height above the warm ground below. Oxygen levels also drop as the density of Earth's atmosphere diminishes with altitude. Technically, there are many definitions of what makes a mountain. In the United States, the criterion is any landform over 1,000 feet high (305m), while anything in the range 501–999 feet high (153–304m) is just a hill.

Rivers

Rivers are channels of fresh water running from elevated terrain, such as hills and mountains, across the land and down to the ocean; they are fed by **rainfall**, seasonal melting of ice, and underground watercourses. There are often flat flood plains running alongside the banks of rivers, areas that have been levelled over time by the flow of water as the river has sporadically burst its banks.

Some rivers that carry a large amount of sediment may form a triangular-shaped 'delta' where they reach the ocean: as the sediment is deposited in a wide fan-shape and gradually built up over time, it re-sculpts the river mouth. An example of this is the mouth of the River Nile in Egypt which has a prominent delta where its waters meet the Mediterranean Sea. Rivers are also partially responsible for the salinity of the oceans, washing salt and minerals from rocks as they flow along.

Valleys

Deep depressions gouged through the landscape, often between hills or mountains, are called valleys. Three main mechanisms work to form valleys. The first is water erosion – a river meandering through the landscape wears away at the rock and soil to carve an ever-deeper channel. Valleys formed like this look V-shaped in cross section. Extreme river valleys can become very deep and spectacular – such as the Grand Canyon in the United States.

Valleys can also be formed through the action of glaciers – huge walls of ice that creep over the terrain, pulverizing rock as they squeeze through the narrowest fissures to form valleys with a cross-section that looks more U-shaped.

The final variety are rift valleys – formed by tectonic processes splitting apart continental plates, to form two new plates with an ocean between. The Great Rift Valley in East Africa is an example of a rift valley forming.

Lakes

A large expanse of water fed by a river but not part of an ocean is a lake. Lakes can form in natural recesses between hills and mountains and in depressions in level ground. Sometimes rivers can form 'oxbow' lakes where a meandering curve in a river tends to deposit silt and sediment preferentially on its outer banks, causing the curve to become cut off leaving a bow-shaped lake behind (see diagram).

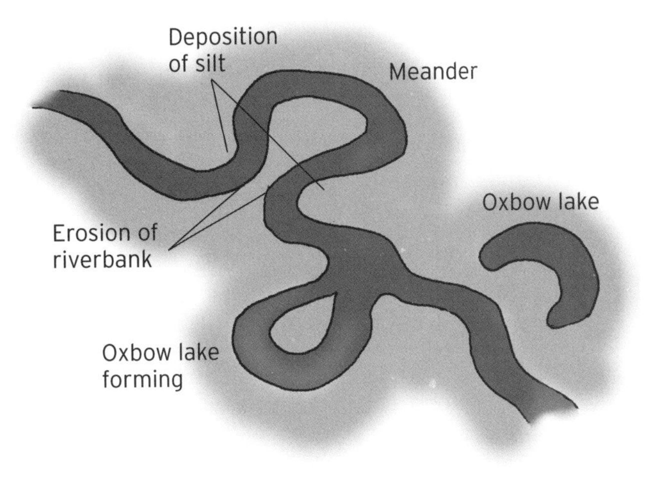

Smaller lakes are known as ponds; the Ramsar Convention on Wetlands of 1971 says that an inland body of water less than 8 hectares (10,000m^2) in size is a pond while anything bigger is a lake. There are over 300 million ponds and lakes worldwide. Canada has the greatest concentration – with 60 per cent of the world's lakes – due mainly to poor drainage.

Wetlands

Lakes and river areas are sometimes surrounded by damp, saturated terrain called wetland; swamps, fens and marshes are all examples and they can be either fresh water, salt water or somewhere in between, depending on their proximity to the coast. The amalgamation of water and land means that wetlands often have rich ecosystems, teeming with many kinds of plants, fish, mammals, reptiles and amphibians.

And yet their limited commercial value has placed them in jeopardy. In 1993, it was reported that half the world's wetlands had been drained to increase the development potential and productivity of the land. For these reasons, wetland areas are now the subject of widespread conservation efforts. In the United States the largest wetland area is the Florida Everglades, and the largest in the world resides in the flood plains of Pantanal, in South America.

Plains

Flatlands make up the bulk of the terrain covering the Earth's land masses. They come in many shapes and sizes – the grassy meadows and fields of Europe, the arid prairies of the United States, the wild African savannah and the desolate steppe and tundra of northern Asia. Each is sculpted by the climate and ecology of its locale. Plains can be formed by erosion, deposition of sediments in water, or by ice or wind, or simply bulldozed flat by the passage of a glacier. Plains can exist on barren expanses of land, in between hills, on valley floors, inland or in coastal areas. As well as playing an important role in natural ecology, plains helped facilitate **agriculture**, and the building of roads and cities.

Forests and jungles

The greatest biodiversity on Earth can be found in the planet's forests and jungles; it has been estimated that 57 per cent of all the planet's living species can be found in jungles. Together, jungles and forests cover around 36 per cent of the Earth's land mass. At high latitudes, above 53° from the equator, where temperatures are lower, forests are usually made up of evergreen conifer trees. Deciduous trees, which lose their leaves in winter, are more common at lower latitudes, while within 10° of the equator rainforests abound – literally, forest areas that receive large volumes of **rainfall**. Jungles and rainforests can be found in South America, Africa and Asia. High temperatures and humidity make rainforests thriving hotbeds of life. And the densest regions of the rainforests are jungle – a living knot of trees, creepers and undergrowth.

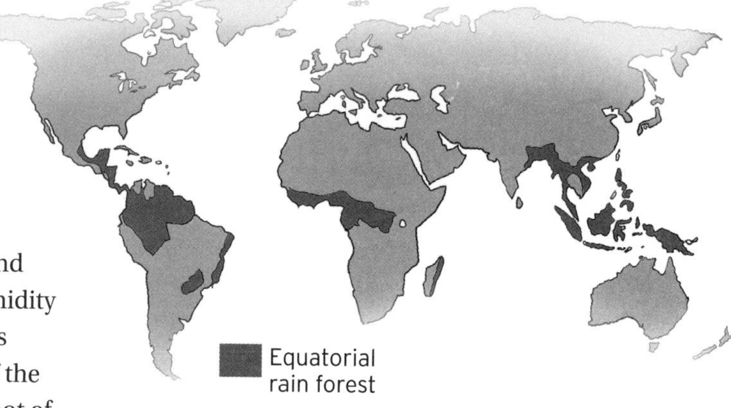

Deserts

Deserts are the driest, most barren places on the Earth. Found on every continent, they form when the rate at which moisture evaporates away from the ground exceeds the rate at which it arrives through **rainfall**. Perhaps surprisingly, the driest desert on the Earth is also the coldest; in the Dry Valleys region of Antarctica, rain hasn't fallen in over 2 million years. Any moisture that does remain is frozen from the air and blasted away by the 320km/h (200mph) winds that routinely scourge the area. Antarctica is also the largest desert in the world, with an area of nearly 14 million square kilometres.

Many green areas of the Earth's surface are being converted to desert through the action of **climate change** and careless agriculture – burning forest and jungle to make arable land, and the over-exploitation of natural watercourses, such as rivers and groundwater. A report in 2008 estimated that a productive area the size of Nebraska is being lost to desert every year.

SUBTERRANEA

Caves

Natural cavities in the Earth's rock formations are called caves; they can be small affairs of just a single chamber, or a sprawling network of tunnels and caverns that take the intrepid explorer deep underground. Caves are formed by a range of different processes. Acids in groundwater can react with and dissolve base-rich rock (see **Acids and bases**), leaving behind a cavity – known as 'solutional caves'. Erosional effects, such as weathering or the constant wear of running water, can cut caves into rock too – as can the lashing assault of the sea, explaining why coastal cliffs are often rich in cave networks. Others, called 'primary caves', are formed at the same time as the surrounding rock.

Caves are found all over the world, and they come in all shapes and sizes: Mammoth Cave, in Kentucky, is the longest at 591km (367 miles) in length; Voronya Cave, in Abkhazia, Georgia, is the deepest reaching more than 2km (1.2 miles) below ground. The largest single cavern is the Sarawak Chamber on Borneo, which measures 700m (2300ft) by 400m (1312ft) by 80m (262ft) – big enough to hold 32 aircraft carriers.

Stalagmites and stalactites

Caves formed by acidic water have streams laden with minerals, such as calcium carbonate, running through them. This water can deposit its rock content throughout the cave, creating spectacular structures such as stalagmites and stalactites. Stalagmites grow upwards from the bottom of a cave through the action of mineral-rich water dripping constantly onto the same spot from above. Just as a stalagmite grows upwards, the drip point above it grows down, as minerals are left behind before each drop falls – and this is a stalactite.

The biggest stalagmite is over 62m (203ft) high and is located in a cave system in Cuba, while the longest stalactite on record measures 20m (65ft) and hangs in Brazil's Gruta Rei do Mato caves. Their growth rates vary wildly, depending on the concentration of the mineral **solution** and the rate of flow. Some may grow by a centimetre in a month, others may take hundreds or even thousands of years to grow by the same amount. Stalagmites and stalactites are types of liquid-deposited rock formations known as speleothems, that also include 'flowstones', which look like an oozing mass of rock, and delicate spiral stalactites called 'helictites'.

Lava tubes

Lava tubes are a kind of 'primary cave' (see **Caves**) formed by the flow of molten lava during a **volcanic eruption**. The exterior of a flowing channel of lava is exposed to the air and so cools more rapidly than the interior, forming a hard crust. Cooling from the outside thickens this crust, while the core of the lava channel continues flowing. Once the volcanic eruption has ceased, the molten lava inside drains away leaving the solidified exterior – and this is a lava tube.

Lava tubes can also grow **stalagmites and stalactites**, and other speleothems. Unlike their limestone cousins, however, these aren't caused by rock dissolved in water but rather the dripping of residual molten lava before the tube has fully solidified. Up to 15m (49ft) wide, lava tubes can reach up to tens of miles in length and are found in Arizona, Oregon, Hawaii and at other volcanic sites around the world.

Underground water

Sometimes a cave system can accumulate liquid in the depressions and recesses of the rock to form underground lakes. These can be supplied by rainfall, natural springs or seeping ground water – porous rocks, such as limestone, can store huge volumes of water in repositories known as 'aquifers'. Caves beneath Sweetwater, Tennessee, have the largest underground lake measuring 240m (787ft) by 70m (230ft).

When the flow of water through caves is sufficiently great, lakes and other stationary bodies of water give way to subterranean rivers. The world's longest underground river stretches for 153 kilometres (95 miles) in caverns running under Mexico's Yucatán peninsula.

Crust

If you could saw planet Earth in half, you would find the inside looked rather like the layers of an onion, with distinct stratified zones of liquid and solid rock stretching from the planet's exterior right down to the core. Outermost of these layers is the Earth's crust, which is a thin veneer of solid rock covering the planet's surface; it is a mixture of **igneous**, **metamorphic** and **sedimentary** rock that in total accounts for about 1 per cent of the volume of the planet.

The thickness of the crust varies greatly – beneath the oceans it is as little as 5km (3 miles) thick, whereas the continental crust, which makes up the Earth's land masses, is much thicker – between 30km (19 miles) and 50km (31 miles). This is why the continents stick up out of the ocean, while the sea floors are submerged. Forming the crust is an interlocking mosaic of **tectonic plates** that float at the Earth's surface because the density of the crust material is lower than that of the underlying layer – the mantle.

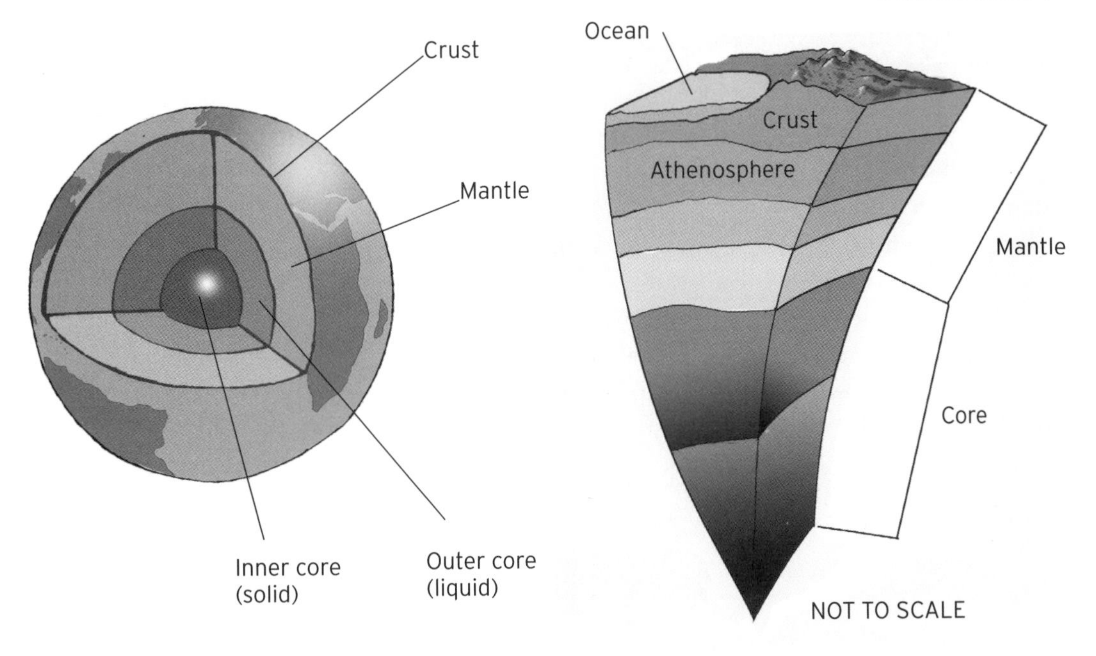

Mantle

Beneath the Earth's crust is a viscous layer of semi-molten rock – nearly 3,000km (1864 miles) thick and accounting for more than 80 per cent of the planet's total volume – known as the mantle. It splits broadly into upper and lower layers; the 'upper mantle' reaches down to a depth of around 400km (248 miles) and this layer is divided into upper and lower zones. The 'lithosphere' is the rocky top part of the upper mantle plus the crust above it. Beneath this is the 'asthenosphere', which is less solid owing to the higher temperature at this depth. The entire upper mantle is coupled to the crust at a boundary called the 'Mohorovicic discontinuity', or 'Moho' for short.

The 'lower mantle', on the other hand, covers the range of depths 660–2900km (410–1800 miles). Here, the rock has become solid once again as the high pressure caused as the weight of the overlying planet compresses it. Between the upper and lower mantles is a layer called the transition zone, which interpolates between the two regimes. Temperatures within the mantle range from a few hundred degrees Celsius at the base of the crust to an estimated 4,000°C (7,232°F) where the mantle meets the next layer down – the core.

Core

The central, hottest part of the Earth is the core, which has an inner and an outer layer and is responsible for the planet's **geomagnetism**. The outer core is liquid metal, predominantly molten iron and nickel, which stretches from a depth of 2,890km (1,806 miles) below the planet's surface down to 5,150km (3,218 miles). Temperatures range from just over 4,000°C (7,232°F) to more than 6,000°C (10,832°F) at the boundary with the inner core. At the inner core, diameter 2,440km (1,525 miles), the pressure is great enough to squash the nickel and iron back into solid form – despite the temperature here approaching 7,000°C (12,632°F).

Geomagnetism

The phenomenon whereby the Earth generates its own magnetic field (see **Magnetism**) – and the study of this field's effects – is known as geomagnetism. At the Earth's surface the field has a strength of between 30 and 60 microteslas, depending on location, which is about 1,000 times weaker than the field of a fridge magnet. The field is generated by the dynamo effect of electrical currents within the Earth's outer core of swirling molten metal (see **Induction**).

Earth's magnetic field is a dipole, meaning it has two poles of opposite magnetic polarity – labelled north and south. Other magnetic dipoles, such as a bar magnet, tend to align themselves with the Earth's field and this is how a navigational compass works, using a needle-like bar magnet that can swing freely to indicate the direction of magnetic north.

OCEANS

Coasts

Coastlines offer a transition zone between the land and the ocean depths, and present some of the most varied natural habitats in the world. From cliffs and beaches to bays and estuaries to exotic coastlines such as fjords and lagoons, the coast has been shaped by the ocean over millions of years through erosion, deposition of sediment and the shifts in sea level brought about by **climate change**. Coasts have transpired to be important for the development of our civilization, facilitating fishing and sea travel.

Oceanography

The total volume of the world's oceans has been estimated at 1.3 billion cubic kilometres. There are five major oceans in the world: the Arctic, which freezes over to create the north polar cap; the Atlantic, which is flanked by Europe and Africa to the

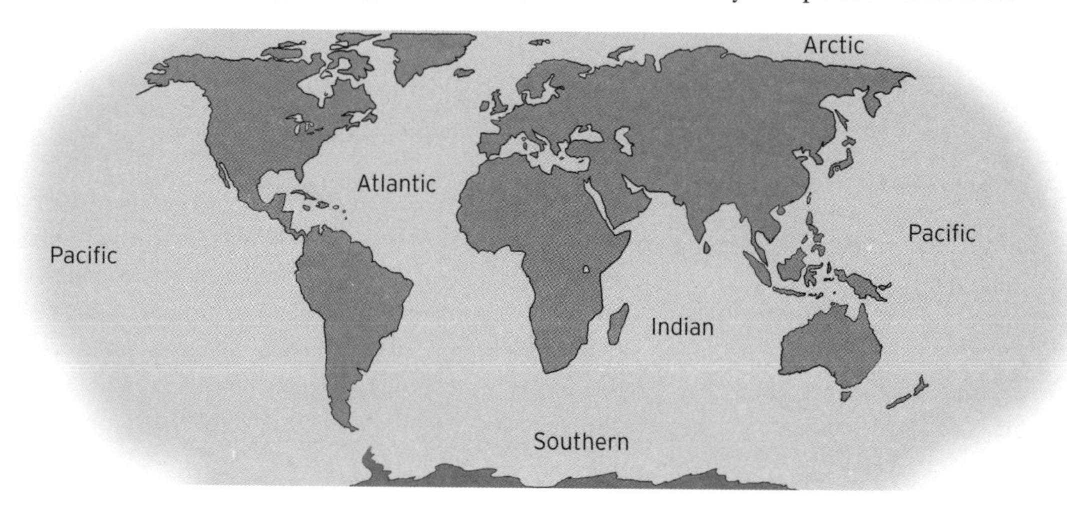

east and the Americas to the west; the Indian, which sits between East Africa and Australia; the Pacific, flanked by Asia and Australia to the west and the Americas to the east; and the Southern Ocean, which laps around Antarctica and the southern polar cap. The divisions are arbitrary, however, as all are connected, making it possible to sail to any ocean from any other. Ocean water is salty, with a salinity of around 3.5 per cent – due to salts washed from rocks by rivers and the production of sodium and chlorine (the components of salt) by other geological processes beneath the sea floor.

Earth is quite literally the blue planet, with 71 per cent of its surface area covered by ocean. Yet the blue tint is not caused by oceans reflecting the colour of the sky – rather, it is because water absorbs a small amount of red from any white light passing through it, leaving an overbalance of blue. The effect is very slight and can only be seen when the volume of water is large.

Ocean ridges

Oceans are not simply large areas of low-lying land that have filled with water. The **crust** making up the ocean floor is much thinner than the continental crust, and whereas the same chunks of continental crust have sat at the Earth's surface for millions of years, ocean crust is continually being created by the process of sea-floor spreading. This is how the oceans themselves were formed and new oceans form where new spreading sites open up – sites called rift valleys, which are located on fault lines in the crust where two **tectonic plates** meet. When the plates move apart they form a rift valley and the effect is to split continents asunder, creating an expanding basin of sea floor between the two pieces, which in time fills with water, giving birth to an ocean. The submerged remains of the rift valley becomes what is known as an 'ocean ridge'.

Such a process is underway at the Great Rift Valley in East Africa where the African continental plate is splitting into two – called the Nubian and Somalian plates. As they move apart, they are opening a rift that will ultimately turn Somalia, Kenya, Tanzania and Mozambique into an island separated from continental Africa by ocean.

Ocean layers

Oceanographers divide the ocean into distinct zones. The uppermost layer is known as the 'pelagic' zone and covers all the ocean's surface. Extending further down to 200m (656ft) is the 'epipelagic' zone within which there is sufficient light for **photosynthesis**, allowing plants to grow. From 200m (656ft) to 1,000m (3,280ft)is the 'mesopelagic' zone, also known as the 'twilight zone' because of its partial light levels. It is home to creatures that like cold, dark water – such as squid, octopus and cuttlefish. Beneath this is the 'bathypelagic', or midnight zone, extending down to 4,000m (13,120ft); the water is pitch black, and the pressure is measured in tons per square inch. Some squid and eels live here. Below this is the 'abyssopelagic' which reaches all the way to the ocean floor. Most creatures at this depth are blind – having no need for eyesight in the unrelenting dark.

The final ocean layer is the 'hadopelagic', which is reserved for the depths of **oceanic trenches**. Over 10km (16 miles) below the surface, the pressure down here reaches 8 tons per square inch. The few creatures that live this deep feed on 'marine snow' – detritus drifting down from the waters above – or on the heat and nutrients spewed out by **hydrothermal vents**.

Seas

In addition to the five major oceans of the world there are numerous smaller bodies of water known as seas. Like the oceans, most are connected to one another by water. The largest of them is the Arabian Sea, but at nearly 3.9 million km^2 it is still dwarfed by the smallest full-scale ocean, the Arctic (13.2 million km^2). Some large lakes are classified as seas, such as the Dead Sea in Jordan, while others don't even bear the name 'sea' – despite being connected to the oceans – such as the Bay of Biscay and the Persian Gulf.

Tides

High and low ocean tides result from the combined gravitational pull of the Moon and the **Sun** acting on the huge mass of liquid water sloshing across the planet's surface. Gravity is an attractive force that increases in magnitude the closer you are to the source of the gravitational field, which means that water on the side of the Earth nearest the Moon feels a greater gravitational pull than the planet itself and is thus raised up relative to the planet to create a high tide. But by the same reasoning, the planet also experiences greater pull than the water on the side opposite the Moon. Relative to the Earth's frame of reference, the effect is to raise a high tide on this side of the Earth too and this is why we get two high tides every day.

It is not only the Moon that raises tides – the Sun's gravity exerts roughly half as much force as the Moon, its colossal mass offset by its great distance. When the Moon and Sun both align in the sky especially high tides can result – called 'spring tides'. But when Sun and Moon are aligned at right angles to each other, the difference between high and low tide is at a minimum – these are known as 'neap tides'.

Tidal bores

Waves that travel from the ocean and up estuaries and rivers – against the flow of the current – are known as tidal bores. They tend to occur in coastal regions where there is a wide margin between the heights of low and high tides. On its own this is not enough to produce tidal bores. A big factor is that rivers hosting tidal bores normally have a funnel-like shape along their length, so that the further you head inland the narrower and shallower the river gets, which concentrates the energy of the water into a small volume, causing a wave to rise up – often by several metres – which then surges upstream. Even so, the best bores only occur when the ocean tide is at its highest, such as during the spring tides when the Sun and Moon align. The low-frequency rumble of a bore wave approaching can be heard sometimes from several kilometres away. Prime locations to see them are the Bay of Fundy in Canada, and the Qiantang River in China.

Ocean currents

The water in the world's oceans is swept this way and that by a maelstrom of circulating global currents. The best known of these is the Gulf Stream, which

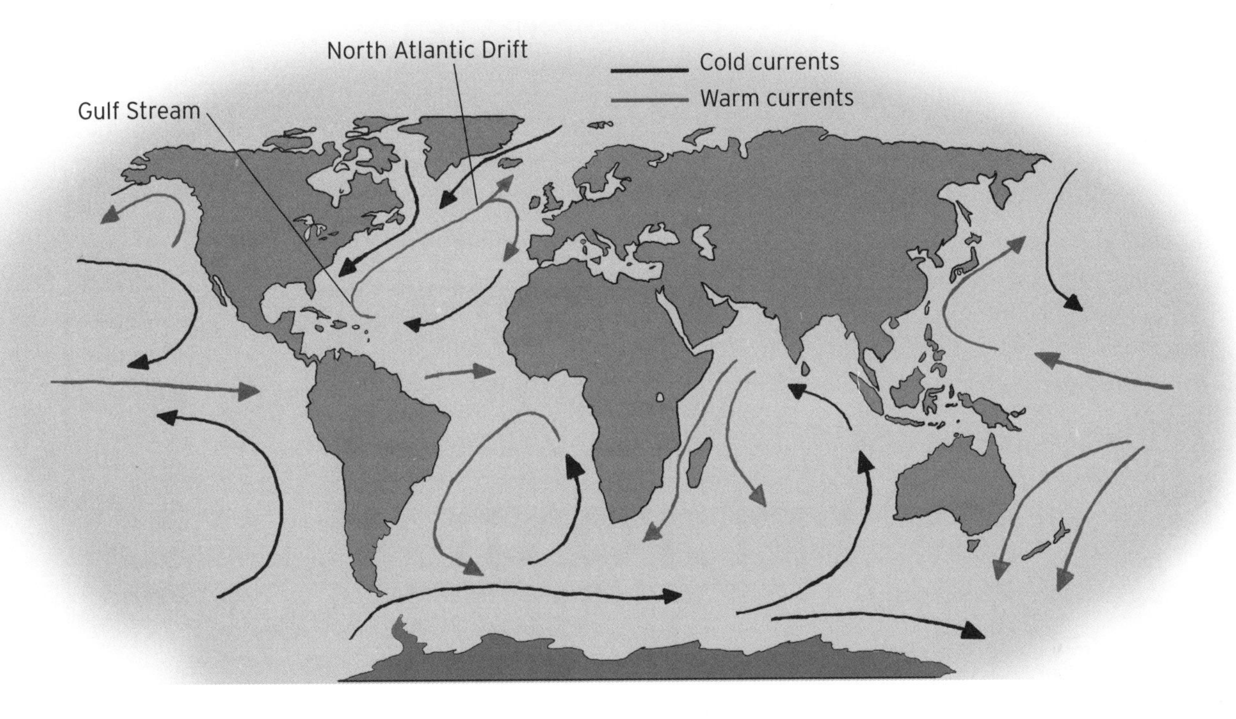

carries warm water from the Gulf of Mexico to the mid-Atlantic, where another current – the North Atlantic Drift – funnels it across to the west coast of Europe. Scientists believe that the heat delivered by this current is what keeps the United Kingdom and France warmer than other locations at the same latitude (see **Latitude and longitude**). Other currents form giant loops in the Pacific, Indian and South Atlantic oceans – known as 'gyres' – and another that skirts around Antarctica.

These currents run at the ocean's surface, in the epipelagic zone and the upper mesopelagic (see **Ocean layers**). They are driven by factors such as wind, the **Coriolis effect**, and temperature. Deeper currents exist too, and usually result from differences in the density and temperature of the deep ocean water around the world.

Oceanic trenches

Oceanic trenches are the deepest, darkest depths of the sea floor, forming at subduction zones (see **Convergent faults**), where the oceanic **crust** is being pulled back into the bowels of the planet by churning convective motions of the semi-molten **mantle** rock below. A trench's bottom is typically several thousand metres below the level of the surrounding sea floor. The deepest known is the Mariana Trench in the Pacific Ocean, west of the Philippines, which reaches a maximum depth of 11km (6.8 miles) – enough to swallow Mt Everest with 2km (1.2 miles) to spare.

The deepest part of floor of the Mariana Trench is a small valley on its floor is known as Challenger Deep, after the British navy vessel first sent to study it in the 19th century. Since then, it has been explored by a number of submersibles – both crewed and robotic.

Hydrothermal vents

Cracks in the ocean floor near an oceanic ridge – where two tectonic plates are moving apart to create new areas of sea floor – can give rise to hydrothermal vents, which spew superheated seawater rich in sulphur and other minerals. The water is heated by the **geothermal energy** of the hot rocks beneath. They are also known as black smokers after the dark clouds of particles they belch out.

Hydrothermal vents are often epicentres of life on the deep ocean floor, thousands of metres below the surface. With no natural sunlight reaching these depths, the vents provide a much-needed source of energy and nutrients for creatures living at these depths. Temperatures around a hydrothermal vent can reach as high as 400°C (750°F). Organisms have adapted to this, developing extreme levels of heat tolerance which has earned them the name 'extremophiles'.

TECTONICS

Magma

Magma is molten rock from the mantle and crust layers of the Earth's churning interior. Most of the crust and upper mantle are solid or semi-molten, but liquid magma forms when this melts in areas where the temperature becomes elevated by geological forces, areas known as hotspots.

Molten magma has a temperature between 750°C (1,380°F) and 1,400°C (2,550°F), depending upon its exact composition. Most magma is a mixture of elements such as silicon, oxygen, iron, sodium and potassium. As the magma cools these elements combine to form different sorts of igneous rock. Magma can ooze up from fault lines where tectonic plates meet – in particular at divergent faults where it solidifies to create new sheets of oceanic crust. Magma pressure is the driving force behind **volcanic eruptions**, where it seeps from the **volcano** as 'lava', or is cast out explosively as fragments known as 'tephra'.

Hotspots

Hotspots are areas on the Earth's surface where either there is a concentration of **volcanoes** or where considerable numbers of **earthquakes** and other forms of tectonic activity are experienced. Convection currents (see **Conduction and convection**) within the Earth's mantle cause hotspots. Rather like the convection in water boiling in a pan on the stove, convection in the mantle transports hot rock to the Earth's surface while cooler rock sinks further down. Where two upward rolling convection cycles mesh together, like cogs in a gear system, heat is focused at a point on the surface creating a hotspot; the stream of hot material dredged up from below is known as a 'mantle plume'. Major hotspots exist under Hawaii, Yellowstone National Park and Iceland.

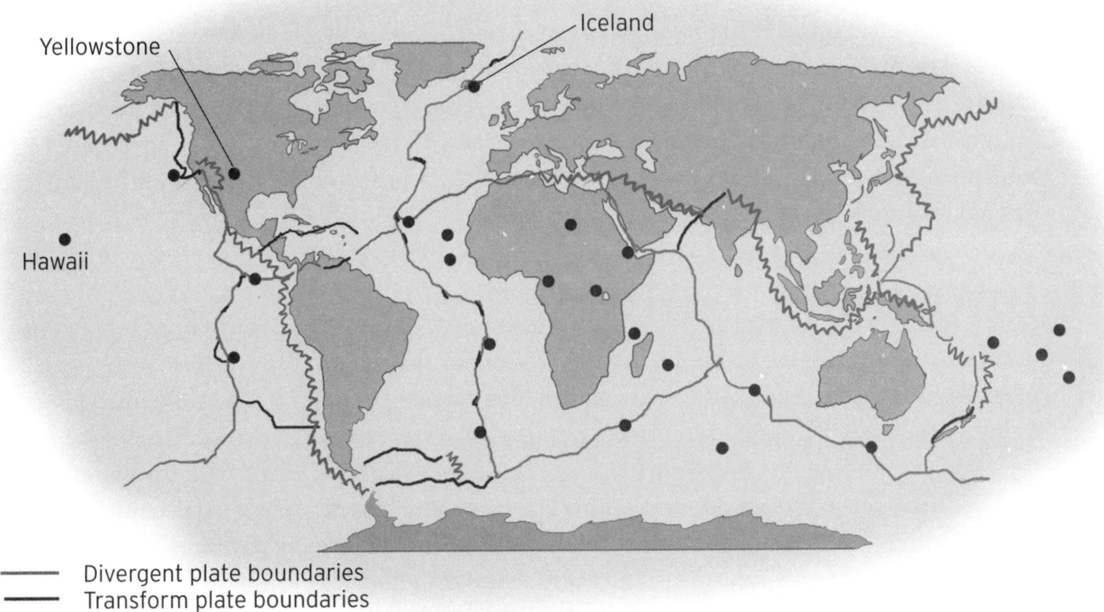

Tectonic plates

The Earth's lithosphere – that is, the crust together with the uppermost shell of the mantle – is broken up into a number of interlocking slabs, called tectonic plates. They jostle and rub together in response to the roiling motion of the molten rock layers beneath. Boundaries between plates are known as fault lines and can take three different forms depending on the relative motion of the plates: divergent, convergent and transform faults.

The chafing together of tectonic plates at fault lines is responsible for **earthquakes** and **tsunamis** as well as important geological processes such as subduction, sea-floor spreading and **continental drift**. Volcanoes are also normally situated along fault lines. There are seven major tectonic plates – African, Antarctic, Eurasian, Indo-Australian, North American, Pacific and South American – together with a multitude of smaller ones.

Divergent faults

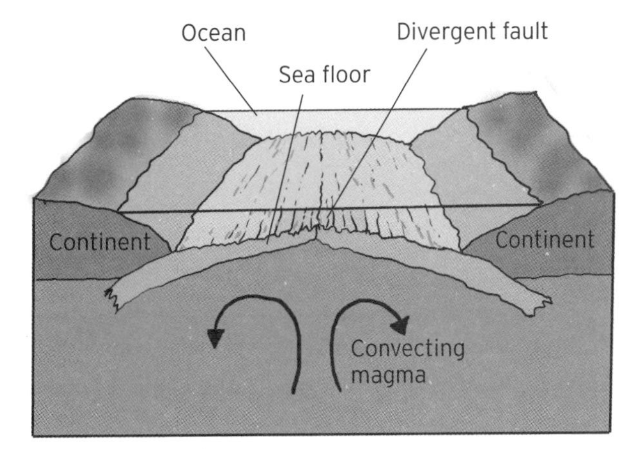

Neighbouring tectonic plates that are moving apart from one another create a so-called divergent fault. Magma is brought to the surface by convection currents, in a similar manner to the formation of a hotspot. But the **viscosity** of the material in the currents causes it to drag on the overlying crust, pulling the crust in opposite directions (see diagram). This can rip open continental land masses, forming rift valleys. At sea, divergent faults in the ocean crust are associated with **oceanic ridges**, where magma rises to fill the gap between the diverging tectonic plates thereby creating new sea floor – a process known as 'sea-floor spreading'. Sometimes, so much magma can be spewed up from an oceanic ridge that it breaks the surface of the water, forming a new volcanic island – an example is Surtsey Island, off the coast of Iceland.

Convergent faults

Convergent faults, the opposite of divergent faults, occur where two tectonic plates are in collision, and usually lead to 'subduction', where one plate slips under the other and is drawn down into the planet's interior. The subducted plate quickly melts once it's under the surface, creating an overdensity of magma and gases, which then burst up through the overlying plate as a cluster of volcanoes. When both the plates involved are made of underwater oceanic crust the subduction carves a deep **oceanic trench** in the sea floor where the plates meet, and the volcanoes created can often break the water's surface to form an arc of volcanic islands behind the fault line. This is the case with the Mariana Trench, Earth's deepest ocean trench, which sits adjacent to the Mariana chain of islands.

When one of the plates involved in subduction is made from continental crust, the continental plate rides up over the other because of its lower density. However, the force with which the plates collide usually rucks up the leading edge of the continental plate to form mountain ranges, interspersed with volcanoes as the subducted plate melts. This is happening along the California coast, where the Juan de Fuca oceanic plate is slipping under the North American Plate.

Transform faults

A transform fault is the boundary between two tectonic plates which are neither diverging nor converging, but instead are sliding past one another. Whereas convergent and divergent faults destroy and create crust, respectively, neither happens at a transform fault; however, transform faults can be destructive in other ways. The motion of the plates past each other is not smooth. Stresses build up over time and when the accumulated force is great enough the plates make a sudden, discontinuous shift, a stop–start motion that can spawn violent **earthquakes**. Perhaps the world's most famous transform fault is the San Andreas fault in California – which was the cause of the earthquake that devastated San Francisco in 1906.

Volcanoes

Convergent faults, divergent faults and hotspots can all lead to the formation of volcanoes – breaches in the Earth's crust that allow hot magma and gases to escape, with sometimes devastating consequences. Volcanoes come in many different types: 'cone volcanoes' resemble the peaked form of ordinary mountains; 'shield volcanoes' are much flatter and less peaked; while 'fissure vents' are little more than holes in the ground. All volcanoes have a central crater, known as the 'caldera', through which they ooze lava – the name given to the pressurized magma from the crust below once it's left the volcano.

Volcanoes formed by hotspots sometimes occur in chains dotted along the Earth's surface, caused by the movement of tectonic plates bringing new areas of crust over the hotspot – which from time to time punches a hole through the new crust, giving birth to a new volcano. The Hawaiian-Emperor seamount chain – a sequence of more than 80 volcanoes, islands and sea-floor mountains in the Pacific Ocean – was created this way.

Continental drift

It doesn't take a genius to spot that the coastlines of South America and Africa look as if they might fit together; and, indeed, they once did. Motion of tectonic plates is constantly reshuffling the layout of the continents; the rate of movement is extremely small – they move at about the same speed human fingernails grow, roughly 10cm (4 in) per year. But the process is inexorable – it tore continents apart in the past and will clash them together again in the future.

The theory of continental drift was first put forward as long ago as the 16th century. However, it wasn't until the theory of plate tectonics was accepted in the 1960s that scientists took the notion of continental drift seriously. Now it's a proven fact; the same fossilized species are found on opposite sides of the Atlantic Ocean, confirming that the two land masses were once linked.

Supercontinents

When two or more continental plates come together they form a supercontinent. In classification schemes where Europe and Asia are considered a single continent, the resulting land mass – Eurasia – is such a supercontinent. Thanks to continental drift, these vast concentrations of land have graced the surface of the Earth many times in the past. As far back as 3.3 billion years ago, all of the Earth's continental crust existed in the form of a supercontinent that geologists have named Vaalbara. The precise layout isn't known but similarities in the **stratigraphy** of rock layers from this period in both Africa and Australia suggest its existence.

The oldest supercontinent for which the layout of the plates is reasonably well understood is called Rodinia, which existed around a billion years ago (during the **Proterozoic aeon**). Others have come and gone since. During the Permian period, 225 million years ago, Earth's continents were conjoined into a land mass known as Pangea. Then in the Triassic Period, 200 million years ago, Pangea broke apart into Laurasia (which would become modern-day North America, Europe and Asia) and Gondwanaland (made up of modern-day South America, Africa, Australia and Antarctica). The continents look set to converge again 250 million years from now into a land mass scientists have named **Pangea Ultima**.

NATURAL DISASTERS

Volcanic eruptions

When the pressure of upwelling magma beneath a volcano reaches breaking point the volcano blows its stack, erupting to blast lava, gas, rock and ash across a wide area. Molten lava spews from the volcano's caldera and from secondary vents that form where the pressure of the rising magma ruptures the volcano's flank. Volcanic bombs – globules of lava hurled through the air which solidify in flight to strike the ground as solid rock – can be scattered over a wide area and be up to several metres across. Meanwhile, clouds of ash and smoke rise into the air and block out the Sun near the eruption, and can even cause dimming on a global scale.

Perhaps the most devastating kind of eruption is known as a 'pyroclastic flow' – a blizzard of gas and partially molten rock at temperatures of nearly 1,100°C (2,012°F) that rolls down the flanks of the volcano and across the surrounding terrain at speeds approaching 750km/h (460mph). Pyroclastic flows can devastate a radius of up to 200km (125 miles).

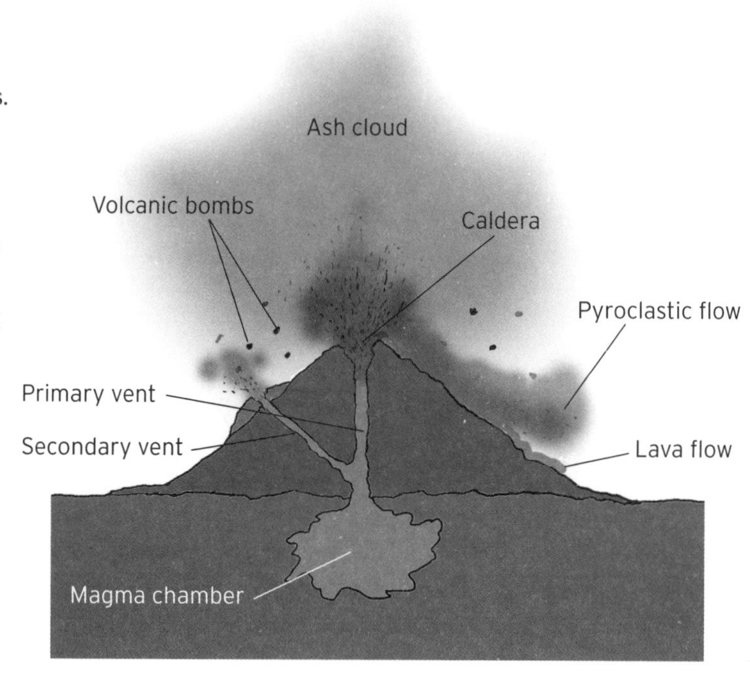

Earthquakes

Earthquakes are a deadly and destructive consequence of tectonic plates grating against each other at fault lines in the Earth's crust. Boundaries between plates are not smooth – lumps and bumps in the rock create friction that stops them moving freely. Friction coupled with elasticity in the rock (see **Hooke's law**) leads to the build-up of strain energy. When enough energy has accumulated to overcome the friction the plates suddenly slip, creating a jolt that spreads out as a ripple through the crust and can be powerful enough to knock down buildings and trigger landslides. If an earthquake's epicentre – the point where the quake spreads out from – is located at sea, it can generate a **tsunami**.

There are three varieties of earthquake, depending on the nature of the fault responsible:

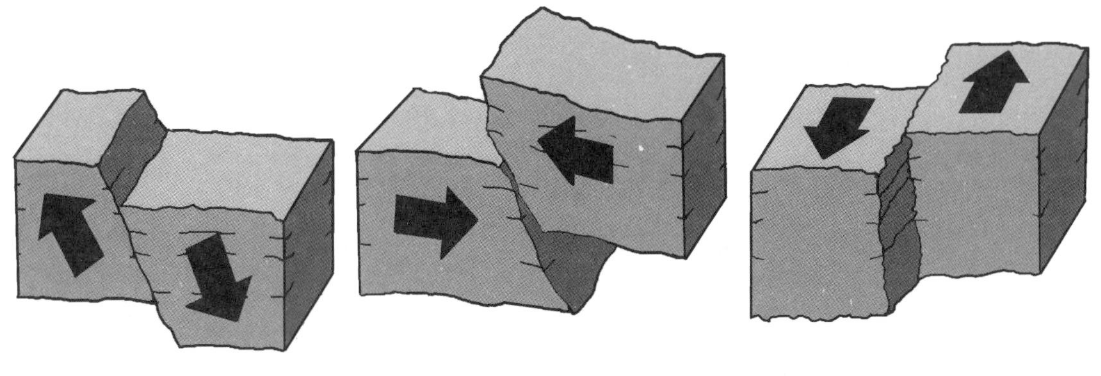

Normal earthquake   Thrust earthquake   Strike-slip earthquake

divergent faults, where two plates are slipping apart, create 'normal' earthquakes; convergent faults, where one plate is being forced under another, create so-called 'thrust' earthquakes; and transform faults, where plates slip past one another, give rise to 'strike-slip' quakes. Earthquakes are measured on the 'moment magnitude' scale, which has now replaced the older Richter scale. Anything with a magnitude of 7 or more poses a serious danger.

Landslides

Loose material on steeply sloping ground is prone to lose its grip and slip to the bottom – the result is a landslide, which can be caused by earthquakes, water action and human activity such quarrying or mining. The effects range from simple inconvenience – for example, when the debris blocks roads – to destruction and loss of life when the debris sweeps away or falls on, inhabited areas. Landslides don't only involve soil and rock. Snow-covered mountains are prone to avalanches – colossal snow-slides – which are a hazard to climbers and skiers. Lethal mudslides can be caused by heavy rains, melting ice and even volcanic activity.

Landslides into the sea can cause **tsunamis** and it is thought this may one day happen on the Spanish island of La Palma, when the western part of the island is expected to slip into the ocean throwing out a tsunami wave 600 metres (656 yards) high that will threaten eastern America.

Limnic eruption

In 1986, Lake Nyos, in the West African country of Cameroon, suddenly and without warning belched up an estimated 80 million cubic metres of carbon dioxide (CO_2) gas. Because CO_2 is heavier than air, the gas cloud settled into low-lying areas within 25km (15 miles) of the lake – driving out all the breathable oxygen. More than 1,700 people suffocated, along with an untold number of wild and domestic animals. The incident is an example of a rare yet terrifying kind of natural disaster called a limnic eruption.

The eruptions are caused by CO_2 dissolved in lake water that is suddenly released. CO_2 can be forced into the water by a number of processes, such as the decay of plant material and the release of the gas from volcanic vents on the lake bed – Lake Nyos sits in a deep extinct volcanic crater. Like a bottle of cola, the water under high pressure at the bottom the lake is able to hold

CO_2 in solution. But when the water is stirred up, for example, by a landslide, earthquake or is volcanically heated, the gas is spontaneously released. Programmes are now underway to gradually siphon CO_2 from the bottom of Lake Nyos to prevent the kind of build-up that causes limnic eruptions.

Floods

Excessive rainfall causing rivers to burst their banks, ocean storm surges breaking down sea defences and the melting of winter ice and snow in spring can all lead to flooding. Despite 21st-century technologies to monitor floods and build defences against them, this form of natural disaster still presents a deadly and devastating hazard in low-lying areas, destroying homes and private property, infrastructure – such as roads and bridges – and causing loss of life through drowning and disease as clean water supplies become contaminated with sewage. The danger from floods is set to increase as **climate change** raises sea levels worldwide, and makes extreme weather phenomena such as cyclones and storm surges more likely.

Tsunamis

Violent events at sea, such as underwater earthquakes and the **cosmic impacts** of comets or asteroids with the ocean, cause huge disturbances in the water that move outwards as a tidal wave, a tsunami that has the power to obliterate coastal regions. The word tsunami derives from the Japanese for 'harbour wave'. Tsunamis carry enormous energy, yet the height of the waves in the deep mid-ocean is small. A boat in the middle of the Atlantic witnessing the passage of a tsunami would barely feel any effect as the swell of deep water passed by. However, once the wave crosses onto a continental shelf, and its energy is concentrated into a shallow layer of water, the wave rears up into a fearsome wall of water often tens of metres high.

On Boxing Day, 2004, a tsunami sped across the Indian Ocean – caused by an undersea 'thrust' earthquake at a subduction site off the coast of Sumatra. The huge wave, in places 30m high, killed over 230,000 people as it washed inland, causing widespread devastation in 11 countries.

Tornadoes

A tornado is a rapidly rotating column of air extending from the ground to a cumulonimbus cloud above; the strongest tornadoes generate windspeeds of over 500km/h (300mph). These savage vortices can tear houses from their foundations, fling cars and trucks through the air and cut a path of devastation across the landscape over a kilometre wide. Tornadoes are formed when air at high altitude is moving faster than the air near the ground, forming a horizontal cylinder of air that rolls over the landscape. If the cylinder rolls over a warm updraught, the upwardly moving air can sweep the cylinder into an upright position and a tornado is born.

They are most common in the Tornado Alley region of the United States from the Midwest down into Texas; the combination of currents of warm, moist air with dry air make the perfect conditions for tornado formation. Radar stations in this region are used to track the course of dangerous tornadoes so that advance warnings can be issued.

Great storms

On 15 October 1987, a terrific storm hit northern Europe, lashing England and France with winds approaching 220km/h (130mph) – uprooting trees, damaging buildings, and killing 22 people. While the storm was not considered a full-scale hurricane (an Atlantic cyclone), it did satisfy several of the criteria, including windspeed and low pressure.

Great storms such as this aren't confined to Europe. In 1962, high winds, rain and snow battered the east coast of the United States, causing 40 deaths and damage running to hundreds of millions of dollars. Thankfully, these storms don't occur frequently because the conditions that cause them – the collision of exceptionally strong hot and cold weather fronts and mergers of smaller storms – are rare.

Cyclones

Cyclones are colossal storm systems that form within the tropics (see **Latitude and longitude**), where the warm ocean stimulates upward convection currents (see **Conduction and convection**), that are then 'spun up' by the **Coriolis effect** to create powerful vortices with a low-pressure core around which extremely high winds circulate. Cyclones in the Pacific Ocean are known as 'typhoons', while those that form in the Atlantic are called 'hurricanes'.

Once formed, a cyclone migrates westwards until it hits land. Mercifully, this usually makes the cyclone start to dissipate, through friction with the land and the removal of its energy source – the heat in the sea. Very violent cyclones are characterized by sustained winds that blow in excess of 250km/h (150mph), and throw ocean waves over five metres high at anything in their path. The most severe winds are found in the so-called 'eyewall', a ring of storms surrounding the cyclone's low-pressure core, which is called the 'eye'. Inside the eye, which can range in size from 8km to over 200km (5 to 125 miles) across, conditions are relatively calm.

Wild fires

Hectares of blazing forests, threatening people's homes and natural habitats for wildlife, have become a common sight on television news. In 2007 flames engulfed woodland around Athens, Greece, killing 65 people; in 2008 it was the turn of southern Australia; while in 2009, blazes devastated 1,300 square kilometres of forest in California.

Many experts blame the recent spate of wild fires on our warming planet, arguing that **climate change** is rendering vegetation tinder dry. Others say that human efforts to suppress wild fires have led to an over-abundance of combustible fuel turning what might otherwise be small, controllable fires into the gigantic conflagrations seen in recent years. Technology is helping to control and manage the fires; monitoring stations across the United States now gather regular data on factors such as temperature, moisture and weather to build detailed predictions of where wild fires are most likely to break out.

Heat waves

As the world warms up, heat waves – prolonged times when temperatures in a region become abnormally high – are becoming more common. They are usually caused by high-pressure weather systems that inhibit cloud coverage, allowing more sunlight to reach the ground and lowering the total amount of rainfall. Cities and urban areas tend to experience some of the worst heat waves, as building materials trap heat more effectively than natural terrain.

Heat waves and associated droughts, can have a catastrophic effect in the developing world. But even in the west, heat is a bigger problem than you might imagine. In 2002, a study by Johns Hopkins Bloomberg School of Public Health concluded that heat waves in the United States kill around 400 people every year.

Cosmic impacts

The collisions of **asteroids** and **comets** with the Earth are terrifyingly destructive events – with the power to devastate cities, continents, and even the planet as a whole. In 1908, a chunk of rock or ice from space struck the Tunguska region of Russia, exploding with the equivalent energy of a 15-megaton **nuclear weapon** – that's 1,000 times as powerful as the bomb which levelled the Japanese city of Hiroshima at the end of the Second World War. Over 2,000 square kilometres of forest were destroyed at Tunguska.

The object that caused the Tunguska event is thought to have been relatively small – about 50 metres across. There are comets and asteroids much larger than this – a kilometre across and more. An impact from a chunk of rock or ice from space a kilometre across could devastate life on Earth, unleashing firestorms, tsunamis and flinging clouds of ash and debris into the atmosphere blocking out the **Sun** for many years, plunging the planet into a severe artificial winter.

Cosmic impacts are believed to be responsible for at least one of the great mass extinctions in Earth's history – the event 65 million years ago when the dinosaurs disappeared from the face of the planet. Astronomers are now combing the skies for these potentially hazardous objects. Meanwhile, other scientists are trying to work out what exactly we're going to do when we find one (see **Asteroid deflection**).

Death rays

Astronomers have suggested that radiation from violent events in space can trigger lethal natural disasters and loss of life on Earth. **Supernovas** and **hypernovas** are powerful explosions marking the deaths of massive stars. If one of these occurred within 100 light-years it could severely damage the Earth's already depleted **ozone layer**, allowing harmful levels of ultraviolet radiation from the **Sun** to poison life on the planet's surface. Evidence for the theory was uncovered by scientists at Japan's RIKEN research institute in 2009 who found signatures of radiation-induced chemical changes in Antarctic ice cores coinciding with the dates – from astronomical records – of known supernovas in Earth's history. Some scientists even speculate that such an event could trigger a mass extinction of life on Earth, arguing that the theory seems to match the characteristics of the extinction at the end of the Ordovician period, in the **Paleozoic era**, around 440 million years ago.

Mass extinctions

Mass extinctions are the ultimate natural disaster, when a large fraction of the species making up life on Earth are wiped from the face of the planet in a short space of time. Scientists studying the **fossil** record of Earth's prehistory have found five major incidents in the past when the number of fossilized species dropped precipitously. Known as the Big Five extinctions, they show up clearly as peaks on a graph of extinction rate plotted against time (see diagram). They are: the Late Ordovician extinction, the Late Devonian extinction and the Permian-Triassic extinction (all in the **Paleozoic era**), the Late Triassic extinction in the **Mesozoic era**, and the Cretaceous-Tertiary extinction which wiped out the dinosaurs. Biggest of these was the Permian-Triassic – with over 90 per cent of marine species and 70 per cent of land vertebrates dying out. Mass extinctions are believed to be caused by powerful events from space, such as cosmic impacts and death rays, as well as terrestrial phenomena such as **climate change** and **supervolcanoes**.

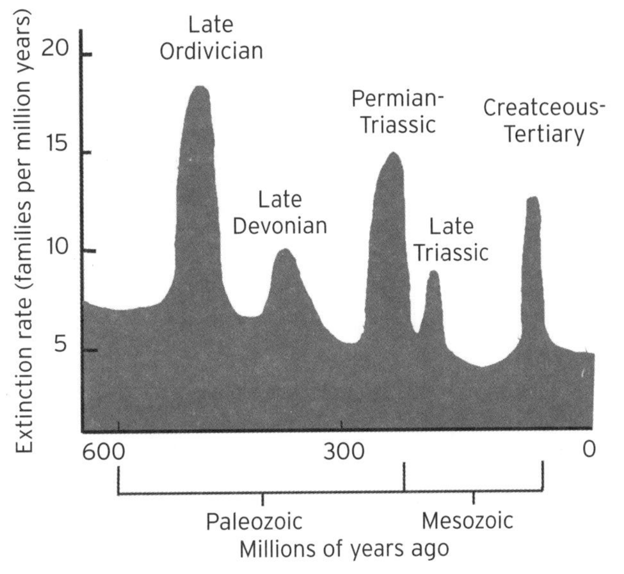

EARTH'S ATMOSPHERE

Atmospheric composition

Earth's atmosphere is a mixture of the gases nitrogen, oxygen, argon and carbon dioxide, in the proportions 78 per cent, 21 per cent, 0.93 per cent and 0.038 per cent, respectively. At sea level on Earth the atmosphere exerts a pressure of around 1 bar, under normal circumstances, where a bar is 100,000 pascals (see **Temperature and pressure**). Meteorologists usually express pressure in millibars (mb), where 1 bar = 1,000 millibars. Weather systems create substantial changes in pressure – for example, in the eye of a cyclone it can drop to as low as 870mb. Even in calm weather, pressure decreases with altitude, halving roughly every 5.6km (3.5 miles) above the planet's surface. Over 99.99 per cent of the atmosphere is within 100km (60 miles) of the surface – and this point is taken to be the edge of space.

The atmospheric composition is what gives the sky its colour; chemicals in the air preferentially scatter blue light in the spectrum of **electromagnetic radiation** from the **Sun** across the sky, giving the sky itself a blue tint.

Atmospheric structure

Scientists divide the Earth's atmosphere into layers. The lowest is the 'troposphere' which stretches up to a height of between 7km

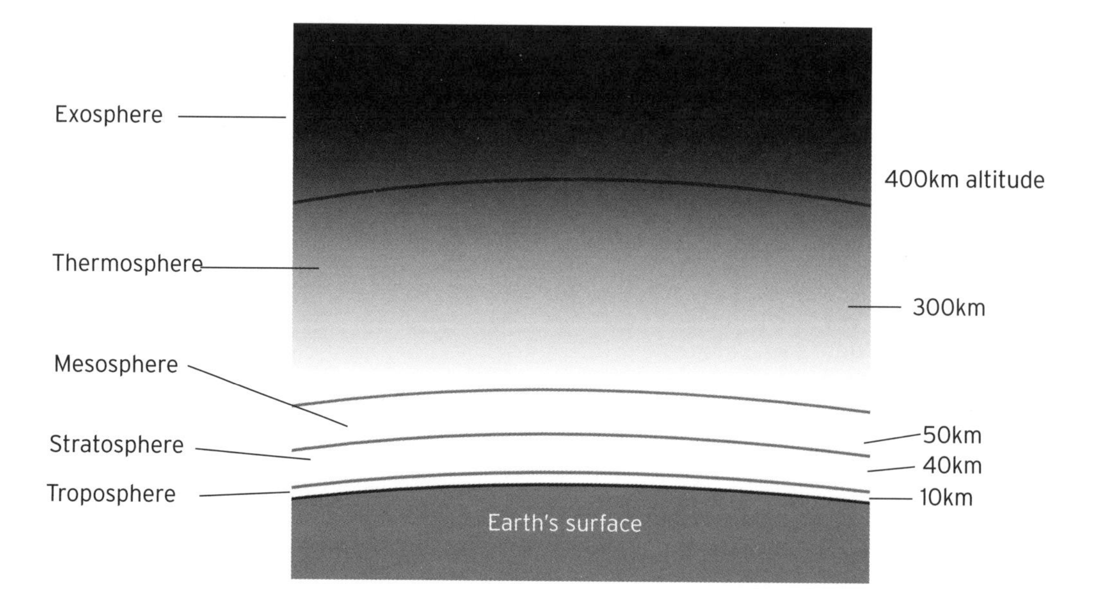

and 28km (4.5 and 17 miles) and is defined as the layer heated by the planet's surface – so temperature decreases with height. At the top edge of the troposphere is a boundary marked by a thin layer called the 'tropopause'. The next level up is the 'stratosphere', extending from the top of the troposphere to 50km (31 miles). Here, convection (see **Conduction and convection**) separates the atmosphere into layers of different temperature, the mercury rising with altitude. Included in the stratosphere is the **ozone layer**, which shields the Earth against harmful solar ultraviolet radiation. Above this is the 'mesosphere', where temperature decreases again with height; this layer reaches up to about 85km (53 miles) and it is the coldest of all the atmospheric layers with the temperature plunging to -100°C (-148°F). Beyond the mesosphere is the 'thermosphere', reaching from 80km (50 miles) up into space to around 400km (250 miles). Here, the dominant source of heat is the Sun, causing the temperature to rise to over 1,000°C (1,830°F). Finally there is the 'exosphere', the atmosphere's rarefied periphery.

Hadley cell

The Hadley cell is a giant convection current (see **Conduction and convection**) that dominates the circulation of air in the Earth's atmosphere. Hot air at the equator rises to heights of over 10km (6.2 miles) and then migrates polewards, to latitudes (see **Latitude and longitude**) of around +/-30° where it cools and falls back to sea level and then sweeps back in towards the equator.

Air in the Hadley cell tends to lose its moisture due to condensation in the low temperatures at high altitude. For this reason the air is dry as it returns to sea level, which is a contributing factor in the formation of deserts around the tropics. Similarly, cool air tends to sink around the poles of the planet, and migrates to latitudes of +/-60° from where it warms up and rises again, a circulation current known as a 'polar cell'. The falling air in the Hadley cell and the rising air in a polar cell set up a third atmospheric convective current between the two, called a mid-latitude, or 'Ferrel cell' which spans latitudes +/-30° to +/-60°.

Jet streams

Jet streams are belts of high-speed air that circle the Earth in an easterly direction in the upper troposphere (see **Atmospheric structure**). There are two principle jet streams – the polar jets and the subtropical jets and they form due to the **conservation law** of angular momentum from **rotational dynamics**. As the Earth rotates, air at the planet's surface is being swept around in a giant circle, but as the air is transported polewards by the Hadley cell then the radius of rotation decreases. Just as an ice skater spins faster by pulling her arms in, as the air moves polewards and its radius of rotation decreases so it speeds up, and the jet streams are the result. They get their easterly direction from the **Coriolis effect**. Aircraft flying eastwards take advantage of the jet streams, using them as a powerful tail wind to lessen both flight times and fuel consumption.

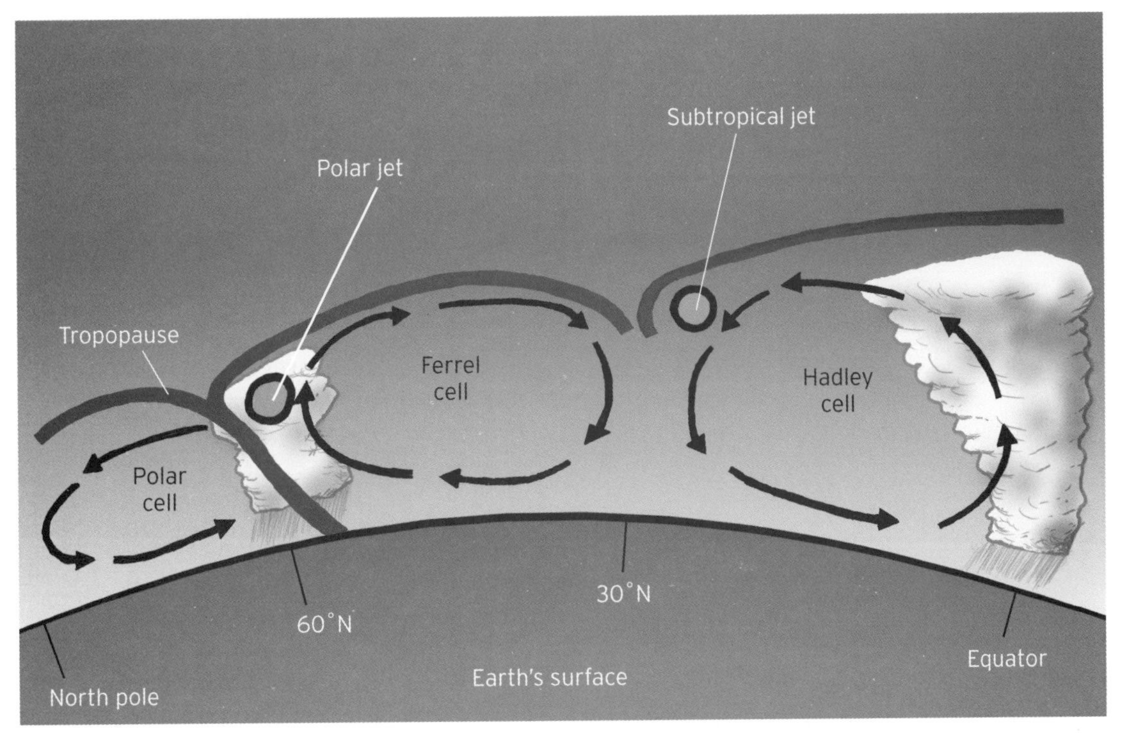

Trade winds

Trade winds are air currents at the surface of the Earth caused by the Hadley cell, a giant convection current in the atmosphere. In the Hadley cell air moves towards the equator along the Earth's surface, but the **Coriolis effect** deflects this straight-line flow, creating air that heads towards the equator from the north-east in the northern hemisphere, and towards the equator from the south-east in the southern hemisphere. These air currents are the trade winds; where they converge around the equator is a belt of still air, known as the 'doldrums', where the Hadley cell scoops up winds to higher altitudes.

Beneath the Ferrel cell (see **Hadley cell**) at latitudes of between +/-30° and +/-60° the same mechanism creates winds, known as 'westerlies', which blow from the south-west in the northern hemisphere and north-west in the southern hemisphere. The polar cells (see **Hadley cell**) generate similar winds in the opposite directions, known as 'polar easterlies'.

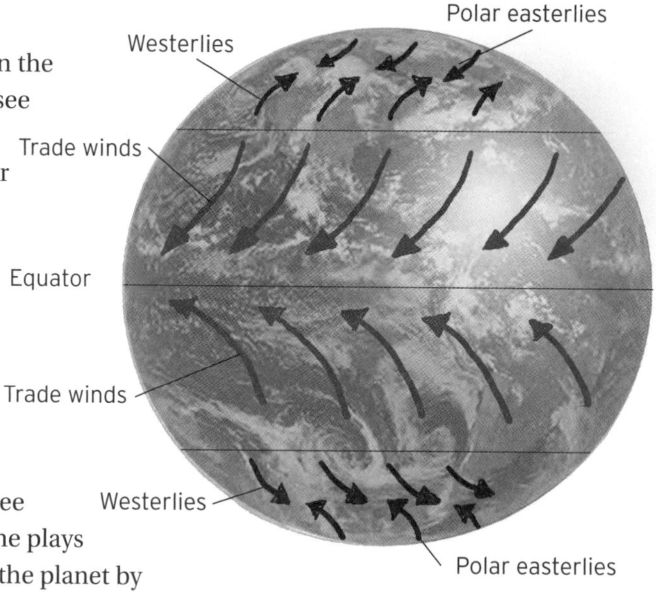

Ozone layer

High up in the Earth's stratosphere is a shell of atmosphere (see **Atmospheric structure**) called the ozone layer, which is rich in ozone – also known as trioxygen, a molecule made from three atoms of oxygen bonded together. Ozone plays an important role in the habitability of the planet by absorbing most of the harmful ultraviolet light from the Sun. Without the ozone layer rates of skin cancer and eye cataracts would soar.

Earth's ozone layer has been depleted by **air pollution** in the form of gases known as chlorofluorocarbons (CFCs), which have been used as propellants in aerosol sprays and as refrigerator coolants. Ultraviolet light acting on these chemicals forms nitrogen oxide **free radicals** that break down ozone, which has led to a serious thinning of the ozone layer over Antarctica – the 'ozone hole'. The use of CFCs is now banned worldwide, and the ozone depletion rate appears to be slowing in response – though it could take as long as 100 years for these pollutants to be flushed from the atmosphere entirely.

Ionosphere

The upper mesosphere and the lower portion of the thermosphere (see **Atmospheric structure**) are home to the 'ionosphere', where the atoms and molecules in the atmosphere are broken into **ions** by the radiation from the Sun. It is in the ionosphere that auroras are formed, as high-energy particles from the Sun collide with atoms and molecules of gas in the atmosphere to create spectacular light shows. The ionosphere is also useful for global communication. Because its gas is ionized and therefore electrically conductive it behaves rather like a metal, meaning it can make **electromagnetic radiation** undergo reflection, an effect that can be used to bounce radio signals between the ionosphere and the ground, channelling them around the planet.

Van Allen belts

The Van Allen belts are two toroidal-shaped belts of high-energy particles from the Sun, trapped by the magnetic field of the Earth (see **Geomagnetism**). The two belts nestle inside one another. The outer belt, which stretches between 15,000km and 25,000km (9,500 and 15,500 miles) above the Earth's equator, is made mainly of electron particles, while the inner belt, located between 1,000km and 5,000km (600 and 3,000 miles), is predominantly high-energy protons. They were first predicted by the American astrophysicist Dr James Van Allen, who was also the first to discover them in 1958 – correctly interpreting the data returned from NASA's Explorer 1 spacecraft. Satellites passing through the belts need shielding to prevent the radiation from damaging their sensitive electronics.

Magnetosphere

Earth's magnetic field doesn't just have an effect at the planet's surface but stretches up through the atmosphere and into space in a giant bubble of magnetism known as the magnetosphere. Close to the planet's surface the magnetosphere is true to its name, being roughly spherical. But beyond, around 60,000km (37,250 miles) up, its shape is influenced by two effects from the Sun: the **solar wind** and the Sun's magnetic field. These sweep the magnetosphere out behind the planet, forming a long tail stretching over 1.25 million km (776,700 miles) into interplanetary space. The Earth's magnetosphere was crucial to the emergence of life on the planet, shielding the surface from electrically charged radiation particles from space.

METEOROLOGY

Wind

Wind is the term given by meteorologists – scientists who study Earth's weather – to the movement of large masses of air across the planet's surface. While certain well-established wind patterns exist, such as the **jet streams** and **trade winds**, most of the atmospheric wind we encounter day to day is less predictable. Wind is caused by differences in atmospheric pressure – the air in regions of high pressure flows to areas where the pressure is lower. The pressure differences themselves are down to thermal effects – updraughts lower the pressure over warm land compared with the cool sea, while cold downdraughts lead to the formation of high-pressure regions. In this way, the Earth's wind systems are powered by the heat received from the Sun – and this is why using the wind to generate electricity is just an indirect form of solar power. Wind speeds are gauged according to the Beaufort scale, depending upon their speed in knots (1 knot = 1.9km/h = 1.2mph).

Weather fronts

We've all seen those lines of blue triangles and red half-circles on television weather maps denoting cold fronts and warm fronts. A cold front is quite literally the edge of a large volume of approaching cold air; its arrival can lead to rainfall as warm, moist air rises over the advancing mass of cold air and its moisture condenses to form clouds. Sometimes thunderstorms even result. Once a cold front arrives it usually bring with it an area of high pressure.

Similarly, a warm front is the edge of an advancing mass of air with relatively high temperature and humidity; these usually bring light rainfall, but nothing as severe as the passing of a cold front. Sometimes, a cold front is preceded by a warm front – known as an 'occluded front'. On the other hand, when a warm and a cold front both of equal strength meet – so that neither gives ground – the result is called a 'stationary front'.

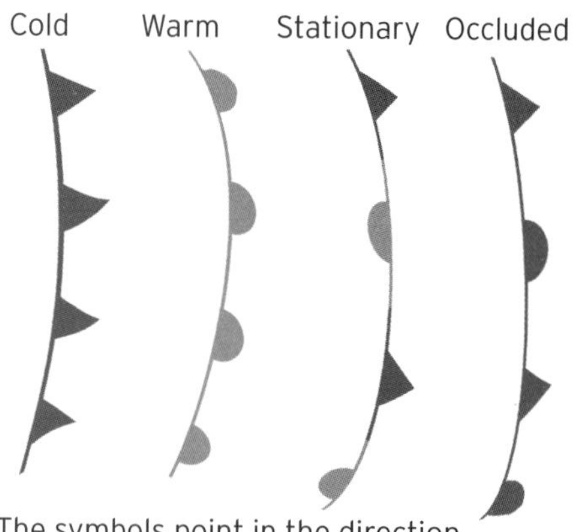

The symbols point in the direction that the front is moving

Clouds

Billowing masses of tiny water droplets in the Earth's skies are known as clouds. They are central to weather phenomena such as rainfall, hail and snow. The water in a cloud diffuses and scatters the sunlight passing through it. For small, rarefied clouds the effect is slight

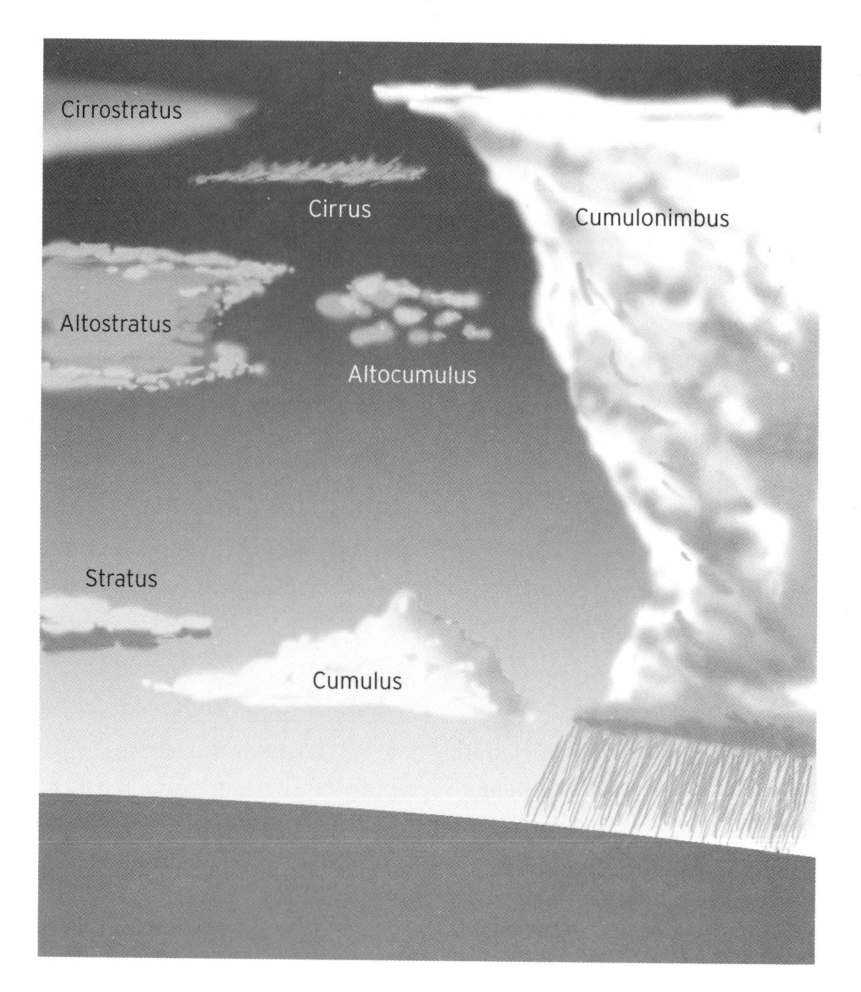

and the clouds appear fluffy and white; however, denser, thicker clouds – laden with water – absorb more of the light, making them appear grey or even black.

Clouds are divided into four families (see diagram). Family A are the high-altitude clouds between 3km and 18km (2 and 11 miles) above ground level (AGL), depending upon latitude (see **Latitude and longitude**). These include wispy 'cirrus' and 'cirrostratus'. Distributed between 2km and 8km AGL (1.25 and 5 miles), the middle clouds form family B, including sheet-like 'altostratus' and fluffy 'altocumulus'. The low clouds found up to 2km (1.25 miles), such as billowing white 'cumulus', make up family C. Finally, there are the vertical clouds of family D, which form at many heights – predominantly the menacing 'cumulonimbus' rain clouds. The scientific study of clouds is called 'nephology'.

Rainfall

When the temperature of a cumulonimbus or nimbostratus cloud is lowered sufficiently that the droplets condense into liquid water, the result is rainfall. Meteorologists refer to the process as 'precipitation' – not to be confused with **chemical precipitation**. Rain is part of the water cycle, the culmination of the process whereby water is evaporated from the oceans to form clouds and then falls back to Earth as rain so that the process can repeat.

Rainfall tends to accompany areas of low pressure because low pressure is caused by rising thermal currents sucking air away from the planet's surface. The currents carry moist air to high altitudes where the temperature is low, allowing the moisture to condense out and fall as rain. When there are prolonged periods of high pressure over a region they can cause reduced rainfall and drought.

Snow and hail

Exceptionally cold temperatures cause the moisture in clouds to condense not into liquid water but into solid ice, in the form of hail and snow. These occur when the updraught of air in a low-pressure region carries water in a cloud up to levels where the air temperature is low enough for vapour and droplets to freeze solid. Cold, wintry conditions help the process along by lowering the altitude at which this freezing can take place. When an ice particle grows past the point where the updraught of air can support it, it falls to the ground under its own weight.

Hail and snow each form in slightly different ways. Snow is made as ice crystals grow directly from the vapour and micro-droplets in a cloud to form delicate flakes. However, when large droplets of liquid water condense first and then freeze the result is hail. Hailstones can be deadly, occasionally measuring up to 15cm across and weighing over a kilogram.

Fog

Fog is a cloud-like mass of water droplets that forms at ground level. It is made by cooling air several degrees below its 'dew point' – the temperature required for the moisture to begin condensing out. The dew point varies with the humidity of the air, with very humid air able to form fog at higher temperatures than dry air. When fog forms, the droplets of moisture cluster around so-called 'condensation nuclei' – usually particles of dust and dirt in the air.

Natural climates in some locations around the world make them especially susceptible to fog. San Francisco Bay is one example, where cold air drifting in off the Pacific cools the air in the bay below its dew point to form thick, rolling fog banks.

Lightning

Lightning is the sudden discharge of electricity from a storm cloud down to the ground – or to another storm cloud of opposite **electric charge**. A lightning strike is an extremely violent process, heating the air in the immediate vicinity to temperatures of up to 30,000°C (54,000°F), while the **electric currents** involved can be tens of thousands of amps. Lightning occurs because the undersides of clouds are negatively charged, which induces a positive charge on the ground. Discharge of lightning proceeds through the formation of a leader – a channel of ionized (electrically conducting) air snaking down from the thundercloud to the Earth's surface. In response, smaller leaders begin to creep up from high points on the ground. When a cloud leader and a ground leader meet, charge rapidly flows up from the ground to the cloud – called the 'return stroke' – emitting a dazzling flash of light accompanied by a massive discharge of sound energy which we hear as a clap of thunder.

El Niño

Every few years an excess of warm Pacific Ocean water flows towards the western coast of America. Low pressure caused by thermal currents rising up from this water leads to increased rainfall, a phenomenon known as El Niño.

Under normal circumstances, the **trade winds** drive water in a westerly direction across the Pacific towards Australia and Indonesia – where sea temperatures are naturally higher. This makes sea levels there higher by a few tens of centimetres but during an El Niño year, the trade winds slacken off allowing the warm water to flow back, creating low air pressure over the Americas – and high pressure at the ocean's western edge. In the intervening years the situation reverses, with low pressure over the western side of the Pacific and highs at the eastern side – sometimes called La Niña. The whole cycle is known as the Southern Oscillation. An analogous phenomenon exists in the northern hemisphere, called the North Atlantic Oscillation.

Weather forecasting

Weather forecasting is the science of using data on current weather conditions, together with models of how weather systems behave, to try to predict conditions several days ahead. Monitoring stations and weather satellites take readings of quantities such as temperature, wind speed, atmospheric pressure and humidity. These numbers are fed to supercomputers – **parallel computing** machines that evolve the data to obtain the most likely future outcome. However, the effectiveness of the process is limited by the fact that the weather is subject to **chaos theory**, making it extremely difficult to obtain reliable forecasts more than a few days in advance – and often much less! Weather forecasting is crucial for agriculture, aviation, shipping and the military.

GEOLOGY

Igneous rock

Petrologists – scientists who study rock formation – classify rocks into one of three types: igneous, sedimentary and metamorphic. Igneous rock is essentially solidified magma – the molten material that spews forth from the Earth's innards during **volcanic eruptions** and at **divergent faults** in the Earth's crust.

There are three different categories of igneous rock: 'plutonic', or 'intrusive', rocks are made from magma that has solidified underground and then burst up through the crust as a rocky outcrop – granite is an example. Volcanic, or 'extrusive', rocks are those which form on the surface from magma that has oozed up in its liquid state – basalt is an example of extrusive rock. 'Hypabyssal' igneous rock is halfway between the two, solidifying at shallow depths to form smaller rocky intrusions at the surface – the dark mineral andesite is a hypabyssal rock.

Sedimentary rock

Minerals formed from layers of sediment deposited and compressed for tens or even hundreds of millions of years are known as sedimentary rocks. Like igneous rock, sedimentary rock is divided into three major categories. 'Clastic' rocks are made of particles from other rocks that have been eroded or crushed and then re-deposited – sandstone is an example. On the other hand, 'chemical' sedimentary rocks are made from sediments laid down in water as a result of **chemical precipitation**. For example, the sedimentary rock gypsum forms when calcium sulphate particles drop out of solution – perhaps as some of the water is evaporated away, increasing the solution's concentration. 'Organic' sedimentary rocks are made from the remains of living plants and creatures. Examples are limestone (formed from the skeletal remains of corals and other marine organisms) and coal (made from the compressed remains of long-dead plants).

Metamorphic rock

Metamorphic rock is a term used to refer to minerals that have changed their properties or structure through the action of **temperature and pressure**. There are three mechanisms to form metamorphic rock: 'contact metamorphism', 'regional metamorphism' and 'dislocation metamorphism'. A contact metamorphism occurs when an uprising channel of magma penetrates a bed of sedimentary rock in the Earth's crust. Heat from the magma spreads into the sedimentary rock causing it to re-crystallize into a new form; for instance, when hot magma rises up through limestone, the result is marble. Whereas contact metamorphisms are localized, regional metamorphisms affect a wide area. An example is when the weight of deep beds of sediments cause the lower levels to sink down into the hotter depths of the crust and re-crystallize. This is the process that converts shale into slate. Dislocation metamorphisms happen at fault lines and tectonic plates meet, and the crushing pressures cause the rock to change form.

The rock cycle

The rock cycle describes the inter-relationship between igneous rock, sedimentary rock and metamorphic rock. Fluid or semi-fluid magma begins the cycle, which solidifies into igneous rock and then can be melted back down into magma or transformed by **temperature and pressure** into metamorphic rock. Or it can be eroded and weathered into particles that are deposited and compressed to form sedimentary rock. Sedimentary rocks can then be weathered and eroded again to form new sedimentary rocks, or melted back into magma (for example, by subduction at a **convergent fault**), or it can be compressed and heated to form metamorphic rock. Metamorphic rocks can go one of two ways – either being eroded to form sedimentary rock or melted back into magma.

The large amount of water present in the Earth's climate plays an important role in the rock cycle, eroding rocks to form new sediment particles, and assisting with the gradual deposition and compression of these sediments to form new rocks.

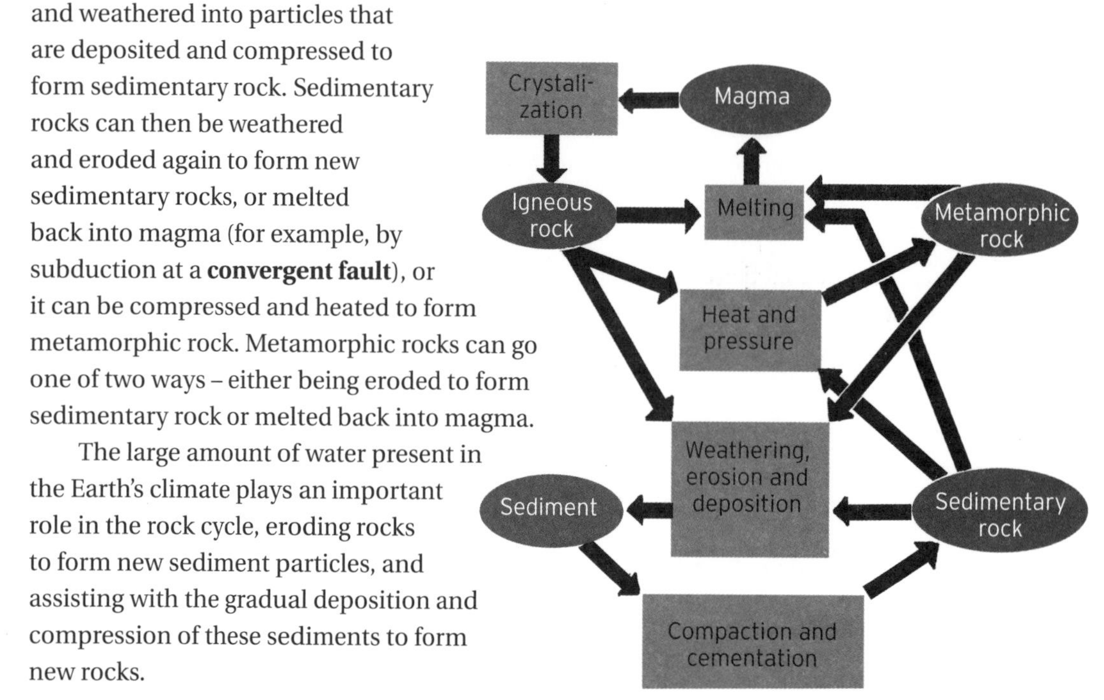

Soil

In most areas of Earth's natural land mass the uppermost layer is composed not of pure rock but of soil – a mixture of eroded rock particles together with organic material, water and air. Soil composition is dependent upon the surrounding area, its mineralogy and climate, and is broadly classified into three types: clay, sand and silt. Clay-dominated soil has small particles and is heavy and sticky; sandy soil has larger grains with less tendency to stick together, but it is more prone to erosion by the wind; silty soil has a constituency midway between sand and clay.

Geologists split soil into four major layers, known as 'soil horizons', the depths of which vary greatly. Topsoil is horizon 'A', containing minerals plus a great deal of organic matter, such as plant roots. Subsoil is horizon 'B' and has a negligible organic component. Horizon 'C' is the 'parent rock' from which the mineral content of the soil has been formed, and below this is horizon 'D', the bedrock. Some schemes also include an upper 'O' horizon, describing the layer of organic detritus above the topsoil.

Mineralogy

Mineralogy is the study of the structure and chemistry of rock. It draws upon aspects of **crystallography** and **materials chemistry** to classify naturally occurring rocks according to a host of measures such as their hardness, strength, density, chemical composition and the arrangement of their **atoms** and **molecules**. The International Mineralogical Association – the field's world governing body – currently recognizes over 4,000 different mineral species.

Stratigraphy

Layers of sedimentary rock and igneous rock, laid down over aeons of **deep time** make up a record of the geological history of the Earth revealing everything from our planet's past climate to the creatures that once lived here. Geologists can examine the stratigraphic record at sites around the world where land movements have exposed a cross section of the accumulated rock – for example, at a cliff face.

Analysis of the different layers in the rock can reveal what minerals are present, whether their sediments were laid down in water or air, what the chemical composition of the atmosphere was at the time, how the Earth's magnetic field was aligned, and what species of plants and animals inhabited our world – revealed by excavating **fossils**. They can assign dates to the distinct layers and what they find in them, by measuring the abundances of different radioactive elements that decay with time – carbon **dating** is an example of this so-called 'isotopic analysis'. Upheavals and interruptions to the **evolution** of life also show up in the stratigraphic record – as dips in the numbers of fossils that match up with deposits of ash corresponding to major volcanic eruptions, or even deposits of extraterrestrial elements (such as iridium) pointing to a calamitous **cosmic impact**.

Geological mapping

Just as land surveyors draw up relief maps to show the **topography** of the Earth's terrain, so geologists construct maps to chart the **mineralogy** of the planet's land masses. Again, contour lines are used but rather than denoting the morphology of the planet's surface, this time they show differences in mineral content across stratigraphic layers, as well as the thickness of the layers themselves. The slope of different rock strata can also be shown by using a mapping convention known as 'strike and dip'.

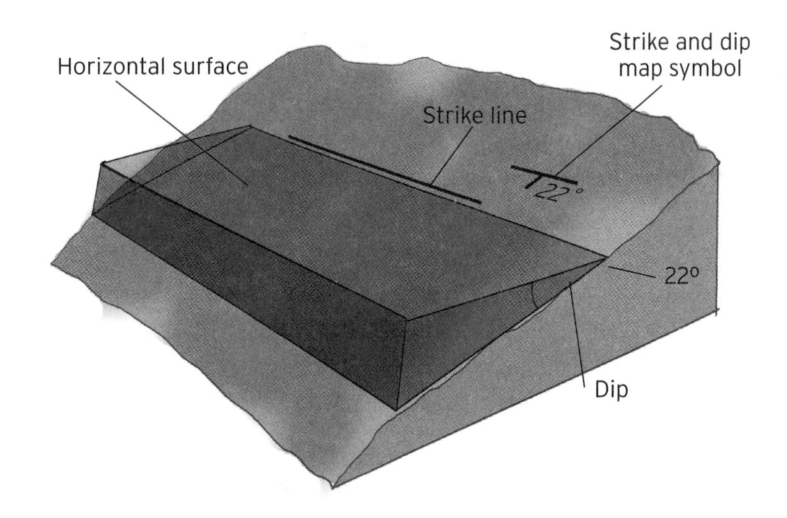

A straight line on the map marks where the plane of the rock layer intersects with the horizontal. Another line at right angles to the first then indicates the downward sloping direction and a number gives the slope in degrees below horizontal

Fossils

The remains of living organisms trapped during the formation of ancient sedimentary rock can be preserved, often in remarkable detail, as fossils. In most fossils the body of the organism is turned to stone, which happens in a number of ways. One is called 'permineralization' where mineral-rich water flows into the cavities and recesses of the organism, gradually depositing its minerals to turn the organic matter to rock. Alternatively, the remains of the organism may decay away, which can leave an exact imprint of its body. Sometimes mineral-rich water flowing through the imprint can make a fossil 'cast' of the organism.

Some of the oldest fossils recovered are of microbe-covered sedimentary rock-forms known as 'stromatolites', which grow in shallow ocean water. The oldest stromatolite fossils date from over 2.7 billion years ago – back in the **Archaean aeon**. Fossils don't necessarily have to be made of rock; the bodies of large animals – such as mammoths – frozen in Arctic ice, and insects preserved in ancient amber (itself fossilized tree resin) are also regarded as types of fossil.

CLIMATOLOGY

Air pollution

An increasing problem of the industrial age is the release of substances into the atmosphere harmful to the environment. Air pollutants can be chemical (such as the CFC gases that have damaged the **ozone layer**), biological (such as methane from the decay of organic waste and sewage) or particulate (such as the tiny flecks of material thought to cause **global dimming**).

Perhaps the worst pollution source is the burning of fossil fuels like **coal** and **petrol**, which generates carbon dioxide contributing to the greenhouse effect; particulate matter adding to global dimming; nitrogen oxides that destroy the ozone layer and produce chemical smog; and sulphur compounds that cause acid rain. There are many other forms of air pollution, including agricultural pesticides, volcanic eruptions, and solvent fumes.

Acid rain

Rainfall that is below the usual pH (see **pH indicator**) for rain – around 5.6 – is known as acid rain. It is a result of air pollution involving compounds such as carbon dioxide, sulphur and nitrogen oxides that react with water vapour in the atmosphere to increase the water's acidity. The pollution can be man-made or derive from volcanic eruptions.

Acid rain erodes buildings, stunts the growth of trees and other vegetation, and poisons lakes – killing fish, impacting on aquatic birds and other creatures further up the food chain. Thanks to measures such as adding catalytic converters (see **Chemical reaction**) to cars to reduce emissions of nitrogen oxides, and the placing of curbs on industry, the control of acid rain now seems to be one of climate science's success stories.

Greenhouse effect

Current concern over climate change centres around the greenhouse effect – which is causing global warming, making our planet warmer year by year. It happens because there are gases in the atmosphere that block a proportion of infrared heat radiation, preventing it from being radiated from the Earth's surface to cool the planet down. If radiation simply arrived from the Sun at infrared wavelengths (see **Electromagnetic radiation**), this wouldn't be a problem – because as much would be prevented from entering the atmosphere as is prevented from leaving. However, sunlight has a range of different wavelengths, many of which pass straight through the atmosphere and are absorbed directly by the ground. This heats the ground up, causing it to re-emit the radiation as extra infrared, which gets trapped – and this is heating the planet up. Carbon dioxide is the major man-made greenhouse gas trapping the heat, enormous volumes of which are churned out by the daily burning of fossil fuels such as **coal** and **petrol**.

Climate change

Estimates made in 2009 by scientists at the Massachusetts Institute of Technology suggest that between now and 2095 average global temperatures could rise by over 5°C (9°F)– as a result of the greenhouse effect, driven by man-made emissions of carbon dioxide (CO_2). If such dramatic global warming were allowed to happen it would lead to catastrophic **sea-level rise**, droughts, famine and an increase in the frequency of great storms and cyclones. These effects, together with the temperature rise causing them, are known collectively as climate change.

Evidence for climate change is summarized in the 'hockey stick' diagram – a graph showing how the world's temperatures have changed over the past hundreds and thousands of years, reconstructed from various sources. A pronounced upturn is shown around the start of the industrial age – when our CO_2 emissions first began to climb.

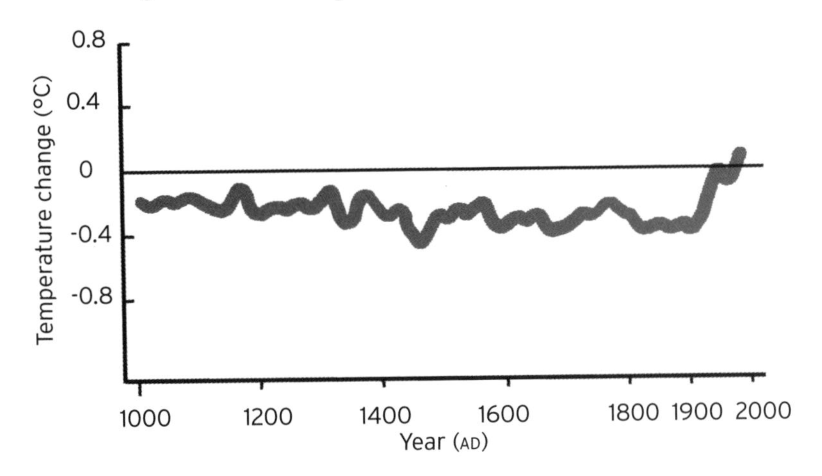

Sea-level rise

Sea-level rise is expected to be one of the most devastating effects of climate change. Recent calculations suggest that the most likely outcome is a rise of between 1 and 2 metres (3 to 7ft) by the end of this century. A one-metre rise would be sufficient to flood the land occupied by some 45 million people, and impact upon the existence of billions more as farmland and fresh water supplies become contaminated by sea water. Rising sea levels are caused mainly by the melting of polar ice as the world warms up; **thermal expansion** of the water itself plays a lesser role. There is enough ice in the world to raise sea levels by 70m (77 yards) – though it's unlikely temperature rises of the levels currently predicted could melt it all.

It's not only the developing world that is at risk from rising sea levels – the flooding would also threaten major cities in the west, including London and New York. A refugee crisis could emerge as stampedes of homeless survivors flee to higher ground. It has even been suggested that nations may be driven to war in the scramble for food and shelter.

Global dimming

While most of the world frets over the planet getting hotter through the effect of global warming, caused by the **greenhouse effect**, other groups of scientists are worried about something that sounds like the complete opposite: global dimming. Whereas global warming deals with the effects of pollution in the form of greenhouse gases trapping infrared radiation and increasing the planet's temperature, global dimming is concerned with solid particles – such as ash and soot – that reduce the amount of the radiation reaching the planet's surface.

It might sound like global dimming is the solution to global warming, but it comes with its own raft of problems for the planet. Climatologists believe global dimming has had a cooling effect on the northern hemisphere – reducing ground and ocean temperatures and thus reducing rainfall. Experts say this could have had a significant effect on Africa, contributing to the droughts that gripped the continent during the 1980s.

Ice ages

Earth sporadically passes through eras when its temperature falls drastically, plunging the world into a so-called ice age, during which much of the surface freezes over. There have been four major ice ages throughout Earth's geological history. These occurred 2.4–2.1 billion years ago, 850–630 million years ago (mya), 460–430mya, and 2.53mya to 10,000 years ago. The second of these was the most severe, during which the expanding ice sheets from both polar caps may have actually met at the equator – a chilly scenario called 'snowball Earth'.

Ice ages are often punctuated by brief warm periods known as 'interglacials' and it is not entirely clear if the most recent ice age is actually over or whether we're just enjoying a relatively temperate interglacial period. Evidence for ice ages comes from the fossil record of planet Earth and its climate locked away in the **stratigraphy** of ancient rock layers, and through the marks left on ancient terrain by passing **glaciers**. Many environmental factors can cause, end or sustain ice ages, including Milankovitch cycles.

Milankovitch cycles

One of the driving factors behind the development of an ice age are the so-called Milankovitch cycles – wobbles in the orbit of the Earth that affect the amount of sunlight arriving at the planet's surface. As the Earth travels around the Sun its orbit is constantly undergoing tiny cyclic changes in tilt and eccentricity, while the planet's own tilt (also known as 'obliquity') and **precession** varies too. In the same way that the temperature on Earth varies through the seasons, the Milankovitch cycles induce temperature changes over much longer timescales of thousands and millions of years.

Milankovitch cycles probably aren't responsible for plunging the world into an ice age – or lifting it out again. However, they are thought to play a part in the onset and retreat of interglacials (see **Ice ages**). The big hitters which make or break the overarching ice ages are thought to be intrinsic changes in the brightness of the Sun, global dimming from volcanic eruptions, and even the layout of the continents – which can inhibit the flow of warm sea water from the equator.

Little Ice Age

The Little Ice Age was a period spanning roughly from the 16th century to the mid-19th century when temperatures cooled by a degree or so. While global ice coverage didn't increase dramatically, there were noticeable effects – winters became longer and harsher and growing seasons shorter, impacting on agriculture and causing famine. During the winter of 1683–4, Britain experienced the worst frosts in its history – with London's River Thames freezing over for two months. The last time the Thames froze at all was in 1814.

The Little Ice Age is thought to have been caused by a double impact of the Sun being at a natural long-term minimum in its activity – known as the Maunder Minimum – at the same time as a number of major volcanic eruptions reduced the amount of light reaching the planet's surface. Other brief eras of both cold and warmth have been discovered from historical data; for example, the Little Ice Age itself was preceded by the Medieval Warm Period, between 800 and 1,300 AD, during which temperatures were slightly elevated.

Climate feedback

There can be little doubt about the reality of climate change or that action must be taken to prevent it. However, uncertainties still exist in the predictions of its future consequences; for instance, the role of water vapour in clouds is still poorly understood. Water vapour is itself a greenhouse gas, yet white clouds help to cool the planet by reflecting heat away into space. This is an example of 'climate feedback' – where a consequence of the changing climate feeds back to either enhance or inhibit global warming. Effects that inhibit climate change are called 'negative feedback', while those that reinforce the warming are examples of 'positive feedback'. Melting ice is a positive feedback phenomenon – ice reflects sunlight back into space helping keep the planet cool. Melt some ice and this process becomes less effective, warming the planet further, melting more ice and so on. For other feedback effects, such as the influence of clouds, the distinction is less clear and until we get a grip on them we don't know how bad the consequences of climate change will be.

Climate modelling

We gauge the future effects of climate change by building models of the Earth's ecosystem inside computers; this is the art and science of climate modelling. The models work by breaking down the atmosphere, oceans and land masses into a three-dimensional grid of boxes. Within each box, the computer applies the laws of fluids, heat, chemistry and the physics of radiation from the **Sun** to calculate properties such as temperature, pressure, sea level and the amount of atmospheric carbon dioxide. The computed properties in each box are then married smoothly to those in the adjacent boxes to build a living, breathing model of the planet's climate.

Climate models are tested by running them backwards to try to predict the known variations in the climate that have taken place over the past few hundred years. Even so, effects such as climate feedback make it an extremely complex task.

Palaeoclimatology

Whereas climate modelling deals with extrapolating the state of the Earth's climate system forward into the future, palaeoclimatology is concerned with tracing it back into the past. Scientists draw upon the thickness of ancient tree rings, which reveal the duration of past growing seasons, the content of sedimentary rock layers and the composition of ancient ice – extracted by drilling out cores from deep below ancient glaciers and ice sheets at the polar caps. The oldest ice core – taken from Antarctica – dates from 800,000 years ago.

These observations enable scientists to work out the past composition and temperature of the atmosphere, as well as the concentration of particulate matter, such as ash – which is the signature of climate upheavals such as volcanic eruptions.

EARTH MYSTERIES

Natural reactor

Mention nuclear reactors, and most people think of the **nuclear electricity** plants that generate power for our homes. But in the Oklo region of Gabon, in Africa, natural deposits of uranium ore have undergone spontaneous **nuclear reactions** in the ground to form a natural nuclear reactor. Nuclear fission reactions (see **Fission and fusion**) at Oklo are not ongoing today, but are thought to have occurred some 2 billion years ago. At its peak the power output of the site is believed to have been as much as 100,000 watts. Scientists were alerted to the natural reactor when uranium prospectors noticed that some of the uranium at Oklo had been converted into other **isotopes** – a process that's only possible through nuclear reactions.

The fission reactions at Oklo are thought to have been made possible by underground water flowing through the uranium deposits. Nuclear fission **chain reactions** require neutrons from one reaction to be slowed down so that they can be absorbed by other uranium nuclei, and the water seems to have served this purpose.

Will o' the wisp

In days of old, folklore was replete with tales of ghostly lights that would lead hapless travellers into treacherous swamps and marshes. Known as will o' the wisp (from the word 'wisp', meaning a burning torch), scientists now know them to be caused by gases given off by decaying organic matter belched up from the ground in **wetland** areas.

Phosphine gas produced in small quantities by marshes undergoes spontaneous combustion upon contact with the air. Once lit, this gas then ignites the far more copious flammable methane gas that's also produced. In laboratory experiments, the addition of other marsh gases has been shown to lower the temperature of the combustion significantly so that it's too low to ignite other nearby fuels – another curious property of will o' the wisp which, for years, defied explanation.

Earthquake light

Like will o' the wisp, earthquake lights are glowing light sources seen hanging in the air. However, their cause seems to be quite separate – as the name suggests, they are triggered by tectonic activity. Appearing as white, red and blue glows over fault zones, where **tectonic plates** meet, they take a range of different forms, from spherical globes to sheets and rays illuminating large areas of sky. Some researchers have suggested the lights could be **electromagnetic radiation** produced by the so-called piezoelectric effect, where some minerals – such as quartz – generate an electric current when they are squeezed. It's the same principle on which gas stove spark lighters work.

In 2007, a NASA team put forward a theory that electrical charges created on the ground by the stress placed on rocks during tectonic activity can create lights and other forms of radiation by interfering with the Earth's **ionosphere**. The theory is yet to be confirmed but if it is correct the electrical signals that accompany the lights could serve as an early warning system, indicating when – and where – an earthquake is about to strike.

Earthshine

Earthshine is the glow of sunlight reflected from the Earth to illuminate the dark surface of a new, or thin crescent Moon. When the relative positions of the Earth, Sun and Moon are just right, sunlight falling on the Earth is reflected onto the Moon's dark side – the side opposite the Sun.

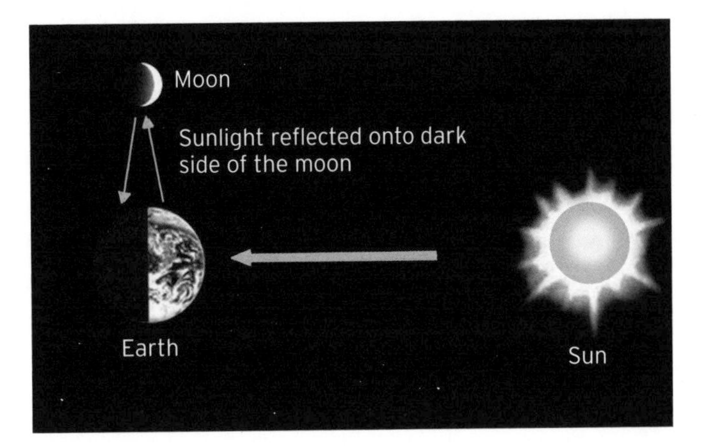

From there it gets reflected back down to the planet, where keen-sighted astronomers can look out for it. Scientists use the measured brightness of earthshine to calculate the total fraction of light falling on the Earth that gets reflected back into space – known as the Earth's 'albedo', which is an important parameter in **climate modelling**.

Other planet–moon systems exhibit the analogue of earthshine; the Cassini **planetary probe** was able to image moons of Saturn illuminated purely by the light reflected from their mother planet.

Noctilucent clouds

Cloud is normally the bane of astronomers, obscuring objects in the sky above. However, one type of cloud puts on a breath-taking night-time display of its own. Called the noctilucent clouds, they are formed by ice crystals high in the mesosphere (see **Atmospheric structure**) around 80km (50 miles) up. For a brief time shortly after sunset, clouds at low altitude fall into the Earth's shadow leaving the undersides of the high-altitude noctilucent clouds beautifully illuminated, lighting up the sky with a magnificent pearly blue glow.

Noctilucent clouds are normally seen at latitudes of between +/-50° and +/-65°, usually in the summer months – between May and August in the northern hemisphere and November and February in the southern hemisphere. In recent years noctilucent clouds have been a more common sight, leading some scientists to wonder whether their formation could be due to processes associated with climate change.

Sprites

Ordinary lightning strikes are caused by the sudden flow of **electrical charge** between thunderclouds and the ground. But lightning phenomena also occur above clouds – known as sprites, which arc between the tops of clouds and the electrically charged **ionosphere** high overhead. They appear as a red glowing halo with tendrils of light dangling beneath and like ordinary lightning they are fleeting phenomena, lasting for just fractions of a second. Their existence was only confirmed in 1994, when they were photographed by high-altitude aircraft.

Sprites aren't the only kind of lightning found at this altitude; another variety, known as 'blue jets', rise up from the tops of clouds to heights of around 50km (30 miles). Each jet lasts for about a second, making them clearly visible to the naked eye as they climb above the clouds.

Ball lightning

The cause of ball lightning is still a matter for debate amongst scientists. It appears during thunderstorms as a crackling ball of electromagnetic energy – between half and several metres across – that moves slowly and steadily before dissipating. Ball lightning's existence was once disputed, but sightings by many witnesses and recordings of electromagnetic disturbances made using scientific instruments have prompted scientists to search in earnest for an explanation.

Current theories ascribe the phenomenon variously to glowing clouds of silicon vapour, tiny batteries formed naturally as nanoscopic particles are charged up during storms, and even the action of small-scale **black holes** wandering in from space. However, at the moment there is scant hard evidence to support any of these theories.

Magnetic field reversal

Every now and then Earth's **geomagnetic** field inexplicably switches polarity – north and south magnetic poles literally changing places. The phenomenon doesn't happen overnight, but on timescales of tens of thousands of years. Scientists know this from examining the magnetic polarity of ancient igneous rocks. As molten magma solidifies its atoms and molecules align their magnetic axes with the prevailing magnetic field, locking in the signature of the geomagnetic field at the time.

The cause of the reversals is unknown, though one idea invokes **chaos theory** in the molten metal currents at the Earth's core – where the field is created. There has been concern that a magnetic field reversal might diminish the strength of the geomagnetic field, leading to an accompanying decrease in the protection it affords us against harmful **cosmic rays**. However, life on Earth seems to have survived many such reversals in the past. Just as well, since scientists think the next one is due to begin in the next few thousand years.

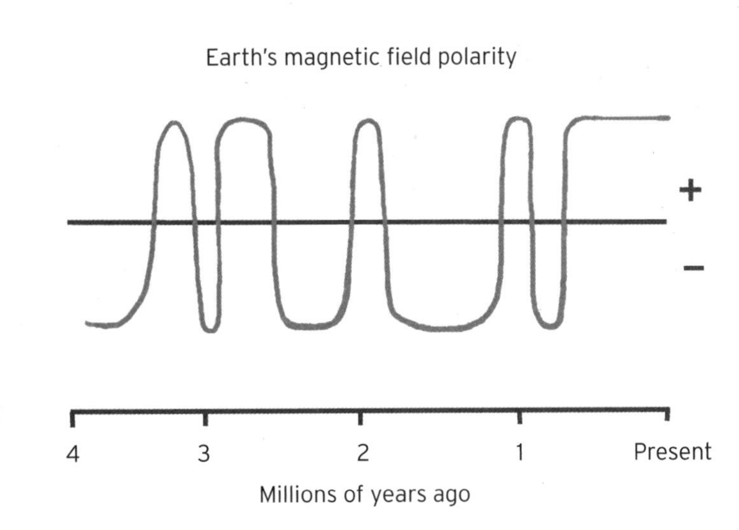

Gaia hypothesis

Gaia was the name given to the Greek goddess of the Earth and the Gaia hypothesis is the theory put forward by prominent British environmentalist James Lovelock that planet Earth, its environment and all its living and non-living elements, behave together as a single giant organism.

Key to the Gaia hypothesis is Lovelock's idea that conditions on Earth regulate themselves to foster the development of life. For example, just as a living animal might regulate its temperature by perspiring, so the temperature of the Earth has remained roughly constant over its lifetime, despite the brightness of the Sun increasing by over 25 per cent. In spite of growing support among environmentalists, many scientists still regard the Gaia hypothesis as controversial. In his 2006 book *The Revenge of Gaia*, Lovelock warned that climate change may ultimately self-regulate itself too – by destroying the human civilization causing it.

SPACE

FROM BACKYARD AMATEUR ASTRONOMY to unpicking the mysteries of the Universe itself, space must be the single most fascinating, most mind-boggling branch of science there is.

The serious scientific study of the Universe began in the year 1609, when Italian polymath Galileo Galilei invented the astronomical telescope, a device as revolutionary to astronomy as the microscope would be to biology. It opened a window on the heavens through which astronomers could accurately measure the motion of distant planets and moons and thus deduce the physical laws governing their orbits. Destroyed was the idea of an Earth-centred Solar System, to be replaced by the Sun-centred view that prevails today.

As telescopes became more powerful, and were ultimately situated in space, above the obscuring haze of the

ANCE • LOOK-BACK TIME • COSMOLOGICAL
LAR SIZE • PARALLAX • MAGNITUDE •
ES • IMAGING • LIGHT POLLUTION • SEEING
TERFEROMETRY • INFRARED ASTRONOMY
COPES • THE SUN • HELIOSEISMOLOGY •
• AURORAE • SOLAR ECLIPSES • LUNAR
AGRANGE POINTS • ASTEROIDS • COMETS
PLANET FORMATION • MERCURY • VENUS
• URANUS • NEPTUNE • DWARF PLANETS
LAR SPACE • NEBULA • STAR FORMATIONS
EQUENCE • BROWN DWARFS • VARIABLE
D GIANTS • PLANETARY NEBULA • WHITE
STARS • HYPERNOVA • INTERGALACTIC

Earth's atmosphere, scientists were able to extend their probing eye even further afield. They analyzed the light from stars to determine what these celestial furnaces are made of. Studying stellar motions revealed the presence of planets in orbit around some stars, just like those orbiting our own Sun. Meanwhile, measuring the brightness and motions of galaxies would start to uncover the workings of the Universe at large.

After centuries of observing, humans are now venturing out into space to experience its wonders first-hand. We've visited Earth orbit and the Moon, and very soon astronauts will turn their eye to a new destination: Mars.

THE NIGHT SKY

Constellations

Most of us can name one or two of the star constellations that decorate the night sky on a clear evening: Orion, the Hunter, with his familiar belt; the pronounced 'W' of Cassiopeia; or the meandering pattern of Ursa Major – the Plough, or Big Dipper. Before telescopes, or other ways to make detailed measurements of the heavens, astronomers categorized the stars into easily identifiable groups, and named them after figures they were thought to resemble – the constellations. They've been with us for millennia; the 2nd-century Greek philosopher Ptolemy makes reference to 48 of them in his book *Almagest.* And a smaller number can be traced even further back, to the 12th-century BC and the Sumerian civilization, which occupied what is now Iraq. Today, there are 88 constellations, including a number of modern additions made during the 16th to 18th centuries, including Fornax, the Furnace, and Antlia – the Air Pump.

Asterisms

Closely related to the constellations are other – unofficial – groupings of stars, known as 'asterisms'. An asterism is an association of stars that form a well-known part of a bigger constellation. Examples include the three stars in the belt of Orion, and the mighty square that dominates the constellation of Pegasus. The stars making up both constellations and asterisms are only grouped together on the 2D celestial sphere – their distances from Earth in 3D space may vary greatly.

Cosmic distance

Here on Earth, we're quite happy using centimetres, metres, kilometres and miles to measure distances but in space these units pale into insignificance. That's why astronomers work with a different set of units. Within the Solar System, the yardstick of choice is the Astronomical Unit (AU), the distance between the Earth and Sun, 150 million kilometres (90 million miles). Venus then orbits the Sun at a distance of 0.7AU, and Jupiter at 5.2AU. But for pacing out distances to stars, this is still too small. Instead, astronomers use the light-year, the distance light can travel in a year – 10,000 billion kilometres (6,000 billion miles).

Look-back time

Measuring cosmic distance in light-years makes it easy to grasp another astronomical concept: look-back time. Light from Earth's nearest star, Proxima Centauri, takes 4.2 years to reach us, so we see it as it was 4.2 years ago. The most distant object visible to the naked eye is the Andromeda Galaxy. That's 3 million light-years away, which means we see it as it was around the time when the first modern humans (genus Homo) were beginning to walk on the Earth.

Cosmological horizons

Our **Universe** is 13.7 billion years old but, perhaps counter-intuitively, this doesn't mean that the furthest objects we can see are 13.7 billion light-years away – as might be expected according to look-back time. This would be the case if the Universe was flat. But on the largest scales, space is curved – in accordance with Einstein's **general relativity**, which distorts distance and means that the furthest objects we can see are actually 46.5 billion light-years distant. This is the distance to our 'cosmological horizon'. It's loosely analogous to the horizon on Earth, in that the Universe continues to exist beyond the horizon – but this is as far as we, in our particular location, can see.

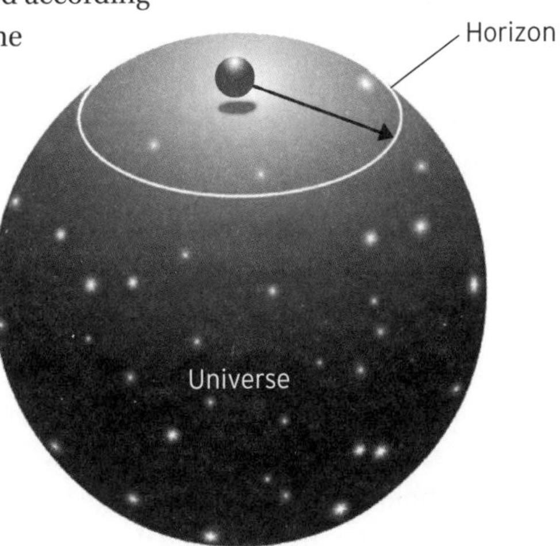

Cosmic coordinates

The night sky forms a sphere around the Earth, and so the natural way to measure positions on it is using degrees of arc. Opposite sides of the sky are 180 degrees from one another. The horizon and the zenith (the point on the sky directly above your head) are separated by 90 degrees.

Astronomers use two main types of angular coordinate system to chart a star's position. The first is altitude-azimuth, or 'altaz', which boils down to measuring two angles – the star's altitude (the angle from zero at the horizon up to 90 degrees at the zenith) and its azimuth (the angle around the horizon with north taken as zero, east as 90 degrees and so on). The trouble with altaz coordinates is that they take no account of the Earth's rotation – so to give an absolute position on the celestial sphere they need to be corrected for the time of day. That's why most astronomers instead use so-called equatorial coordinates: 'right ascension' and 'declination', which record the angular position of a star from a fixed point on the sky – called the First Point of Aries.

Angular size

With a well-established system of cosmic coordinates to map the celestial sphere, astronomers can work out the angular size of objects or the angular distance between them. For instance, the Moon and the Sun are each approximately half a degree across. The Big Dipper in the constellation Ursa Major spans 26 degrees. Using a telescope equipped with a graduated mount it's easy to read off angular coordinates and calculate the angular distances between points.

But for those without telescopes this isn't so simple. Astronomers using the naked eye or binoculars have a rule of thumb – quite literally – for gauging distance on the sky, namely that the width of their thumb held at arm's length spans about two degrees. Similarly, a little finger covers one degree and a full fist about ten dgrees.

Parallax

One way to measure cosmic distance is a technique known as parallax. To get an idea of how it works, place a coffee cup on a table in front of you. Now kneel down so that you're looking on a level with the cup, with your eye-line parallel to the table. If you move your head from side to side, the cup will appear to move relative to objects in the background and the amount of movement decreases the further away the cup is placed. Measure this movement and you can tell how far away the cup is; this is parallax. The same technique can be used to gauge the distance to a star. Only rather than moving their heads from side to side, astronomers measure the apparent shift in the star's position when the Earth is at opposite sides of its orbit around the Sun.

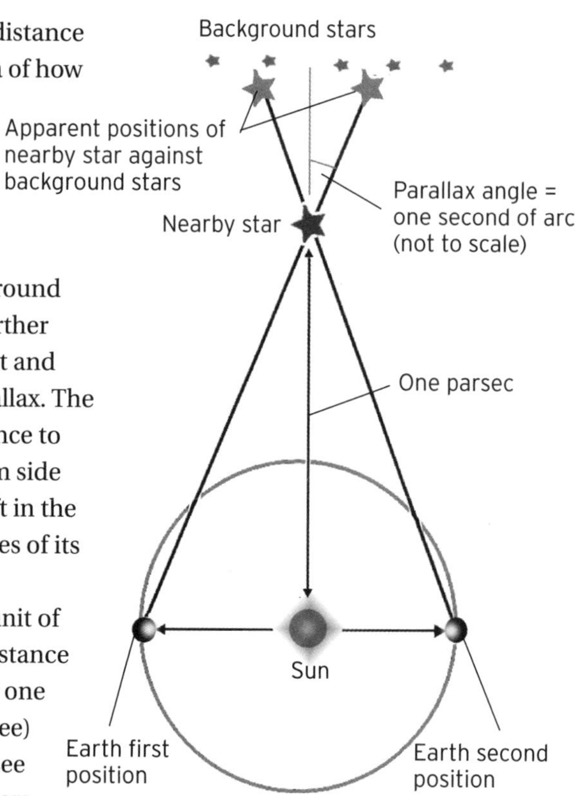

Stellar parallax is used to define a natural unit of astronomical distance: the parsec. A star at a distance of one parsec will appear to shift its position by one 'second' of arc (that's 1/3600 of an angular degree) as the Earth moves by one Astronomical Unit (see **Cosmic distance**). A parsec equals 3.26 light-years.

Magnitude

The brightness of stars and other astronomical objects are measured on what's called the magnitude scale. Magnitude is a number that can be positive or negative; the lower the magnitude, the brighter the object will appear in the sky. For example, the Full Moon has a magnitude of -12.6; Jupiter at its best clocks around -2.9. The faintest objects visible to the naked eye are magnitude 6.5, and faraway Pluto comes in at a very faint 13.7.

The system is constructed so that each full point of magnitude corresponds to a 2.5-fold increase in brightness. That's not quite as mad as it sounds; at the time the magnitude scale was formulated (the mid-19th century) this was the way the human eye was thought to respond to changes in brightness. Astronomers actually deal with two kinds of magnitude – apparent and absolute. An object's absolute magnitude is its brightness as seen from a distance of 10 parsecs; the apparent magnitude, on the other hand, is the observed brightness, as seen from Earth (or wherever the astronomer happens to be). Because objects get dimmer the further away they are, knowing both the absolute and apparent magnitude of a star or galaxy offers another way to calculate its cosmic distance.

Astronomical measurement

The branch of astronomy dealing with the cataloguing of stellar positions is called astrometry. Measuring how position changes with time reveals key information about the motion of stars and the orbits of planets, but there are other fields of astronomical measurement too.

Photometry is concerned with measuring the amount of light coming from a celestial object. Spectroscopy is more complicated, and involves measuring the brightness of individual colours in the rainbow spectrum of the light from a star. Certain **chemical elements** absorb or emit light of characteristic colours, and so respective dips or peaks in the brightness of these colours is a sure indicator of their presence in the star.

ASTRONOMY

Telescopes

The principal tool of the astronomer is the optical telescope, a device using an arrangement of lenses or mirrors, or both, to concentrate light from distant celestial objects at the observer's eye. A telescope's most important specification is its 'aperture', the diameter of its main light-gathering lens or mirror. The bigger the aperture is, the more light the telescope will be able to collect – allowing astronomers to see fainter objects. A typical 100mm-aperture backyard telescope can show objects down to about magnitude 12 – considerably better than the 6.5 you can manage with the naked eye. Meanwhile, a professional-level telescope with a one-metre aperture can see right down to magnitude 17, sensitive enough to pick out faint **galaxies** and **quasars**.

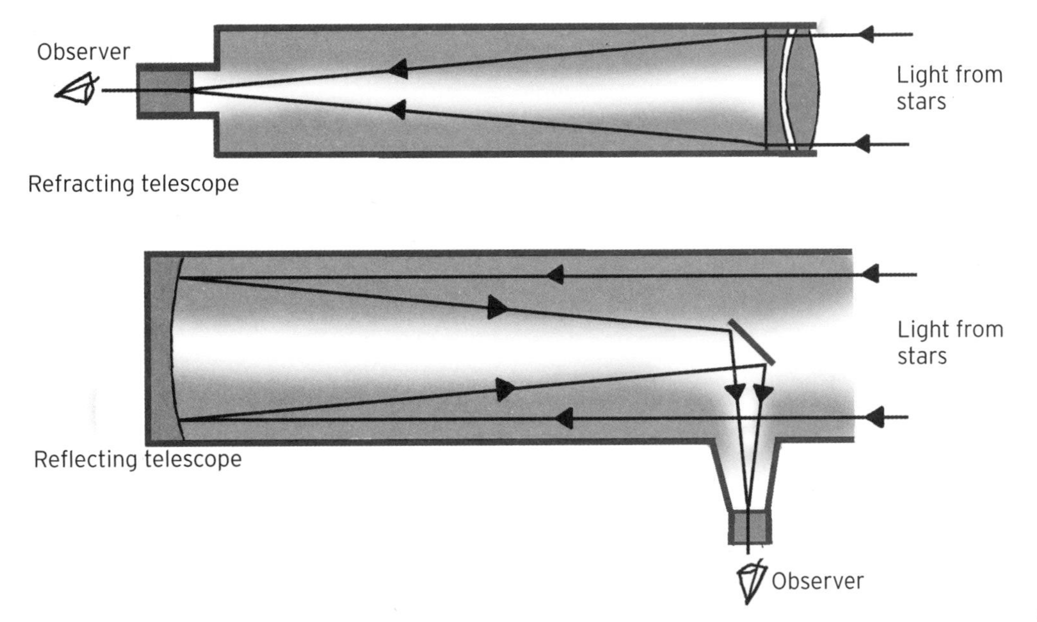

Imaging

Nowadays, it's rare for a professional astronomer to actually put their eye to a telescope. Astronomy research is generally carried out using instruments attached to the telescopes, and perhaps the most common of these are imaging devices – cameras. In the past astroimaging was done using photographic plates, but now both amateur and professional observers record their pictures electronically – using either a consumer digital camera or an astronomical CCD.

CCD is an acronym for 'charge coupled device', a grid of light-sensitive cells that sits in place of the telescope's eyepiece. When the telescope is pointed at, say, a star, the image that would normally be projected into the astronomer's eye is shone instead onto the CCD grid. Each cell in the grid then charges up according to the amount of light falling on it. These charges can then be read out to a computer – the charge in each cell controlling the brightness of the corresponding pixel in the resulting image.

Light pollution

The celestial sphere, observed in the dead of night from an unpopulated area, is a jaw-dropping sight – a teeming mass of stars that simply cannot be seen from an urban setting. The reason the view is so poor from cities is light pollution; light energy from street lamps, car headlights, and other sources must all go somewhere and a large portion of it goes straight up into the sky. Here, it gets scattered by the atmosphere, making the sky glow. The faint stars and cosmic clouds that look so beautiful from the country are simply lost in the glare. Most professional astronomical observatories are situated away from population centres for the purpose of overcoming light pollution.

Seeing

It's not just light pollution that plagues astronomers – another problem is known as atmospheric 'seeing'. This is caused by **turbulence** making the view through the atmosphere move and shimmer – as if you're looking through the heat haze above a fire. It's the turbulence that causes stars to 'twinkle'. Take a look next time you're out on a clear night. It will be the stars low down in the sky that are twinkling the most, because their light has travelled through more of the Earth's turbulent atmosphere than those overhead.

One innovative technique to combat seeing was developed in the 1990s, and is called 'adaptive optics'. A network of servomotors is positioned beneath a reflecting telescope's main mirror. A laser beam is fired up into the sky and the way the beam gets distorted by atmospheric turbulence is measured and fed back to the servomotors, which then deform the mirror in just the right way to correct the distortion in the telescope's images.

Space telescopes

A somewhat brasher solution to the issues of light pollution and seeing is to bypass the Earth's troublesome atmosphere altogether and put your telescope in space. American astronomer Lyman Spitzer was one of the first to seriously investigate this idea, as far back as 1946. The American space agency NASA began researching space telescopes in the 1960s and Spitzer was involved with the development. It took until 1990 for what is now known as the Hubble Space Telescope (HST) to reach orbit.

And yet it would still be several years before the observatory was working as advertised. An **aberration** in the main mirror meant that the optics were flawed – and a set of corrective 'spectacles' had to be built and installed. Hubble has been a workhorse of astronomy ever since, delivering discovery after discovery.

Radio telescopes

Celestial objects don't just give off light; the radiation coming from the heavens encompasses virtually every region of the **electromagnetic radiation** spectrum. At the long-wavelength end, cosmic radio waves have brought momentous discoveries about galaxies, the death throes of stars and the origin of the Universe itself.

Modern radio telescopes consist of a reflective dish that concentrates the waves onto a radio receiver placed at the focal point. The dish is analogous to the mirror in an optical reflecting telescope. Only, because the wavelength of radio is very much longer than that of light (tens of metres for radio, compared with less than a millionth of a metre for light), radio dishes need to be much larger than their optical counterparts. The world's biggest radio telescope – the Arecibo dish, in Puerto Rico – spans 305m (1,000ft), nearly a third of a kilometre across.

Interferometry

Large-aperture telescopes don't just gather more light than their small-bore brethren. The bigger the aperture size, the better the 'resolution' of the telescope – that is, its ability to show fine details, for example craters on the Moon or gaps in the rings of Saturn. But there is a way to cheat, and get high-resolution images without a large telescope using a technique called interferometry. It works by combining the light from two smaller telescopes separated by a large distance – call it 'D' – to produce images with the

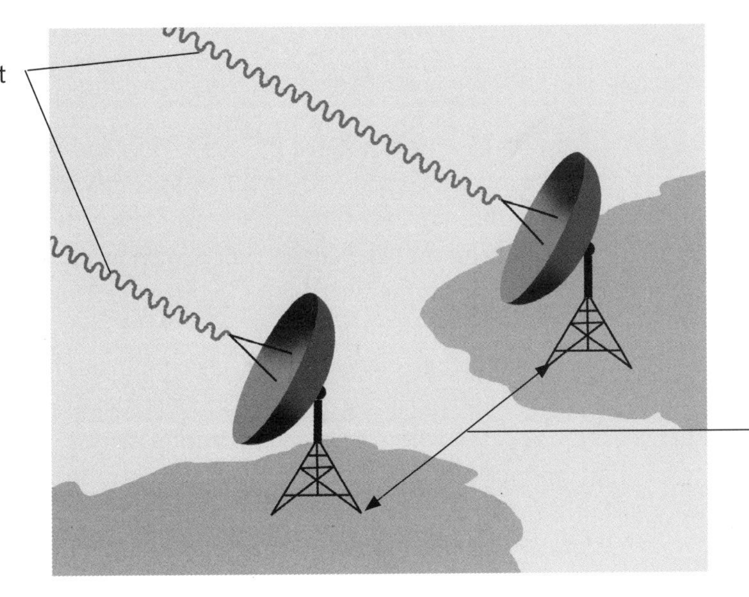

resolution that you would expect from a single telescope with an aperture size 'D'.

Interferometry was pioneered for radio astronomy, where the long wavelength of the waves meant that a huge radio dish was normally needed to achieve good resolution. But the technique has since been hijacked for optical telescopes too – most notably the European Southern Observatory's Very Large Telescope (VLT) in Chile.

Infrared astronomy

In 2003 NASA launched the Spitzer Space Telescope – named after American astronomer Lyman Spitzer. Whereas the Hubble Space Telescope had benefited from amazing clarity of view, the Spitzer telescope demonstrated one of the other great advantages of space observatories. Earth's atmosphere is opaque to certain wavelengths of **electromagnetic radiation**: in particular, a large portion of the infrared spectrum gets absorbed before it can reach the ground, and so can be seen only from space. That what Spitzer does. Other parts of the spectrum that can only be scrutinized from space include X-rays, gamma rays and most of the ultraviolet spectrum. And there are also space telescopes to observe these.

Cosmic rays

Not all the radiation coming from space is electromagnetic in nature. Earth's upper atmosphere also gets battered by particles known, perhaps ironically, as cosmic rays, consisting primarily of protons and **ions** that have been accelerated by **supernova** blast waves, active galaxies and even our own Sun. They pack huge amounts of energy – a single cosmic-ray particle can strike the atmosphere with the same energy as a fast tennis serve.

Although the primary ray particle itself won't make it to the ground, its impact with the atmosphere generates a cascade of fragments – billions of secondary particles that rain down. So-called 'air shower observatories', networks of particle detectors covering large areas of land, can spot the cascade and tell scientists the energy and trajectory of the cosmic ray that caused it.

Extremely Large Telescopes

The biggest optical telescopes at the surface of the Earth today are giant reflecting instruments with primary mirrors around 10 metres across. But there is a new generation of telescopes now on the drawing board that will make these behemoths like midgets. Not named in jest, the 'extremely large telescopes' (ELTs) will have main mirrors up to 40 metres in diameter. The mirrors are segmented, made up from hundreds of interlocking hexagons, each just a metre or two across, not only making them easier to support but also allowing each segment to be controlled separately so the telescope can benefit from adaptive optics to counteract atmospheric **seeing**. Calculations suggest that a telescope of this sort will be able to capture images of the heavens many times more detailed than those taken by space telescopes. The most advanced ELT project is the 24.5-metre Giant Magellan Telescope, expected to begin operations in 2018.

SOLAR SYSTEM

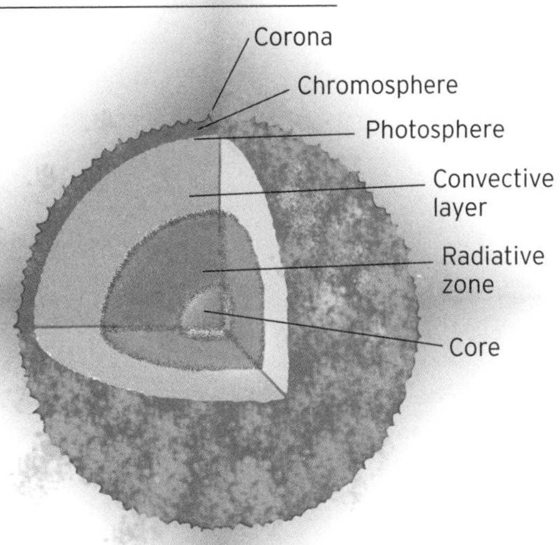

The Sun

Over 100 times the size of the Earth and more than 300,000 times its weight, the Sun is the powerhouse at the centre of the Solar System, cranking out energy at the rate of 4×10^{26} watts (in **scientific notation**) – enough to light 4 million billion billion light bulbs. It derives its energy from fusion **nuclear reactions** taking place in its core; temperatures in the core reach a staggering 15 million°C (2 million°F), hot enough to fuse atoms of hydrogen together to form helium.

The Sun has a complex internal structure, arranged like layers within an onion. Outside the core is the 'radiative zone' within which energy flows outwards as **photons**. Beyond this is the 'convective layer' where the Sun switches to a different method of heat transport: convection. Then there is the photosphere, where the Sun becomes transparent and its energy is free to stream out into space as heat and light. Outside this is a red-coloured layer called the chromosphere. Finally there is the corona, the pearly white atmosphere of plasma (see **Plasma physics**) where temperatures can exceed 5 million°C (1.8 million°F).

Helioseismology

Observing the interior properties of the Sun would be extremely difficult were it not for the field helioseismology, the study of sound waves propagating on the solar surface. Just a like a bell that's been struck, the surface of the Sun rings when it's disturbed and the properties of the 'sound', such as its frequency and loudness, reveal key details about the underlying structure.

Astronomers make helioseismological observations of the Sun by measuring its undulating surface and then plotting the shape of the waves in a computer, from where they can be studied. The applications of the technique are manifold – enabling astronomers to determine the Sun's chemical composition, temperature and pressure, density and interior motion, and even to study sunspots on the far side. The field is closely related to asteroseismology – investigating the internal structure of distant stars by measuring the frequency of vibrations on their surfaces using powerful telescopes.

Sunspots

Sunspots are relatively cool regions on the Sun's surface, where the temperature dips from 5,500°C (9,930°F) to around 4,000°C (7,232°F). They are caused by irregularities in the Sun's magnetism that disrupt its convective layer in small areas. Blocking convection lowers the amount of energy passing through each of these small areas and causes them to cool down temporarily. Sunspots usually appear in small groups which each last for a few weeks. The number of sunspots visible at any time fluctuates over an 11-year cycle.

Counter-intuitively, the presence of cool sunspots is a sign that the Sun is heating up, because the magnetic fluctuations that cause the spots also generate flares and other kinds of intense solar activity. During a 70-year period at the end of the 17th century, sunspot numbers reached a record low. Called the Maunder Minimum, it coincided with a period of extreme cold on Earth, known as the **Little Ice Age**.

Solar activity

Variations in the Sun's **magnetism** can lead to a host of violent effects on the solar surface which astronomers group together as 'solar activity'. Most common are solar flares and coronal mass ejections (CMEs). Solar flares are sudden and intense brightenings of small regions on the Sun's surface and they occur when energy built up in the Sun's tangled magnetic fields is suddenly released. A single flare can give off the same energy as millions of hydrogen bombs.

A CME, on the other hand, is a belch of charged particles cast out by magnetic activity from the Sun's superheated corona over the course of a few hours. During a typical event, the corona will disgorge 1,000 billion kg of material. CMEs often accompany flares but also occur as isolated events. The bursts of radiation that accompany a flare or a CME can cook electronic equipment onboard spacecraft and even here on Earth. More seriously, this radiation poses a lethal hazard to astronauts on interplanetary missions – outside the protective bubble of Earth's magnetic field. Monitoring and forecasting solar activity, and its effect on near-Earth space, goes under the umbrella term of 'space weather'.

Solar wind

Even when the Sun isn't especially active, there is a steady flow of particles streaming outwards into space from its corona – called the solar wind. It exists because the solar corona is so hot. According to the **kinetic theory** of gases, the warmer a gas is, the faster its constituent particles are moving. The corona is so hot that a small fraction of particles in the plasma it's made up of – typically electrons, protons and ionized atoms – will be moving faster than the Sun's escape velocity.

The solar wind travels out from the corona at a speed of 400km/s (almost 1 million mph) and its effects can be seen on comets' tails (making them 'blow' away from the Sun) and in the formation of spectacular auroras on Earth.

Auroras

Solar wind particles slamming into Earth's upper atmosphere produce the breath-taking lightshows known as the Northern and Southern Lights – the auroras. The solar wind carries with it some of the Sun's magnetic field. When conditions are just right, this field can hook up with the magnetic field of the Earth, funnelling high-speed solar wind particles down onto our planet's magnetic poles. These particles then collide with atoms of gas in the atmosphere, causing electrons in the atoms soak up energy from the wind particles, raising them up to a higher energy level. Over time the electrons drop back down, giving off **electromagnetic radiation** in the form of light with a characteristic colour, depending on the kind of gas involved. For example, oxygen atoms give off red and green light while nitrogen gives off blue and purple. Auroras don't just happen on Earth; astronomers have also observed them on the gas-giant worlds of Jupiter and Saturn.

Solar eclipses

A total eclipse of the Sun is perhaps the most spectacular event in space that anyone can hope to see from Earth (though take care to use the correct eye protection – observing the Sun with the naked eye can damage your eyesight). The Moon crosses directly between the Earth and the Sun, and casts a shadow that sweeps across the Earth's surface. Anyone in the shadow's path will see the Sun's face gradually eaten away by the Moon's dark disc, leaving just the ghostly glow of the solar corona. Observers not quite in the shadow's path may still see a 'partial eclipse', where just part of the Sun's disc is obscured.

A total eclipse is visible somewhere on Earth roughly once every 18 months. We are extremely fortunate that the angular size of the Sun and the Moon as seen from Earth are identical or such a perfect eclipse wouldn't be possible.

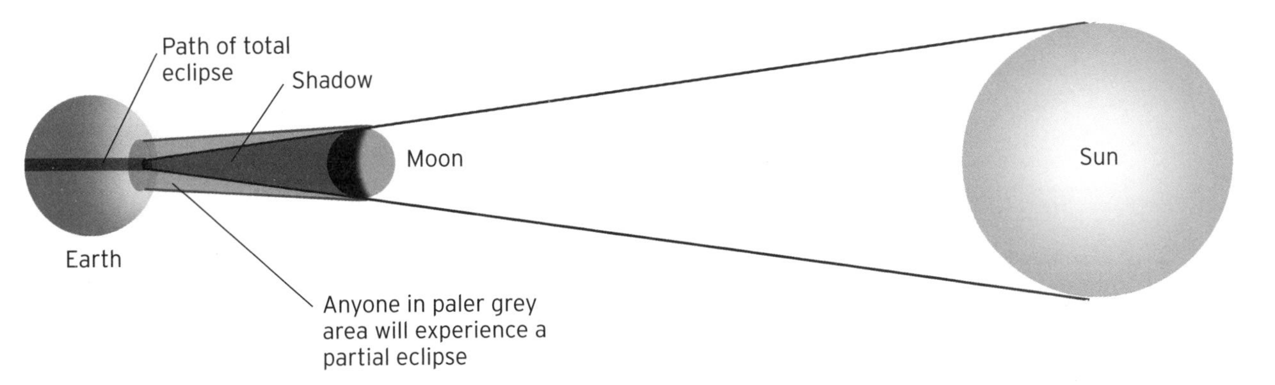

Lunar eclipses

When the Earth passes between the Sun and the Moon, astronomers at the right place on the planet can observe the Earth's shadow cut across the lunar surface – a spectacle known as a lunar eclipse. There are both partial and total lunar eclipses – depending how much of the Moon's surface falls into shadow. The darkness level of a total lunar eclipse can also vary greatly; sometimes the Moon will be virtually invisible during totality, while at other times it simply assumes a coppery-red hue – a colouring caused by sunlight that has passed through the Earth's atmosphere. There is a great deal of dust floating in our planet's atmosphere which scatters away blue light, leaving just reds and orange. We see the same effect when the Sun is low in the sky – for example, at sunset.

Transits

A transit takes place when a planet passes in front of the Sun. Observers see the planet's dark silhouette as a black dot creeping across the solar surface. Because the transiting object has to pass *between* the observer and the Sun, we on Earth can only hope to see transits of Venus and Mercury, but a future astronomer on Mars could also see transits of Earth.

The word 'transit' is sometimes also used in astronomy to describe the crossing of a planet or moon in front of celestial objects other than the Sun. So it is possible to watch transits of Jupiter's moon Io across the face of the giant planet, for example. If the object in transit is bigger than the object it's transiting – so that the more distant object is completely hidden – then the event is referred to as an 'occultation'. So, shortly after Io has transited Jupiter's face it will be occulted as its orbit takes it behind the planet.

Titius–Bode law

A mathematical formula for predicting the orbital radii of the planets in the Solar System is the Titius–Bode law. An empirical formula, meaning it has no basis in the laws of physics, it has simply been constructed to fit the observed data. The modern formulation of the law says that the distance of each planet from the Sun, measured in AU (see **Cosmic distance**), is given by $0.4 + (0.3 \times k)$, where k is a whole-number. If k is allowed to be the sequence of powers of the number 2 – 0, 1, 2, 4, 8, 16, 32, and so on, then it accurately gives the orbital radius of every planet (with the exception of Neptune, which is nearly 9AU closer to the Sun than the law predicts).

The law – named after German astronomers Johann Daniel Titius and Johann Elert Bode – was first published in 1768, 13 years before the discovery of Uranus, the orbit of which it predicts correctly.

Planet	*k*	Titius–Bode law orbit (AU)	Actual orbit (AU)
Mercury	0	0.40	0.39
Venus	1	0.70	0.72
Earth	2	1.00	1.00
Mars	4	1.60	1.52
asteroid belt	8	2.80	2.80
Jupiter	16	5.20	5.20
Saturn	32	10.0	9.54
Uranus	64	19.6	19.2
Neptune	-	-	30.1

Lagrange points

It is a relatively simple matter to use **Newtonian gravity** to calculate the gravitational field produced by a single star or planet but what happens when there are two gravitational sources? This is a complex problem but the Italian–French mathematician Joseph Louis Lagrange worked out the answer in 1772. A key feature of his solution was the existence of five points in space where the gravitational attraction of the two sources to some extent cancels out. For example, consider the Earth–Sun system. An object placed on the line connecting the Earth and Sun feels the force of the two bodies pulling it in opposite directions. The object's exact position is crucial – too close to the Earth and it will fall towards the planet, too close to the Sun and it will fall that way instead. But in between is a spot where the object will stay at the same distance from the Sun, orbiting in lockstep with the Earth. This is known as the first Lagrange point, or L1 for short.

Another four of these points exist. Labelled L2 – L5, they form a cross shape through the Earth-Sun system. L1 – L3 are unstable. In other words, shove an object sitting at any of these points and it will carry on moving away. But L4 and L5 are stable. Populations of asteroids often lie at L4 and L5 – for example, the Trojans in the Jupiter–Sun system.

Asteroids

Rocky bodies that prowl the Solar System are known as asteroids. Too small to be classed as planets, they range from boulders just tens of metres across to flying mountains spanning hundreds of kilometres. Millions of them can be found in the main asteroid belt, which circles the Sun between Mars and Jupiter; but other populations also exist. Trojan asteroids gather at the L4 and L5 Lagrange points of Jupiter. Meanwhile, Vulcanoids are a hypothetical asteroid population thought to skirt the Sun inside the orbit of Mercury. Some asteroids have even been found to have tiny misshapen moons in orbit around them. The first of these was Dactyl, the moon of main-belt asteroid Ida; it measures just over a kilometre across and was discovered in 1994 by NASA's Galileo **planetary probe**.

Then there are Earth-crossing asteroids, those whose trajectories take them across the orbit of the Earth. These objects pose a danger to our planet; 65 million years ago, a 10km (6 miles) long asteroid (or possibly a comet) is believed to have struck Mexico's Yucatán peninsula. The blast, resulting tidal waves and environmental damage are generally believed to have brought about the extinction of the dinosaurs. For this reason, a number of astronomical observatories around the world run projects to detect and catalogue Earth-crossing asteroids. Operating under the umbrella name of Spaceguard, they aim to ultimately track 90 per cent of asteroids bigger than 1km (0.6 miles) in size.

Comets

To complement the rocky asteroids, the Solar System is also home to a class of icy wanderers known as comets, which vary in size between 100 metres and a few tens of kilometres. They are divided between short-period comets, orbiting the Sun once every 200 years or less, long-period comets which orbit less frequently, and single-apparition comets which make just one passage through the inner Solar System before disappearing back out into space never to be seen again.

When a comet enters the inner Solar System the heat of the Sun starts to evaporate its icy surface, creating both a nebulous 'coma' surrounding the comet and a stream of particles that sweep out behind it as a spectacular tail. In fact, most comets have two tails – a gas tail, which is blown directly away from the Sun by the solar wind, and a dust tail pointing somewhere between the gas tail and the direction the comet has come from.

Kuiper belt

The Kuiper belt is a disc of icy planetoids orbiting the Sun out beyond Neptune and named after Dutch-American astronomer Gerard Kuiper, who postulated its existence in 1951. Astronomers in Hawaii in 1992 discovered the first Kuiper belt object (KBO) – or rather, the first to be recognized as such. Most astronomers now consider Pluto, discovered in 1930, to be a large KBO, and indeed this realization led to Pluto and the belt's other largest members being reclassified as **dwarf planets** in 2006. Today over 70,000 KBOs are known.

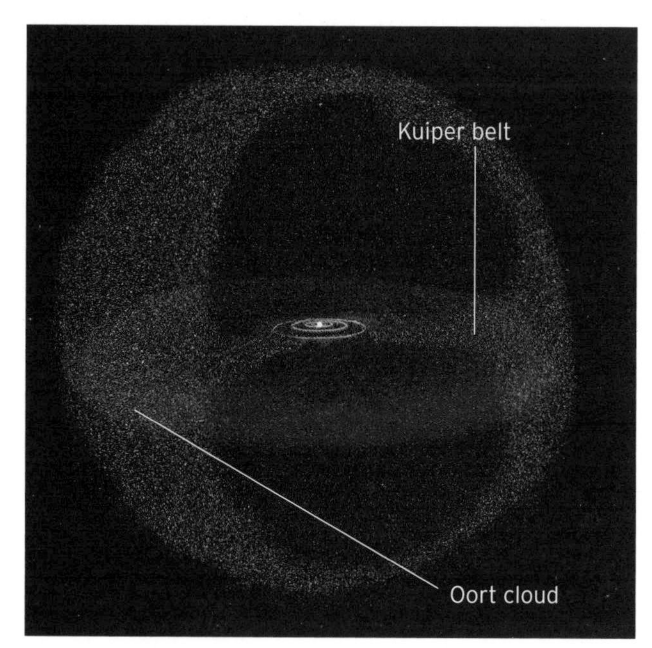

The gravity of the giant planets, mainly that of nearby Neptune, occasionally yanks KBOs out of the Kuiper belt and flings them into orbits arcing high above the Solar System's mid-plane. These 'scattered disc objects' are thought to be the primary source of short-period comets. Long-period and single-apparition comets originate even further out in a swarm known as the Oort Cloud which surrounds the Solar System at a distance of 50,000AU.

Meteors

Small chunks of material in space, measuring anywhere between the size of a dust particle and a small boulder, are known as 'meteoroids'. Amazingly, millions of them slam into Earth's atmosphere and burn up every day, creating short-lived bright streaks across the night sky – called shooting stars, or meteors. Occasionally a meteor is too big to burn up entirely, and some of it reaches the ground below – this is a 'meteorite'.

Meteors come in two varieties – sporadic and showers. Sporadic meteors occur randomly, whereas showers are caused as the Earth's orbit carries it through well-known streams of dusty debris that have been left behind by comets. For example, the Perseid meteor shower that takes places every August is caused by material shed in the wake of Comet Swift-Tuttle.

Heliosphere

The solar wind carves out a giant bubble surrounding the whole Solar System, which protects it from the rigours of **interstellar space**. This is called the heliosphere. The stream of charged solar wind particles and their magnetic field serve to bat away high-energy **cosmic rays,** as well as fast-moving wind particles from other stars.

There are three principal components to the outer boundary of the heliosphere: the 'termination shock' is where the solar wind begins to run out of steam, dropping from supersonic

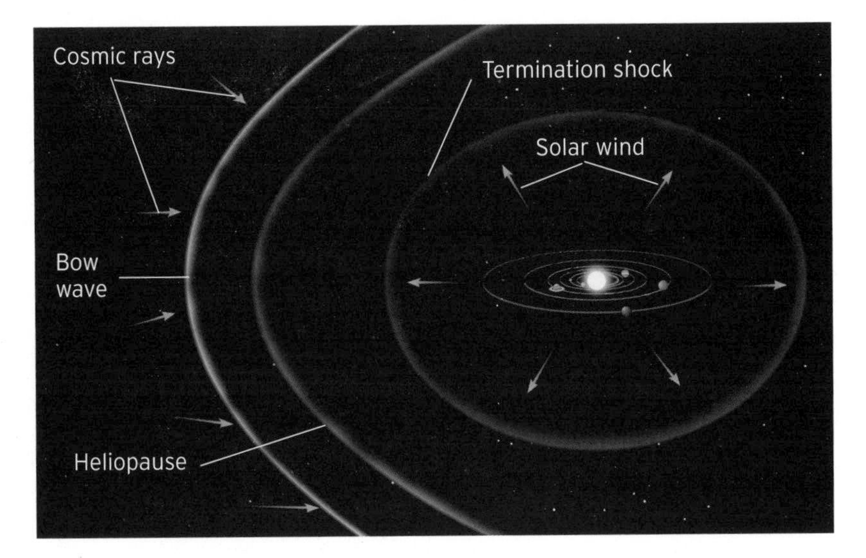

to subsonic speed; where the wind collides with the matter of the interstellar medium is called the 'heliopause'; and finally there's the 'bow shock', a pressure wave preceding the Solar System's motion through the galaxy – rather like the bow wave in front of a ship.

PLANETS

Planet formation

Some 4.6 billion years ago the Sun and Solar System are believed to have condensed from a giant molecular cloud, a vast cloud of hydrogen molecules, which collapsed under its own gravity. It would have been rotating ever so slightly and speeded up as the cloud got smaller. The centrifugal force generated by the rotation eventually halted the cloud's collapse in two of its three dimensions to create a flat, spinning 'protoplanetary disc'. As the embryonic Sun continued to grow at the centre of the disc so planets began to take shape within it as dust particles collided and stuck together, formed rocks and then boulders and then grew further by their gravity.

Close to the Sun, with all gases blasted away by the young star's heat, the rocky terrestrial planets formed. But further out the temperature was cooler; gases and icy chunks were able to coexist and merge, and the giant planets grew true to their name. This outline for the formation of the Solar System is known as the solar nebular theory. Astronomers see the same process taking place in planet-forming clouds many light-years away.

Mercury

The innermost planet of the Solar System, Mercury, dashes round the Sun once every 88 days. It is a baked world, with daytime temperatures reaching as high as 430°C

(806°F), and yet because Mercury has no atmosphere to speak of, it cannot retain this heat, meaning night-time temperatures are bitterly cold, plunging to -170°C (-274°F). Mercury's surface resembles that of the Moon – a bleak and pocked landscape. It was first photographed close-up by the Mariner 10 space mission in the mid-1970s, and more recently by the MESSENGER probe in 2008.

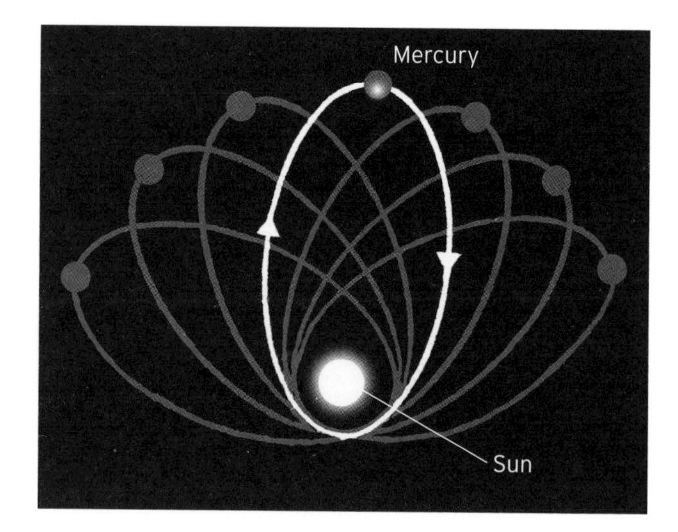

Mercury's orbit provided one of the early experimental tests for Einstein's theory of **general relativity**. Astronomers in the 19th century had noticed that as Mercury circles the Sun, the overall shape of its elliptical orbit also rotates, so that over time the planet's path through space traces out a petal-like pattern. **Newtonian gravity** was unable to explain this so-called 'perihelion precession', but calculations using Einstein's theory matched the observations exactly.

Venus

Venus is often described as the Earth's sister world, being four-fifths the weight and almost exactly the same size as our planet. Many scientists believe the similarities run deeper still, and that Venus once had a temperate climate with liquid water on its surface. The reality today is very different though – present-day Venus is swathed in a thick atmosphere of carbon dioxide creating a runaway **greenhouse effect**, trapping radiation from the Sun and warming the planet's surface to a blistering 460°C (860°F). That's hot enough to melt lead and, indeed, even hotter than it gets on Mercury – despite Venus being further from the Sun. The planet's surface pressure is 93 times that of Earth – comparable with our deepest ocean trenches.

Even so, NASA scientist David Grinspoon has speculated that there could be life on Venus. High in the planet's cloud decks, some 50km (31 miles) up, the temperature and pressure drop to Earthly levels. In fact, while the atmosphere remains toxic – and you'd need a mask to breathe – it's the only place in the Solar System beyond Earth where you wouldn't need a spacesuit.

Earth

Our planet is the third in line from the Sun. As the world that we know best, Earth's surface temperature, pressure, diameter (12,700km/7,890 miles) and mass (6×10^{24} kg), and the length of our day and year, have become the benchmarks against which other planets are gauged. The Latin name for Earth is 'Terra', and planets similar to the Earth – including Mercury, Venus and Mars – are known as terrestrial worlds.

Earth is the only place in the Solar System where we know for sure that there's life. Our planet's habitability is down to a host of factors being 'just right'; the surface temperature is in the correct range for water to exist as a liquid, it has a moon that stabilizes the planet's rotation, in turn stabilizing the environment, and the planet has a magnetic field that fends off harmful radiation from space.

Moon

Orbiting at a distance of around 610,000km (380,000 miles) the Moon has a diameter of just over a quarter of the Earth's and gravity about one-sixth of what you're feeling as you read this. The surface terrain is split between bright, rugged highlands and the dark, smooth maria – the lunar 'seas', large areas of solidified magma. Our Moon is believed to have been formed during a colossal impact event called the Big Splash. In this scenario, a Mars-sized body struck the Earth a glancing blow around 4.5 billion years ago and the Moon coalesced from the debris.

When it was born, the Moon was spinning rapidly; the Earth's gravity has since put the brakes on this rapid rotation, giving us a Moon that is 'tidally locked' – keeping the same face pointing towards our planet at all times. That is why when the Moon's full – that is, directly opposite the Earth from the Sun – you always see the same familiar 'man in the Moon' markings.

Mars

Mars circles the Sun at a distance of around 1.5 AU (see **Cosmic distance**). Known colloquially as the Red Planet, its crimson hue is caused by the large proportion of iron in its soil and surface rocks – the planet is quite literally rusty. It is just over half the diameter of the Earth and has one-third as much gravity at its surface. Mars has an active climate with seasons, and weather phenomena such as dust storms and tornado-like 'dust devils' have been photographed. The planet has two orbiting moons: Phobos and Deimos.

NASA's Phoenix Mars lander spacecraft found water-ice on the planet's surface when it touched down in 2008. The planet's low temperature and pressure make it unlikely liquid water exists there today, though chemical signatures and erosion marks on the surface suggest that Mars was once a damp place in the past, fuelling the debate over the existence of past or even present life on Mars. However, Mars's cold and arid environment means nothing more advanced than microbes could realistically survive there now.

Jupiter

The giant of the Solar System, Jupiter is ten times the size of the Earth and weighs over 300 times as much. It is a 'gas giant' planet – structurally very different from the more Earth-like 'terrestrial planets' that orbit closer to the Sun, and composed mostly of the gases hydrogen and helium.

Jupiter is perhaps most famous for its Great Red Spot – a swirling storm system big enough to swallow up several planet Earths – and smaller storms are also seen raging within the planet's counter-rotating belts of gas. The planet has 63 known moons; the four biggest ones – Io, Europa, Callisto and Ganymede – are known as the Galilean moons, as they were discovered by Italian polymath Galileo using one of the first astronomical telescopes 400 years ago. They can be seen from Earth through binoculars as pinpricks of light straddling the planet. Io is the innermost of the four; orbiting so close to the giant planet, it is subjected to constant squashing and squeezing by the tidal forces of Jupiter's gravity, producing a heating effect that makes Io the most volcanic place in the Solar System, with over 400 active **volcanoes** on its surface.

Saturn

The sixth planet out from the Sun – and the second largest, after Jupiter – is the gas giant world Saturn, orbiting the Sun at a distance of 10AU (see **Cosmic distance**). It is mostly made up of the gases hydrogen and helium.

Saturn's spectacular system of rings, 360,000km (223,694 miles) in diameter and just 20km (12.4 miles) thick, appear like a knife edge at Saturn's great distance; the rings are composed of small icy chunks in orbit around the planet. There are 61 known moons of Saturn. Titan is the largest at just under half the size of the Earth and shrouded by a dense nitrogen atmosphere. There has been much speculation about what lies beneath Titan's murky haze – including lakes of liquid methane and even life. When the European Space Agency's Huygens probe landed there in 2004 it found neither, though it did see evidence that rocks near the landing site had been eroded by liquids in the past.

Uranus

Uranus was discovered by the British astronomer Sir William Herschel in 1781. Like Jupiter and Saturn, it is also a gas giant, though, unlike those worlds, its hydrogen-helium atmosphere is supplemented with a proportion of methane, which absorbs the red component in the light from the Sun, giving the planet its pale blue colour.

Uranus is an awfully long way away – 20 AU (see **Cosmic distance**). At -224°C (-371°F), it is also the coldest planet in the Solar System – colder even than its outer neighbour Neptune – though the reason for this is not well understood. The low temperature means there is insufficient energy in the planet's atmosphere to support anything like the swirling weather systems or coloured bands that cross the faces of Jupiter and Saturn. Perhaps the strangest feature of Uranus is that its rotation axis is tipped over by 98°, making it orbit the Sun 'on its side'. Because of this and the fact that the planet takes 84 years to complete a single orbit of the Sun, its north and south poles each experience 42 years of continuous light followed by 42 years of darkness.

Neptune

By the time you venture 30 times further than the Earth is from the Sun, the Solar System is beginning to get very dark and chilly. But that's where eighth planet, Neptune, can be found. It was discovered in 1846 by German astronomer Johann Gottfried Galle, acting on calculations made by French mathematician Urbain Le Verrier, who had deduced that tiny perturbations observed in the orbit of Uranus were due to the gravity of a new world.

Neptune is nearly four times the diameter of the Earth and of similar size and chemical composition to Uranus. Unlike Uranus, though, Neptune has activity visible on its surface, including several large, dark spots, thought to be **hurricane**-like storms similar to Jupiter's Great Red Spot. Neptune presents a problem for models of planet formation – which currently suggest that there would have been insufficient material this far from the Sun to create such a large planet. One possible solution is a theory known as 'migration', which argues that Neptune would have formed nearer to the Sun, where the density of matter was greater, and then migrated outwards to its present position.

Dwarf planets

Many of us were taught that Pluto is the Sun's ninth 'planet', but in 2006, the International Astronomical Union – the world governing body of astronomy – tightened its definition of the word. Under the revised scheme, Pluto became one of a new group of objects known as 'dwarf planets'. The rethink was motivated by the discovery of new bodies in the Solar System's outer reaches larger than Pluto. If Pluto was a planet, then so were these new discoveries – and there could be many of them out there.

There are currently five dwarf planets known. Pluto, Haumea, Makemake and Eris are chunks of rock and ice orbiting out beyond Neptune. The fifth, Ceres, was formerly the Solar System's largest asteroid, lying in the main belt between Mars and Jupiter. Before the reclassification, Eris – at the time carrying the informal name 'Xena' – was regarded by many astronomers as the tenth planet. Pluto and Eris each have a single moon, called Charon and Dysnomia respectively. Haumea has two: Hi'iaka and Namaka; Ceres and Makemake have none.

Exoplanets

It's an amazing fact that we now know of more planets orbiting other stars than we do in our own Solar System. At the time of writing, there are 374 of these so-called exoplanets known. The first was detected in 1995 by Michel Mayor and Didier Queloz, at the University of Geneva, Switzerland. They discovered a planet-sized companion orbiting the star 51 Pegasi, 50 light-years away in the constellation Pegasus. The planet was too small to observe directly, but they were able to infer its presence by measuring the star's motion through space. As the unseen planet orbits it causes the star to wobble back and forth ever-so-slightly, and sensitive equipment enabled them to measure the wobbles. The '51 Peg' planet is a gas giant world, like Jupiter, but recently planet hunters have turned up terrestrial worlds – with masses just a few times larger than our own, and which reside in their star's 'Goldilocks zone', the range of orbits for which the surface temperature is just right for liquid water to exist.

Rogue planets

Perhaps more bizarre than **exoplanets** is the possibility that there may exist a population of planets roaming freely through deep space – not bound to any star. Theoretical models predict that a small number of these rogue planets should have been flicked out of their host star systems as matter clashed together during the violent process of **planet formation**. However, although candidate rogue planets have been found by astronomers, there is yet to be a confirmed detection.

Professor David Stevenson, at the California Institute of Technology, believes that if rogue planets are out there they could even bear life. He says these worlds would have been formed when their star systems were rich in hydrogen gas and this would have given them thick atmospheres capable of trapping heat, making them warm enough to sustain liquid water oceans and a temperate climate, despite lacking the warmth of a nearby star.

STARS

Interstellar space

Outside the protective bubble of our Solar System lies the harsh cosmic wilderness that is interstellar space. Extremely empty, punctuated by rarefied clouds of gas and dust – the densities of which are so low they are measured not in grams, but atoms per cubic centimetre – the principle components of interstellar space are hydrogen clouds. These come in three forms: neutral clouds of hydrogen are made of simple hydrogen atoms; molecular clouds comprise molecules, each made from two hydrogen atoms bonded together; while so-called H II regions are created where radiation from a nearby star has ripped the electrons from hydrogen atoms to leave a cloud of positively charged hydrogen **ions**. In addition, these all come with a smattering of cosmic dust. It is thought these clouds account for around 15 per cent of the mass of a galaxy like the Milky Way.

Nebula

Clouds of cosmic gas visible from Earth are known to astronomers as nebulas. Broadly speaking, there are three types. 'Emission nebulas' are gas clouds that give off their own light, usually because the atoms of gas in the nebula have lost one or more electrons to become ions and energy is released as the ions and electrons gradually recombine. H II regions seen in interstellar space are examples of emission nebulas. A 'reflection nebula', meanwhile, is a cloud of dust that generates no light of its own but instead reflects that of nearby stars. Finally, there are 'dark nebulas', conspicuous not by the light they emit or reflect, but by the light from sources behind them which they blot out. Perhaps most famous is the Horsehead in the constellation of Orion.

Star formations

Stars form inside cold molecular clouds in interstellar space. The initial density of the cloud is in the range of a few to a hundred particles per cubic centimetre (compare with Earth's atmosphere where each cubic centimetre of which holds 30 billion billion particles). However, because the biggest clouds can span hundreds of light-years molecular clouds are still home to a great deal of material – hundreds of thousands of solar masses. Tiny density irregularities within a cloud act as the seeds for regions to collapse under their own gravity and as the gravitational attraction of these regions pulls more matter in, so their density goes up and their gravity increases further. It's a runaway process.

When one of these 'protostars' condenses, the gas inside is squashed, causing the temperature to rise, until the temperature in the core reaches around 15 million°C (27 million°F), then fusion **nuclear reactions** ignite and a star is born. Radiation pouring from the new star blows a bubble around it in the surrounding molecular cloud. The total time needed to form a star this way (like our Sun) is around 50 million years.

Hertzsprung–Russell diagram

Astronomers map out the evolution of stars on what's called a Hertzsprung–Russell (H–R) diagram, after the two astronomers who drew the first one in 1910. Essentially it is a scatter plot of stars' **magnitude** against their colour. Most stars are found to lie on a strip known as the **main sequence**. Star formation creates protostars which evolve across the diagram and on to the main sequence via routes known as Hayashi tracks and Henyey tracks, which interpolate the starting state of a collapsing cloud of gas onto that of a fully formed star. When a star has completed its main sequence life, it can evolve in a number of directions. Stars like our Sun, for example, become redder and brighter, climbing the so-called 'giant branch' of the diagram as **red giant** stars – before ending their lives as faint, hot **white dwarfs**.

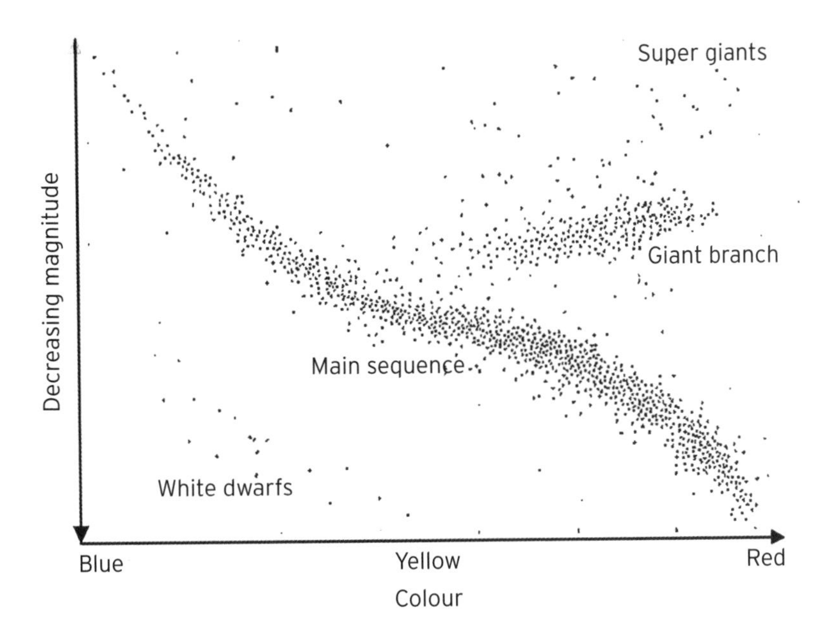

Main sequence

Average stars like our own Sun occupy what's called the 'main sequence' – a diagonal strip on the **Hertzsprung–Russell** diagram. Internally, a main sequence star burns hydrogen fuel in its core through fusion **nuclear reactions** to generate energy; also produced is helium, which is the 'ash' from the nuclear burning. Stars on the main sequence are classified into 'spectral types' according to their temperature. For historical reasons the categories are labelled O, B, A, F, G, K, M – in order of decreasing temperature, from 50,000°C (90,000°F) down to 3,000°C (5,4300°F). The Sun is a G-type star at 5,500°C (9,930°F). The main sequence stage of a star's life lasts until it runs out of hydrogen fuel and begins burning helium instead, at which point it becomes a red giant.

Brown dwarfs

Protostars that don't quite make it out of star formation to the point where nuclear reactions switch on inside them are known as brown dwarfs. They are, to all intents and purposes, supermassive versions of the planet Jupiter – spheres of gravitationally bound gas, though not quite so massive that their gravity can create the temperature needed to kick-start nuclear fusion in their cores. Most brown dwarfs are about the same size as Jupiter but weigh anywhere in the region of a few to 90 times as much. Their existence was put forward in the 1970s but the first one wasn't discovered until 1995; there are many now known.

Variable stars

Some stars display drastic changes in brightness – these are the variable stars, and there are a number of different types. Eclipsing variables are binary stars, each component of which periodically moves in front of the other causing the net brightness of the pair to briefly drop. Algol and Beta Lyrae variables fall into this category.

Other variables periodically swell and shrink in size. Shrinking causes the star to become denser and more opaque, meaning it traps more radiation, which in turn re-inflates it, and so on. Mira and Cepheid variable stars exhibit this behaviour. The pulsation period of Cepheid variables is directly linked to their maximum brightness; knowing their true brightness and measuring their apparent brightness then allows their distances to be calculated.

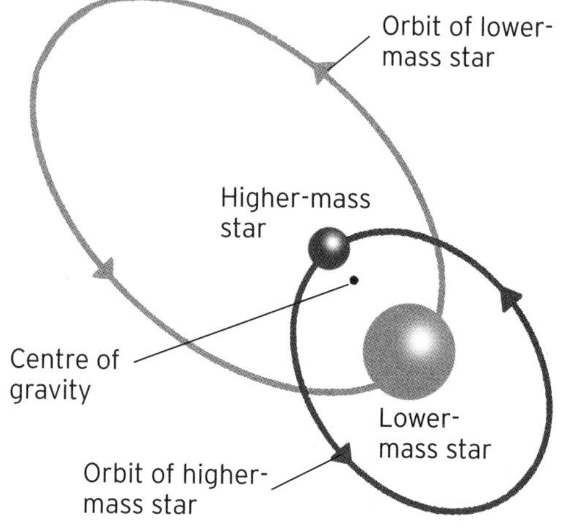

Binary stars

Just as planets orbit the Sun, so stars can orbit around one another. A binary star consists of two stars that have formed close to each other and whose motion has brought them into orbit around their common centre of gravity. There are various types, depending on how their binary nature is detected; for example, 'visual binaries' can be seen through a telescope as two stars. 'Spectroscopic binaries' show up because of the

Doppler effect shifting in the spectrum of their light as they move back and forth. While in an eclipsing binary, the two stars move in front of each other to create a variable star.

Binary stars are one way to detect a **black hole**, which cannot be observed directly – because it's black – but if it's part of a binary system then its presence can be inferred from the gravitational effect it has on its bright companion.

Star clusters

Stars usually form in groups, as regions within a star-forming cloud shrink down and then fragment to spawn a brood of young stars. But stars can occupy even larger groups, called star clusters. Our **Milky Way** has two types: open clusters and globular clusters. Lying in the plane of the galaxy, open clusters are typically a few tens of light-years across and home to several hundred stars. Open clusters are only loosely bound by gravity, hence the name 'open'; for this reason they are easily disrupted, and so their member stars are usually quite young.

Globular clusters, on the other hand, are spherical concentrations of stars that orbit in a galaxy's outer halo. Extremely dense, they pack up to a million stars into a region a few tens of light-years in size. The density makes their gravity stronger, meaning these star systems are tightly bound; accordingly, their members include some of the oldest stars known.

Red giants

When a main sequence star runs out of hydrogen fuel in its core, nuclear reactions are temporarily halted and the core begins to cool. But the cooling leads to a reduction in pressure that makes the core contract and start to heat up again. The heating continues until the temperature is high enough – around 100 million°C (180 million°F) – to ignite nuclear fusion reactions burning helium; then the high temperature swells the star's outer envelope to hundreds of times the diameter of the Sun. At the same time, the expansion brings about a cooling of the outer layers – taking it from being a yellow, Sun-like star to a cooler, redder one, known as a red giant.

Our Sun is destined to become a red giant around five billion years from now. It will engulf the planets Mercury and Venus, and bake the Earth's surface to a crisp. Interestingly, conditions on Saturn's moon Titan will become warm enough for life when the Sun is a red giant – but, sadly, this phase won't last long.

Planetary nebula

What happens when a star reaches the end of its life? Medium- and relatively low-mass stars, such as the Sun, go out rather gracefully – as a billowing gas cloud known as a planetary nebula. The star will already have become a red giant, burning helium at high temperature in its core. However, red giants are unstable stars – tiny changes in core temperature produce enormous variations in the star's brightness leading to pulsations in the outer layers, which grow in size until they are ultimately large enough to cast off the outer envelope entirely.

Planetary nebulas measure about a light-year across. Even at stellar distances from Earth this makes them appear disc-like – resembling a planet rather than the point-like dot of a star and that is the origin of their name, coined by British astronomer William Herschel in the late 18th century. The star's core – a glowing ember of carbon and oxygen called a white dwarf – remains at the nebula's centre.

White dwarfs

A white dwarf is a stellar corpse – a remnant state, left behind after a sun-like star has ended its life; it is, in fact, the core of a red giant star that has flung off its outer layers of gas to become a planetary nebula. The core is extremely hot, sometimes over 100,000°C (180,00°F) – making them appear white in colour. However, their small size means they radiate their heat slowly, and so they are relatively faint. With no internal power source, they fade over time – cooling to become a so-called 'black dwarf'.

White dwarfs are extremely dense, packing a solar mass of material into a sphere roughly the size of the Earth. Their thermal pressure cannot resist the gravitational pull created by such high density. Instead, white dwarfs are held up by **quantum theory**. Whole atoms don't exist inside a white dwarf – the material has been squashed down to become a soup of atomic nuclei and electrons. As gravity tries to squash the electrons down, the **exclusion principle** kicks in to prevent them all from being forced into the same state. This so-called electron 'degeneracy pressure' is what stops a white dwarf folding in on itself to become a **black hole**.

Supernova

Stars weighing more than about ten times the mass of the Sun end their lives with a bang, blowing themselves apart in a spectacular explosion called a supernova. When such a big star suddenly runs out of the fuel it needs for nuclear fusion, the effect is like opening a trapdoor under an elephant. With the star's source of support removed, its outer layers go into free fall, and the star implodes, compressing the star's core, turning it into a neutron star and halting the collapse. This in turn triggers a sudden 'bounce' that flings the collapsing layers back outwards and drives the resulting supernova explosion.

Neutron stars

The core of a star ending its life in a supernova explosion remains at the centre of the debris cloud left by the blast. Called a neutron star, it is formed as the increasing pressure within the dying star squashes together electrons and protons, turning the core into a giant ball of neutrons. But that's not all; for every electron–proton pair that merges, a **neutrino** is given off. Billions upon billions of neutrinos stream from the doomed star and out across space – and can warn astronomers of an impending supernova before it becomes visible to telescopes.

Neutron stars cram a solar mass of material into a ball just 12km (7.5 miles) across; this is so dense that a teaspoon of the stuff would weigh as much as a mountain. Like white dwarfs, neutron stars are held up by quantum pressure – although here it's caused by applying the quantum **exclusion principle** to neutrons rather than electrons.

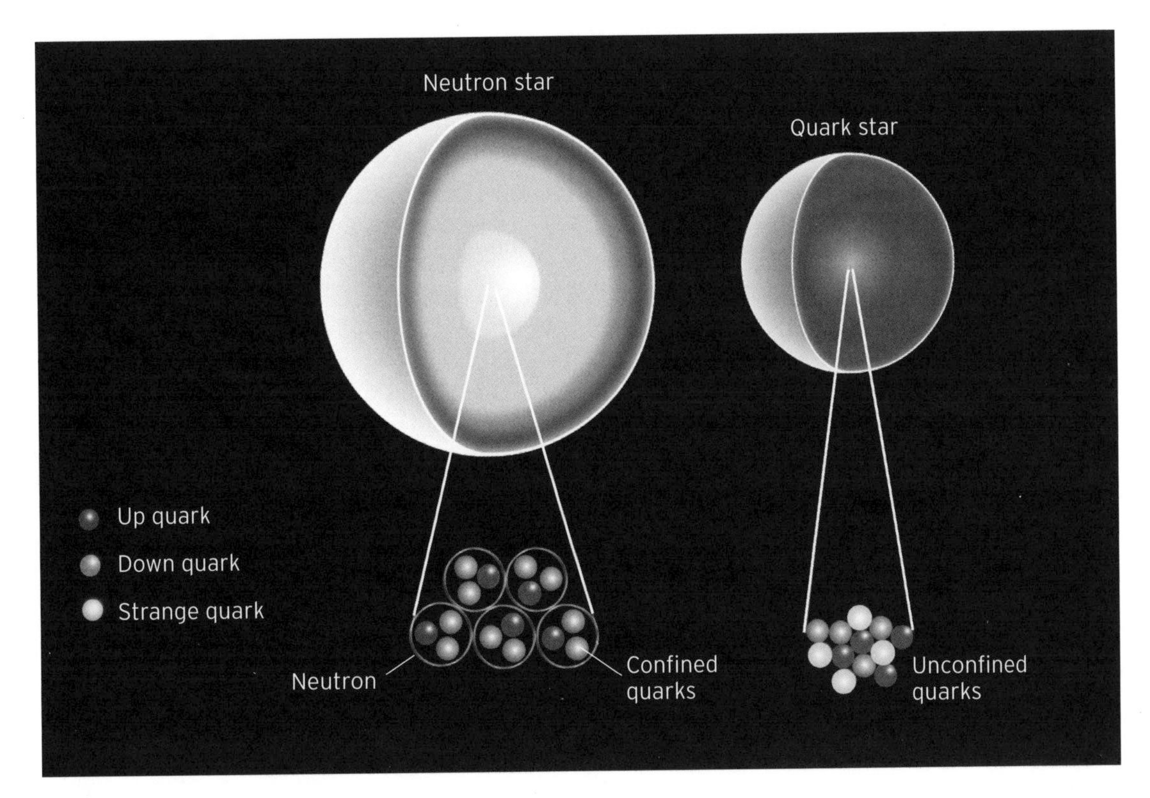

Quark stars

If a neutron star is heavy enough its gravity can shrink it down even further – melting its neutrons into their constituent **quarks**. In 2002, astronomers led by Jeremy Drake at the Harvard-Smithsonian Center for Astrophysics, in Cambridge, Massachusetts, announced that they had discovered a candidate quark star in the constellation Corona Australis. Called RXJ1856, the object appeared too small to be a neutron star and yet was too big to be a stellar-mass **black hole**.

Both neutron stars and quark stars exhibit intense magnetism which, if the star is spinning, can interact with nearby charged particles to produce beams of radiation from its magnetic poles, that sweep around like a cosmic lighthouse as the star turns. Such objects are called pulsars.

Hypernova

When a dying star is so big that not even the formation of a quark star can stop the gravitational collapse, the core swallows itself to become a **black hole**, releasing energy equivalent to 100 supernovae in the process. This is a hypernova. Hypernovas are believed to be the source of 'gamma-ray bursts', flashes of high-energy **electromagnetic radiation** discovered in the late 1960s by satellites. Roughly one a day is observed but their origin was a mystery until 1997 when the detection of optical counterparts to the gamma-ray bursts enabled their sources to be traced to exploding stars in distant galaxies.

GALAXIES

Intergalactic space

If you thought interstellar space was dull, then little can prepare you for the stark cosmic hinterland that lies in between the galaxies. Intergalactic space is about as close to a perfect vacuum as nature gets, with just a scraping of matter amounting to a **density** of around one hydrogen atom per cubic metre.

The major inhabitants of intergalactic space are the galaxies themselves – these giant gatherings of stars are like islands in the cosmic vacuum separated from one another by a few million light-years, which is about 20 to 40 times the typical size of a galaxy. Compare that with the average distance between stars within a galaxy – typically tens of millions of times their size – and it can be seen that intergalactic space is, in fact, much more densely populated than interstellar space.

Spiral galaxies

The most commonly occurring galaxies in the **Universe** are the spirals, which have a bright disc, within which spiral arms sweep around. In fact the arms are a spiral density wave that circles the galaxy once every few tens of millions of years. Ordinarily this pattern would be invisible, but where it squashes the disc material, star formation takes place, lighting up the spiral arms with new generations of hot, bright stars. The relatively young stars found in the disc are known as 'population I' stars.

The disc is surrounded by a spheroidal 'halo' of material normally home to older stars, known as 'population II'. Dynamic studies of galaxies also suggest the halo harbours a large quantity of **dark matter**, which makes up the bulk of the galaxy's mass. Finally at the hub of the galactic disc is the so-called 'bulge', where stars cluster densely with clouds of cosmic gas and dust. At theheart of this most galaxies are thought to contain a supermassive **black hole** weighing millions of times the mass of the Sun.

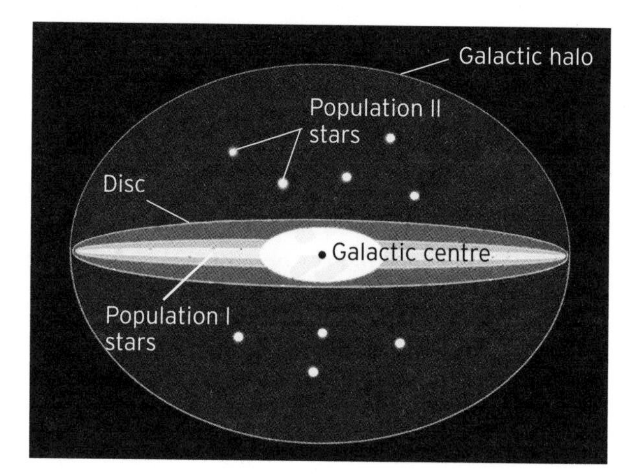

Barred spiral galaxies

The structure of some spiral galaxies is made more complicated by the presence of a bright 'bar' running through their cores. Rather than emanating directly from the core, the galaxy's spiral arms trail from the ends of the bar. The physics of galactic bars is not well understood. However, scientists believe their formation is the result of gravitational interactions with other galaxies, coupled to density waves similar to those that form the spiral patterns found in normal spiral galaxies. Roughly a third of all known spiral galaxies are barred, including the Milky Way.

Milky Way

Our own galaxy is called the Milky Way and believed to be a barred spiral, of classification SBb-SBc on the **tuning fork diagram**. The disc of the galaxy is 100,000 light-years across, around 1,000 light-years thick, holds some 300 billion stars and weighs 600 billion times the mass of our Sun. The Milky Way has two major spiral arms that trail from the ends of a bar, along with a small number of shorter arms called 'spurs'. The spiral pattern revolves around the Milky Way disc once every 50 million years.

Lying about 26,000 light-years from the Milky Way's centre on the edge of a spur known as the Orion Arm are the Sun and Solar System. The Sun completes one orbit of the galactic centre every 220 million years. Judging from the age of its oldest stars, the Milky Way is thought to have formed around 13.2 billion years ago, when the **Universe** itself was just 500 million years old.

Elliptical galaxies

Not all galaxies have the ornate structure of the spirals; some, known as elliptical galaxies, have plain old ellipsoidal shapes, with their stars distributed smoothly inside them. The bulk of the stars in an elliptical galaxies are population II – the stellar old-timers (typically billions of years in age) that are also found in the halos of spiral galaxies. There's also very little structure in the orbits of stars in an elliptical galaxies. In a spiral galaxy they all circulate in one direction in the plane of the galaxy's disc but in an elliptical galaxies, stellar orbits criss-cross this way and that. Elliptical galaxies also encompass a large range of masses – anything between 10 million and 10,000 billion times the mass of the Sun. Their sizes vary dramatically too – ranging from a few hundred light-years up to hundreds of thousands. They are a minority group in the Universe, accounting for just 10 per cent of known galaxies though that may change as the **Universe** grows older. Ellipticals are thought to be formed when two spiral galaxies merge together. Indeed, the Milky Way may one day become an elliptical; it is on a collision course with the nearby Andromeda Galaxy – the two are due to crash together in 3 billion years' time.

Galactic misfits

Some galaxies don't fit the description of either spiral galaxies or ellipticals. Lenticulars – meaning 'lens shaped' – have a disc, just like a spiral galaxy, but none of the intricate spiral structure within it. These are thought to be spiral galaxies that have used up all of their star-forming material so that there's nothing left to make the bright, new stars that delineate spiral structure; they have also avoided the galactic collisions that transform many spirals into ellipticals. On the other hand, so-called 'peculiar galaxies' haven't been so lucky. These have unusual, misshapen appearances that defy classification and are thought to be the result of especially violent galactic smash-ups.

Tuning fork diagram

Astronomers sum up the shapes and appearances of different galaxies on what's called the tuning fork diagram, conceived by astronomer Edwin Hubble in 1926. The first sequence, on the far left of the diagram, shows the elliptical galaxies, which range left to right from spherical (E0) to highly flattened ellipsoids (E7). Then there are the lenticulars (see **Galactic misfits**), labelled S0. At this point the diagram splits to form the two 'prongs' of the fork: the upper prong charts the spiral galaxies, which range left to right, from those with tightly wound arms and large central bulges (Sa) to those with loosely wound arms and small bulges (Sc); the lower prong denotes barred spirals, which range from SBa to SBc by the same rules.

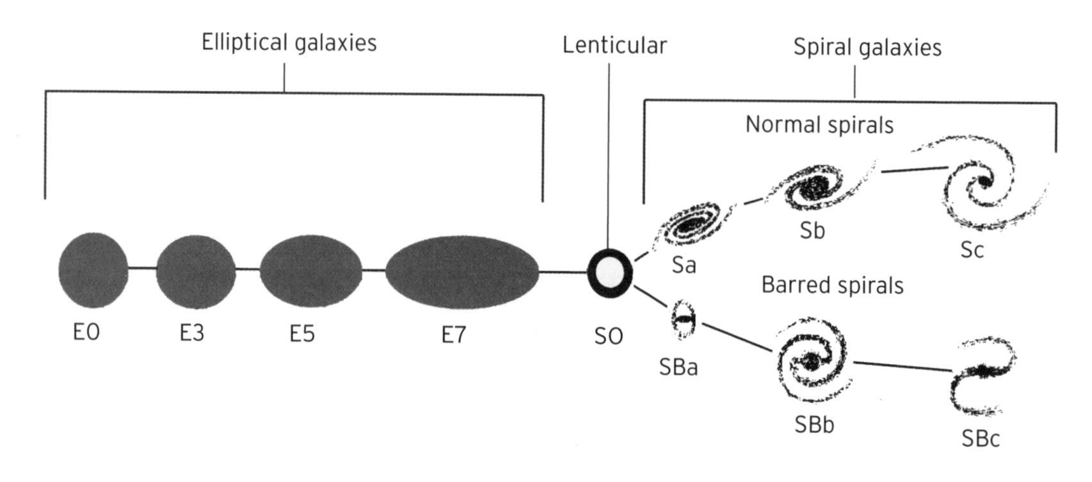

Messier catalogue

In 1771, the French astronomer Charles Messier published a catalogue of galaxies visible to the telescopes of the day. Distant galaxies often appear faint and fuzzy through the telescope, much like comets, and Messier, a keen comet hunter, was fed up with mistaking known galaxies for new cometary discoveries – so he decided to make a list of them.

Originally numbering 103 objects, the Messier list has since been expanded to 110 – as astronomical historians have discovered evidence for objects that Messier was clearly aware of but, for whatever reasons, he neglected to include. Messier's catalogue doesn't just cover galaxies, but also star clusters, **nebulas** and **supernova** remnants. Because most of the objects on the list are visible through binoculars or small telescopes, the Messier catalogue remains to this day a definitive list of targets for amateur astronomers.

Dwarf galaxies

Rather like planets and **binary stars**, some galaxies are orbited by miniature brethren. Perhaps the archetypal examples are the Large and Small Magellanic Clouds that skirt the periphery of the Milky Way. The Large Mellangenic Cloud, for example, is home to around 10 billion solar masses of material – broken down into some 30 billion stars – and around 14,000 light-years across and 157,000 light-years from the Milky Way. Astronomers believe it was once a small barred spiral galaxy before it strayed too close to the Milky Way, whose gravity distorted it into its present, irregular form.

The Milky Way has 14 dwarf galaxies in orbit around it. Other dwarf galaxies, such as the Phoenix Dwarf and the Tucana Dwarf are isolated and wander deep in intergalactic space.

Galaxy formations

Dwarf galaxies grew in a similar way to the process of star formation – clouds of material shrunk by virtue of their own gravity. Cosmic clouds generally have some degree of rotation, and the centrifugal force generated by this rotation supports the galaxies' collapse in two dimensions while letting it continue in the third – making the resulting galaxies flattened, or disc-shaped. Larger galaxies are believed to have then formed via what's called a bottom-up process, where smaller units – namely the dwarf galaxies – merged via their gravity to create the large galaxies that we see in the modern Universe.

Computer simulations show that **dark matter** plays a crucial role in galaxy formation – without it, the process couldn't take place. Galaxies are still forming today, although the formation rate in the past was much higher.

Galaxy evolution

Like lifeforms under a microscope, galaxies evolve over time. Left to its own devices, a galaxy will change its chemical make-up over its lifetime as stars process hydrogen gas into helium and heavier elements, through nuclear reactions, and then return this enriched material to interstellar space via **supernova** explosions and the ejection of **planetary nebulas**.

Edwin Hubble's tuning fork diagram was previously thought to represent an evolutionary sequence – a process by which galaxies evolve from ellipticals into spirals. This is now known not to be true. If anything, the evolutionary path operates in the opposite direction, converting spiral galaxies into ellipticals as galactic collisions turn ordered spiral structure into the chaotic tangle of an elliptical. Individual stars don't merge during such collisions – the gaps between them are too large for them to collide – but gravity pulls the stars from each galaxy into one consolidated swarm. The compression waves seeded by galaxy mergers trigger bursts of intense star formation – these are sometimes referred to as 'starburst galaxies'.

Active galaxies

Most galaxies are thought to have enormous **black holes** lurking at their cores. In active galaxies these have become fierce sources of radiation, turning the galaxies into bright cosmic beacons. Active galactic nuclei – the cores of active galaxies – are powered by black holes devouring stars, gas and dust. The gravity of the black hole pulls the material in and accelerates it to high speed, but as it converges on the black hole's event horizon it bunches up and collides with other infalling material, causing it to heat up and emit electromagnetic radiation.

There are two main varieties of active galaxies: radio-loud and radio-quiet, depending on how much radiation they emit in the radio region of the electromagnetic spectrum. Radio-loud active galaxies spawn energetic jets of plasma that can stream thousands of light-years into space – at right angles to the disc of the galaxy, like cosmic spindles. Material in the jets travels at close to the speed of light and, where it slams into the gas pervading **intergalactic space**, creates shock waves that billow out to form large radio-emitting lobes. No one is quite sure how the jets from active galaxies are powered, though **frame dragging** is one possible explanation. In radio-quiet galaxies, it is believed that the galaxy is oriented so that intervening material – such as the galaxy's own disk – is blocking our view of the radio-emitting core.

Quasars

Quasars are active galaxies seen at vast distances from the **Milky Way**; the most distant quasar known is about 28 billion light-years away, meaning its light was emitted when the Universe was just a tiny fraction of its present age. That quasars can even be seen at this colossal distance makes them the brightest objects in the universe. Hundreds of thousands of quasars are known, though none are closer than 3 billion light-years, which has led some astronomers to suggest that quasars are embryonic galaxies, tempestuous youngsters that would later settle down to become the more modest galaxies we see in the local Universe today. Under this scheme, even our own Milky Way could have undergone a quasar phase in its past.

Astronomers first began to spot the objects that would become known as quasars in 1960. But it wasn't until 1963 that they would be recognized as such – by Dutch-American astronomer Maarten Schmidt. The name 'quasar' is a contraction of 'quasi-stellar object' – a name they were given because, although effectively galaxies, their great distance made them appear as star-like (or 'stellar') points of light in the sky.

THE EARLY UNIVERSE

The Universe

What exactly is the Universe? Everything that exists? Just the bit you can see? Or is there some altogether more spiritual definition? As far as scientists are concerned, the Universe is the full extent of the three-dimensional space we all live in. Some scientists might extend that to include time, so that by their reckoning the Universe is 3D space plus its entire past and future.

Cosmologists place special emphasis on our 'observable universe' – that is, the portion of the Universe from which light has had time to travel to the Milky Way since the Big Bang. Or, in other words, the part of the Universe that lies within our **cosmological horizon**. Other universes, exterior to our own, may also exist – for instance, those making up the **Multiverse** predicted in the **many worlds** view of **quantum theory**.

The Big Bang

Around 13.7 billion years ago, the biggest explosion there has ever been brought our Universe into existence – except it wasn't an explosion. There was no blast wave tearing through space – the material just sat there while space itself expanded. And there was no single 'point' where the Big Bang took place – it happened everywhere at the same time. There is, however, one thing the Big Bang has in common with proper explosions: a fireball. The Universe was born extremely hot and dense before the expansion of space brought it down into the more clement cosmos we see today. What caused this big bubble of matter, radiation, space and time to pop into existence in the first place? The truth is nobody knows, though fields such as **quantum cosmology** and the **ekpyrotic theory** are offering insights.

Microwave background

Anyone reading for the first time that our Universe spontaneously sprang into existence billions of years ago and then expanded and cooled to form galaxies, stars, planets and ultimately us might, quite rightly, ask how we know all this. The Big Bang theory rests on two solid pillars of evidence. One concerns the abundance of the chemical elements in space. According to the theory, the early universe was filled with hydrogen which then got cooked up in nuclear fusion reactions in the Big Bang fireball into helium plus a smattering of heavier elements. Calculations based on the Big Bang theory predict that the Universe should have emerged from this hot fireball with about 25 per cent of its hydrogen converted into helium; and this is exactly the balance observed in isolated cosmic gas clouds. But there's another, more astonishing piece of evidence. The theory predicts that there should be a detectable echo of the Big Bang still travelling through space. It is not a sound echo, but one made up of electromagnetic radiation with a very specific wavelength – microwaves with a **black body radiation** temperature 2.7°C (4.85°F) above absolute zero.

In 1964, Arno Penzias and Robert Wilson, two radio astronomers at Bell Labs, New Jersey, found this cosmic microwave background radiation, a discovery made completely by accident. Indeed, initially, they tried to get rid of this unwanted 'noise' plaguing their detector but it confirmed that our Universe really was born in a hot Big Bang.

Inflation

The theory of inflation says that at one hundred-million-billion-billion-billionths of a second after the Universe was born in the Big Bang it underwent a tremendous growth spurt, expanding by a factor of 10^{26} (see **Scientific notation**). There are a number of motivations for the theory of inflation. It is one way of getting the Universe up and out of the quantum realm after the Big Bang when, without inflation, it should have just recollapsed back in on itself by gravity; it also neatly solves two other niggling cosmological conundrums – the flatness problem and the horizon problem.

Moreover, inflation offers a way to explain how structures such as galaxies first formed, saying that tiny density fluctuations generated by the quantum **uncertainty principle** were magnified to astrophysical scales by the rapid expansion, and that these formed the seeds around which galaxies could later grow. Indeed, the pattern of fluctuations observed in the cosmic **microwave background** radiation observed by satellites seems to match well with inflation's predictions.

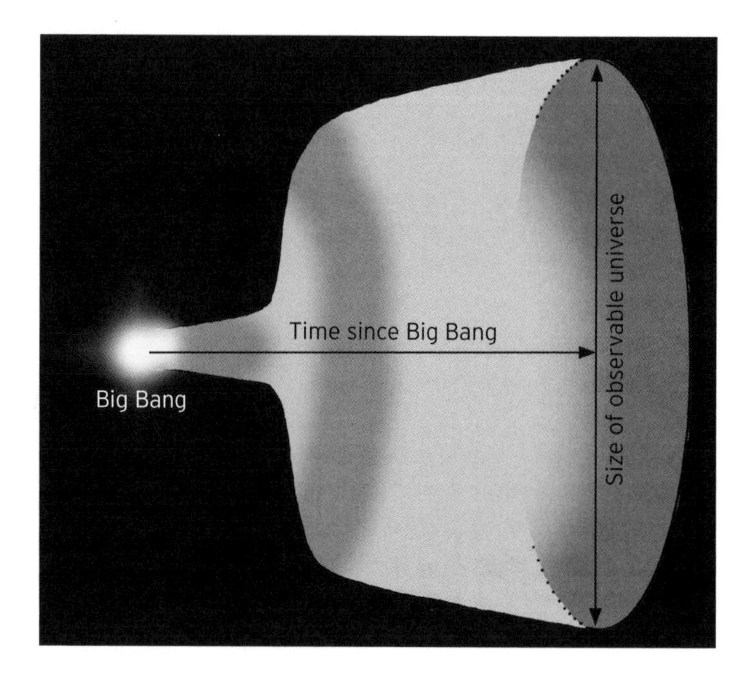

Flatness problem

When inflation was first put forward in 1980, by MIT physicist Alan Guth, it cleared up a couple of problems with the Big Bang theory that had been worrying theorists for ages. First was the flatness problem. Put simply, worrying astronomical measurements of the present-day Universe show that space on the largest scales is incredibly flat, which is surprising given that **general relativity** – our best theory of space and time – allows for all manner of different curvatures. The Big Bang theory could not explain this flatness.

Inflation solves the problem by blowing up the Universe to terrific size, making any curvature negligible. Imagine you were perched on top of a beach ball – its curvature would be very apparent as you tried to keep your balance. But inflate the beach ball to the size of the Earth and – as our daily experience tells us – the small portion of the ball that you're standing on would seem flat.

Horizon problem

There's another big glitch with the Big Bang theory that the theory of inflation clears up – the horizon problem: why do opposite sides of the Universe appear more or less the same? True, the constellations on opposite sides of the sky are different and there are different galaxies and clusters to be found from one side of the sky to the other. But there are no large qualitative differences; for example, we don't see one half of the sky blazing brilliantly while the other is pitch black. There's no reason why this should be. The furthest we can see – our **cosmological horizon** – is 46.5 billion light-years. And so opposite sides of the sky are 93 light-years apart. Even travelling at the speed of light, there simply hasn't been time since the Universe began – 13.7 billion years ago – for the two sides to come into contact.

Inflation gets round this by making the size of the very young Universe much smaller than in the standard Big Bang model. The matter content of the Universe then had time to even out its differences before inflation blasted it up to the size we see today.

Chaotic inflation

One problem with inflation is explaining how it got going in the first place – the conditions for this to happen had to be just right. A version of the theory put forward by physicist Andrei Linde, called 'chaotic inflation', offers a natural mechanism. Linde supposed that the early Universe just after the Big Bang was a seething, 'chaotic' tangle of quantum fluctuations – **virtual particles**. All it took, reasoned Linde, was for the conditions in one small corner of this tangle to become conducive to inflation and that tiny corner would then grow rapidly to dominate the volume of the Universe.

Linde went on to suggest that this process continues today, but so far away – thanks to the inflationary expansion of our own local bubble of the Universe – that we can't see it. If he's right, then new universes are constantly budding off from our own in a process he calls 'eternal inflation'.

Life of the cosmos

The collapse of dead stars into **black holes** might be causing new universes to sprout from our own. One theory says the black holes form **wormhole**-like bridges into these new domains. American theoretical physicist Lee Smolin has suggested that the production of new universes is analogous to biological reproduction, and that successive generations of universes might evolve according to a kind of cosmological version of Darwinian

natural selection. Like biological organisms, each new universe would have traits – such as flatness and expansion rate – that are slightly mutated from those of its parents. And, as in Darwinian evolution, only those universes with the 'fittest' traits survive.

Cosmic defects

Twisted knots of energy left over from the Big Bang, cosmic strings, monopoles, textures and domain walls form a family of weird objects in space that scientists call 'cosmic defects'. Defects were first proposed in the 1970s by Tom Kibble, of Imperial College London; he pointed out that they are an unavoidable consequence of **spontaneous symmetry breaking** in the Universe. The unified, symmetric Universe can be thought of as rather like a lattice of needles, all perched on end; symmetry breaking is analogous to what happens as the needles inevitably start to topple over. Crucially, there's no reason for all the needles to topple in the same direction – in some regions they will fall pointing one way, while in others they will fall pointing in other directions entirely.

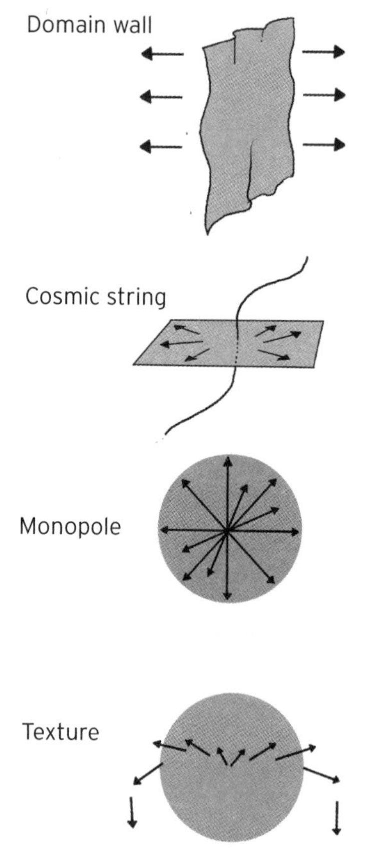

The simplest kind of cosmic defect forms where two regions of space in which the needles have fallen differently meet up – a two-dimensional boundary known as a 'domain wall'. Where three or more regions come together, so that the needles point outwards from a central line-like core, the result is a defect known as a 'cosmic string'. And if all the needles point outwards in three dimensions, they define a spiky-looking point, a defect called a 'monopole'. A most complicated kind of defect is an abstract object known as a 'texture', forming when all the needles point outwards in four dimensions.

Multiply connected universes

Imagine if you could travel so far out into space that you came back to where you started. Einstein's theory of **general relativity** does a good job explaining the behaviour of our cosmic neighbourhood, but it says nothing about the overall shape, or 'topology', of space. Is the Universe an infinite flat sheet, a closed sphere, a ring-shaped 'torus', or something even more bizarre? Mathematicians call universes that wrap around on themselves 'multiply connected'. In 2003, scientists led by Jean-Pierre Luminet at the Paris Meudon Observatory, in France, carried out a study of the **microwave background** radiation. The group of scientists found patterns in the radiation that suggested space may have a complex wraparound structure based on a 12-sided polygon called a dodecahedron. Exit on one side of the dodecahedron and you re-enter through the opposite face. They are currently waiting for space probes to gather new, more accurate maps of the microwave background to test the idea further.

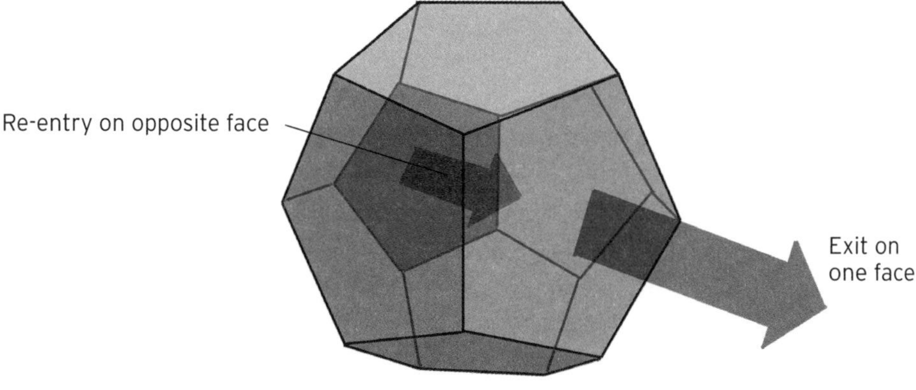

Quantum cosmology

According to **general relativity**, the Big Bang in which our Universe was born should have started out as a gravitational singularity, a point of zero size and infinite density. As is the case in black-hole physics, scientists view the presence of a **gravitational singularity** as a sign that the classical theory of general relativity is breaking down – it doesn't provide a good description of the Universe's first moments. Instead, what's needed is a quantum treatment of gravity.

Quantum cosmology is an attempt at this, and was first proposed in the 1980s. Whereas other theories of **quantum gravity** take an existing quantum system (i.e. **particle physics**) and try to build gravity in, quantum cosmology works the other way, by quantizing the existing curved space and time of general relativity itself. The formulation makes it possible to calculate the quantum probabilities of different kinds of universe springing into existence from nothing. However, subjecting the theory to rigorous scientific tests will require better astronomical data than is currently available.

The Multiverse

Parallel universes have long been a staple of science fiction but, in 1957, physicist Hugh Everett put forward a new philosophy of **quantum theory**, known as the **many worlds** interpretation, which brought parallel universes into the here and now. If correct, it means that our Universe is just one in a sprawling network of many, known as the Multiverse. In the Multiverse there are universes where every conceivable possibility is played out – universes where you don't exist, where you wrote this book, or where you are the Emperor of Poland. Scientists have even toyed with the possibility that there may be parallel universes where the laws of physics themselves are radically different, so that gravity and the interactions between subatomic particles take very different forms.

Holographic universe

What if the whole Universe was just a hologram, a 3D projection of events taking place on a 2D surface, just like the hologram on your credit card? That's the basic premise of the 'holographic principle'. It was first formulated in response to a problem in **black hole** physics, namely that information stored on material falling into a black hole seems to get destroyed – something that's at odds with the laws of **information theory**. The holographic principle says that the information doesn't fall in, but sticks at the hole's event horizon.

But then physicists realized that if the information content of a black hole is encoded on its horizon, then this must also be true for the Universe too, which is bounded by its own **cosmological horizon**. It may be possible to test the idea; space in a holographic universe should have an unusually coarse-grained structure, making detailed small-scale measurements impossible. In 2007, scientists using the GEO600 **gravitational wave** experiment in Germany detected a tentative signature of just such graininess, but more data is still needed.

Ekpyrotic universe

In the standard Big Bang theory, it makes little sense to talk about what happened before our Universe began; the Big Bang marked not just the beginning of all the matter, but also space – and time. Asking what happened before the Big Bang is like asking what's north of the North Pole.

But a theory put forward by physicists Neil Turok and Paul Steinhardt in 2001 could change that. Called the ekpyrotic theory (after the Greek word for 'conflagration'), it imagines the four-dimensional space and time of our Universe as a sheet-like 'membrane' sitting in a 5D **extra-dimensional** hyperspace. A short distance away in the fifth dimension is another membrane with which ours periodically collides, and each collision ignites what we would term a Big Bang. The cycle repeats every 30 billion years, so that an endless succession of Big Bangs stretches back into the past and ahead into the future.

The ekpyrotic model is motivated by ideas in **string theory**, offering its own mechanism for forming structure in the Universe, such as galaxies and clusters – and this makes the ekpyrotic universe a challenger theory to **inflation**.

COSMOLOGY

Olbers' paradox

Look outside on a clear night and you'll see a smattering of stars with dark sky behind. And yet if the Universe is infinite, no matter which direction you look in, then surely your line of sight must ultimately fall on a star, meaning the night sky should be a blazing mass of starlight. This apparent contradiction was highlighted in the 18th century by German astronomer Heinrich Olbers, and is known as Olbers' paradox.

The paradox is resolved by a combination of two factors. Firstly, our Universe was born a finite time ago, meaning that only light from a finite number of stars – those lying within our **cosmological horizon** – has had time to reach us. Discovery of Hubble's law and the expansion of the Universe in the 1920s added a new twist. The expansion stretches light out as it tries to cross cosmological distances, reducing its energy and making distant stars and galaxies even fainter than they would ordinarily be.

Cosmological principle

Cosmology is a strange science; from a single vantage point in space we are trying to unpick the workings of the entire Universe. We've taken what amounts to a single snapshot – a few centuries of observations, compared to billions of years of cosmic evolution – to try to piece together the Universe's entire history, and predict its future. Indeed, cosmology would be impossible if we didn't make a few assumptions, and paramount among these is the cosmological principle. It says that on large scales, the Universe is both homogeneous and isotropic. Homogeneous means it has the same properties from point to point throughout space; isotropic means it looks the same in all directions. The cosmological principle is closely related to the Copernican principle, which states that the Earth does not occupy a special location in space – our view of the Universe is typical.

Hubble's law

Albert Einstein's theory of **general relativity** made a startling prediction: space is expanding. Only, Einstein found this so unpalatable that rather than announcing it as something for astronomers to go and look for, he fiddled his equations to remove the effect.

But in 1929, American astronomer Edwin Hubble and his assistant Milton Humason found the evidence that space really is getting bigger. They had collated many observations of the spectra of distant galaxies – that is, how the brightness of the galaxy's light varies with wavelength. Galaxies can be moving towards or away from the observer, and this changes the wavelength of their light because of the **Doppler effect**. Astronomical measurements, in particular spectroscopy, can reveal how fast galaxies are moving. Hubble and Humason found that most galaxies are receding – their light is red-shifted, moved towards the longer-wavelength red end of the spectrum. By concentrating on galaxies with well-known distances, they saw that the recession speed increases with distance; the speed of a galaxy was just given by its distance multiplied by a number that's since become known as Hubble's constant. This was solid evidence for the cosmic expansion Einstein had predicted yet ignored.

Cosmic expansion

Although astronomer Edwin Hubble was on the right track when he calculated the recession speeds of galaxies according to the **Doppler effect** of **wave theory**, in fact there was a deeper physical mechanism behind his observations. It is this expansion that gives rise to Hubble's law. Imagine space as the surface of a balloon that's being inflated. Draw dots on the surface of the balloon and then watch the position of the dots as you blow air into it. Every dot moves away from every other dot as the balloon gets bigger. And, for a steadily expanding balloon, the speed that any two dots move apart increases with their separation – just as Hubble predicted.

Cosmic expansion shifts the light from distant galaxies to longer wavelengths – the so-called red-shift effect. However, this is quite different to the redshift produced by the Doppler effect. Cosmological redshift is caused by the expansion of space stretching out the light to longer, redder wavelengths.

Dark matter

There's more to galaxies than meets the eye – that's the basic gist of dark matter. The theory says that space is pervaded by a field of invisible material. What's more, this 'dark matter' weighs around five times as much as all the Universe's visible material put together. Astronomers got their first sniff of dark matter in the 1930s, in the form of discrepancies between the mass visible in galaxy clusters and the gravitational mass inferred from the motion of their constituent galaxies.

More evidence followed in the 1970s when astronomers started measuring the 'rotation curves' of spiral galaxies. These are graphs that plot the orbital speed of stars in the galaxy against their distance from the galactic centre. **Kepler's laws** predicted that for galaxies with their mass concentrated at the centre, the speed gradually diminishes to zero as you move outwards, but astronomers found that once clear of the galaxies' central bulge, the orbital speed was roughly constant. One way to account for the problem was if the galaxy was embedded within a massive ellipsoidal halo of hidden material. Although there is much astronomical evidence, individual dark matter particles have yet to be detected experimentally.

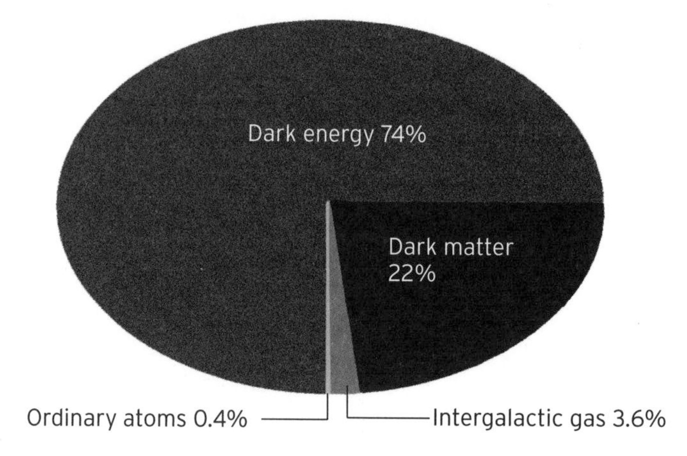

Dark energy

Poor Albert Einstein. When he discovered that **general relativity** predicted the Universe should expand, he added a term called the 'cosmological constant' to the mathematics of the theory to cancel the expansion. But when Edwin Hubble discovered cosmic expansion, Einstein promptly removed the cosmological constant from his equations, referring to it as the 'biggest blunder' of his life. Now astronomical observations indicate that the cosmological constant exists after all. Reincarnated with the name 'dark energy', it is now believed to act in the opposite direction – accelerating the expansion of the Universe rather than damping it down. The accelerating effect of dark energy on cosmic expansion was first noticed in the late 1990s by astronomers measuring the redshifts of supernova explosions in far-off galaxies.

The latest experimental data from NASA's Wilkinson Microwave Anisotropy Probe spacecraft suggest that dark energy accounts for about 74 per cent of the mass of the Universe, dark matter makes up 22 per cent and, last of all, are normal atoms at around 4 per cent.

Galaxy clusters

Like stars, galaxies tend to not live alone. Most are members of associations known as galaxy clusters measuring millions of light-years across. Each is home to anything up to 1,000 galaxies. The smallest clusters are believed to have merged to form the largest ones – cluster formation proceeded via a 'bottom-up' process, rather than the opposite 'top-down' scheme. Clusters group together into larger 'superclusters' that span hundreds of millions of light-years and make up great sheets and filaments of matter that surround vast empty voids.

The Milky Way is a member of a galaxy cluster known as the Local Group. Other clusters near to us include the Great Attractor – an unseen concentration of mass in the constellation Centaurus that's pulling the Milky Way towards it at 600km/s (350 miles/s). Our Local Group is part of the Local Supercluster, which also includes the nearby Virgo and Ursa Major clusters.

Gravitational lensing

Much like the **bending of star light** around the Sun – which provided the first experimental test of general relativity during a solar eclipse, in 1919 – light from distant galaxies can be bent by the gravity of intervening galaxy clusters.

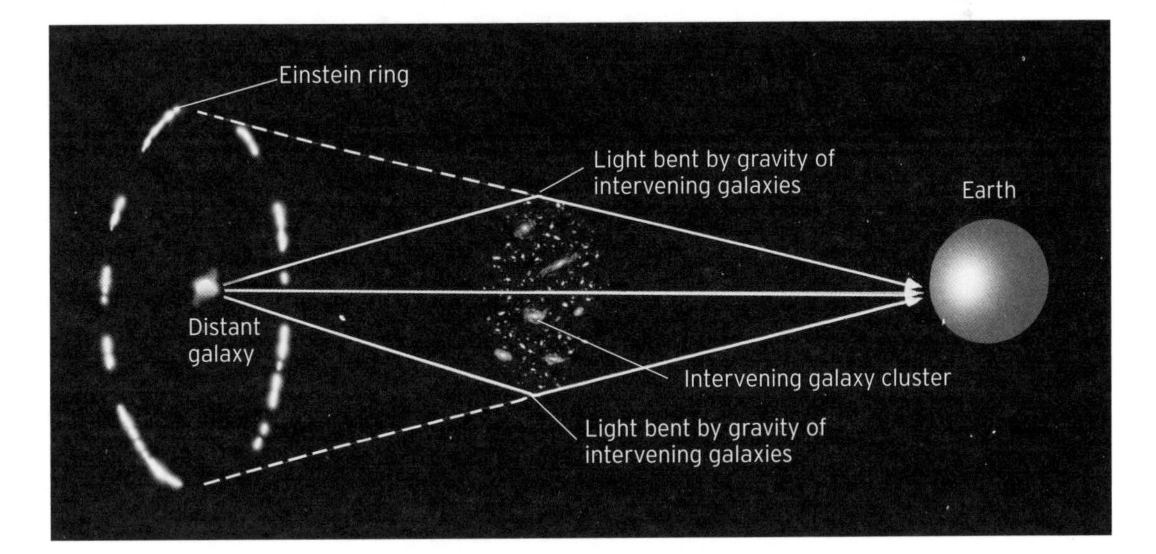

The effect can amplify the light from faraway galaxies, which is why the process is known as gravitational lensing. In two dimensions, it curves the light from the distant galaxy so that the observer sees two mirror-image copies of it (see diagram). In the three dimensions of our real Universe, a galaxy cluster perfectly aligned between the observer and a distant galaxy will smear the light from the galaxy into a perfect ring around it. Such an alignment is known as an 'Einstein ring', and the first one was discovered in 1998. Gravitational lensing has a small-scale cousin known as microlensing, used to detect low-mass dark objects (such as **exoplanets** or **brown dwarfs**) by the blip they produce in the light from a star behind them.

End of the Universe

How will our Universe finally end its days? It won't happen for many billions of years, but a number of possible scenarios have been put forward. If the density of matter within the Universe is higher than a certain critical value, then its gravity will be enough to one day halt **cosmic expansion** and pull space back in on itself, in a cataclysmic anti-Big Bang, which cosmologists have dubbed the Big Crunch.

If, on the other hand, the Universe has less than this critical density then it will never recollapse and its expansion will continue for ever. Eventually, the stars run out of fuel and die, particles of matter all decay and the Universe fades into nothingness. This is sometimes called the 'heat death' scenario because, according to the laws of heat transfer, the Universe has reached a state of **thermodynamic equilibrium**, and so no more useful work – including the conversion of mass into energy inside stars – can take place. With **dark energy** to help cosmic expansion on its way, the heat death scenario seems likely. But dark energy also raises a third possibility – if the dark energy is potent enough, taking a form called 'phantom energy', it could eventually accelerate cosmic expansion to such a degree that it literally tears the Universe apart in a somewhat extreme scenario known as the Big Rip.

SPACE TRAVEL

Rockets

Space rockets are a classic demonstration of the third of **Newton's laws of motion** – that is, for every action there is an equal and opposite reaction. Rockets work by accelerating large volumes of exhaust gas to high speed, the reaction to which is a force that pushes the rocket in the opposite direction. On Earth, propulsion is easy because there's always something for a vehicle to push against. Boats push against the water – the propeller drives a mass of water backwards, the reaction to which is a force that accelerates the boat forward. But in the vacuum of space, there's nothing; spacecraft have to take their own 'reaction mass' to push against – namely, rocket propellant. This comes in two parts – fuel plus an oxidant, a chemical containing the oxygen that's needed for combustion to take place. On Earth there is oxygen aplenty in the atmosphere but in space you must take your own.

Earth orbit

Some of the first trips into space were on what are called suborbital trajectories, meaning the flight paths were giant arcs through the sky – essentially scaled-up versions of throwing a ball in the air and watching it come back down. America's first manned spaceflight – made by Alan Shepard in 1961 – was a 15-minute suborbital trip, landing Shepard 480km (300 miles) from his launch site.

Putting a rocket into orbit around the Earth requires much more power than the quick up and down of a suborbital flight. Orbit is a special feature of gravity where the gravitational pull of a planet provides the **centripetal force** needed to keep a spacecraft circling around it. Rockets need to reach extremely high speed to attain Earth orbit, typically around 28,000km/h (17,00mph), making orbital flight inherently more dangerous. Re-entering the atmosphere at this speed generates temperatures of thousands of degrees Celsius, meaning spacecraft need effective heat shields so as not to burn up. Russian Yuri Gagarin was the first human to orbit the Earth – in April 1961.

Escape velocity

Sometimes space missions require a spacecraft to go beyond suborbital flight and Earth orbit – and leave our planet's gravity behind altogether. Scientists can use the laws of **Newtonian gravity** to calculate how fast a rocket needs to travel to do this. For the Earth, it comes out to around 40,000km/h (25,000mph), known as Earth's escape velocity. The Moon, with its lower gravity, is easier to get away from – with an escape velocity of just 8,600km/h (5,300mph).

Earth's escape velocity is enough to send a rocket off on a journey to the planets. However, embarking upon a voyage to the stars will require a craft that can travel even faster – as it'll need to break away from the Sun's gravity as well. Thankfully, at Earth's orbital distance the gravity of the Sun isn't as strong as it is close up, but a spacecraft that has already escaped Earth's gravity will still need to find an additional 150,000km/h (93,000mph).

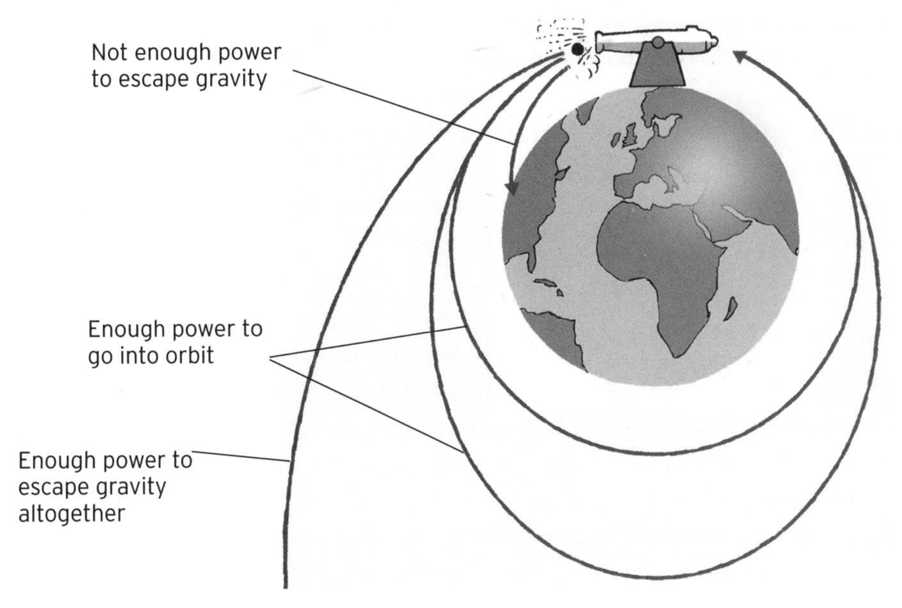

Artificial satellites

Artificial satellites are man-made spacecraft that circle in Earth's orbit. The first was the Russian Sputnik 1, launched in 1957, which made 1,440 orbits of the Earth before burning up in the atmosphere early in 1958. Most artificial satellites are uncrewed, performing remote-controlled functions in space such as Earth-observing, astronomy, sending navigation signals (as is the case with the GPS satellites) and acting as communication platforms. A small number, however, have human crews; at present the only permanently occupied human outpost in orbit is the International Space Station (ISS). This is by far the largest satellite in Earth orbit, measuring over 100 metres (325 feet) across. It is a scientific platform for studying the effects of space on the human body and for conducting other research in space's zero-gravity environment – which can't be carried out on Earth.

Space tourism

Space tourism has now arrived, with the first commercial charters to the high frontier – rocket ships loaded with tourists, each paying hundreds of thousands of dollars to experience the thrill of outer space. The leading space tourism operator is Virgin Galactic, run by British entrepreneur Richard Branson, offering suborbital flights into space aboard a winged rocket ship that glides back down to land on a conventional runway. Tickets cost $200,000 each, though this hefty price tag is expected to fall considerably in the decades to come. Future space tourism projects promise orbital flights and possibly the chance to make extended stays, of days or weeks, in space hotels – circling the Earth or even on the Moon.

Clarke orbit

The time it takes for an Earth-orbiting body to make one complete circuit of the planet depends on the altitude of the orbit. Low-Earth orbit, relatively close to the planet, has a period of about 90 minutes but this time span increases as you move outwards. There comes a point where the orbital period equals the time it takes for the Earth to spin once on its axis, i.e. 24 hours. An artificial satellite placed in the Earth's equatorial plane at this altitude – 35,786km (22,236 miles) – will appear to hang in the sky above the same spot on the planet's surface.

In 1945, a little-known science fiction author called Arthur C. Clarke published an article in the British electronics magazine *Wireless World* suggesting that satellites occupying such orbits could be used to bounce radio signals around the world. Hundreds of such communication satellites now encircle the Earth in these 'Clarke' orbits.

Planetary probes

A small number of spacecraft have been placed in orbit around other bodies of the Solar System. Launched in 1959, the Soviet Luna-1 was the first man-made space probe to leave Earth orbit; it made a fly-by of the Moon before entering orbit around the Sun. Its sister craft Luna-10, launched in 1966, was the first to orbit around the Moon. Orbiting another planet would take a further five years, but was achieved by NASA's Mariner 9 spacecraft in 1971 when it orbited Mars and returned with the first close-up photography of the Red Planet.

Other spacecraft have been even bolder – not content with merely orbiting other bodies, they've landed on them. Russia's Luna-3 was the first on the Moon's surface in 1959, although this was more of a crash than a landing. Luna-9 made the first soft landing there in 1966 – the same year that Russia's Venera-3 crashed into the surface of the planet Venus. Robotic spacecraft have since orbited Jupiter, Saturn, and made fly-bys of Uranus and Neptune. Meanwhile, controlled landings have been made on Mars, Venus and Saturn's moon, Titan.

Ion engines

One alternative to rockets that has recently made the jump from experiment to proven technology is ion propulsion. Ion engines work on the same basic principle as rocket propulsion – they carry fuel and create motion by throwing it out behind them as fast as possible. But where rockets do this by burning the fuel, ion engines give it an **electric charge** and then accelerate it using an electric field.

Spitting out the fuel atom by atom in this way may seem painfully slow – and it is. Ion engines deliver a tiny thrust, meaning that it takes days or weeks of continual running to build up a head of speed. However, the advantage is that they are extremely efficient; each gram of fuel in an ion engine will deliver up to 20 times as much thrust as the same mass of combustible chemical fuel. Eventually.

Solar sails

Solar sails are a completely different propulsion technology to rockets. These large sheets of silvered Mylar quite literally hitch a ride on the light from the Sun. As was demonstrated by Einstein when he worked on the photoelectric effect in 1905, light is not just a wave – it can equally well be thought of as particles, known as **photons**. Sunlight is full of high-energy photons and, just like particles of air beating against the sail of a yacht, photons raining against a solar sail impart momentum which pushes the sail along. Angling the sail enables a pilot to steer it, tilting the sail to accelerate the craft's and so vary the size of its orbit around the Sun.

Humans in space

In the 1960s, human beings finally left the Earth. First we flew outside the atmosphere, then we orbited the Earth and finally, as the decade was drawing to a close, we went to the Moon. Yet since then, humans have done relatively little to advance the crewed exploration of space. After the Moon, the plan was to forge on to Mars, and land the first crews there by 1986. But propelling humans across the gulf of interplanetary space has proven harder than we first thought. Studies have shown that radiation in deep space poses a lethal hazard to astronauts. And adding bulky shields to spacecraft is impractical, as this adds greatly to the weight that has to be lifted into space. Even so, scientists at Rutherford Appleton Laboratory, in England, have come up with a possible solution. Most of the harmful radiation in space consists of electrically charged particles and the RAL team believe that a spacecraft could be wrapped in a magnetic bubble that can bat away this subatomic threat. Better still, the magnets required to do this would be small and light enough to launch into space.

LIFE IN SPACE

CHON

Carbon, hydrogen, oxygen and nitrogen are the elements upon which life on Earth is based and they go under the collective acronym CHON. Hydrogen emerged in huge quantities from the Big Bang. The other three CHON elements were forged in the nuclear furnaces inside stars – binding elements together to make progressively heavier atomic nuclei. When these stars ended their lives in **supernova** explosions, the elements were scattered across the Universe to form clouds in interstellar space from which future generations of stars and planets would grow. Experiments have shown how CHON chemicals in water subjected to sparks of electricity form amino acids – the building blocks of organic life (see **Miller–Urey experiment**).

Life in the solar system

Earth is the only place in the Solar System where life has been found, though many scientists suspect that primitive microbial life may one day be discovered on Mars and on Jupiter's moon Europa. On Mars, a number of recent discoveries have led scientists to believe that life may be lurking deep underground – water-ice has been discovered there and observations from orbit have detected methane gas in the planet's atmosphere. Atmospheric methane is easily broken down so its detection suggests that a source is present on Mars. This could be volcanic, although the lack of obvious active vulcanism on Mars makes this implausible; a biological origin seems more likely to many.

Meanwhile, some planetary scientists believe Europa has an ocean of liquid water tucked away beneath its surface. Heat generated by the squashing and squeezing of nearby Jupiter's gravity keeps the water from freezing and could sustain it at a temperature warm enough to support life.

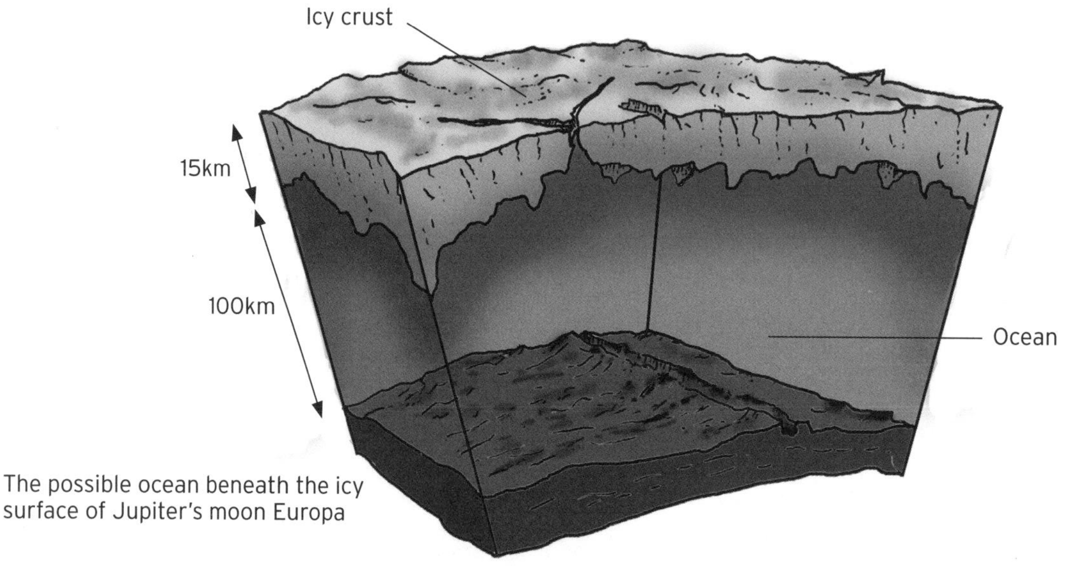

The possible ocean beneath the icy surface of Jupiter's moon Europa

Panspermia

The possible transfer of microbes between planets in the Solar System is known as panspermia. It's an idea that was first put forward by Swedish chemist Svante Arrhenius in 1903 and later developed by British astronomers Fred Hoyle and Chandra Wickramasinghe.

Organisms are believed to be able to drift through space inside rocks, where they are shielded from the rigours of the extraterrestrial environment. Some scientists believe that micro organisms inside tiny specks of dust can be blown out through interplanetary space by the radiation from the Sun – the same method of propulsion used by **solar sails**. In November 1969, the Apollo 12 mission landed on the Moon just a stone's throw from Surveyor 3 – a robotic lander that had touched down two years earlier. Astronauts from Apollo 12 recovered microbes from inside Surveyor 3's camera which, when returned to Earth, were found to have survived their stay in space. The discovery gave the theory of panspermia a massive boost.

Earth-like planets

Astronomers are starting to detect planets orbiting other stars that are similar in mass to our own, and that orbit in the 'Goldilocks zone' – where the temperature is just right for liquid water to exist. The first planets beyond our Solar System – called **exoplanets** – were detected in 1995. Most exoplanets found so far are Jupiter-like but as detection technology improves, lower-mass worlds are coming into view. The most Earth-like planet known today is called Gliese 581d with a mass around seven times that of the Earth. Although it orbits at just 0.2 Earth orbital radii, its parent star is a cool red dwarf – making the temperature very similar to that of the Earth.

In 2009, NASA launched its Kepler space mission – a space telescope that's watching for the minute, tell-tale dimming caused by small planets making **transits** across the faces of distant stars. A further space mission, called Darwin and proposed by the European Space Agency, will use optical **interferometry** to analyze the atmospheres of Earth-like worlds. The simultaneous presence of water, carbon dioxide and ozone (a **molecule** made of three oxygen **atoms**) would be a sure sign of life.

Drake equation

In 1960, American astronomer Frank Drake wrote down a mathematical equation to predict the number of extraterrestrial civilizations that exist in the Milky Way. Known as the Drake equation, it's essentially a long list of multiplicative factors, each encapsulating the probability of particular events that would be necessary – first, for the emergence of life, and then for it to evolve intelligence. They include terms such as the rate of star formation,the fraction of these stars that have planets, the fraction of these planets that can support life, and so on.

When Drake originally formulated the equation and estimated the numerical values of its terms, he obtained an answer of about ten intelligent extraterrestrial civilizations in our galaxy. The best values for the terms according to modern science give the somewhat more conservative result of two.

SETI

The search for extraterrestrial intelligence (SETI) is an effort by astronomers to detect signals broadcast by intelligent life elsewhere in the Universe. Searches are typically conducted at radio wavelengths close to 21centimetres, the wavelength of radio waves given off by clouds of atomic hydrogen gas in the Milky Way. An alien civilization might broadcast near to this wavelength in the hope that its signal would be picked up accidentally by radio astronomers studying the hydrogen. Scientists have been carrying out SETI searches since 1960, using some of the biggest radio telescopes available – including the 300m Arecibo dish and the new Allen Telescope Array, an **interferometry** network of radio dishes funded by Microsoft co-founder Paul Allen. And yet no confirmed alien signal has ever been detected.

Some have sent deliberate transmissions into space. In 1974, astronomer Frank Drake used the Arecibo telescope to beam a 'hello' to the Hercules globular star cluster. Other scientists, including Cambridge's Professor Stephen Hawking, have warned of the dangers of giving away our location to potentially hostile alien races.

Fermi paradox

In 1950, Italian-American physicist Enrico Fermi posed the question 'Where is everybody?' – if there are intelligent lifeforms elsewhere in the Universe, then why don't we see them? This has become known as the Fermi paradox. Fermi's reasoning was simple. Given the huge number of stars in the Universe – over 10,000 billion billion – even the smallest chance of an intelligent civilization emerging must be realized somewhere. Once an advanced civilization emerges in our Milky Way it will expand, spreading across the entire galaxy in a few tens of millions of years – the blink of an eye compared with the 13.7-billion-year age of the Universe. Yet we see no evidence for such interstellar empires.

SETI enthusiasts resort to a number of arguments to explain the Fermi paradox – for example, speculating that the aliens may choose not to interact with Earth owing to some kind of non-interference policy. Scientists of a more sceptical outlook take it as evidence that life in the Universe is just very, very unlikely – a school of thought that has become known as the 'rare Earth hypothesis'.

Kardashev scale

How advanced an extraterrestrial civilization has become is measured on what's called the Kardashev scale, drawn up by Russian astronomer Nikolai Kardashev in the 1960s. Kardashev split advanced alien civilizations into three categories – which he labelled types I, II and III – according to the amount of energy that they have access to. A type I civilization is able to harness all the energy on its home planet, requiring the construction of a vast solar panel to capture the light arriving from the planet's host star – as well as putting every available atomic nucleus through nuclear fission and fusion. It's been calculated that the energy available to a type I race would be equivalent to what you'd get by detonating a large thermonuclear bomb every second. Type II civilizations have extended their mastery further, to capture every scrap of energy given out by their home star; it would involve building an energy-gathering structure around the star, such as a Dyson sphere. Meanwhile, an alien race of type III is able to make use of all the energy given off by every star in its native galaxy – around 100 billion times the energy output of the Sun. Human civilization is still some way below type I.

Dyson spheres

Suggested by the physicist Freeman Dyson, a Dyson sphere is a structure built around a star to capture and harness its energy. The sphere could take the form of a rigid framework or, more likely, would comprise a swarm of energy-gathering spacecraft that together enshroud the star. A civilization that succeeded in building a Dyson sphere would attain type II status on the Kardashev scale.

Some astronomers say Dyson spheres might be detectable – a sphere enclosing a star would heat up and re-radiate energy at infrared wavelengths, which could in principle be picked up through infrared astronomy. Others have wondered whether alien-built structures orbiting a star – maybe parts of an incipient Dyson sphere – could be detected optically as they **transit** across the star's face. With sufficiently sensitive instruments, their sharp, angular profile could be distinguished from the smooth, round silhouette of a planet. Hunting for such objects has been dubbed SETT – the 'search for extraterrestrial technology'. It has even been suggested that NASA's Kepler space telescope, launched in 2009, might be up to the task.

Anthropic principle

The anthropic principle is a method of scientific reasoning that works by constraining the laws of physics, chemistry and biology by the simple fact that life has arisen here on Earth; any candidate scientific theory that expressly forbids life in the Universe must therefore be wrong. British astronomer Fred Hoyle famously used the anthropic principle to predict that a particular carbon-producing nuclear reaction must take place inside stars. If it didn't there wouldn't be enough carbon around for organic life, including ourselves, to emerge. Experimental physicists went away and looked for Hoyle's prediction – and duly found it.

Some scientists have commented how incredible it is that we are here to observe ourselves at all – so improbable is it that the laws of physics should be tuned just right for life. Others invoke the **many worlds** idea of **quantum theory** to argue that it's really no surprise at all – if many worlds is correct, they say, then every possible universe exists somewhere. And we must – by virtue of our existence – find ourselves in one of the small number where life is permitted.

HEALTH AND MEDICINE

PUT SIMPLY, MEDICINE IS THE SCIENCE OF HEALING. Primitive forms of medical practice were a feature of many ancient civilizations, including the Chinese, Egyptian and Indian. But it is the healing hands of the ancient Greeks to which we owe the biggest debt of gratitude – not necessarily for their knowledge, but for the methodology they instigated. The Greek physicians Hippocrates, generally revered as the Father of Medicine, and Galen laid the foundations for a scientific approach to medical care, where treatments are prescribed based on evidence, rather than folklore and superstition. And it was the Greeks who introduced the Hippocratic oath, a vow to practise medicine ethically which even today all new doctors must swear by.

RT • BLOOD • LUNGS • MUSCULOSKELETAL
ER • SKIN • URINARY SYSTEM • IMMUNE
OLOGY • PHYSIOTHERAPY • PAEDIATRICS
OTOLARYNGOLOGY • OPHTHALMOLOGY
CCUPATIONAL THERAPY • EMERGENCY
Y • BACTERIOLOGY • VIROLOGY • PRIONS
AIDS • CANCER • OBESITY • DIABETES •
E • BLOOD PRESSURE • TEMPERATURE
ON • BLOOD AND URINE TESTS • BIOPSY
DEFIBRILLATION • KIDNEY DIALYSIS •
-RAYS • ULTRASOUND • ANGIOGRAPHY •
TROENCEPHALOGRAPHY • TOMOGRAPHY
ECULAR IMAGING • PHARMACOLOGY

While the Dark Ages following the fall of the Western Roman Empire in AD476 halted medical progress for hundreds of years, it picked up pace again in the 18th century with a string of discoveries – including antiseptics, anaesthetics, vaccines and, later, antibiotics – which would pave the way for the development of modern medicine.

Nowadays, the genetics revolution that dominated biology in the 20th century is making its presence felt in the world of medicine with many promising new treatments for deadly illnesses, including cancer and AIDS, based on gene therapies that reprogram or reinterpret patients' DNA.

THE HUMAN BODY

Human anatomy

Anatomy is the study of the structure and layout of the organs making up a living being. Human anatomy breaks down into a number of areas. 'Superficial anatomy' concerns the layout of surface features and structures – arms, legs, head, genitalia, and so on; internal anatomy divides into particular systems of the body – for example, the gastrointestinal system, the musculoskeletal system and the immune system. This study of large-scale organs in the body is also known as 'gross anatomy'. At the other end of the scale is 'microanatomy'; it breaks down into areas such as cytology (the anatomy of **cells**) and histology (the anatomy of **tissue**).

Human physiology

Physiology concerns the organs and other components of the human body – with emphasis on the functions that they perform, such as respiration, circulation, digestion and excretion. Many of the functions of these systems overlap – for example, the heart and lungs both contribute to the 'cardiovascular system' that circulates oxygen around the body. Different systems of the body are integrated and the interactions between them mediated by communication systems such as **hormones** and the nervous system (see **Neurobiology**).

The study of physiology is one of the oldest medical sciences, dating back to the time of the Greek philosopher Hippocrates in the 5th century BC. Physiology took off in the 19th century with the development of the theory of **cell biology**, which enabled scientists to understand the functions of the human body.

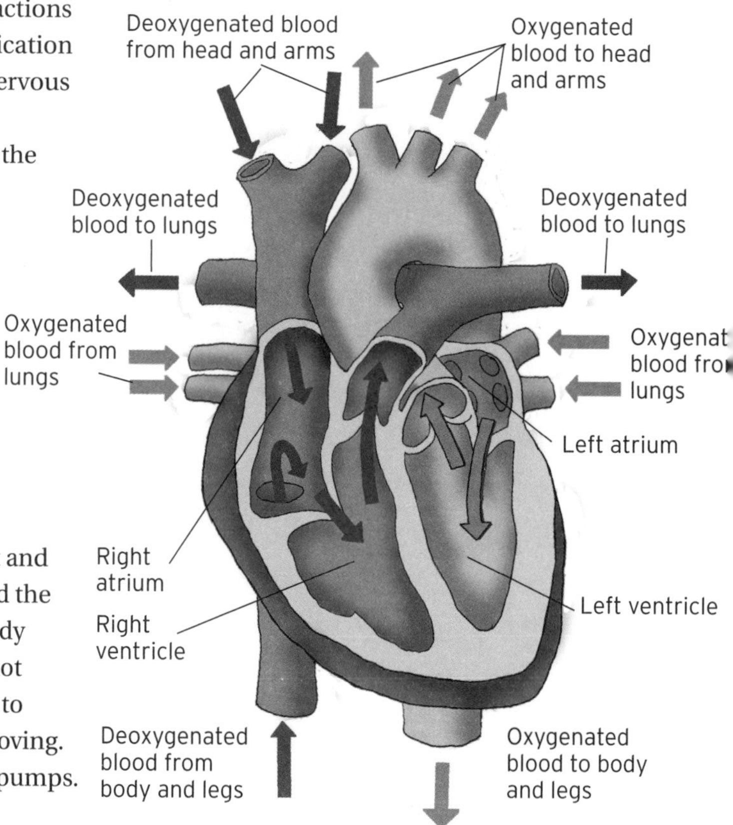

Heart

The workings of the heart and the vessels that circulate blood around the body fall under the area of clinical study known as cardiology. The heart is a knot of muscle that expands and contracts to work as a pump, keeping the blood moving. In fact, it is divided into two separate pumps.

Half the heart – the right ventricle and right atrium – drives the 'pulmonary system', a network of vessels that carries blood through the lungs in order to oxygenate it. The freshly oxygenated blood then enters the heart's second pump – the left ventricle and left atrium – which forces the blood through the 'systemic circulation', carrying it around the whole body. Blood vessels are divided into those that carry blood away from the heart – the arteries – and those that return it to the heart – the veins. The systemic circulation of the human body contains a staggering 96,000km (59,650 miles) of blood vessels, enough to go more than twice around the Earth.

Blood

Blood is liquid tissue that is pumped through the circulation system by the heart and transports oxygen, nutrients and chemical messengers known as **hormones** around the body, as well as carrying away waste products. The medical study of blood is known as haematology.

Blood has four principal components. Red blood cells, also known as 'erythrocytes', carry oxygen around the body – when blood passes through the lungs, a protein in the cells called haemoglobin binds to oxygen gathered by lung tissue. White blood cells, also called 'leucocytes', are part of the immune system that helps the body to fight infection. Platelets, or 'thrombocytes', are the third blood component and they play a role in blood clotting – essential in the healing of wounds. The final component of blood is plasma, the fluid component in which the cells are suspended. Blood cells are manufactured in the marrow inside bones, and a cubic millimetre of human blood contains around 5 million erythrocytes, 4,000–10,000 leucocytes and 150,000–400,000 platelets.

Lungs

All air-breathing animals use lungs to filter oxygen from the air and transfer it to the blood – and, conversely, to expel waste carbon dioxide back to the atmosphere. Oxygen is used by body cells for aerobic **respiration** – a process for converting nutrients into energy.

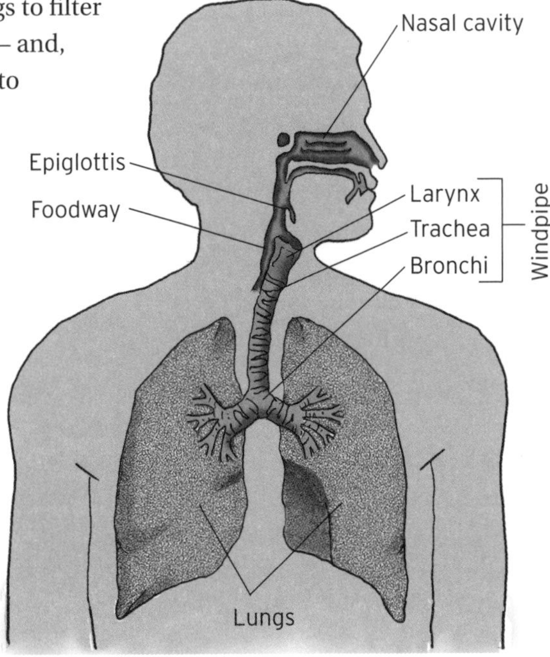

Humans have a pair of lungs situated in the thorax – the main cavity within the ribcage – consisting of a membrane through which the gases are exchanged. The membrane has a total area of around 70 square metres – about the same size as a tennis court – but is folded up into a compact organ. Air enters through the 'windpipe'; because the airway and foodway in mammals both cross over, the windpipe is closed off by a flap of tissue, called the 'epiglottis', while food is being swallowed, to prevent choking. Breathing is controlled by muscles around the lungs that contract and expand to expel and draw in gas. The clinical study of the lungs and airways is known as pulmonology.

Musculoskeletal system

The human musculoskeletal system, the linkage of bones and muscles that work together to produce movement and give the body its rigidity, consists of not just bones and muscles, but also the ligaments and cartilage joining the bones together, and the tendons which connect muscle to bone. Muscles are fibrous bundles of cells that are capable of contracting in response to nerve impulses; they come in three types. 'Voluntary muscle' moves in response to a consciously produced signal from the brain – walking or using your arms involves voluntary muscles. 'Involuntary muscle' is what powers internal organs, such as shuffling food through your digestive tract, and it operates subconsciously. Lastly, 'cardiac muscle' is the special kind of muscular tissue found in the heart.

Bone, on the other hand, consists of a scaffold made from fibres of collagen protein into which are woven minerals, such as calcium phosphate (which give rigidity), and bone cells which regulate the **biochemistry** and behaviour of the body's skeletal structure.

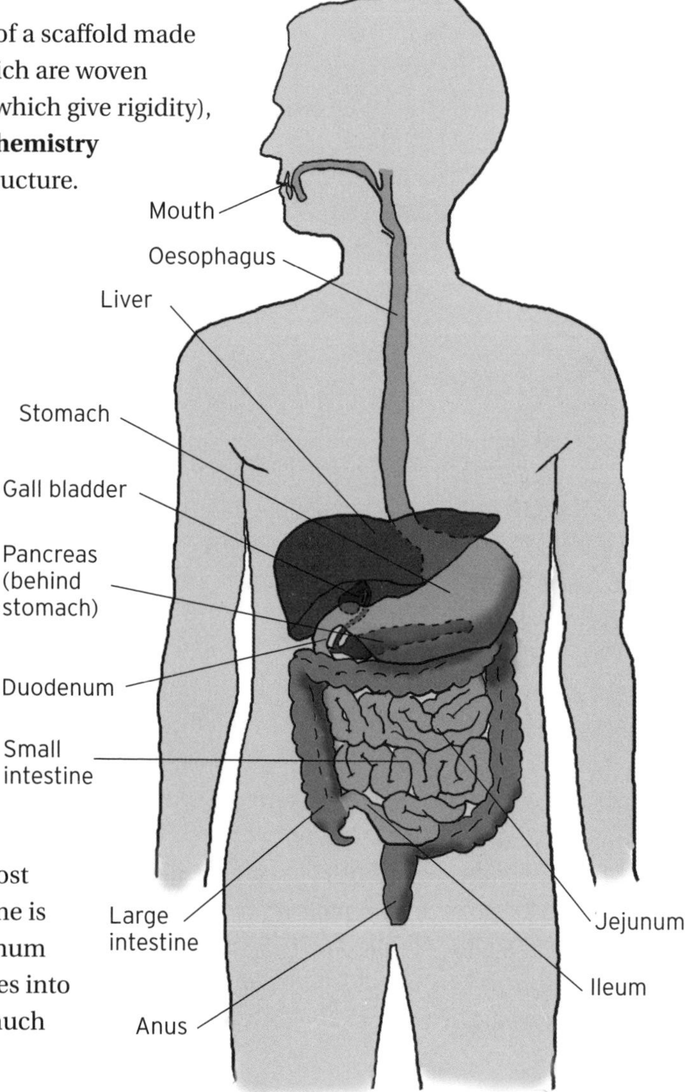

Gastrointestinal system

The chain of organs found in mammals for the digestion and absorption of food is known as the gastrointestinal system. Food enters the gastrointestinal system at the mouth and then passes into the oesophagus, a passage leading down to the stomach, where acidic gastric juices are mixed into the food by contraction of the stomach muscles. It then passes out of the stomach and into the small intestine, which is the powerhouse of the human digestive system – here, food is converted into nutrients via a host of digestive **enzymes**. The small intestine is split into three parts – duodenum, jejunum and ileum. Next the residual mass moves into the large intestine, which removes as much water as possible before it reaches the anus for excretion. Other organs – such as the liver and pancreas – contribute to the gastrointestinal system by secreting further digestive enzymes. The gastrointestinal tract in an adult human, from mouth to anus, can be up to 9 metres (30 feet) in length. Doctors who specialize in treating conditions of the gastrointestinal system are known as gastroenterologists.

Liver

Besides the heart, lungs and kidneys (see **Urinary system**), the liver is another of the body's essential organs. It plays a major role in the digestive system, producing chemicals essential for the processing of food – such as fat-digesting bile. Bile is stored temporarily in the gall bladder, and enters the digestive tract at the duodenum, at the top of the small intestine (see **Gastrointestinal system**). The liver also plays a part in removing toxins from the blood (such as alcohol) and, along with muscles, stores glycogen – a chemical used by the body for short-term energy reserves. The study of liver diseases is known as hepatology.

Skin

Technically it's known as the integumentary system – the skin that covers the exterior of the body, together with the hair and nails. The clinical study of skin is known as dermatology. The outer layer of the skin, the epidermis, is made of a type of **tissue** known as epithelial cells, which also make up the protective covering for most of the internal organs and ducts in the body. Beneath the epidermis is the dermis, a layer made from collagen and other proteins. The innermost layer is the subcutaneous tissue, composed mainly of fat. Hair and nails, meanwhile, are made of tough protein called keratin.

Skin plays an important role in protecting the body from injury and infection, and for retaining water. It also mediates the body's interactions with its environment, being laced with nerve receptors (see **Neurobiology**) providing sensitivity to touch, pain and temperature. And it plays a part in **homeostasis**, helping to regulate body temperature through the sweat glands embedded in the subcutaneous tissue, which deliver moisture to the surface through channels called pores.

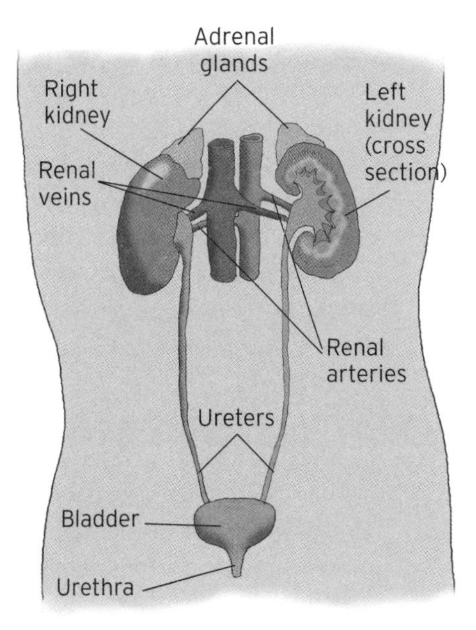

Urinary system

Excretion of waste chemicals from the body's **metabolism** is dealt with by the urinary system. The urinary system's principle organs are the kidneys, which filter waste products from the blood – primarily urea (a by-product of breaking down protein) and uric acid (produced in the breakdown of nucleic acids). The kidneys also play a role in **homeostasis**, by regulating blood pressure, blood acidity (see **Acids and bases**), electrolyte levels and by the production of **hormones** such as erythropoietin, which regulates the production of red blood cells. From the kidneys, liquid waste is transported along a tube known as a 'ureter' and into the bladder where it is stored prior to excretion through the 'urethra'.

There are two kidneys in the human body, located towards the back of the abdominal cavity. The study of kidney function is called 'renal physiology' and the clinical study of diseases of the kidney is known as 'nephrology'. Meanwhile, the clinical study of the urinary system taken as a whole is 'urology'.

Immune system

The immune system is a complex network of processes throughout the body that swing into action in response to the presence of an infection – usually in the form of viruses or bacteria (see **Bacteriology** and **virology**). One of the primary mechanisms of immune response consists of the production of 'antibodies' by white blood cells known as 'lymphocytes'. Antibodies are proteins that stick to invading cells – known as 'antigens' – and neutralize them. However, there is no 'catch all' antibody; instead, they must be tailored to each specific antigen, which is done by lymphocytes known as B cells that recognize the surface structure of particular antigens. With the aid of other lymphocytes, called 'helper T cells', the B cells generate a flood of antibodies tuned to seek out and destroy the invading antigens.

Once the infection is destroyed the antibody levels return to normal, although some degree of memory of the antigen is retained by the immune cells so that in the event of re-infection their response is much swifter, known as 'acquired immunity'. Some diseases can turn the immune system against the body – causing antibodies to attack healthy cells. Called 'autoimmune' diseases, they include lupus and rheumatoid arthritis.

Reproductive system

The set of organs that deal with the production of **gamete** cells, the mechanics of sexual intercourse and the gestation of offspring through to birth, is known as the reproductive system. In females, this includes the ovaries, fallopian tubes and uterus, as well as the vagina and cervix; in males, the reproductive organs are the penis, testicles and the sperm ducts connecting them. There are a number of other body functions that can also be considered as indirect contributors to the reproductive system such as the **hormones** produced by the 'endocrine system', and the generation of pheromones – odour chemicals which are thought to boost the sexual arousal of a mate.

Medical treatment of reproductive disorders is known as 'gynaecology' in females and 'andrology' in males. Gynaecology is often grouped together with the clinical care of pregnant women – 'obstetrics'.

Neurobiology

Study of the nervous system is known as neurobiology. The systems includes the network of nerve fibres that thread throughout the body that relay senses such as touch, taste, smell, sight and sound, and carry the electrical impulses triggering our muscles to work (see **Musculoskeletal system**). It also includes the brain itself – a central mass of nerve cells, called neurons, that coordinates all of the body's functions and activities and which serves as a central processing unit housing our thoughts, our emotions and our very consciousness.

Neurons have the ability to communicate with other neurons via gateways called 'synapses'. Some types of neuron have extended structures called axons, to which many other neurons can form electrical connections, allowing complex networks to be built up. The 'peripheral nervous system', the mass of nerves that run through most of the body, is made up of thin bundles of axons. On the other hand, the 'central nervous system' is made up of neurons and comprises the core structures of the brain and the spinal cord (see **Vertebrates**).

MEDICAL CARE

Physiotherapy

Physiotherapy is a branch of medical care concerned with improving and restoring body function – usually following injury, illness or age-related debilitation – to maximize quality of life. Often physiotherapy will involve the treatment of musculoskeletal system disorders by the manual manipulation of muscles, bones and joints – alongside the development of exercise routines – to improve mobility. This is known as 'orthopaedic' physiotherapy. But the techniques of physiotherapy are also used to address conditions of the heart and lungs, and the nervous system (see **Neurobiology**) – where they are effective for relieving the symptoms of neurological diseases such as cerebral palsy and multiple sclerosis. Physiotherapy is even applied to skin problems – particularly in rehabilitation following burns. Unlike other forms of manipulative therapy – such as **chiropractice** and **osteopathy**,which are areas of complementary medicine – physiotherapy is a branch of mainstream medical science.

Paediatrics

The area of medicine that deals with the treatment of injury and illness in children (up to 18 years of age in the UK and 21 in the United States) is known as paediatrics; practitioners are called paediatricians. It is a specialized discipline owing to the physiological differences between adults and children and the different range of illnesses that children are prone to. Paediatricians are usually present at caesarean births and at other high-risk deliveries.

In 1802 the first children's hospital opened in Paris – and is still open today. Later on in the 19th century other countries followed suit, with the opening of London's Great Ormond Street Hospital in 1855, and the Children's Hospital of Philadelphia in the United States in 1855.

Fertility

Fertility treatment is an area of medicine that seeks to assist couples to conceive children. There are a number of factors that can cause fertility problems and these can arise in both men and women. Common causes in women include difficulty ovulating, which is the process whereby an egg cell (see **Gametes**) is deposited into the uterus; damage to the fallopian tubes, which can cause an egg to lodge there (which may become fertilized and develop – a dangerous condition called 'ectopic pregnancy'); or simply age, as female fertility drops sharply past 35 years. The principle cause in men is low sperm count (oligospermia).

A number of medications are available for women, which stimulate ovulation. There are, however, no drugs to effectively treat oligospermia. In this case, in-vitro fertilization (IVF) is an option, involving the extraction of eggs from the female which are then manually fertilized with sperm from the male and the resulting embryo implanted back into the uterus and brought to full term. Studies into the effectiveness of IVF around the world indicate success rates of between 20 and 40 per cent.

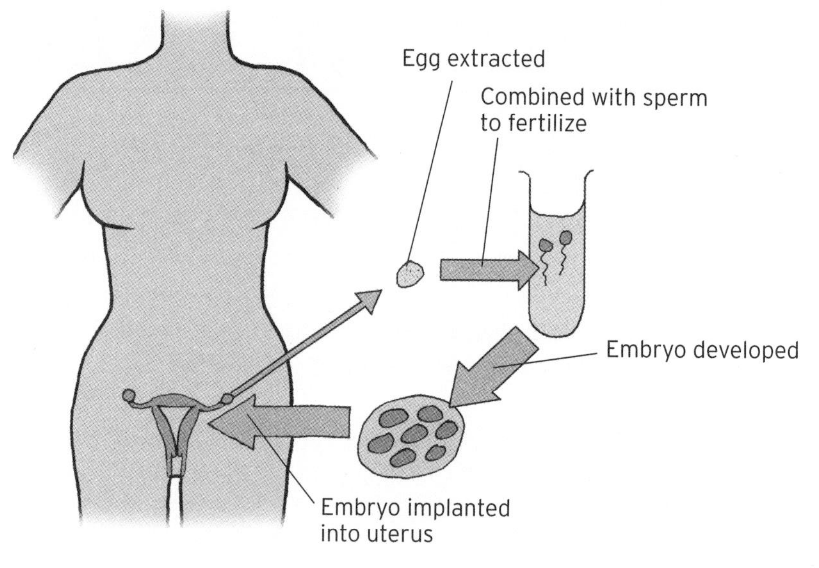

Geriatrics

At the other end of the medical scale from paediatrics is geriatrics – the care of the elderly. This may involve the treatment of age-related illnesses, as well as the specialized care needed to account for physiological changes brought about by age. The job of the geriatrician is made harder by the fact that many elderly patients may be suffering simultaneously from a range of conditions. Treating them all runs the risk of adverse effects arising from interactions between different drugs, meaning that in some cases treatment of the most debilitating conditions must be prioritized. Common problems that occur in the elderly are immobility, incontinence and impairment of mental faculties, particularly memory (see **Dementia**).

As healthier lifestyles and better medical care continue to drive up life expectancy, geriatric care looks set to become increasingly important. In 1900 there were 3.1 million Americans aged 65 and older (around 4 per cent); by 2030, 20 per cent of all Americans alive are expected to fall into this age group. In the UK, figures published in 2008 by the Office of National Statistics revealed that there are now more over-65s in Britain than there are children under 16.

Dentistry

Dentistry is concerned with medical care of the teeth, gums, tongue and the interior of the mouth cavity. Much dental work is preventative, administered through regular check-ups and cleaning; though, from time to time, dentists need to perform surgery. The most common form of surgery is to fill dental 'caries' – the cavities left by tooth decay. Caries is caused by bacteria in the mouth which feed on the sugars in food and then excrete acid, which eats away at tooth enamel. Left untreated, the rot will eat its way down to the soft pulp at the core and cause the tooth to die. Dentists drill away rotten tooth material and fill the cavity either with a metal alloy, called 'amalgam', or a resin-based composite. Other dental procedures can be more involved, such as fitting a crown (a complete synthetic cap over the top of a tooth); root canal surgery (where the pulp inside the tooth is drilled out and filled); and extractions of troublesome teeth.

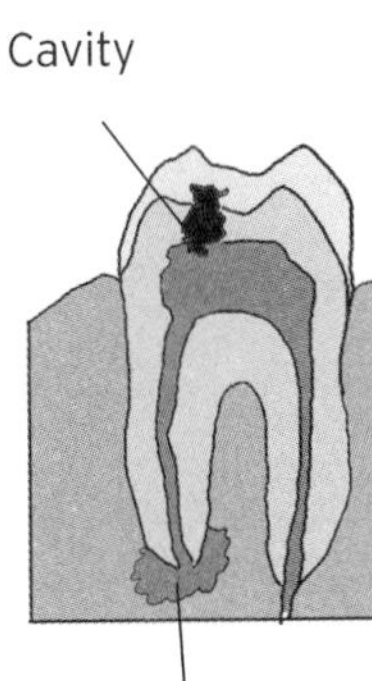

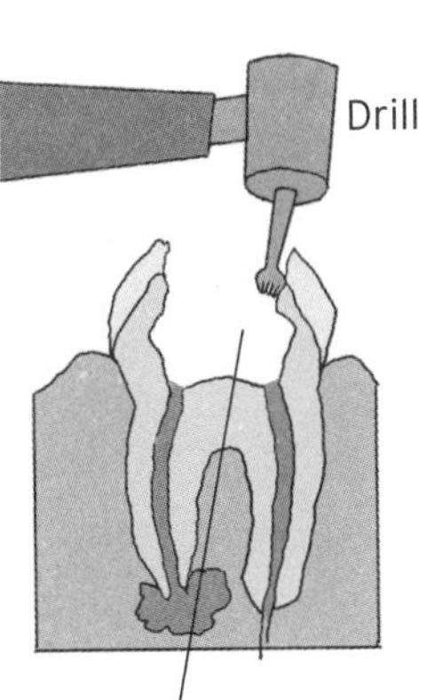

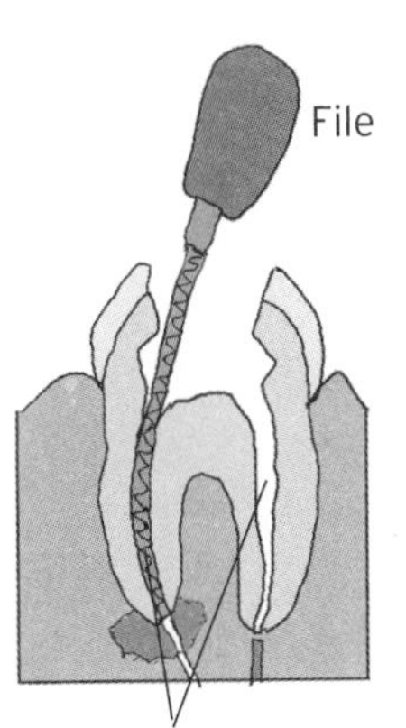

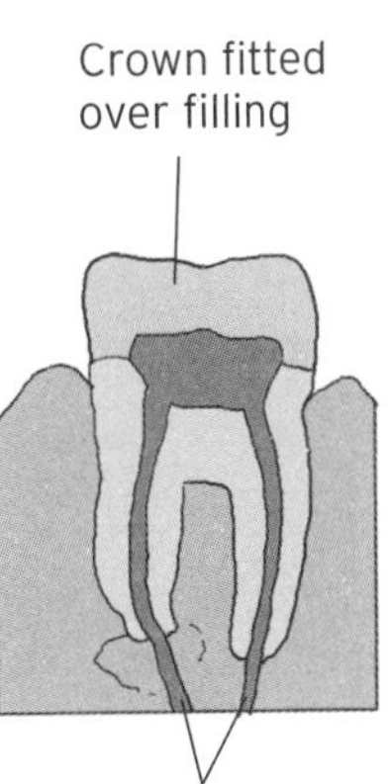

Otolaryngology

Medical disorders of the ears, nose and throat are treated by specialists in the field of otolaryngology; some otolaryngologists broaden their remit to treat general disorders of the head and neck. Conditions of the ears include otitis – inflammation of the ear canal – as well as hearing loss and dizziness caused by disruption of the fluids in the inner ear which contribute to our sense of balance. Nasal conditions include rhinitis, which causes a continual runny nose in response to inflammation of the mucus membranes inside the nose, and sinusitis which is inflammation in the sinus cavities of the skull that lead into the nose. In the throat, common disorders are laryngitis, inflammation of the larynx (the voice box), and nodules – growths on the vocal cords inside the larynx.

Ophthalmology

The clinical specialism dealing with diseases and disorders of the eye and the surrounding tissue is ophthalmology. Ophthalmologists treat conditions such as glaucoma (damage to the optic nerve), infections and injuries to the cornea (the transparent front of the eye), diseases of the retina (the light sensitive region inside the eye) and cataracts. Cataracts can cause partial or total blindness and result when the protein making up the lens of the eye 'denatures' – a process similar to the cooking of an egg white – turning the lens a milky colour.

There are three other categories of eye care: 'orthoptics' deals with eye movement and defects in the ability of both eyes to work together; 'optometry' is concerned with eye testing, measuring patients' visual acuity and issuing prescriptions for corrective aids. These aids usually consist of contact lenses or spectacles, and are fitted by the final class of eye-care specialist – 'opticians'.

Rheumatology

Rheumatology is the clinical science handling the treatment of disorders of the joints, connective tissues between the bones – such as cartilage and ligaments – and the muscles. These ailments are referred to collectively as 'rheumatism'; they are 'autoimmune' diseases – in which the immune system mistakenly identifies healthy tissue in the body as a foreign invader and attacks it.

Perhaps the most common form of rheumatism is rheumatoid arthritis, caused by the swelling up of tissue in the joints as a result of the immune response. Symptoms begin with discomfort and eventually lead to pain and immobility. Rheumatoid arthritis is treated symptomatically, using analgesics to alleviate the pain, and anti-inflammatory drugs to bring down the swelling. Sufferers also report that physiotherapy is helpful. There is, as yet, no cure.

Palliative care

The alleviation of the symptoms experienced by sufferers of serious illness is a sphere of medicine known as palliative care. It may mean providing medical care during the final months of a terminal condition, such as cancer or AIDS, or providing long-term support for those with chronic, yet non-life-threatening illness. The goal of palliative care is to relieve pain, fatigue, nausea and immobility – so as to improve patients' quality of life. Palliative care for the terminally ill is normally provided by organizations called hospices; the care may either be administered in specialized residential centres or to patients living in their own homes. As well as offering medication and other clinical treatments, palliative carers are often also skilled counsellors able to help patients come to terms with emotional distress.

Occupational therapy

Occupational therapy is a branch of medical care with the goal of improving the quality of life of patients through occupational pursuits – helping them overcome disabilities resulting from illness, injury or age so that they can carry out the activities they need to do in order to lead healthy and fulfilling lives. This means both giving them the means and mobility to pursue occupations, and selecting activities that are therapeutic to their particular condition. For example, a patient learning to get around in a wheelchair might need structural changes to their environment (widening of doorways, fitting of ramps, and so on) as well as a programme of physical exercise to build strength in their arms and shoulders. Another patient, who has suffered a stroke, might be encouraged to try arts and crafts as a way to rebuild their manual dexterity skills. Occupational therapy has a long history, dating back to the 1st century BC and the physicians of Ancient Greece.

Emergency medicine

For some people being admitted to hospital, their first point of contact will be the accident and emergency department (A&E) or emergency room (ER). Anyone who's sustained a serious accidental injury, or is taken suddenly and severely ill, will present themselves at the A&E (ER) for treatment (or be taken there by ambulance). The primary goal of emergency medicine is to stabilize patients so that their condition can be treated by specialists. Emergency medicine became a speciality in its own right in the United States during the 1960s. Prior to this time, the department would be staffed by physicians spared from the other parts of the hospital. In the UK the first full-time A&E consultant was at Leeds General Infirmary in 1952.

Intensive care

The constant monitoring and attention required by patients who are in an unstable condition, or critically ill, is known as intensive care. Many of the patients in intensive care may not be able to breathe on their own, and so require an artificial lung or 'ventilator', and may be suffering from failure of other organs too. Particular departments of a hospital may have their own intensive care wards to treat specific conditions – such as burns, nervous system disorders and cardiology (heart disorders).

Because of the equipment required – and the quality and quantity of staff needed to operate it – intensive care is expensive and places great demands on a hospital's resources. In the year 2000, for example, annual spending on intensive care medicine in the United States is estimated to have totalled well in excess of $70 billion (around £45 billion). Annual expenditure on intensive care in the UK in 2006 was £1 billion.

PATHOLOGY

Epidemiology

The medical science of diagnosing disease, and the study of its causes, is called pathology. Epidemiology is the subset of pathology dealing with the spread of diseases through large populations of organisms – in the case of medical science, human beings. Historically, the role of the epidemiologist was to chart the spread of infectious diseases – such as influenza or smallpox – as they moved through a population during an epidemic. Nowadays, epidemiologists have a broader remit that includes diseases which aren't spread by infectious agents – such as heart disease and lung cancer. Rather than being caused by infection, these conditions are due to 'environmental factors' – such as diet or toxic chemicals.

Infectious diseases, on the other hand, are normally passed through the population by either viruses or bacteria (see **Virology** and **Bacteriology**). Though there are other agents of human infection, such as fungi (see **Mycosis**), and rogue proteins known as prions (see **Prions**).

Bacteriology

Study of bacterial infections and the diseases that they produce goes under the banner of bacteriology. Bacteria (see **Prokaryote microbes**) that are potentially harmful are known as 'pathogenic'. Bacterial infections are often recognized because they tend to produce symptoms localized around the site of the infection. For example, a wound which develops a bacterial infection may become red, inflamed and painful; but symptoms are typically confined to that area. In extreme cases, however, a condition known as 'sepsis' may result when the infection spreads through the blood leading to a whole-body response of the immune system, causing fever, vomiting and – if not treated – organ failure and death. Notable diseases caused by bacterial infection include anthrax, E. coli, MRSA, tuberculosis, salmonella, tetanus, typhoid and leprosy. Bacterial infections are treated using **antibiotics**.

Viral infections (see **Virology**) can lead to 'secondary' bacterial infections – such as the sore throat and nasal congestion that accompanies influenza. The virus weakens the immune system, so allowing hostile bacteria to flood in.

Virology

Pathogenic viruses are dealt with by a branch of medicine known as virology. Some of the deadliest diseases known are spread by viral infection – such as smallpox, AIDS, hepatitis, Spanish influenza and ebola. Viral infections can be distinguished from infections due to bacteria (see **Bacteriology**) by the fact that the symptoms of a viral infection are normally 'systemic' – that is, they are not localized to the site of the infection.

Not all viral infections will lead to full-blown illness. The classic case is polio – where many people become infected with the polio virus but only around 5 per cent actually develop the disease. Polio is said to be highly contagious but not very 'virulent'. Compare that with the Spanish influenza virus that broke out in 1918, which was both contagious and virulent (see **Pandemic**). The only effective medication against pathogenic viruses are **vaccines and antivirals**.

Prions

The word prion is a contraction of 'proteinaceous infection' – prions are pathogens made mainly of protein. They are responsible for deadly diseases including bovine spongiform encephalopathy (BSE or 'mad cow disease') and its human form, Creutzfeldt–Jakob disease (CJD) – both of which are incurable and fatal. Prions were predicted theoretically in the 1960s but not discovered until the early 1980s, by Stanley Prusiner at the University of California, San Francisco.

Prions are believed to be proteins that have folded into the wrong shape. So-called prion proteins (PrPs) are produced normally by nerve cells (see **Neurobiology**) in the body. Whereas natural PrPs are destroyed over time, the misfolded ones aren't. As a result, they collect in the central nervous system – the brain and spinal cord – where they destroy cells, creating microscopic holes that give the tissue a spongy appearance when viewed under a microscope. The result is irreversible brain damage and death.

These diseases are spread by ingesting prion-infected tissue. Although prions contain no genetic material they are able to replicate, causing other protein molecules they encounter to assume the same misshapen form. A variant of CJD is believed to have arisen in humans through the consumption of BSE-infected beef.

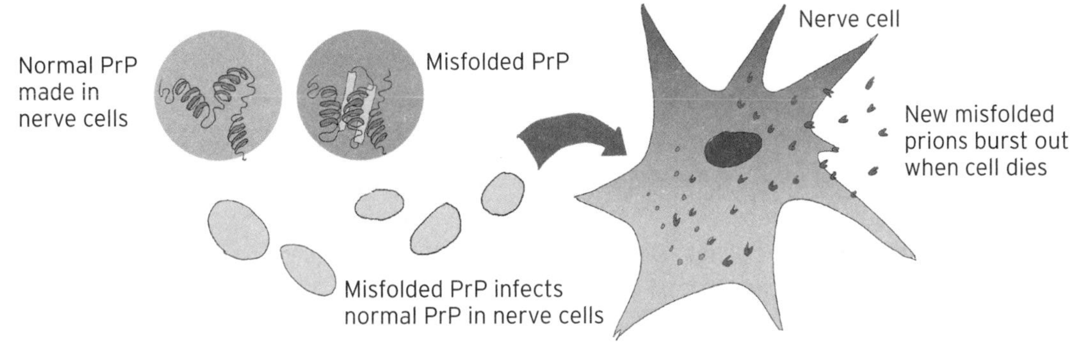

Mycosis

Fungal infections (see **Fungi**) of the human body are known as mycosis. In mild cases, mycosis affects just the outer layer of skin, the epidermis (called 'superficial mycosis'), but it can extend further down into the dermis layer and the tissue beneath ('subcutaneous mycosis'). In extreme cases, deep fungal infections can be life-threatening.

A common form of mycosis in humans is thrush, an infection of the mucus membranes, which can affect the mouth, the gastrointestinal system, urinary system and reproductive system. Symptoms are an itching or burning sensation in the infected area caused by fungus of the Candida **genus**. Another common fungal infection is athlete's foot, which causes flaking and cracking of the skin on the soles of the feet. Mycosis is treatable using antifungal drugs – also known as 'antimycotics', which exploit differences between fungal and human cells, triggering chemical processes that break down the outer membranes of fungal cells and cause them to die.

Pandemic

A disease epidemic that spreads to a significant portion of the human population – and sometimes a significant portion of the world – is known as a pandemic, from the Greek *pan* (meaning 'all') and *demos* (meaning 'people'). Perhaps the worst pandemic disease in history was smallpox, which killed 300 million people through the 20th century before it was eradicated. Meanwhile, the Black Death in the 14th century, wiped out a third of the population of Europe – an estimated 75 million. And the Spanish flu virus of 1918–1920 killed between 50 million and 100 million in the space of just two years – more than the number killed during the First World War.

Today, improved medical care has dramatically reduced the number of deaths resulting from the outbreak of harmful pathogens. At the start of 2010, the swine flu pandemic had killed just over 14,000 people worldwide – tragic deaths, but substantially fewer than have succumbed to the pandemics of the past. However, technology has proved to be a double-edged sword, with air travel spreading diseases around the globe overnight, and turning localized outbreaks into pandemics almost instantaneously.

MALADIES AND MORBIDITIES

Heart disease

The leading cause of death in the Western world is heart disease; in the United States alone, it is estimated to kill over 100 people every single hour. Heart disease is a general term for disorders of the heart muscle and the circulation system responsible for transporting oxygenated blood around the body. The biggest single killer is coronary heart disease, which accounts for the deaths of nearly 460,000 Americans every year, caused by failure of blood vessels to deliver sufficient oxygen to the heart muscle itself. This can lead to chest pains (called 'angina') and ultimately a heart attack (a 'myocardial infarction') as cells in the heart muscle die through oxygen starvation. Coronary heart disease is usually caused by atherosclerosis – fatty deposits inside the arteries – creating a blockage in the vessels that feed blood to the heart. Lifestyle is a big factor in bringing on heart disease, with contributory causes being smoking, excessive alcohol consumption and obesity. Age and family history also play a part.

AIDS

Acquired immune deficiency syndrome, or AIDS, is a disease of the immune system resulting from a viral infection by the human immunodeficiency virus (HIV). The virus attacks and destroys 'helper T cells' in the human body which play a crucial part in the **immune system**'s production of antibodies that fight off invading microorganisms. Effectively, this opens the door to infections by other bacteria, viruses, fungi and parasites, leading to serious illnesses such as pneumonia, tuberculosis, meningitis and cancer. Cancer is the biggest killer of HIV-positive patients.

HIV is spread by body fluids, such as blood, semen, vaginal fluids and breast milk. Unprotected sex is the most common form of transmission – followed by blood transfusions, the sharing of hypodermic needles among drug users, and transmission of the virus to infants from infected mothers. Once infected, a patient normally develops full-blown AIDS within ten years; however, the use of antiviral drugs can slow this process (see **Vaccines and antivirals**).

Cancer

A cancer is a malignant tumour that grows uncontrollably in regions of healthy tissue, inhibiting their function and ultimately causing death. Cancer was responsible for 13 per cent of all human deaths in the world in 2004. The disease is caused by genetic mutations – changes introduced into the DNA structure of a cell by factors such as radiation (see **Radiobiology**); harmful chemicals (called carcinogens) including tobacco smoke, asbestos and organic compounds known as polycyclic aromatic hydrocarbons; cancer-causing viruses (called oncoviruses), such as the human papillomavirus (HPV) that transmits cervical cancer; and inherited genetic defects.

Genetic errors cause new cells to grow abnormally, and at an accelerated rate – leading to a region of swollen tissue known as a tumour. Some tumours are benign. But when the abnormal tissue grows aggressively, invading the surrounding healthy tissue, and even spreads to other parts of the body through the blood (a process called 'metastasis'), then it is said to be malignant and cancer results. Common forms of the disease are lung cancer, bowel cancer and malignant brain tumours. Not all cancers form tumours – an exception is leukaemia, cancer of the white blood cells. The clinical study and treatment of cancer is known as oncology.

Obesity

People who have accumulated so much body fat that their weight poses a serious hazard to their health are known as obese. Strictly speaking, you are obese if your 'body mass index' – calculated by dividing your weight in kilograms by the square of your height in metres – exceeds 30. (Though there is criticism of this definition as it takes no account of muscle, which is heavier than fat.) While technically not a disease in its own right, obesity increases the risk of developing many serious illnesses – including diabetes, heart disease, chronic lung disease, arthritis and kidney failure. So much so, that the World Health Organization now considers the condition to have reached epidemic proportions. It is estimated that around 10 per cent of world's entire population could be considered obese.

The principle causes of obesity are poor diet and lack of exercise (though other factors such as genetic inheritance and mental health also contribute). Basic treatment usually means correcting these principal causes first. Medication can be prescribed for obesity, and in extreme cases, surgery is an option – such as the fitting of 'gastric bands' to limit stomach volume.

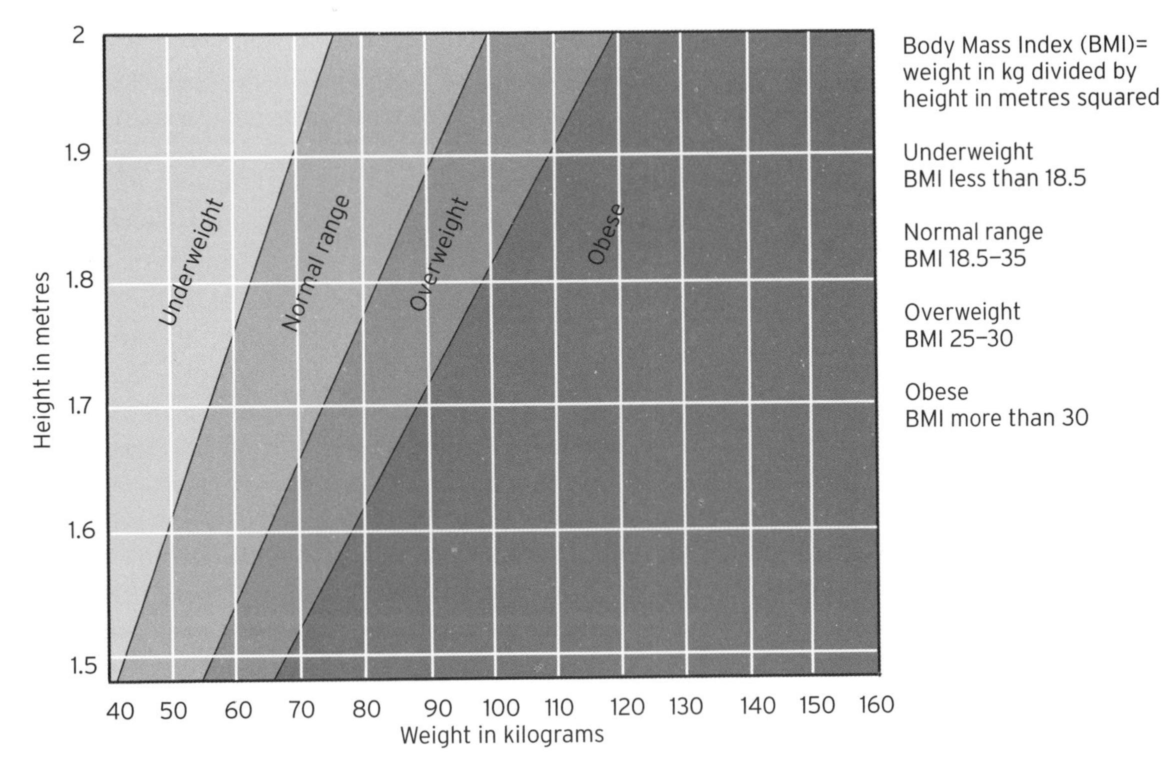

Diabetes

Diabetes – or, diabetes mellitus, to use its full name – is a condition where the body is unable to remove glucose from the blood. If untreated it can lead to hyperglycaemia, where blood sugars become dangerously high. Blood glucose levels are normally controlled by the **hormone** insulin, which is released by the pancreas, stimulating the liver and muscles to absorb glucose and store it as the molecule glycogen. Diabetes disrupts this process.

The disease manifests itself in one of two forms. In type 1, the pancreas makes insufficient insulin – this form of diabetes is congenital (present from birth) through a combination of genetic and developmental factors. In type 2, on the other hand, the body's cells become resistant to insulin – this is by far the most common form of the disease, affecting 90 per cent of the world's 220 million people with diabetes. Type 2 can be caused genetically; however, it can also be acquired later in life through poor diet and lack of exercise – which is why it is often brought on by obesity. Both forms of diabetes can be medicated with regular insulin injections.

Influenza

Influenza, commonly known as flu, is a viral infection (see **Virology**) that spreads through the human population, often in winter. Symptoms include sore throat, runny nose, aches and pains, and a fever. The virus is extremely contagious, transmitted through particles ejected by coughs and sneezes. For most people the symptoms abate after a week or two without the need for medical attention, although in the especially young, elderly or those with pre-existing respiratory conditions, such as asthma, it can be more dangerous.

Every now and again, a virulent strain of flu comes along that is potentially lethal to everyone. The swine flu pandemic of 2009, caused by a strain known as H1N1, was such an outbreak. H and N refer to types of protein forming the envelope surrounding the virus, typically made up of a combination of haemagglutinin (H) and neuraminidase (N) proteins, with the numbers following the letters giving the exact form of each. Influenza is treated with **vaccines and antivirals**. However, the virus mutates so rapidly that vaccines need to be constantly modified.

Superbugs

Normally a bacterial infection is treated with antibiotics. However, some strains of bacteria have started to become resistant to antibiotic drugs, making them extremely difficult to treat – the superbugs. One of the most notorious is the strain methicillin-resistant Staphylococcus aureus – MRSA. Symptoms begin with a fever and small red spots that eventually grow and become filled with pus. In some cases, the infection spreads to internal organs leading to sepsis (see **Bacteriology**) and death. MRSA killed 17,000 people in the United States during 2005, and nearly 1,600 in the UK in 2007. The disease can, in some instances, be treated with the antibiotic vancomycin. Though this can have toxic side effects, and MRSA is even beginning to evolve resistance to this drug as well – leading to a new strain known as VRSA.

The emergence of superbugs is largely down to the overuse of antibiotics, which has stimulated the bacteria to evolve a defence mechanism in order to survive – through **natural selection**. This is a manifestation of the **Red Queen hypothesis** – the evolutionary 'arms race' between two competing species, in this case the antibiotics and the bug.

Coma

Coma is a state of deep unconsciousness that can result from brain injury, toxins in the body or oxygen deprivation to the brain caused by conditions such as a stroke (where the brain's blood supply is interrupted). A coma may also be induced chemically as a medical treatment when a patient is suffering severe and painful injuries that need time to heal.

Comas normally last for a matter of days or weeks, but sometimes for longer periods. Forty years is the longest a person has been in a coma for – Edwarda O'Bara fell into a coma as a result of diabetes in 1970. As of 2010, she was still unconscious, but alive. The longest coma that anyone has awoken from ended after 19 years, when Terry Wallis – who had entered the coma following a car accident in 1984 – suddenly started speaking in 2003.

MEDICAL PROCEDURES

Pulse

One of the primary vital signs for checking the immediate health of a patient is their pulse. This can be taken where an artery – a vessel carrying blood away from the heart – passes close to the surface of the skin so that a finger placed over it can feel the pressure waves in the blood flow generated by the rhythmic pumping of the heart. It's normally done at the neck (the carotid artery) or the wrist (radial artery).

In most cases, the pulse is a direct reflection of the patient's heart rate. A healthy adult's resting pulse should normally be somewhere in the range of 60 to 100 beats per minute. Though patients with a high physical fitness – such as athletes – may have resting pulse rates considerably lower than this. A trained medical professional can estimate a patient's systolic (peak) blood pressure by feeling their pulse. A pulse meter – usually attached either to a finger or an earlobe – is a standard piece of hospital bedside monitoring equipment.

Blood pressure

Blood pressure is the pressure exerted on the walls of the blood vessels by the beating of the heart. The strength of the pressure decreases the further from the heart it is measured. In order to obtain comparable readings, it is usually measured at the brachial artery, situated on the inside of the elbow. Blood pressure is normally quoted as two numbers, such as 112/64: the first number is the peak (or 'systolic') blood pressure, produced during each beat of the heart; the second is the minimum (or 'diastolic') blood pressure present between each beat. Blood pressure is measured in units of mmHg – millimetres of mercury, the height that the pressure is able to raise a column of liquid mercury, measured on a device called a 'manometer'.

Readings are usually taken using an inflatable cuff; the pressure in the cuff is released until blood can be heard by **auscultation** just starting to flow through the artery. At this point the pressure in the cuff is equal to the systolic blood pressure. Releasing the cuff pressure further until no sound is heard gives the diasystolic pressure. Excessively high blood pressure is known as 'hypertension'; low pressure is called 'hypotension'.

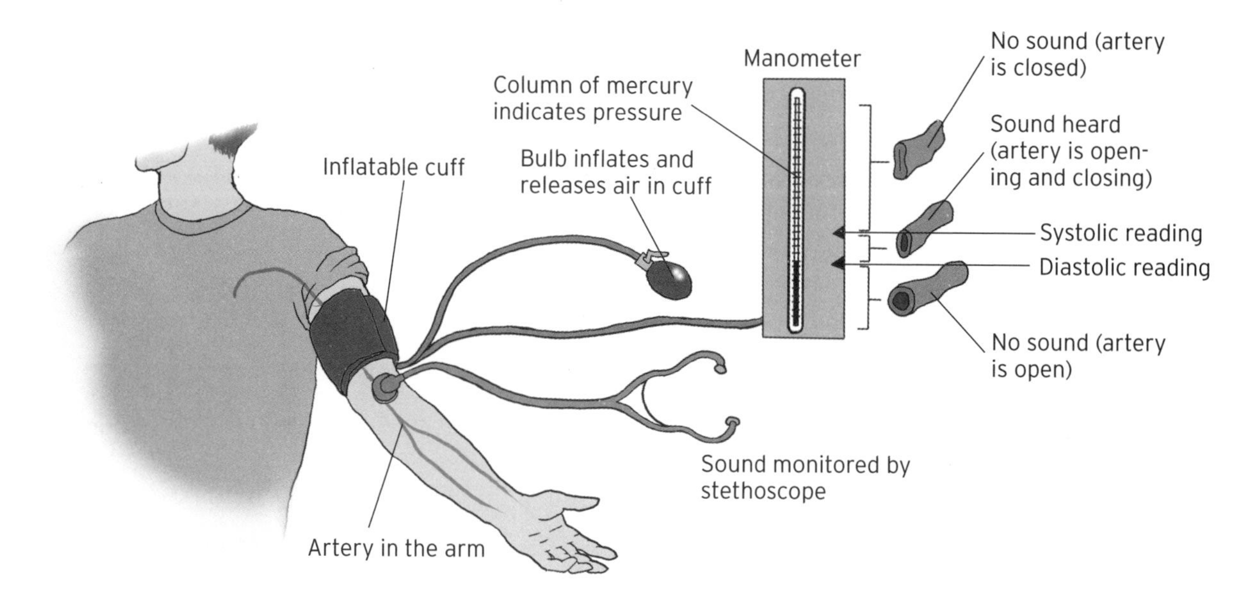

Temperature and breathing

In addition to pulse and blood pressure the two other key vital signs used in medical care are body temperature and respiratory rate (breathing). Body temperature is taken by placing a thermometer under the patient's armpit or in their mouth for 30 seconds. A healthy human should have a temperature of around 37°C (98.4°F). More than 40°C (104°F) (hyperthermia) or less than 35°C (95°F) (hypothermia) is cause for concern. Raised temperature can be a consequence of heat stroke, or the presence of toxins or infection. Hypothermia can be caused by exposure, toxic poisoning and certain injuries.

Respiration rate can either be counted manually by a doctor or monitored using a bedside device called a 'respirometer'. A typical adult should take around 12 breaths every minute.

Auscultation

Auscultation is the medical practice of diagnosing illness by listening to the body's internal sounds, usually with a stethoscope. The stethoscope was invented in 1816 by French physician Rene Laennec. Early models resembled ear trumpets, consisting of a rigid tube, one end of which was placed on the patient and the other to the physician's ear. Modern stethoscopes have flexible tubes and fittings that enable doctors to listen with both ears. The end of the device that's placed against the patient also has two sides in modern instruments – one optimized for listening to the respiratory system and the other for the heart and circulation. Some of the latest stethoscopes are electronic, and have inbuilt amplifiers and noise-reduction systems to mask background sounds.

When performing an auscultatory examination of the respiratory system, doctors listen for wheezing and signs of any blockages. Checks to the heart and circulation can pick up any irregularities in the heart's rhythm. While auscultation of the gastrointestinal system can show up irregularities and obstructions in the bowels.

Palpation

Palpation is conducted as part of a physical examination, and involves the physician examining the patient's body with his or her hands looking for abnormalities and clues to aid diagnosis. This might mean detecting lumps and swellings and estimating their size; gauging the patient's response to having areas touched (e.g. is it painful?); assessing mobility of joints in rheumatological complaints (see **Rheumatology**); or determining the position of the fetus in a pregnant woman. Palpation is often used on the abdomen and upper torso as a precursor to other diagnostic techniques such as biopsy or medical imaging.

A related kind of physical examination technique is 'percussion', where the doctor strikes an area of tissue lightly with the fingers to try to determine its nature. A resonant, hollow sound indicates an air-filled cavity below, whereas a staccato 'thud' is suggestive of hard tissue.

Blood and urine tests

Physical and chemical tests on body fluids are a powerful diagnostic tool in medicine; blood and urine tests are two of the most common types. Blood tests are performed on samples drawn from a vein using a hypodermic needle. Tests are carried out to determine the patient's blood group prior to a transfusion; blood acidity; the number of cells of different type (low numbers of red cells indicate anaemia and abnormal numbers of white cells can be a sign of infection); the presence of harmful viruses or bacteria; and testing for concentrations of various chemical compounds.

Similar analysis can be carried out on a patient's urine, and tests can determine the presence of glucose (an indicator of diabetes); the presence of proteins (which can be a sign of kidney infection); and the presence of drugs and other chemicals. Meanwhile, taking a culture of the urine – that is placing a sample in a Petri dish lined with nutrients and seeing if any microorganisms grow in it – can reveal the presence of a urinary infection.

Biopsy

Sometimes the information returned from a palpation and a blood and urine test is insufficient to make a concrete diagnosis of a medical condition. In this case, a biopsy may be required – where a sample of tissue is taken for analysis. Tissue for a biopsy can be taken in various ways. In the case of internal organs, a 'core needle' may be used

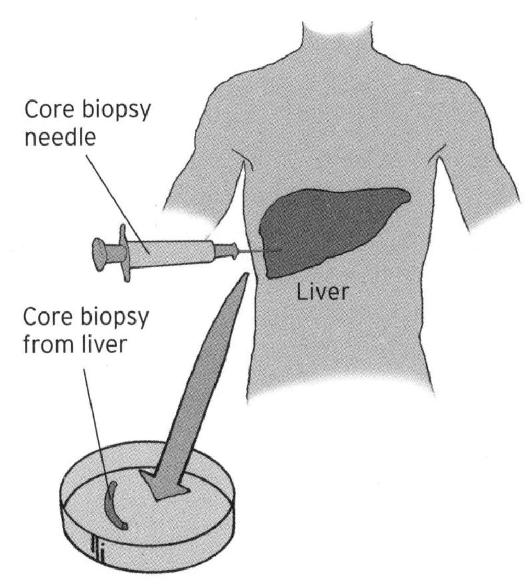

– a hollow needle is pushed through the skin and into the organ beneath to remove a small core of tissue from it. Open surgery is another option, where a surgeon removes a small piece of suspect tissue through an incision in the skin. In the case of suspected cancer tumours, the whole tumour may be removed for analysis. This has the advantage of preserving its overall shape, which can be important for diagnosis. Samples are then examined under a microscope to determine their cell structure and whether they are normal – and, if not, to diagnose the nature of the disorder.

Cardiopulmonary resuscitation

When a patient has stopped breathing and has no pulse, cardiopulmonary resuscitation (CPR) is the first course of action. The procedure involves compressing the chest to artificially operate the heart, passing blood around the body, and breathing hard into the patient's mouth to deliver oxygen to their lungs. CPR is not, in itself, a life-saving medical technique – victims of heart attacks rarely survive on CPR alone. But it is an effective method to keep a patient alive until proper life-support equipment, such as defibrillators (see **Defibrillation**) to stimulate the heart, and ventilators to assist with breathing, can be used.

Defibrillation

Defibrillators are devices used to treat arrhythmias of the heart, or full-blown heart attacks, by delivering an electric shock to reset the beating of the heart muscle. The first defibrillators entered service in the 1940s. These devices could be used only during heart surgery and were connected directly across the heart organ to deliver their electric current. In the late 1950s an alternative method was developed that didn't require the chest cavity to be open. It used a circuit of capacitors (see **Capacitance**) and inductors (see **Induction**) that could be charged up to deliver a jolt of electricity powerful enough to reach the heart from pads placed on the surface of the chest. These defibrillators are now a key piece of equipment in emergency medicine. Today, there are even implantable defibrillators available, which are placed in the chest via a surgical procedure, where they monitor heart rhythm and deliver low-power shocks automatically to correct irregularities – and a high-power shock in the event of a heart attack.

Kidney dialysis

The kidneys are the body's natural filters, sifting through the blood, removing waste chemicals produced by the metabolism and then funnelling them to the bladder where they are excreted as urine (see **Urinary system**). There are two kidneys in the body but it is possible to survive with just one. However, in cases of kidney disease so severe that both organs have to be removed, a machine is used to do the filtering work instead – a dialysis machine.

A line is inserted into an artery on the patient's arm and blood passes from this into the dialysis machine, where it flows over a semipermeable membrane. On the other side of the membrane is a fluid called the 'dialysate'. Impurities in the blood flow across the membrane and into the dialysate by **osmosis**. Holes in the membrane are large enough to allow these chemicals through but are too small for blood cells and proteins to cross over. The cleaned blood is then fed back into a vein. Artificial kidney dialysis is usually carried out three times a week, and each session takes three to four hours.

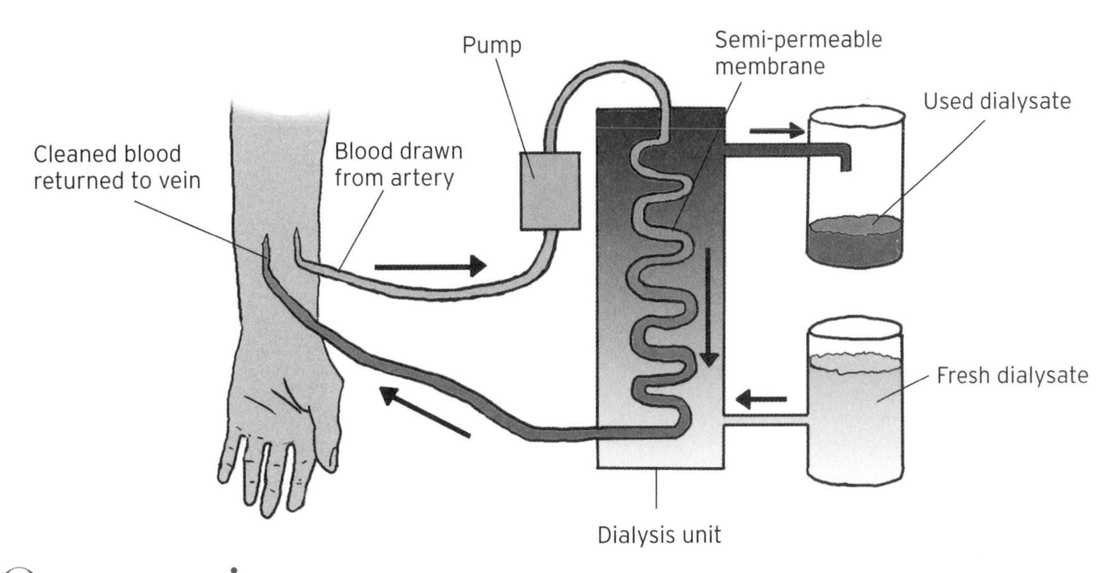

Quarantine

When a group of people or animals are suspected of carrying an infectious agent (see **Bacteriology** and **virology**) they may be placed in isolation to prevent the pathogen from spreading to the rest of the population. Quarantine may last for a predetermined period to see if any of the group develops symptoms, or until decontamination or treatment can be administered. During the 2009 swine flu pandemic, Japan quarantined 47 airline passengers in a hotel for a week after three people who had been on the same flight tested positive for the virus.

Quarantine is sometimes enforced as a purely precautionary measure, even when there's no particular pathogen involved. This was the case when the Apollo astronauts returned from the Moon. They were forced to spend several days living in a sealed environment upon landing for fear they may have brought extraterrestrial bugs back with them.

Autopsy

An autopsy is a procedure carried out on a corpse to determine the cause of death. Usually it will involve dissection of the body so that internal organs can be examined. Autopsies can be conducted from a legal standpoint – a so-called **forensic** autopsy – to provide evidence in a criminal investigation, or done for medical purposes when a patient has died for no established reason. A medical autopsy can reveal medical errors and improve the overall process of medical diagnosis to help save the lives of future patients. Following an autopsy, the body can be reconstituted, with organs replaced and incisions sewn back up neatly so that the body can be viewed during burial if the relatives so desire.

MEDICAL IMAGING

Microscopes

A microscope is an optical device used for magnifying microorganisms. It is used in medical microbiology – for diagnosing microbial infections – and in a host of other applications, including histology (the study of living tissue), forensics and in metallurgy.

One of the first microscopes was built in 1590, by lensmakers in the Netherlands; this was a simple device with one lens. In 1625, Italian scientist Galileo Galilei made the first compound microscope – with one lens to create an image of the target and another to magnify it. The magnified view appears as a so-called 'virtual image' behind the target (see diagram); many microscopes today also have a light source behind the target to illuminate it. Modern microscopes often dispense with the need for the user to actually look through the instrument. Instead, they make use of charge couple devices (CCDs) that gather an image and transfer it electronically to a computer screen, which also allows scientists to apply onscreen processing techniques to enhance the image.

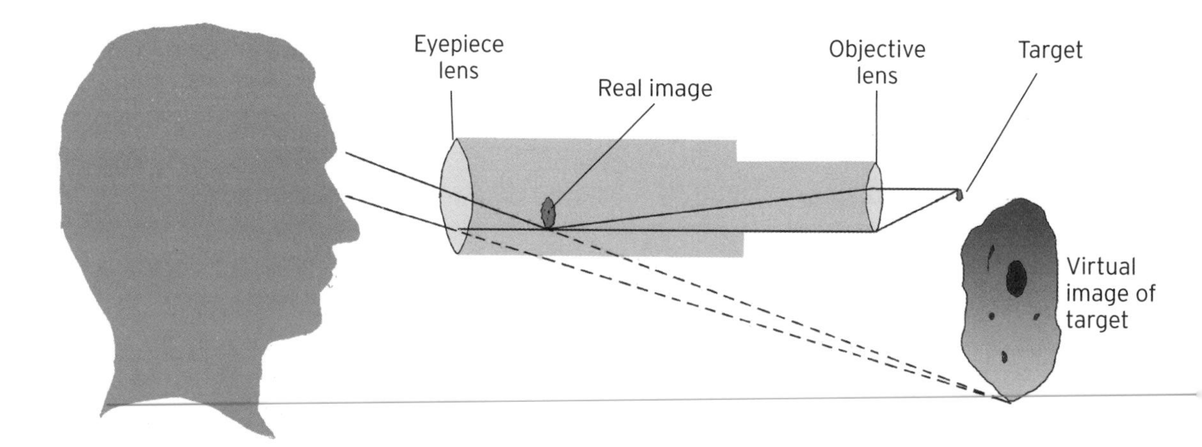

X-rays

X-rays are a class of **electromagnetic radiation** with a shorter wavelength than ultraviolet light. They can penetrate matter, which is why doctors can use them for carrying out internal examinations of patients without the need for surgery. These high-energy waves were discovered by accident in 1895 by German physicist William Roentgen who was investigating the properties of electrons given off by heated negative electrodes – so-called cathode rays. He found that when certain metals are bombarded with cathode rays they give off X-rays in response. The power of the X-rays produced can be altered by varying the voltage of the electrode.

In medicine, X-rays pass through soft body tissues and are blocked by denser tissues to create an imprint of the internal structure when the emerging beam is projected onto either a photographic plate or, more likely nowadays, a **semiconductor**-based detector.

Ultrasound

X-rays can be harmful to sensitive tissues or if applied in high doses, which is why in some cases – such as surveying the progress of an unborn fetus – a safer option is ultrasound. Ultrasound uses sound waves with a frequency above 20kHz, too high to be detected by the human ear. The technique works in a similar way to submarine **sonar** – by bouncing sound waves off objects. Measuring the time taken for each wave to bounce back reveals the distance to the various layers of tissue in the target, enabling an image to be built up.

In recent years, more sophisticated forms of medical ultrasound have become available. 'Doppler ultrasound' measures the speed of moving parts in the target through the **Doppler effect**, and can be used to map blood flow – for example, to check the heart function of an unborn fetus. Meanwhile, '3D ultrasound' is able to display a three-dimensional image rather than the two-dimensional slice revealed in an ordinary scan. Ultrasound is also occasionally used in therapy, where the high cycle rate of the sound can be employed, for example, to break up kidney stones.

Angiography

Angiography is a medical imaging technique used to show up veins and arteries, and the cavities inside internal organs, in order to diagnose circulatory problems. The technique works by injecting a so-called 'contrast agent' into the patient's blood that is opaque to X-rays, so that the blood vessels show up in an X-ray image with similar clarity to bone structure as in an ordinary X-ray.

Angiographs usually consist of video images, so that changes in blood flow can be monitored over time. This is done using 'fluoroscopy', where the patient lies between an X-ray source and a fluorescent screen – rather like the screen in a tube-based television – on which the moving image can be observed.

Endoscopy

An endoscope is a device used for performing minimally invasive examinations of patients. It consists of a flexible tube incorporating a viewing device that can be inserted into orifices such as the mouth and anus. The endoscope tube has a tiny camera mounted to the end, which relays images to a computer screen. All kinds of medical conditions may be assessed by endoscopy, including gastric (see **Gastrointestinal system**), respiratory (see **Lungs**), urinary-system and reproductive-system complaints. The first endoscopy was performed in the early 19th century, by German physician Philip Bozzini.

Sometimes endoscopes are used not just for observations but to carry out procedures such as cleaning – by blowing water or air down the endoscope tube – or even biopsies, where surgical instruments can be passed down the tube to gather tissue samples.

Electrocardiography

Otherwise known as ECG, electrocardiography is a non-invasive way of measuring a patient's heartbeat by monitoring the electrical impulses of the heart muscle tissue. The electrical signals are normally plotted on a graph that shows a peak each time the heart beats. In 1942 the modern ECG system, involving 12 leads attached at various points on the body, was introduced and this provides the most accurate picture of the heart's electrical activity.

An ECG examination can reveal things such as glitches in the heart's rhythm indicative of minor disorders in the heart rate or identifying the degree of heart muscle damage sustained in a heart attack. ECG equipment has become considerably smaller over the years. Some doctors speculate that in the future, ECG systems may become so compact that people with heart disease will be able to wear one continuously and be alerted at the first sign of irregularities.

Electroencephalography

Electrical signals produced by brain cells, or neurons, as they fire indicates brain activity that can be measured by electroencephalography, or EEG. EEG is used in the monitoring and diagnosis of brain conditions such as stroke, epilepsy, sleep disorders and brain injury. They are also used to determine the level of brain function in **coma** patients. A patient undergoing an EEG has a large number of electrodes (up to 25) placed around their scalp to pick up the brain signals, which are then plotted as waves on a computer screen. The different types of brain waves are classified by their frequency and correspond to different cycles of neural activity. Alpha waves, for example, have a frequency between 8Hz and 12Hz while beta waves are higher, between 12Hz and 30Hz. Each wave type offers diagnostic insights into different aspects of brain function.

Tomography

Tomography is a medical imaging technique that uses a scanner to take virtual slices through a patient's body. The slices reveal extremely detailed internal images which can then be inspected by a doctor for the purpose of diagnosis, or to monitor the progress of a known illness. There are several different types of tomography used in hospitals. 'Computerized tomography' (CT) works by rotating an X-ray machine around the patient and taking an image every few degrees. These images are then assembled in a computer to build up a virtual reconstruction of the patient's insides.

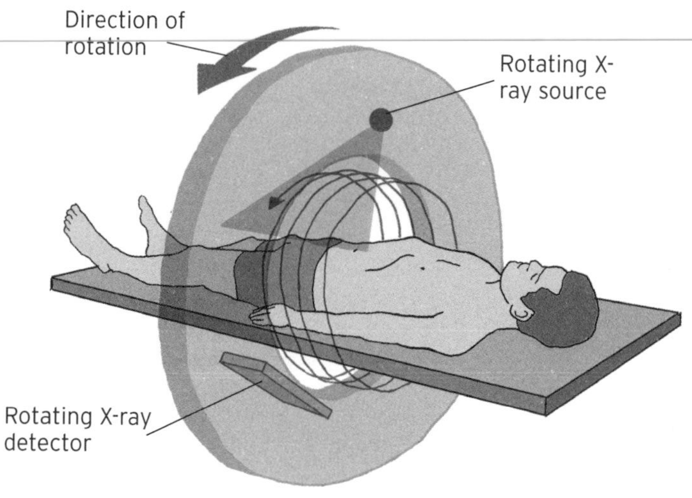

In 'positron emission tomography' (PET), patients are injected with a radioactive 'tracer' substance which emits positrons – the **antimatter** counterpart to electrons. As these positrons meet with and annihilate electrons they give off gamma rays. The tracer also mimics the action of glucose, meaning the resulting gamma-ray images map out areas of high metabolic activity. 'Single photon emission computed tomography' (SPECT) is similar to PET but uses a tracer that emits gamma rays directly.

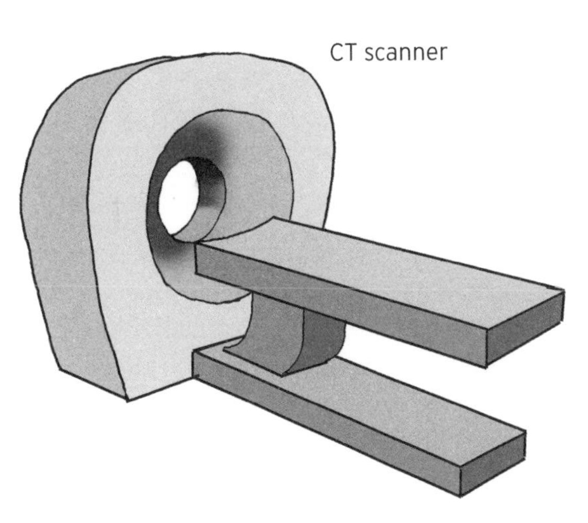
CT scanner

Magnetic resonance imaging

MRI is a scanning technique that works by coaxing radiation from hydrogen atoms inside the body's water molecules by passing a strong magnetic field through the body. This causes the electrically charged nuclei of the hydrogen atoms to snap into alignment with the field. Next, a pulse of radio waves is passed through the area to be scanned, which briefly flips some of the hydrogen atoms in the area out of parallel with the field. As the atoms flip back they give off radio waves, but different types of tissue flip back at different rates, and this is what reveals the structure of the body within. The magnetic fields used are intense – up to 3 tesla, or 100,000 times the strength of Earth's magnetic field.

Functional MRI (fMRI) looks in particular at blood flow, which shows up the parts that are using the most oxygen. Used predominantly as a brain scanning technique, it reveals which areas of a patient's brain are the most active. Sometimes, a 'contrast agent' is given to the patient via an intravenous line. This boosts the contrast of the fMRI image by enhancing the magnetic properties of hydrogen nuclei in the blood.

Molecular imaging

Imagine being able to use imaging techniques to see not just bone, skin, blood and tissue but the fundamental molecular processes in the cells of a living patient. This is what the relatively new field of molecular imaging aims to do. It works using 'bio-markers' similar to the radioactive tracers used in PET and SPECT scans (see **Tomography**), that will show up on scans in response to various **biochemistry** processes. That allows doctors to test for different chemical and metabolic pathways, which often appear as the precursors to disease before any physical symptoms have presented themselves.

MEDICATION

Pharmacology

The study of how drugs and medication interact with living tissue for the purpose of treating disease, injury and other medical conditions is called pharmacology. It includes understanding the **biochemistry** of cell behaviour and how **chemical compounds** can be used to alter that behaviour for the benefit of the patient.

The field of pharmacology began to take off in the mid-19th century as knowledge of chemistry and biology enabled the **scientific method** to be applied to the development of drugs – where before there were just potions based on superstition and shaky reasoning (see **Evidence-based medicine**). Today, pharmacology is divided into 'pharmacodynamics' (the effects of drugs on the body) and 'pharmacokinetics' (the effects of the body on the drugs – how the drug is absorbed, metabolized and excreted).

Antiseptics

Before the days of antiseptics, surgery used to be a dirty business with many patients dying from bacterial infections (see **Bacteriology**) contracted while their internal organs were exposed during the procedure. That all changed in 1867, when English physician Joseph Lister – building on the research into germs by French biologist Louis Pasteur – sterilized surgical instruments, and patients' wounds, using carbolic acid. After doing this, he noticed a marked reduction in cases of infection.

Lister had realized that in order to grow and thrive, bacteria (see **Prokaryote microbes**) need a source of food, moisture and oxygen. Antiseptics are simply chemicals that prevent these essential nutrients from reaching the bacteria. He went on to introduce the wearing of clean surgical gloves and insisted that surgeons wash their hands in antiseptic solution prior to operating. Common antiseptics in modern use are ethanol, boric acid, hydrogen peroxide and iodine. The antiseptic mouthwash Listerine is named in Joseph Lister's honour.

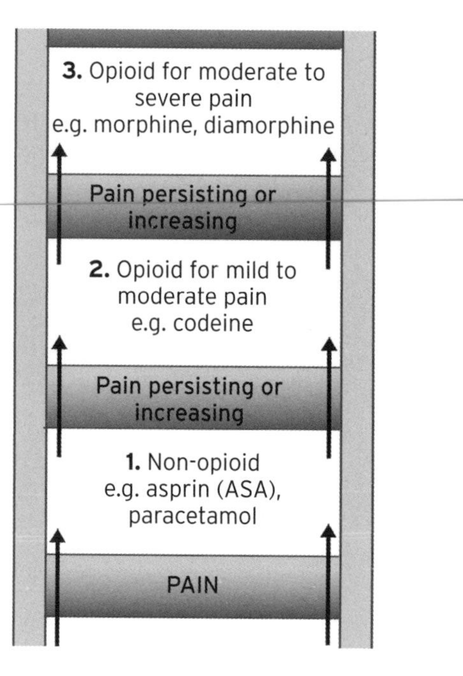

Analgesics

Drugs that relieve pain caused by illness or injury are known as analgesics. In everyday parlance, they are called 'painkillers'. Analgesics work either by blocking the path of pain signals to the brain or altering the brain's interpretation of them so that they aren't recognized as pain. They are different from **anaesthetics**, which either produce a complete loss of feeling in a localized area or induce unconsciousness.

The World Health Organization classifies analgesics on a three-rung 'pain ladder': the bottom rung is for mild pain and comprises over-the-counter medications such as paracetamol and ibuprofen; on the middle rung are the weak opium-based drugs, such as codeine; while the top rung of the ladder, reserved for severe pain, contains the strong opium-derived drugs including morphine and diamorphine, also known as heroin. Painkillers on the middle and top rungs bring with them a number of side effects, such as vomiting and constipation, and have the potential to become addictive.

Antibiotics

Antibiotics are chemical compounds that destroy or inhibit the growth of bacteria (see **Prokaryote microbes**), especially those that cause disease. Antibiotic processes were first described by 19th-century French microbiologist Louis Pasteur and German physician Robert Koch, who noticed how some bacteria could inhibit the growth of anthrax. The first natural antibiotic, penicillin, was discovered by Scottish biologist Alexander Fleming in 1928 (see **Serendipitous discoveries**). Later, in 1939, Australian pharmacologist Howard Florey and German-born biochemist Ernst Chain worked out how to purify penicillin into a form that could be administered clinically. This was just in time for the Second World War, and is thought to have saved millions from death by infected wounds. The term 'antibiotic' was coined by American biologist Selman Waksman in 1942.

Antibiotics work by interfering with the chemical pathways that bacteria rely on, while leaving those in healthy human cells untouched. This might mean disrupting the processes by which bacteria feed or by which they build their cell membranes. Recently there has been concern that some bacteria strains are becoming resistant to antibiotics, creating **superbugs**.

Vaccines and antivirals

Antibiotics are usually an effective way to clear up a bacterial infection (see **Bacteriology**) but they are no use against viral infections (see **Virology**) – the biochemistry of viruses is very different. Instead, different drugs, known as vaccines and antivirals, must be used. Vaccines work by injecting the patient with a virus that's either dead – killed by heat or chemicals – or has had the dangerous genetic material removed, leaving just the surrounding protein envelope. This inactivated virus material then trains the body's immune system to recognize that particular viral strain and attack it.

Vaccines are a preventative measure that must be administered in advance of infection, to give the immune system time to adapt. If a patient is already infected with a virus they must be treated with antiviral drugs. Like antibiotics, these selectively disrupt chemical processes on which the virus relies. For example, the antiviral drug Tamiflu works by blocking the action of an **enzyme** that the influenza virus uses to release copies of itself from infected cells, slowing the spread of the virus and thus giving the immune system longer to destroy it.

Immunosuppressants

Whereas vaccines and antiviral drugs try to give the **immune system** a helping hand in staving off infection, sometimes the body's immunity is its own worst enemy – and needs to be reined in using drugs called immunosuppressants. Some medical conditions, such as rheumatoid arthritis (see **Rheumatology**) and lupus, are 'autoimmune' diseases – that is, they trigger the immune system to misidentify healthy cells in the body as foreign invaders, and attack them. Following **transplant surgery**, the immune system will try to destroy the foreign tissue in the new organs, so these drugs are vital.

Immunosuppressant drugs can counteract the damage done by these unwanted immune responses by inhibiting the body's production of lymphocytes – white blood cells through which the immune response is mediated. The danger with immunosuppressants is that they work non-selectively – meaning that as well as weeding out the unwanted immune response, they cripple the ability to fight genuine infections. In addition, they can cause side effects such as high blood pressure and damage to the liver and kidneys (see **Urinary system**).

Chemotherapy

Chemotherapy is the use of chemical drug treatments to fight cancer – as opposed to **radiotherapy**, which uses radiation. Cancer chemotherapy involves a cocktail of toxic chemicals that selectively kill any cells that undergo rapid cell division and growth – as cancer cells do. However, some healthy cells in the body also divide rapidly – such as hair follicles (which is why chemotherapy causes hair loss), as well as bone marrow (causing a reduction in blood cell count) and the gastrointestinal system (leading to swelling of the digestive tract). Other side effects include nausea, fatigue and suppression of the immune system.

Chemotherapy treatment for cancer began during the Second World War, when it was noticed that exposure to the chemical weapon mustard gas resulted in a reduced white blood cell count. When doctors subsequently injected the active toxin from mustard gas into bloodstreams of patients suffering with lymphoma – cancer of the lymphocyte white blood cells – it improved their condition.

Euthanasia

Deliberate medical intervention to end a patient's life in order to spare them from unnecessary suffering is known as euthanasia. Active euthanasia – where a doctor performs the procedure on a patient – is, for humans at least, illegal in most countries. In Switzerland, however, it is permitted for a physician to assist a patient in committing suicide. In other words, the physician is allowed to furnish the patient with the means to end their life, but it must be the patient who administers it.

Dignitas, a Swiss non-profit organization, offers assisted suicide to the terminally ill. Their protocol involves the patient orally ingesting a lethal dose of the barbiturate drug pentobarbital, which first induces drowsiness, followed by full-body anaesthesia, then coma and death within 30 minutes. As of March 2008, they claimed to have helped 840 people to end their lives.

SURGERY

General surgery

Any medical procedure that involves entering the body of the patient to remove, modify or manipulate the tissue beneath is known as general surgery. Surgery has a surprisingly long history, dating back many thousands of years. The oldest form of surgery is 'trepanation' – where ancient people would drill holes through the skulls of the mentally ill, believing it would cure them by 'releasing the demons'. Only in relatively recent times has surgery become based on more solid science. A real breakthrough was the discovery of anaesthetics in the 19th century, though other major developments were antiseptics, which reduced the risk of infection, and **blood transfusion**s, which enabled doctors to control blood loss during surgery.

Anaesthetics

Anaesthetics are drugs given to patients prior to undergoing surgery to deaden their sense of feeling – either in a localized area, in the case of minor operations (local anaesthetic), or across the whole body for major surgical procedures (general anaesthetic). The first surgical procedure under anaesthetic was carried out by William Morton, an American dentist, who in 1846 performed a tooth extraction having first anaesthetized his patient with the general anaesthetic diethyl ether.

Modern general anaesthetics are administered in several stages. One drug, usually injected, induces the patient into an unconscious anaesthetized state – the barbiturate sodium pentathol is a common choice. Then another chemical – such as 'sevoflurane', usually inhaled through a mask – will be used to maintain the anaesthetized state while the operation is carried out. Amazingly, no one knows for sure quite how general anaesthetics work – though few people would be prepared to go through major surgery without one. Local anaesthetics, on the other hand, are understood. Injected through the skin, they function by inhibiting the flow of electrical signals across nerve cells around the injection site.

Amputation

Before the properties of anaesthetics were discovered, surgery was truly grim. Speed was the order of the day to minimize suffering, meaning that only the simplest operations were possible – and most of the time these were amputations.

Surgical amputation is a procedure used to remove limbs or other parts of the body which have become infected to a point when they cannot be saved and they threaten the health of the rest of the body – through conditions such as sepsis (see **Bacteriology**). Amputations may also be performed on areas where cancer has taken hold, or on body parts causing chronic pain.

Today, the procedure is normally carried out by tying off arteries to prevent bleeding, cutting through the muscle with a scalpel, and using an electric saw to cut the bone. Finally, the skin is stitched over the stump. An artificial, or **prosthetic**, limb may then be attached to minimize disability.

Keyhole surgery

Keyhole surgery – or, to use its technical name, laparoscopic surgery – is a minimally invasive surgical technique where surgeons work through small incisions, around a centimetre long. Long-handled surgical tools are inserted through the incisions together with a camera device called a laparoscope, similar to an endoscope, to guide their actions. Carbon dioxide gas is often used to inflate the abdomen and give surgeons space to work. The technique is often used to remove gallstones, stomach tumours and to repair hernias. The benefits are manifold – lessening blood loss, hastening recovery times (patients are often discharged the same day) and reducing the quantity of post-operative analgesics (see **Analgesics**) required.

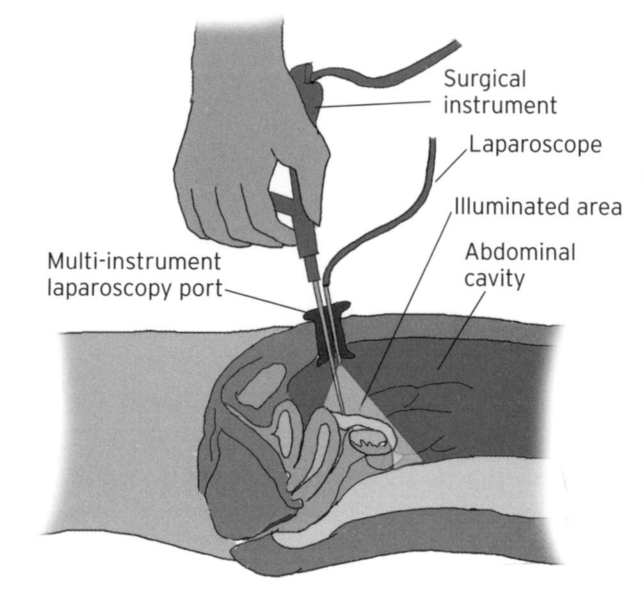

The first laparoscopic procedure on a human was carried out by Swedish physician Hans Christian Jacobaeus in 1910. Strictly speaking, 'laparoscopy' is confined to the abdominal cavity; a keyhole technique applied to the upper body – the thorax – is known as a 'thoracoscopy'.

Plastic surgery

Surgical procedures that involve reconstructive techniques to rebuild tissue destroyed or disfigured by disease or injury fall under the umbrella term of plastic surgery. No use is necessarily made of plastic – rather the word refers to reshaping of the body. Basic reconstructive techniques have been carried out for thousands of years. In India in around 600 BC, the Hindu surgeon Sushruta practised techniques for reshaping the noses of patients – a cosmetic technique called 'rhinoplasty' (commonly known as a 'nose job'). And the ancient Romans had procedures for removing scars.

In more recent history, the first person generally acknowledged to have benefited from plastic surgery is British naval sailor Walter Yeo, who lost both eyelids at the Battle of Jutland in the First World War. He was given skin grafts to repair the damage by surgeon Harold Gillies – now regarded as the father of plastic surgery. Today, there is a booming industry in cosmetic plastic surgery dealing with everything from breast surgery to facelifts.

Eye surgery

Eye surgery is a specialist branch of **ophthalmology** and can include procedures to correct cataracts in the eye, to repair nerve damage (known as glaucoma) or to alter the optical properties of the cornea at the front of the eye in order to correct vision defects – removing the need for the patient to wear corrective devices such as glasses.

The kind of eye surgery people are most familiar with is the 'laser eye treatment' used to correct sight defects. Also known as 'photorefractive keratectomy', it works using a laser beam to literally burn away tiny areas of the cornea, reshaping it to change the way it refracts light rays passing through it. This can be used to treat conditions such as myopia (short-sightedness), hyperopia (long-sightedness) and astigmatism (where the cornea is oval instead of round, making focusing more difficult).

Brain surgery

Brain surgery is the name given to any surgical procedure carried out within the skull. Usually brain surgery is preceded by extensive **medical imaging** of the head so that the procedure can be planned meticulously. The surgeon then places the patient under general anaesthetic and performs a craniotomy – where a bone flap is opened in the patient's skull, usually by drilling small holes and then cutting between them with an electric saw. This gives the surgeon access to the brain. Brain surgery can be performed for various reasons – for example, to remove brain cancers, to stem bleeding, or to remove blood clots caused by conditions such as a stroke. The field of brain surgery is a sub-discipline of neurosurgery, which encompasses all surgery performed on the central nervous system – including the spinal cord.

In the early 20th century, Portuguese surgeon Egas Moniz developed a type of brain surgery to treat psychiatric disorders, known as 'lobotomy'. It involved parts of patients' brains being either destroyed or physically severed by inserting sharp implements through holes in the skull. Thankfully, lobotomies are rarely performed today.

Robot surgery

An amazing development in keyhole surgery has been the use of remotely operated robots to perform minimally invasive procedures. Keyhole surgery means that the surgeon's hands don't need to be inside the patient – all cuts and stitches are carried out by delicate instruments inserted through keyhole incisions. And this has made it possible for these instruments to be operated mechanically by a surgeon at a console that does not necessarily need to be in the same room, the same hospital – or, indeed, the same country.

In 2000, the US Federal Drug Administration approved a robotic system called 'da Vinci' for clinical use. In conventional keyhole surgery, the surgeon must stand and transfer his gaze between the instruments and the display screen. However, da Vinci allows the surgeon to be seated – for greater comfort during lengthy procedures. Two endoscopes give full stereoscopic vision, while hand and foot controls are used to operate the instruments simultaneously – and with greater range of movement than permitted by human hand and wrist joints. In the future, the system could be used by surgeons to conduct operations over great distances. For example, a top surgical specialist in the United States could operate on an injured soldier on a faraway battlefield.

TRANSPLANT SURGERY

Blood transfusion

Blood transfusions are a life-saving medical technique allowing blood lost by a patient as a result of injury or surgery to be replenished through a donation from a healthy individual. Attempted blood transfusions have a long history, dating back to the 15th century and even involving the transfer of sheep's blood into the veins of humans. The first human-to-human blood transfusion to actually work was carried out in 1818 by English obstetrician James Blundell. However, it wasn't until 1901 – when blood groups were discovered – that transfusions became a reliable form of therapy. Mixing certain groups of blood, it was found, led to dangerous immune system reactions.

A little over ten years later it was discovered that adding anticoagulants to blood could increase its shelf life, thus enabling 'blood banks' – where supplies of donated blood could be stored and used to treat sick patients on demand. Today, donated blood is screened for pathogens, such as HIV and hepatitis, and then separated into red cells, platelets and plasma – which are dispensed separately according to each patient's needs.

Recipient's blood group	Allowed donor blood groups
A	A or O
B	B or O
AB	A, B, AB or O
O	O

Organ transplants

The first successful human organ transplant took place in 1954 in Boston, Massachusetts, when a patient became the recipient of a kidney transplanted from his twin brother. Since then doctors have successfully transplanted hearts, lungs, livers, pancreases, as well as tissue such as skin and corneas (see **Eye surgery**). In the Boston transplant of 1954, recipient and donor were genetically identical twins – so there was no chance of the transplanted kidney being rejected by the recipient's immune system. Modern transplant surgery – where donors can be almost anyone – normally involves the recipient taking **immunosuppressant** drugs. In recent years, more exotic types of transplant surgery have become possible – such as limb transplants, **face transplants**, and stem cell therapy to grow transplant tissue that will not be rejected.

Cybernetic implants

The development of organ transplant surgery, to transfer organic body parts from one person to another, has brought about a revolution in medicine. But another revolution is hot on its heels – the development of artificial organs, made from plastic and metal, that can do the job of their organic counterparts. They are known as cybernetic implants. Some of the first cybernetic systems to be implanted inside the body were pacemakers, electronic devices that regulate the beating of the heart by delivering rhythmic electrical pulses that stimulate the heart muscle to contract in time.

Recently, Arizona-based company SynCardia has developed the implantable CardioWest Total Artificial Heart. In 2004, the device was approved as a temporary solution for patients awaiting an organic heart transplant and in early 2010 the company claimed over 800 patients have had the device fitted. Meanwhile, in other patients artificial cochleas have been implanted to repair damaged hearing and much research has been carried out into cybernetic eyes, by linking cameras to the optic nerve – with some success.

Limb transplants

In 1998, New Zealander Clint Hallam became the recipient of the world's first successful hand transplant, having lost his right hand in an accident involving a circular saw. Over a period of two years, Hallam learned to use the new hand and could even use it to write. Ultimately, however, he was not comfortable with it and it was amputated at his request in 2001.

The first hand transplant to be a long-term success was performed in Louisville, Kentucky, on American Matthew Scott in 1999. Since then, there have been many successful hand transplants – including a number of double procedures, grafting new left and right hands onto a patient. A hand transplant operation takes between 8 and 12 hours. Like other transplants, patients must take immunosuppressants for as long as they wish to keep the new body parts.

Xenotransplants

Since 1975, hundreds of thousands of patients have received replacement heart valves from a rather unusual source: pigs. The valves are not made of living material, and are pickled prior to use so there's no chance of them introducing any kind of infection. But what if a human being could receive a complete heart or liver transplanted from a pig? Advances in genetic modification are bringing this possibility – known as xenotransplantation – closer to the realms of possibility.

Some doctors are concerned that xenotransplants could throw open the door for animal viruses to jump to humans – especially since the recipient of a xenotransplant would need to be heavily immunosuppressed. It has happened in the past with AIDS and vCJD (see **Prion**). And yet the practice of transplanting animal organs into humans could potentially end the desperate shortage of transplant organs which prevails today – currently, half of all transplant patients die while on the waiting list.

Stem cell therapy

Embryonic stem cells are a type of mutable cell that can grow into any other kind of tissue in the body – such as blood or skin or brain. They are the first type of cell present in an unborn embryo, before the embryo starts to acquire the structure that turns it into a fetus (see **Cellular differentiation** and **developmental biology**).

One of the principle dangers of transplantation surgery is organ rejection. Embryonic stem cells could theoretically be used to grow transplant tissue that is an exact genetic match to the patient and so could never be rejected by their immune system. Stem cell therapy works by cloning the patient. Known as 'therapeutic cloning', this would be done by somatic cell nuclear transfer – the same process used to create cloned mammals, such as Dolly the sheep. The process is halted at the embryo stage, at which point embryonic stem cells are harvested from the embryo. These can then be injected into the patient and coaxed to grow into whatever cell type is needed to patch up the damage done by serious conditions such as cancer, heart disease and brain injury.

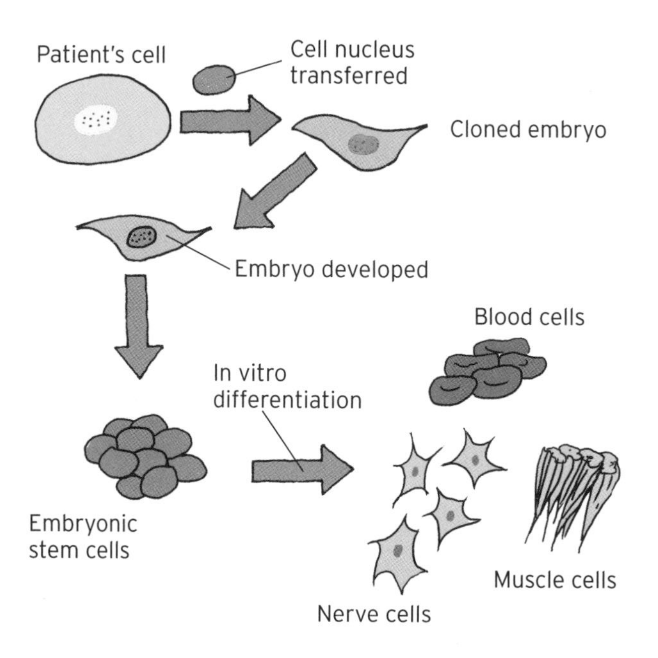

Face transplants

For people horribly disfigured in accidents, it is a way to bring some semblance of normality back to their lives. In 2005, French woman Isabelle Dinoire found herself in just that situation after having been savagely mauled by her dog, leaving her lower face in tatters. In November that year, French surgeon Jean-Michel Dubernard reconstructed Dinoire's face by grafting on a large triangle of facial tissue – including the nose and mouth – that had been taken from a recently deceased donor. In 2008 – while admitting it had been difficult coming to terms with having someone else's face – Dinoire reported that she had a full sense of feeling in her new skin. That same year, another team in France announced the first successful full-face transplant, on a man disfigured by facial tumours.

Following surgery, the recipient's new face neither resembles their old face nor the face of the donor. It's somewhere in between – with the donor's features stretched over the recipient's bone and muscle structure. As with all transplants, the recipient must continually take **immunosuppressants** to prevent the new tissue from being rejected.

MODERN MEDICINE

Evidence-based medicine

In an age of science, you might reasonably expect all of the medicines and clinical procedures prescribed by doctors to be based on sound scientific evidence. And yet up until recently, many doctors would still make their decisions based on personal preference and results from limited statistical samples – namely, what's worked before on the relatively small number of patients they themselves have treated.

Now, reason is taking over, thanks to the easy access doctors have to data from clinical trials and records of the experiences of other doctors around the world. The drive to assess the effectiveness of different treatments scientifically is known as evidence-based medicine. Health services operate a system of rating different forms of medical evidence based on its reliability – ranging from the results of rigorous clinical trials down to opinions of doctors based on their clinical experience.

Clinical trials

In 1747, Scottish naval surgeon James Lind conducted a historic experiment. He took a group of 12 sailors all suffering from the disease scurvy and divided them into six pairs. Each pair was given a supplement to their daily rations – some were given cider, some vinegar and some garlic mixtures. But it was the group given citrus fruits who fared best and indeed recovered from the scurvy within six days. Lind had conducted the first clinical trial – scurvy is now known to be caused by deficiency of vitamin C, which citrus fruits are rich in.

Today, clinical trials are carried out on much larger scales – assessing the effects of new candidate drugs and treatments on thousands of volunteers to test their efficacy and to assess the risk of side effects. The best clinical trials assess a particular drug against a **placebo**. Which people get the real drug and which get the placebo is determined at random – it's known as a 'randomized' trial – and neither the doctors administering the treatments nor the patients are told which is which, known as 'double-blinding'. The power of clinical trials can be improved using a technique known as 'meta-analysis', where the results from several trials of the same drug are combined to increase the sample size.

Laser medicine

Lasers are well known for their applications in treating sight defects through laser **eye surgery**, but they are finding a raft of new applications. Laser scalpels are used in surgery to make incisions with zero risk of infection and on a very fine scale – sometimes smaller than the size of a cell. They are particularly useful in brain surgery, for burning away tumours without handling the delicate surrounding tissue. Surgery with a laser scalpel is also self-cauterizing and therefore bloodless. And in dentistry, lasers are used for removing cavities and for tooth whitening.

In cosmetic surgery lasers are effective for the removal of scars, tattoos and unwanted hair – and have also found uses in the treatment of acne, cellulite, moles and postnatal stretch marks. Recent studies have even shown that lasers can be effective in pain relief, due to the ability of light sources to alter cell function – an emerging field called 'laser therapy'.

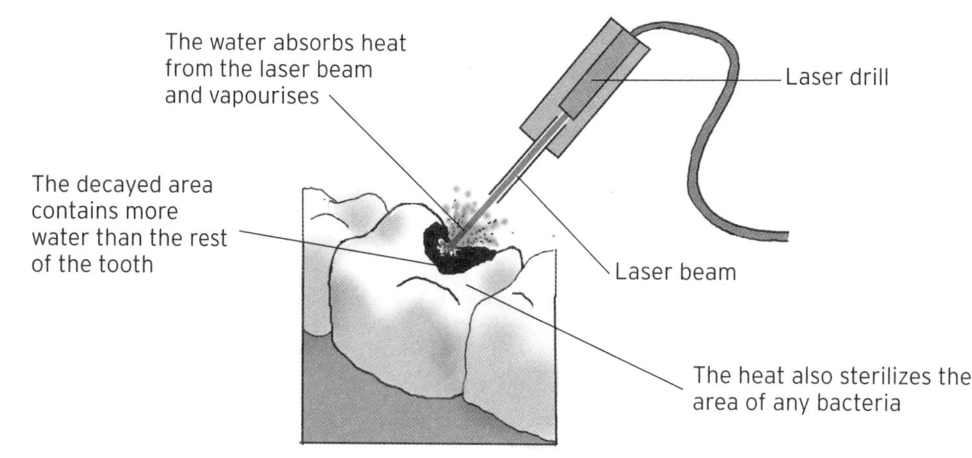

Radiotherapy

Radiotherapy, or radiation therapy as it's also known, exploits the deleterious effects of ionizing radiation (see **Radioactivite decay**) on the human body (see **Radiobiology**) to kill harmful cells such as those in cancer tumours. It works by smashing apart the DNA in the cell nucleus. As the cells divide and grow the radiation damage is passed on from generation to generation and accumulates until eventually their growth slows and they die.

The procedure can either be administered externally, as a beam of radiation that is directed onto the infected area, or the source may be internal, through the ingestion of a radioactive solution or surgical implantation of radioactive source material. Potential side effects of radiotherapy are hair loss in the infected area, fatigue and, in some rare instances, secondary cancers caused by the therapy.

Telemedicine

We've all been on the Internet now and again to check out our aching shoulders, blotchy skin, and other strange symptoms – rather than trudge all the way to the doctor for a proper diagnosis. Now, it seems you can do both. The Internet is being used to deliver professional medical care – this is telemedicine. It can involve having a real-time consultation with your doctor via video chat, or uploading medical readings you've gathered yourself for a healthcare professional to examine, or even remote monitoring – where a doctor can keep track of a patient's vital signs, transmitted in real time via a wearable device. It's especially useful for the treatment of patients in isolated locations, such as on oil rigs or remote research stations.

A related field is e-health, dealing with the centralization of patient records on databases that doctors can access from anywhere, providing Internet services to patients (such as ordering repeat prescriptions), and posting reliable health information online.

GENETIC MEDICINE

Gene therapy

The science of using **genetic modification** techniques to alter and repair defective genes in humans – of the sort that cause hereditary diseases such as cystic fibrosis – is known as gene therapy. One method is to introduce the new genes, as a piece of **recombinant DNA**, into a virus that is then used to infect the patient. As the virus spreads by injecting its own DNA into each cell, it actually injects the new genes, modifying the patient's genetic make-up cell by cell.

Gene therapy has had a chequered history. In 1999, Jesse Gelsinger took part in a clinical trial at the University of Pennsylvania to investigate gene therapy for a type of liver disease. However, his immune system reacted severely to the virus being used to deliver the treatment and he died shortly after. Since this tragedy, further trials have been conducted – with greater success. These have used gene therapies to treat blood disorders, defective vision and even HIV infection.

Personal genomics

Many companies are now springing up around the world that offer to analyze your DNA sequence for you to see whether you are genetically predisposed to develop certain diseases later in life. The idea is that if the tests showed, for example, that you are susceptible to heart disease then you can make the effort to eat healthily and get plenty of exercise during early life in order to offset the extra risk later on. This form of predictive medicine is known as personal genomics.

Some commentators, however, have argued that personal genomics may bring about the downfall of the health insurance industry. Those who find they are at low risk will buy less insurance while those at increased risk will buy more – but of course they won't tell their insurer the results of their gene scan as this would push up their premium. With no way to tell the healthy from the soon-to-be-sick, insurance companies would have to raise premiums for everyone, deterring even more of the low-risk cases, pushing premiums higher still – until, ultimately, the market collapses.

Reprogenetics

The merger of reproductive medicine with genetics is known as reprogenetics. The principle application is to be able to screen embryos for heritable genetic diseases, such as Huntington's and cystic fibrosis, as well as genetic conditions such as Down's syndrome. A couple would use in-vitro fertilization (see **Fertility**) to create a number of embryos; the gene sequences of these embryos are analyzed and any found to contain potentially disease-causing **genetic mutations** are rejected. The couple are then free to select which embryos are implanted back into the mother and brought to full term, which is where some of the controversy surrounding reprogenetics begins. It is just as easy to reject those embryos with bad genes as it is to select the ones that have especially good genes. You might want to choose embryos likely to grow into offspring that have a high IQ, or outstanding athletic prowess, or that will be especially attractive. Critics have compared this possible future practice, dubbed 'designer babies', with **eugenics**.

Pharmacogenomics

Why is it some of us have a bad reaction to certain foods while others can eat as much as they like with no ill effects? The same is true of medicines – some of us respond well to a certain treatment, others will feel no effect, while a few will suffer adverse reactions. The key to understanding the effects **substances** have on our bodies lies in our **genes**. Pharmacogenomics – or 'designer drugs', as it's sometimes called – seeks to predict our reactions to medications before we take them, in a bid to concoct clinical drugs tailored to the needs and body **chemistry** of the individual. In the case where the patient is particularly under- or over-sensitive to certain drugs, it might mean adjusting the dosage or prescribing alternative medication entirely. At some stage in the future, pharmacogenomics coupled to powerful **computing** resources could even be able to formulate new drug molecules on demand that produce a specific effect when combined with that patient's body chemistry, as determined by their **DNA**.

Cloning

Cloning is a technique for creating genetically identical copies of an animal. It works by taking genetic material from an adult animal and inserting it into the nucleus of an egg cell (see **Gametes**) taken from a female animal. The cell is then zapped with electricity to stimulate it to grow, and then implanted into the female and brought to full term. This technique is known as 'somatic cell nuclear transfer'. In 1996 it was used to produce Dolly the sheep – the world's first-ever cloned mammal. Dolly died prematurely in 2003, from what appeared to be age-related illnesses. Some have suggested Dolly's cells were already six years old when she was born – the age of the donor animal that her cells were taken from. For this reason, it's unlikely the production of human clones by this technique will be approved for many years. Nevertheless, there has been considerable interest in cloning from the point of view of **stem cell therapy**.

RNAi

RNAi, short for RNA interference, is a way of silencing **genes** responsible for illness in an organism's DNA sequence – by blocking **gene expression**. It works using short lengths of so-called 'double–stranded RNA' – rather like the double-stranded DNA of which the **chromosomes** in cell nuclei are made. Injecting these RNAs into a cell interferes with the action of the messenger RNAs that carry genetic information from the nucleus to the **ribosome**, where proteins are manufactured. The double-stranded RNA can be tuned so that particular genes are targeted and silenced.

Researchers believe that once the technique is perfected it will have applications in halting the expression of genes in viruses including influenza, the liver disease hepatitis and even cancer and HIV. The researchers who pioneered the technique – American biologists Andrew Fire and Craig Mello – received the 2006 Nobel Prize for Medicine.

COMPLEMENTARY MEDICINE

Dietary supplements

From a vitamin tablet a day to the more questionable benefits of 'natural fat-buster' pills, people spend $52 billion (2006 figures) every year worldwide on dietary supplements; in the United States alone, Americans spend $23 billion. These include herb extracts, vitamins and minerals, and even biochemicals such as amino acids and enzyme supplements to aid digestion. Many of us take them believing that they will enhance our health and well-being. In the case of simple vitamins this is no doubt true, but for other supplements the benefits are less obvious.

Unlike the pharmaceutical drugs industry, the market for supplements is relatively unregulated. In the United States, supplements are classed as foods not drugs – meaning that they do not require approval by the FDA before being brought to market. Indeed, there's no requirement for the manufacturer to even prove that their products have the claimed effect – as is the case with clinical trials of pharmaceutical drugs. On the other hand, proponents argue that the vast number of people who take the various types of supplement available actually constitute some of the largest clinical trials ever performed.

Homeopathy

Homeopathic remedies are treatments derived from the so-called 'law of similars' – treating illnesses with chemical compounds that induce similar symptoms to those of the ailment being treated. The compounds are administered in a form heavily diluted in water – sometimes so much that not even a single molecule of the active compound remains, which has led most medical professionals to consider homeopathy a **pseudoscience**. And, indeed, this seems to be backed up by the lack of scientific evidence or **clinical trials** demonstrating any benefits – at least, beyond the **placebo effect**. The closest was a piece of research published in the reputable science journal *Nature* in 1988 where the researchers claimed to have demonstrated that water has a kind of 'memory' for substances that it has previously come into contact with. However, attempts by other independent groups to reproduce the results have not been successful.

Acupuncture

Acupuncture is a form of alternative medicine that works by inserting thin needles a few millimetres through the skin at specific points on the body, usually as a method of pain relief but also to treat other conditions. With acupuncture, which has its origins in China, it is claimed that the body's lifeforce, or 'chi', flows along channels called 'meridians' in the body. Interrupting the flow, or relieving the pressure, at 'acupoints' along these meridians can have an effect on body function – or at least so the theory goes. There is no known mechanism in biology that can place these ideas on a scientific footing.

Indeed, there is little scientific evidence on the effectiveness of acupuncture either, leaving many medical practitioners sceptical of its benefits. Nevertheless, a great many patients report positive effects from the treatment. Although this may be due to the placebo effect, MRI imaging does suggest that acupuncture has an effect on brain activity.

Placebo effect

If you take a pill that you are told will help treat a medical condition and it does, even though the pill contains no active ingredients, then you've just experienced the placebo effect. Placebos are treatments that work purely through the power of suggestion – mind over body. Scientific evidence confirms that the placebo effect is real, but there is little consensus over the exact mechanism by which it works. Some research has suggested that endogenous opioids – the body's natural painkillers, such as endorphins – may be involved.

Interestingly, the opposite of the placebo effect has also been observed. Called the 'nocebo effect', it makes patients feel worse after taking an inert drug that they are told is bad for them. Modern clinical trials of new drugs are designed to take account of the placebo effect.

Chiropractic

This is a form of alternative therapy that involves manipulation of the joints and muscle tissues to treat and prevent disorders of the musculoskeletal system. Often combined with the use of massage, exercise and dietary supplements, clinical studies have found it to be especially effective in the treatment of lower back pain. However, claims that chiropractic manipulation of the spine can treat other disorders not necessarily related to the musculoskeletal system – in particular, conditions in children, such as colic, feeding problems, ear infections and asthma – have been mired in controversy. In 2008, the science writer Simon Singh publicly criticized chiropractors making these claims, resulting in a libel action being brought against him by the British Chiropractic Association. Founded in 1890, today chiropractic is the third largest medical profession in the world after conventional medicine and dentistry.

Osteopathy

Another branch of complementary medicine that involves manipulation of the joints and soft tissues of the musculoskeletal system is osteopathy. However, the techniques of manipulation used by an osteopath are quite different from those used in chiropractic. Osteopaths, especially in the United States, normally undergo a more conventional medical training compared with a chiropractor, making them more akin to a medical doctor specializing in bones and muscles – indeed, the field has been largely rebranded as 'osteopathic medicine' to reflect this.

Like many other forms of complementary medicine, osteopaths take a holistic approach to treatment, considering the effects on the whole body. Unlike chiropractors, however, osteopaths confine their attention to disorders of the musculoskeletal system and don't believe that the therapy they offer can necessarily cure other ailments. Osteopathy was founded by American physician Andrew Taylor Still in 1874.

Neurolinguistic programming

NLP, short for neurolinguistic programming, is a form of alternative **psychotherapy** based on the field of **neurolinguistics**. The idea is to draw upon the connection between brain function and our learning and use of language to alter the way we think and, as a result, change our behaviour. Practitioners claim it can be used to effectively treat phobias, **depression**, **anxiety disorder** and other psychological problems – and does so more rapidly than would be possible through conventional psychotherapy.

Therapists believe that examining the language a patient uses to describe their problems, and watching their gestures and expressions – even the direction their eyes move in when they talk (see diagram) – enables them to build up a picture of the patient's mindset which they can then work to change. NLP is frequently a feature of life coaching and self-improvement classes, and increasingly crops up in corporate training where course leaders claim it can improve subjects' interpersonal skills and powers of persuasion. As with so many forms of alternative medicine, NLP has attracted harsh criticism from scientists who claim there is scant evidence that it offers any real benefit.

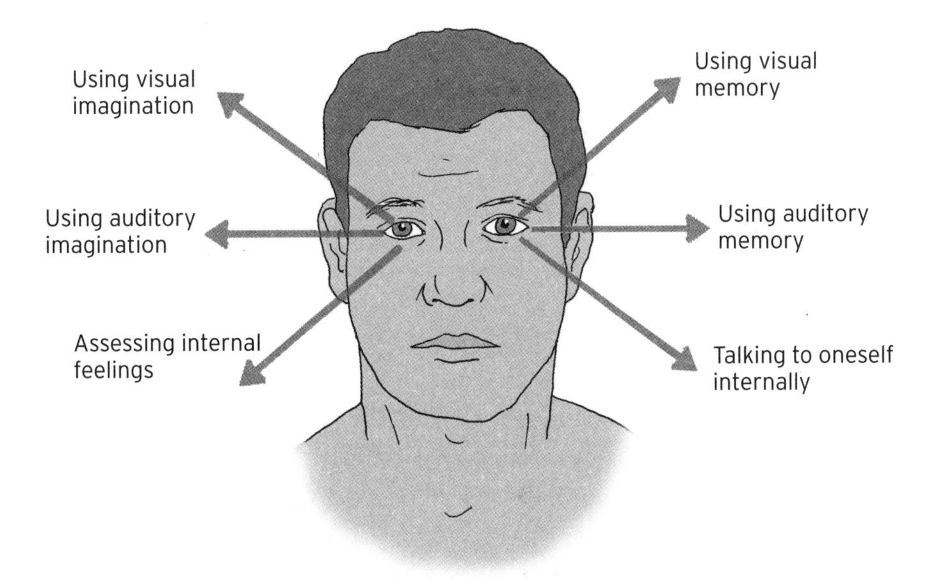

SOCIAL SCIENCE

WHAT IS IT THAT MAKES SOCIETY TICK – that binds populations of people together into communities and civilizations? Answering this, and the raft of other questions it throws up, is the province of social science.

It is a broad church of disciplines encompassing linguistics (how we talk and communicate), psychology (how our brains make sense of reality and, in turn, dictate our reactions to it), economics (how society functions through the exchange of goods and services) and politics (how communities of people arrive at the decisions that shape their future).

Some of the first social scientists were Islamic scholars in the Middle Ages. For example, the 11th-century Persian scientist Al-Biruni produced some of the first writings

GUAGE ACQUISITION • STRUCTURE OF
ITIVE LINGUISTICS • SOCIOLINGUISTICS
• PSYCHOANALYSIS • BEHAVIOURISM •
L PSYCHOLOGY • PERSONALITY TRAITS
CHOLOGY • INTELLIGENCE • EMOTIONAL
EMORY • CONFABULATION • ILLUSIONS •
IMENT • SYNAESTHESIA • KAPPA EFFECT
Y • CROWD CRAZING • SLEEP PARALYSIS
CONTROL DISORDER • PERSONALITY
DISORDER • AUTISM • DEMENTIA •
SENTIENCE AND SAPIENCE
CONSCIOUSNESS •

comparing different societies, in which he looked at people from cultures across the Middle East and Asia.

The phrase 'social science' was first used in the early 19th century by Irish writer William Thompson, in a book arguing why workers should receive better pay for their toil. The idea was taken further by Frenchman Auguste Comte, who was among the first to apply the scientific method to social problems.

Today, through the scientific study of cause and effect upon society, social scientists have helped bring about some of the great changes that have benefited our civilization over the years – such as nationalized health care, education, welfare, and the minimum wage.

LINGUISTICS

Semiotics

The scientific study of languages – how they emerged and evolve, and how humans use them to communicate – is called linguistics. Perhaps the most fundamental subdiscipline of linguistics is semiotics – the science of signs, their use and meaning. A sign can be a gesture made with the hands or face, a picture or symbol, or a piece of written text. Semiologists break signs down into a surprisingly complex structure: the sign itself splits into the 'signifier', that is the actual hand gesture or words, and the 'signified', which is the concept that the signifier refers to. Different types of signs are classified in a number of ways. For example, 'iconic signs' are those for which the signifier resembles the signified (such as pictures); 'indexical signs' occur when there is cause and effect between signified and signifier (such as smoke and fire); while 'symbolic signs' are where the relationship between signifier and signified is determined by convention (such as words).

Origin of language

At some point in the evolution of modern humans, our brains became complex enough to acquire language – developing a form of semiotics whereby they assigned distinct vocal sounds ('signifiers') to particular concepts (the 'signified'). There are several schools of thought as to how this actually happened. The 'ding-dong' theory supposes that early humans constructed sounds that imitated the concept they were trying to describe, rather like a verbal iconic sign (see **Semiotics**). And the 'pooh-pooh' theory says that early words and phrases were expressions of emotions and feelings, such as hunger or anger. Alternatively, the 'bow-wow' theory suggests that our first attempts at speech were imitations of animal noises.

Modern humans – *Homo sapiens* – evolved in Africa around 200,000 years ago, and by about 30,000 years ago they had replaced all other forms of early humans, such as Neanderthals and *Homo erectus*. Some palaeontologists ascribe this to our superior communication skills, primarily through language.

Language acquisition

Language acquisition is the process by which human infants first learn to talk. The field has been something of a battleground in the **nature vs nurture** debate. Nowadays, it's generally accepted that both aspects contribute: nature gives human babies the aptitude to understand human grammar, but nurture plays a part by exposing babies to the sounds of humans talking which they then imitate.

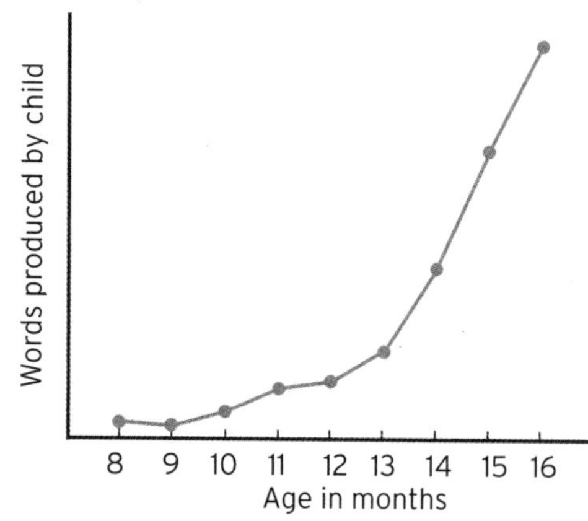

Linguists have named the 'nature' component of language acquisition as 'universal grammar' (see **Generative grammar**), describing a kind of grammar common to all human languages and which human babies are born with the innate ability to understand. All they need to learn are the grammatical quirks of their own specific language. Language skills typically begin to emerge in infants between the age of nine months to a year.

Structure of language

Language is made up of a number of different components. 'Phonetics' deals with the sounds of particular words, and how they are produced and understood. Linguists have developed a 'phonetic alphabet' of speech sounds, that is independent of any particular language. There are 107 distinct sounds – or 'phonemes' – in the alphabet, plus a large number of modifiers that can be applied according to accents and intonation. 'Semantics' is concerned with the particular meanings of words. 'Morphology' is the study of the internal structure of words and the language-specific rules by which they are formed – for example, 'happy' can join with 'ness' to give a word for how happy someone is, but in doing so we swap the 'y' for an 'i'. 'Pragmatics' addresses how context contributes to the meaning of words and sentences – for example, does the sentence 'I am in the red' mean I am in debt, or that I am wearing a crimson suit? Meanwhile, 'syntax' is the name linguists give to the rules of grammar – how individual words are put together to form sentences. 'Nouns', 'adjectives' and 'verbs' are all aspects of syntax.

Generative grammar

Generative grammar is a theory of syntax (see **Structure of language**) developed by American linguist Noam Chomsky. In its most basic form the theory splits grammar into 'deep grammar' – a kind of universal grammar (see **Language acquisition**) which all humans are born with – and a 'surface grammar' that differs from language to language. In generative grammar, Chomsky developed a mathematical framework for grammar in which he set out the fundamental rules for what sentences were grammatically possible in any language and which were not. He argued that these rules are ingrained in the human brain from birth.

Generative grammar is a powerful and versatile tool for analyzing language structure. So much so, American composer Fred Lerdahl has even used its principles to deepen his understanding of music. But generative grammar is by no means universally accepted amongst linguists – an alternative theory is cognitive linguistics.

Cognitive linguistics

At odds with the theory of generative grammar, cognitive linguistics is a theory that says language emerges as part of the natural human capacity for thought and that +there are no 'pre-programmed' grammatical rules in the human brain. Proponents argue that our knowledge of language and vocabulary is no different from our knowledge in any other area and that data is stored and retrieved in the brain in much the same way.

The two theories differ in the emphasis each places on meaning. Generative grammar is independent of meaning – it's a theory of grammar and the rules of syntax (see **Structure of language**) – whereas in cognitive linguistics, meaning, and our interpretation of it, is the true driver.

Sociolinguistics

Sociolinguistics is a sub-field of the science of language looking at how language changes and evolves through the action of the society that speaks it. A good example might be how different classes use different versions of a language. Upper classes might use a very traditional 'proper' dialect, whereas lower classes might use a clipped version that's laden with slang. Some people might use different language in different social situations – say, more casual among their friends than they are with their work colleagues. This is known as 'code switching'.

The related field of 'evolutionary linguistics' examines how language evolves over time, to reconstruct how languages emerged and changed in the past and how they will develop in the future. With the advent of the Internet and other communication technologies, language now seems to be evolving faster than ever. It's also moving in new directions – as new forms of slang and shorthand become popular for **instant messaging**.

Neurolinguistics

The field of science that attempts to identify the processes taking place in the brain (see **Neurobiology**) accompanying the formation and use of language is called neurolinguistics. Language skills emerge in the frontal lobe of the brain in a part known as Broca's area, after the French physician Pierre Paul Broca (1824–80), who identified damage to this area as the reason why one of his patients had lost the ability to talk. Meanwhile, our ability to understand language is housed in another brain region, located towards the rear of the brain's left hemisphere – known as Wernicke's area, after German neurologist Carl Wernicke.

Closely related to neurolinguistics is the sphere of 'psycholinguistics', dealing with the influence not just of brain anatomy and chemistry on language processing, but also psychological factors, such as state of mind and past experiences. Neurolinguistics has led to a field of alternative psychotherapy, known as **neurolinguistic programming**.

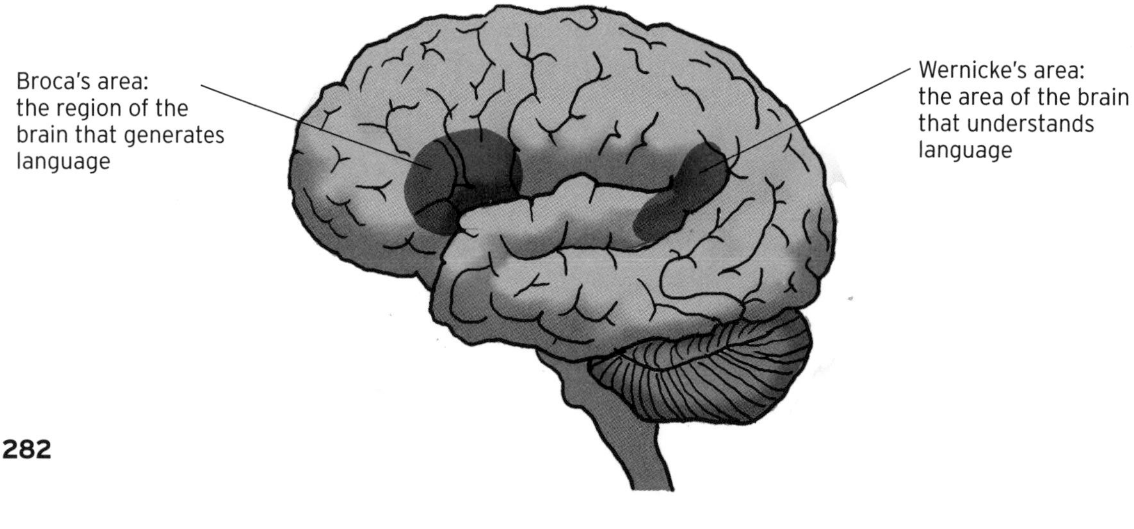

PSYCHOLOGY

Psychotherapy

Psychology is a branch of science dealing with the workings of the mind, its quirks and defects and how it influences human behaviour. Psychotherapy attempts to use our understanding of psychology to benefit people with mental health issues, or full-blown mental illness. A psychotherapist will talk with the patient to try to get to the root of their thought processes, their mindset and their emotional state. The procedure is purely verbal; no use is made of medication, a reflection of the psychotherapist's belief that psychological problems arise emotionally, not because of defective **brain function**. And this is where psychotherapy diverges from psychiatry, which is a branch of mainstream medicine specializing in the treatment of mental disorders. A psychiatrist can prescribe psychotherapy for a patient – as well as other treatments, such as medication.

Psychoanalysis

The Austrian psychologist Sigmund Freud developed psychoanalysis, a branch of psychotherapy, in the late 19th and early 20th centuries. Freud's central idea was that processes in the unconscious mind are responsible for people's behaviour, in particular that emotional energy locked up in the mind is responsible for psychiatric disorders. He believed that counselling could release this pent-up energy, and he did this by encouraging patients to recount their fantasies and dreams, and through 'free association' exercises – where patients are asked to say the first thing that comes into their mind following a prompt.

Freud also theorized that there are three primary areas of the mind: the 'ego', the 'superego' and the 'id'. The ego is the rational part of the mind that makes our level-headed decisions; the superego is the angel on our shoulder, ever moralizing about what we should do; the id is the devil on the other shoulder tempting us with what we'd most like to do. The human mind, said Freud, is torn between these three influences.

Behaviourism

Behaviourism was a **positivist** view of psychology holding that all that really matters in the evaluation of a patient's mental state is what can be measured and quantified, i.e. their behaviour. What's more, the theory says that all of our behaviours are acquired through our past experiences.

American psychologist John B. Watson introduced the theory of behaviourism in 1913. He based his idea on an experiment carried out on a group of dogs by Russian scientist Ivan Pavlov. Each time Pavlov fed the dogs, he would ring bell. Eventually, he found, the dogs would salivate upon hearing the bell, even if there was no food present – their behaviour had been conditioned by their experiences. Similarly, argued proponents of behaviourism, human behaviour is guided by 'operant' conditioning – where we learn through punishment and reward. Although an influential theory until the mid-20th century, behaviourism leaves little room for inherited behaviours (see **Sociobiology** and **Nature vs nurture**) and for this reason has now fallen from favour.

Cognitive psychology

Probably the best psychological model that scientists have to work with today is cognitive psychology, a term coined by American psychologist Ulric Neisser in 1967. In essence, the theory says that our thought processes are the root of all psychological and behavioural phenomena.

Central to the idea of cognitive psychology is information processing. Our brains take in a huge volume of information through the senses and processes it – our behaviour, so the theory goes, is determined by what we make of that information. In this view, it's rather like the brain is a computer, taking in data from the world around, and the mind of each individual is analogous to the particular piece of software that's running on that computer. Now, cognitive scientists are trying to figure out the exact 'specs' of the brain computer – and the form of the program each person has running on it.

Developmental psychology

The changes that the mind goes through as we grow older is known as developmental psychology. Practitioners in this field chart how the psychological outlook of human beings varies with age. For example, as we mature out of infanthood, we gain the ability to experience complex emotions – such as embarrassment and pride. As we get older we also become better equipped to empathize with the emotions of others. For many people, their outlook on life also changes as they move from childhood through adolescence and adulthood – usually becoming wiser and more balanced.

A big question in developmental psychology is the role of **nature vs nurture** – how much of this change in behaviour is determined by life experiences (see **Behaviourism**), and how much is pre-programmed by genetics (see **Sociobiology**)? This is an area of ongoing research.

Personality traits

In an effort to classify the different psychology of individuals on a scientific basis, psychologists have come up with a range of major personality traits that most people exhibit to varying degrees. There are five key traits, and assigning scores for each of these five factors, psychologists believe they can capture the mindset of most people. The five factors are: 'openness to experience', how readily a person will try new things; 'conscientiousness', how self-disciplined and careful a person is; 'extroversion', the degree to which a person enjoys the company of others; 'agreeableness', how easy to get along with the person is; and finally, 'neuroticism', how easily stressed out and upset they are.

The work of Swiss psychologist Carl Jung produced a similar kind of classification which is often used in psychometric tests – written exams designed to assess someone's personality. Known in the business as 'Myers-Briggs' tests, these rate candidates in four dimensions: extroversion–introversion, sensing–intuition, thinking–feeling and judging–perceiving.

Social psychology

Social psychology is the application of principles from psychology to describe not the behaviour of individuals, but that of groups of individuals. Social psychology draws upon scientific methods to understand society, shedding light on phenomena such as the persuasiveness of advertisements to the policy tactics of our leaders (see **Social and political engineering**).

Sociologists and psychologists tend to differ in their approach to social science. Psychologists tend to take a 'bottom-up' approach, focusing on the individual and then trying the aggregate the behaviour of many individuals to explain the behaviour of the group. Sociologists, on the other hand, tend to work 'top-down' – primarily concerned with group behaviour, giving the actions of the individual secondary consideration.

Environmental psychology

It's a simple fact that no one likes being in a dark, depressing room for too long. Now this phenomenon has its own field of science – the interaction between people and the immediate locale around them is an area of study known as environmental psychology. It is leading architects and planners to build offices in which workers are more productive, and hospitals where – simply by the addition of a nice view from the window – patients have been found to heal more quickly.

Town planners are also catching on, using environmental psychology to gauge the interplay between people, buildings and locations to design new urban spaces that minimize crime and boost happiness levels. They achieve this through the use of practical yet aesthetic street layouts, innovative architecture and plenty of parkland. In the UK, environmental psychology is already being used to rejuvenate the built environment in some of the more dismal inner cities.

BRAIN FUNCTION

Intelligence

Our capacity for learning, reasoning, problem-solving, and for remembering knowledge and drawing upon it at a later time, is all grouped together in a property of the human mind called intelligence. The oldest theory of intelligence is the idea of 'general intelligence' put forward by British psychologist Charles Spearman in 1904. He believed general intelligence could be quantified by a single number – the intelligence quotient, or IQ, given by a person's mental age divided by their actual age, multiplied by 100. British-American psychologist Raymond Cattell later refined this, suggesting that there are two kinds of intelligence. 'Fluid intelligence' is a person's inherent ability to learn and reason, while 'crystallized intelligence' refers to the knowledge and abilities they acquire through life.

In 1983, American developmental psychologist Howard Gardner went further still, proposing the theory of multiple intelligences. This divided an individual's intellectual capacity into seven areas – linguistic intelligence (ability with words and languages); logical-mathematical intelligence (numerical problem solving skills); spatial-visual intelligence (spatial judgement and coordination skills); musical intelligence (hearing and sense of rhythm); bodily kinaesthetic intelligence (manual skills and practical aptitude); interpersonal intelligence (ability to interact with others); and intrapersonal intelligence (understanding of the self).

Studies have shown that inheritance accounts for around 75 per cent of a person's intelligence – intelligent parents generally do have intelligent children – with the other 25 per cent shaped by the person's experience (see **Nature vs nurture**). However, attempts to isolate genes for intelligence have failed. It seems our natural intellectual ability is the product of complex interactions between many different areas of our DNA code.

Emotional intelligence

As a supplement to the theory of multiple intelligence introduced by psychologist Howard Gardner, American student Wayne Payne, in his doctoral thesis of 1985, put forward the idea of emotional intelligence. It is loosely defined as the quality of an individual to be aware of the emotional feelings of themselves and others – and to incorporate emotional considerations into their reasoning processes.

People with high emotional intelligence are generally more successful in relationships and have better ability as team workers, and indeed as team leaders – something companies are becoming increasingly aware of, sometimes using psychometric tests (see **Personality traits**) to screen the emotional intelligence of potential employees.

Intuition

Intuition is the ability of some people to simply know the right decision to make, without stopping to analyze the facts in detail. In Myers–Briggs psychometric testing (see **Personality traits**) intuition is considered to be the opposite of 'sensing' – with intuitive individuals more likely to rely on gut feeling and instinct rather than concrete evidence and analysis. The best model of intuition that psychologists have is that experience in a particular subject or activity helps us build an instinctive grasp of it so that the subconscious is able to process incoming data much faster than the conscious mental faculties ever could.

Malcolm Gladwell, author of the book *Blink: The Power of Thinking Without Thinking*, believes experts in particular areas become able to filter irrelevant information from the flood of data pouring into our senses. He calls this ability 'thin slicing' and cites the example of the American relationship psychologist John Gottman, who – from watching a couple talk for 15 minutes – can predict with 90 per cent accuracy whether they will still be together 15 years later.

Creativity

What is it that enables some of us to pluck innovative solutions and new ideas readily from thin air, while others just don't have the creative spark. Scientists have found that creativity has no particular centre in the brain – it seems to be more about mental flexibility and the ability of the brain to make unusual connections between seemingly unrelated areas. In this view, creativity seems to be a facet of the whole brain working together – and indeed, **split brain** personalities generally seem to be far less creative.

Other experiments in animals have shown that new behaviours seem to emerge from a mixing up of existing ones – a process reminiscent of the mixing up of existing genes during sexual reproduction (see **Gametes**). So it may be that new ideas evolve in a similar way to the appearance of new species through evolution. Many psychotherapists recommend creative pursuits as a way to boost happiness levels, even to the point of suggesting it as a form of therapy for **depression**. Then again, this may be a double-edged sword. Psychiatrist Kay Redfield Jamison of Johns Hopkins University, in Baltimore, has confirmed the long-held suspicion that creativity and mental illness go hand in hand – finding that successful artists are statistically more likely to suffer from mood disorders.

Memory

The precise location in the brain of the neural databanks where our long-term memories are stored isn't well understood. It may well be that, like creativity, our capacity for remembering isn't localized to any particular part of the brain, but is spread across it. What is known is that certain brain areas play a part in memory processing. We have two principle types of memory – short-term and long-term. Short-term memory holds information for up to about a minute before it is transferred to the long-term memory. It seems the hippocampus, an area buried deep inside the brain, is the main arbiter of this transfer. Sleep is also thought to be important for reinforcing information that's written to our long-term memory.

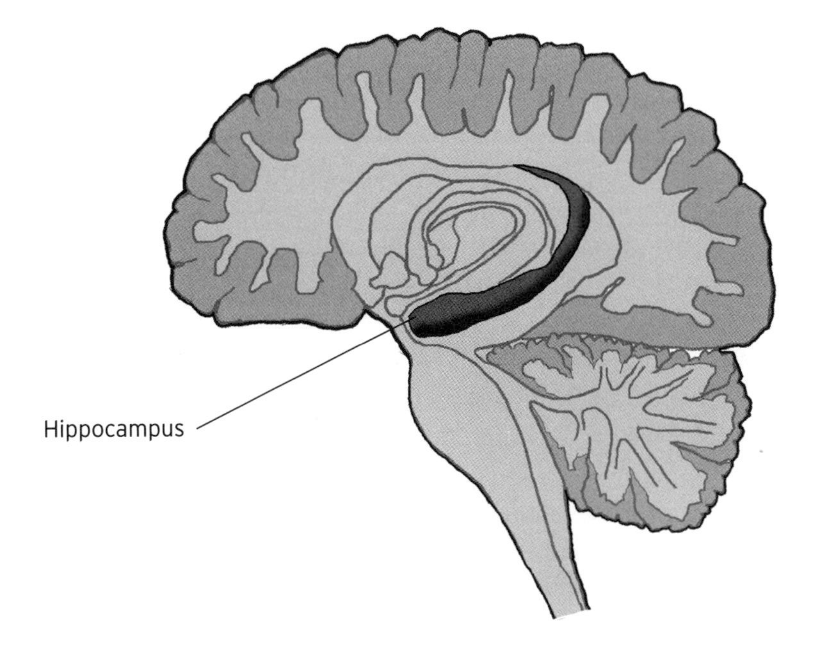

PSYCHOLOGICAL PHENOMENA

Confabulation

In 2003, psychologists Elizabeth Loftus and Jacqueline Pickrell at the University of Washington carried out an experiment that calls into question the human ability to remember past events accurately. They gave 24 volunteers booklets containing accounts of childhood events that had happened to them. Most of these stories had been provided by relatives, but one was completely bogus – a made-up story about getting lost at a shopping mall and being reunited with their parents through the help of a kind elderly lady. When Loftus and Pickrell interviewed the participants about their stories, a quarter of them reported being able to remember vividly the day they got lost at the mall, some even providing extra details beyond what was in the booklet.

There is no well-established theory to explain why we humans are so prone to embellishing our memories in this way – a phenomenon psychologists call confabulation. One possibility is that our memory is tempered by our rational thoughts. The subjects in the Washington experiment were told that trusted relatives had provided the stories in the booklet – and this may have convinced the subjects that the stories were true.

Illusions

It is not only false memories introduced through confabulation that can fool the brain. Human beings are also susceptible to optical illusions – where certain arrangements of shapes and colours can deceive the visual cortex. A good example is known as the cafe wall illusion (see diagram). Focus in on any particular region and you'll see that all the rows of bricks and lines are in fact parallel, but look at the big picture and some appear to be slightly wedge-shaped. The cafe wall illusion is thought to be caused by something psychologists call 'border locking'. In simple terms, this is to do with the way the brain uses changes in colour to detect edges. At the top and bottom of the picture, where the pattern is very regular, it's easier to make out the edges of the bricks, marked out by

the mortar lines between them. But in the middle the edges are harder to see, especially when the eye isn't looking directly at them. And that's what makes the lines appear to move around.

Top-down processing

Illusions demonstrate how easy it is for the brain's visual system to be fooled; but it is possible to pull the wool over not just our eyes, but our ears too. One manifestation of this is through a phenomenon called top-down processing, which effectively makes us hear what we're told to hear.

A good example is the Led Zeppelin song 'Stairway to Heaven', which is supposed to contain a secret message when played backwards – a phenomenon called 'backmasking'. If you first play the song backwards, it's nonsensical; but if you read what the backwards lyrics are meant to be, then listen to the backwards audio again, suddenly the message jumps out as plain as day.

Top-down processing is a term from **cognitive psychology**, referring to the way information is organized in the brain – with basic information at the bottom and the most complex, processed information at the top. Top-down means that processed information in the form of memories is influencing the basic information coming in through the senses.

Milgram's experiment

Illusions and top-down processing alter our perception of the world, yet it's possible to alter our behaviour as well – sometimes in frightening ways. A chilling experiment carried out by the Yale University psychologist Stanley Milgram in 1961 demonstrates this. Milgram recruited a group of volunteers for what was advertised as a memory experiment. Each participant was told they would take the role of a 'teacher' and have to test a subject in another room – the 'student' – on their memory of a list of word pairs. Each time the subject got a question wrong the participant was told to press a button administering an electric shock of progressively increasing voltage.

In fact, the electric shocks were not real and the student was an actor. At first, he complained that he was in pain. Later he would be heard screaming as the voltage was supposedly increased. And finally there was an ominous silence. Whenever a participant voiced their concern, they were told by the experimenter that it was essential they continue

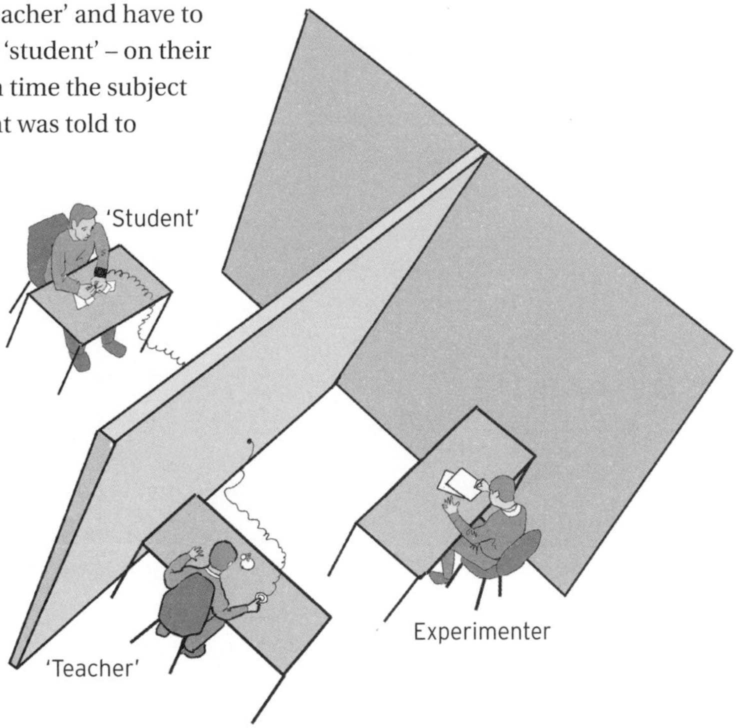

'Student'

Experimenter

'Teacher'

and that they wouldn't be held responsible. Remarkably, over 60 per cent of them completed the experiment, administering what would have been a fatal shock. Milgram's experiment demonstrated how willing many of us are to do as we're told.

Synaesthesia

Synaesthesia is a condition where a person's senses become jumbled up. It is experienced by roughly 1 in every 2,000 people. A common example is the association of particular letters or numbers with colours – a person will literally see the letters in different colours. Others can hear colours, or taste words.

Medical imaging studies show that those who experience synaesthesia – referred to as 'synaesthetes' – have different brain activity from those who don't. Those who see colours in response to words show activity in both colour areas and word-related areas of the brain, suggesting that a literal cross-wiring of synaptic connections could be responsible. Some scientists think this linking of different brain areas could also give synaesthetes enhanced creativity. Famous synaesthetes include Tori Amos, Leonard Bernstein and physicist Richard Feynman.

Kappa effect

Albert Einstein once famously remarked: 'Put your hand on a hot stove for a minute, and it seems like an hour. Sit with a pretty girl for an hour, and it seems like a minute.' His point was that human perception of time is subjective.

In 1953, a group of researchers writing in the science journal *Nature* came to much the same conclusion – although in slightly more prosaic terms. The group looked at a car journey split into two parts, each taking an equal amount of time but in one half the car was travelling considerably faster and thus covered more ground. They found that passengers would perceive the faster leg of the journey, covering more ground, as taking longer – this they called the Kappa effect.

Cognitive dissonance

Cognitive dissonance is a major field of study in **social psychology**, and illustrates how people's attempts to rationalize contradictory facts can lead to the most irrational of beliefs and behaviours. For example, a supporter of animal rights given a pair of leather shoes as a gift might wear them – rationalizing the contradiction by arguing that they didn't buy the shoes, therefore wearing them is acceptable.

In the 18th century, American politician Benjamin Franklin once used the technique to win over a political enemy by borrowing a book from him, and then returning it promptly with thanks. In doing his enemy a good turn, the other man was plunged into cognitive dissonance. He resolved the conflict by deciding that he must like Franklin after all – and the two subsequently became great friends.

Herd mentality

Human beings don't have to obey orders (see **Milgram's experiment**) to lose command over their own actions. Like sheep and other herd animals, we are prone to follow the actions of others – when we think the other person knows better, or when we simply can't be bothered to think for ourselves. This is a phenomenon in social psychology known as herd mentality. It can lead to group behaviours such as the 'bandwagon effect', where increasing numbers of people copy what everyone else is doing. An example is when investors all frantically buy or sell certain stocks, which can create stock-market bubbles and crashes. The bandwagon effect can also lead to the collapse of banks, as a small loss of confidence causes ever more customers to withdraw their deposits.

Another type of dangerous herd behaviour is known as 'groupthink', where members of a group tasked with making a decision often end up opting for a course of action that simply minimizes conflict within the group. This happens because members who disagree with the apparent group consensus don't want to look foolish or alienate themselves. It leads to what James Surowiecki, author of *The Wisdom of Crowds*, describes as a 'loss of cognitive diversity'.

Crowd crazing

In 2004, when a new branch of IKEA opened in Jeddah, Saudi Arabia, offering large discounts to the first customers through the doors, managers were expecting and prepared for crowds of up to 5,000 people. But when 20,000 turned up the result was a stampede in which 16 people were injured and 3 were killed. Many events like it have happened at stores around the world, including WalMart on Long Island, in the United States in 2008, and in Japan in 2002. This behaviour is known as crowd crazing. Psychologists put it down to a form of herd mentality when people are possessed by a primal instinct to follow the rush, either towards a potential benefit or away from danger, once the crowd situation has become threatening. The effect is thought to be triggered by lack of information, when people aren't clear where they should go or queue and are worried that they might miss out as a result.

Sleep paralysis

When a person is asleep, the body's muscles are naturally paralysed – a condition known as atonia. But sometimes the symptoms of paralysis can persist briefly into wakefulness leading not just to immobility but also hallucinations and feelings of terror caused by the momentary disconnection of mind and body. The phenomenon is known as sleep paralysis, and normally abates after a few minutes. Some psychologists have suggested that sleep paralysis could be a reason for some accounts of paranormal experiences and alien abduction.

MENTAL ILLNESS

Depression

Feelings of sadness, mood swings, low motivation and in some cases thoughts of suicide or death are all symptoms of the psychiatric condition known as depression. It is thought to affect as many as 17 per cent of the population at some point during their lives. Depression can be brought on by stress, illness, and change in personal circumstances such as the breakdown of a relationship or the death of a family member. It also tends to run in families, suggesting a significant genetic component too.

The strongest scientific evidence links depression to disturbances in brain chemicals; in particular, deficiency of the neurotransmitter serotonin. This deficiency is believed to be why antidepressants such as Prozac – known as selective serotonin re-uptake inhibitors (SSRIs), which block the absorption of serotonin by the brain – have proven so effective in treating the disorder. Some evolutionary biologists have suggested that depression may even have emerged through **natural selection** as a beneficial adaptation that turns our interest away from fruitless pursuits.

Anxiety disorders

Anxiety disorder is a mental state in which sufferers experience inordinate levels of fear and angst. There are various forms of anxiety disorder: 'generalized anxiety disorder' is a state of perpetual worry with no apparent cause or trigger; 'social anxiety disorder' is an irrational fear of social situations; 'obsessive-compulsive disorder', or OCD, is anxiety that leads to obsessive rituals, such as repeated hand-washing; while 'post-traumatic stress disorder' is anxiety that's experienced as a result of one or more past traumatic experiences.

Anxiety disorder is thought to have a neurobiological basis in a brain neurotransmitter called GABA (gamma-aminobutyric acid), which dampens down activity in the central nervous system; those prone to anxiety disorder seem to have low levels of GABA. Antidepressants of the selective serotonin reuptake inhibitor (SSRI) class seem to help with anxiety disorders. Scientists suspect this is because the drugs agitate GABA-receptors in the brain, mimicking the effect of the neurotransmitter and thus having a calming effect on the patient. Cognitive behavioural therapy (CBT), a form of **psychotherapy**, has also been found to be effective. Proponents of the treatment believe that mental disorders are caused by the thoughts and behaviour of the individuals themselves; CBT seeks to address that by changing the way those with a disorder think and act.

Impulse-control disorders

A set of psychological conditions characterized by compulsive behaviour are impulse-control disorders. There are five principle forms of recognized impulse-control disorder: intermittent explosive disorder (inability to control violent outbursts), pathological gambling (compulsion to gamble), kleptomania (the irresistible urge to steal), pyromania (the impulse to start fires) and trichotillomania (the urge to pull out one's hair).

Impulse-control problems seem to arise alongside other psychiatric conditions such as anxiety disorders, depression, eating disorders and personality disorders. As with depression, evidence indicates that the neurotransmitter chemical serotonin plays a big part in controlling impulses, which is why Prozac-related antidepressants are effective treatments.

Personality disorders

Aspects of someone's personality that may cause problems in relationships with others, and in some cases lead to behaviour considered unacceptable to society, are known as personality disorders. Psychologists divide personality disorders into three clusters. 'Cluster A' is made up of the 'eccentric' personality disorders (PD): paranoid PD (distrusting of others), schizoid PD (socially detached) and schizotypal PD (having odd ideas and beliefs). 'Cluster B' are the 'dramatic, emotional or erratic' disorders: antisocial PD (showing lack of empathy or conscience), borderline PD (emotionally unstable), narcissistic PD (self-obsessed) and histrionic PD (needing to be the centre of attention). Finally, 'Cluster C' forms the anxious disorders: avoidant PD (having extreme social inhibition), dependent PD (relying overly on the help of others) and obsessive-compulsive PD (obsessed with perfection). However, obsessive-compulsive personality disorder is different from obsessive compulsive anxiety disorder (see **Anxiety disorders**).

In 2005, a study found a higher incidence of histrionic, narcissistic and obsessive-compulsive personality disorders in a sample of UK company executives than among the inmates of Broadmoor – a UK hospital for the criminally insane.

Bipolar disorder

Formerly known as 'manic depression', bipolar disorder is characterized by wholesale swings between episodes of severe depression and times of almost delusional elation, known as 'mania'. Each episode may last for several weeks or more at a time. The condition is probably brought about by chemical imbalances or anatomical defects in the brain, though isolating the exact cause has proven extremely difficult. One study found that sufferers have lateral ventricles – C-shaped structures within the brain that join up with the spinal cord – that are 17 per cent larger than those in non-sufferers.

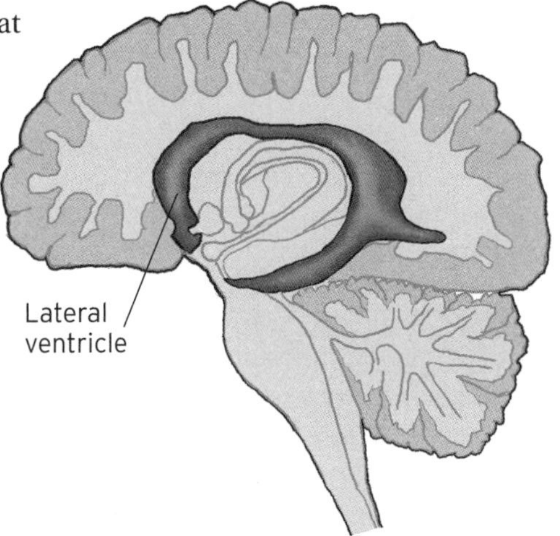

Bipolar disorder is thought to affect around 2 per cent of the population and can be passed genetically from parents to their offspring. In some people who are genetically predisposed to the illness, it seems a trigger is needed to activate the first manic or depressive episode – this is usually a stressful event or series of stressful events, often during late adolescence.

Eating disorders

Mental health problems that result in abnormal eating patterns are known as eating disorders. The three most common are anorexia nervosa, bulimia nervosa and binge eating. Anorexia nervosa, usually known simply as anorexia, is a fear of gaining weight that manifests itself through extremely limited food intake. Some anorexics will eat so little as to almost starve themselves, which can result in organ damage and even death. People with bulimia nervosa, or bulimia, will eat to excess and then rid themselves of the food either by vomiting or taking laxatives. The constant cycles of eating and purging can cause damage to the gastrointestinal system and, if the purging is by vomiting, the teeth can be damaged by repeated exposure to stomach acids. Binge eaters gorge themselves like bulimics, but do not purge the food afterwards, leading to considerable weight gain that can cause **obesity** and **diabetes**.

There are a host of possible causes in **neurobiology** for eating disorders including genetic predisposition from parents, and imbalances of neurotransmitter chemicals in the brain such as serotonin, norepinephrine and dopamine.

Autism

Autism is a learning disorder usually manifesting itself within three years of birth and staying with the person for life. It severely impairs communication and social interaction skills – and the imagination. Children with autism are typically uninterested in any kind of play that involves imagination or pretending in preference to repetitive activities such as arranging their toys in a line. The evidence suggests that autism symptoms are caused by a disorder of the brain's synapses, the links between neurons that allow signals to pass from one neuron to another. In people with autism it is believed some of these synaptic channels become disrupted. It seems that autism is caused by abnormal changes in brain development occuring early in the development of the fetus, which may be triggered by an unwanted **immune system** response, chemical pollutants in the environment or genetic factors inherited from parents. At present, autism is not understood well enough to say for certain.

Autism is sometimes referred to as a 'spectrum disorder', meaning that people with the condition experience symptoms to differing degrees. Some areas of the autistic spectrum are given their own names, such as 'Asperger's syndrome'. Those with this condition exhibit obsessive behaviour patterns, but retain a workable degree of communication and interaction skills.

Dementia

This is the degradation of mental faculties that accompanies ageing, and is considered part of the field of **geriatrics**. The most common form is Alzheimer's disease, where clumps of misfolded protein gather around brain cells – forming 'plaques' – and inside the cells themselves, forming 'neurofibrillary tangles'. Plaques and tangles inhibit the functioning of the cells, causing them to die – leading to gradual degeneration of the brain.

Symptoms usually begin with difficulty remembering recently learned facts and then worsen to more severe memory loss, difficulty talking, followed by lack of mobility and the requirement of full-time care. Roughly 0.4 per cent of the world's population – 27 million people – have dementia. And with medical science set to prolong old age ever further, this figure looks set to rise.

Schizophrenia

Schizophrenia is a serious mental disorder in which sufferers are unable to distinguish between imagination and reality. This may take the form of sensory hallucinations or delusional beliefs – symptoms collectively known as 'psychosis' – as well as disordered thinking and catatonia, where the person becomes silent, still and unresponsive to stimulus. The condition affects roughly 1 per cent of Americans. Schizophrenia is treated using antipsychotic drugs, together with psychotherapy – often cognitive behavioural therapy (see **Anxiety disorder**).

CONSCIOUSNESS

The mind

Most people would agree that they are conscious. But despite this – and after many centuries of active research by scientists and philosophers – the processes inside the head responsible for it have proven remarkably elusive. Consciousness glues together our perceptions of reality into a unified view of the world that allows us to recognize our place within it.

Researchers investigating consciousness are quick to draw a divide between the brain – the 1.4kg of soft grey matter inside your head – and the mind, the conscious entity within and its emotional states, experiences, beliefs and desires. They talk about 'philosophy of mind' – studies of the nature of our conscious minds, in particular the relationship between mind and body. In the 17th century, French philosopher René Descartes believed in a theory called 'dualism' – that the ethereal mind and the physical body are made from different materials. Today we know this cannot be true because of the effect of **anaesthetics** – physical substances that are able to induce an unconscious state.

Self-awareness

Key to consciousness is the idea of self-awareness, the ability to distinguish your thoughts from the entity producing them. As Catholic philosopher Augustine of Hippo put it in the 5th century AD: 'I understand that I understand.'

Zoologists use a technique called the 'mirror test' to gauge whether an animal is self-aware. This involves marking the animal with a dye spot, and then placing it in front of a mirror so that the spot is visible to it. If, after seeing its reflection, the animal makes an attempt to investigate the spot on itself then it has recognized the reflection as its own and is considered to be self-aware. There are ten species known that pass the mirror test – humans, orangutans, chimpanzees, bonobos, gorillas, bottlenose dolphins, orcas (killer whales), pigs, elephants, and magpies. Human babies usually pass the test after the age of about 18 months.

Qualia

Experiences perceived by the minds of conscious beings are known as qualia (singular 'quale'). Examples are the 'redness' of the colour red, the taste of an apple or the pain of stubbing your toe. These are subjective sensations which are hard to describe – one person's experience of the colour red may differ from another's. Indeed, describing qualia to other people is impossible – they are ineffable experiences that can only be conveyed by analogy, for example pointing to something that is red. Qualia play a part in the formation of intentions in the conscious mind – you like the colour red so you decide to turn your head and look at the fire truck.

Sentience and sapience

Philosophers and psychologists say that creatures that can experience qualia have a quality of mind known as sentience. Qualia include the experiences of pain and suffering, which has led to the sentience of all animal life forms becoming a central argument used in their defence by the animal rights movement. A related concept to sentience is sapience, meaning the ability to demonstrate higher cognitive powers – manifested through human-like (or better) intelligence and capacity for thought and reasoning.

Neural correlates of consciousness

Neural correlates of consciousness (NCCs) are the physical processes that happen in the brain when we experience qualia in the mind. The study of NCCs took an enormous step forward with the development of brain scanning techniques such as **magnetic resonance imaging** (MRI). Functional MRI relays real-time images of the activity inside the brain. A test subject could now have their brain scanned while reporting what qualia they were experiencing to a researcher. And comparing the scan to the subject's report could reveal what patterns of brain activity produce certain qualia. For example, when looked at under a scan, an area of the brain called the anterior cingulate cortex lights up when we feel the emotion of envy.

Some of the most productive brain-scanning experiments in the search for NCCs have been carried out on Buddhist monks, because of their ability to create and hold a given mental state long enough for the scanner to build a clear picture of the corresponding brain function.

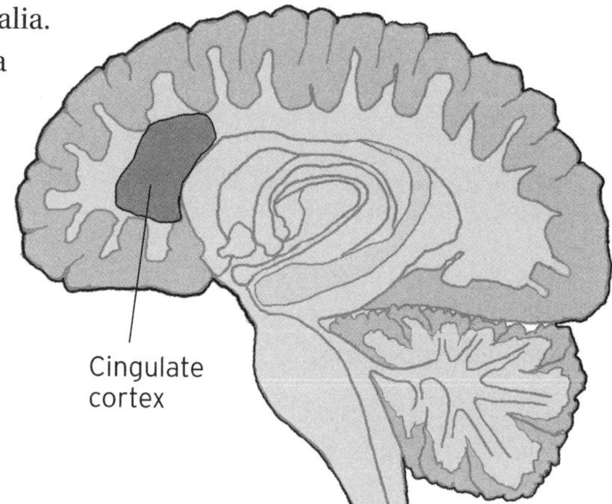

Hard problem of consciousness

The simple question 'Why is it that we experience qualia?' is known as the hard problem of consciousness, which is in contrast to many of the other questions in consciousness research, which are easy by comparison. Even linking all of the qualia we experience in the mind with neural correlates of consciousness – the physical processes between neurons in the brain that explain *how* and *where* qualia are formed – will not solve the problem of *why* we experience them. The term was coined by Australian philosopher David Chalmers.

Split brains

Some fascinating insights into the nature of consciousness have been gathered by researchers carrying out studies on so-called 'split brain' patients. These patients have had the 'corpus callosum', which connects the brain's left and right hemispheres, severed as a treatment for severe epilepsy. Because stimuli to one side of the body are processed by the opposite side of the brain this has enabled researchers to investigate consciousness in each side of the brain separately. For example, the behaviour of a patient in response to information presented to their left visual field will be exclusively due to the right-hand side of the brain.

The studies seem to show that the nature of consciousness in the two hemispheres is very different. Patients shown an image on the right of their vision can describe verbally what they see. Those shown an image on the left cannot – but they can later pick out the image they saw from a group. It seems that whereas the brain's right side can remember what it sees, the left-hand side houses all of our capacity for language and creativity – and seems to have an overall higher level of consciousness than the right.

Antedating

An astonishing, yet baffling, breakthrough in consciousness research was made in 1979 by the neurobiologist Benjamin Libet, at the University of California, San Francisco. He and his colleagues performed experiments in which they directly stimulated the brains of a number of test subjects. First of all they found that the brain only records the stimulus if it lasts for half a second or more. Next, they stimulated the brains of the patients again but followed this a quarter of a second later with a small stimulus applied to the surface of the skin. They were expecting the detection of both stimuli to be delayed by half a second so that the patients should feel the brain stimulus first, followed a quarter of a second later by the skin stimulus. In fact, the patients all reported feeling the skin stimulus first. It was as if the brain was editing out the half-second delay in the skin response to give the illusion that both the stimulus and the perception of it were taking place at the same time.

Libet reasoned that the brain does this same 'antedating' process with all sensory inputs, to account for the time taken for the inputs to travel along the nerves, and to then be processed by the brain. In this way, the brain stitches together the inputs, and the qualia they produce, to create a consistent picture of reality.

Free will

We all like to think that we decide to do things because we want to – a facet of conscious thought known as free will. But an experiment carried out by Benjamin Libet at the University of California, San Francisco, as a follow-up to his work on **antedating**, suggests that we are, in fact, simply slaves to our subconscious.

In the 1980s, Libet wired a sample of volunteers up to brain monitoring equipment that measured the brain's so-called 'readiness potential' – activity in the motor cortex that coordinates muscle movement. He asked each of them to move their wrist and while doing so, report the time, as measured on a clock in front of them, at which they took the conscious decision to make the movement. As was already well known, the detection of a readiness potential preceded the actual wrist movement, because of the time taken for the signal to travel from the brain, down the nerves and to the wrist. What Libet wasn't expecting to find was that the readiness potential also preceded the intention to move – actually beginning to build around half a second before the subjects reported that they had made the conscious decision. Libet's experiment seemed to show that free will is an illusion.

Quantum mind

In 1989, Roger Penrose – a professor of mathematics at the University of Oxford, who has made seminal contributions in physics to the fields of **general relativity** and **quantum theory** – put forward his own theory of consciousness. Working with anaesthesiologist Stuart Hameroff, of the University of Arizona, Penrose argued that consciousness arises as a result of quantum processes taking place between the proteins making up microtubules – tubular structures that form part of the scaffold-like cytoskeleton around which cells in the brain are built. What's more, Penrose and Hameroff suggested that it's these quantum processes that enable the brain's capacity for creativity, intuition and innovation – in particular, the problem-solving powers of mathematicians and scientists. These are abilities that far outstrip anything that can be achieved with modern computers which, Penrose says, are limited in their scope by **Gödel's incompleteness theorem**. Penrose set out the theory of the quantum mind, known as 'orchestrated objective reduction', or simply Orch-OR, in his book *The Emperor's New Mind*.

Out-of-body experience

Even though the theory of dualism (see **The mind**) has been disproven, many people still cling to the idea that consciousness can exist outside of the body. One in ten people claim to have had an out-of-body experience (OBE), in which they are able to view their surroundings from a perspective exterior to their physical self.

Despite a number of scientific studies of OBEs, there is little to no compelling evidence that they represent a real detachment of the mind from the physical brain. British psychologist Susan Blackmore believes OBEs may be a kind of wakeful dreaming state, in which a person's consciousness becomes disconnected from their physical senses and distorted by dream-like images from elsewhere in the brain.

SOCIAL TRENDS

Population dynamics

One species that doesn't seem in need of conservation is *Homo sapiens*: us. The human population on Earth is currently doubling around once every 40 years, with the growth put down to medical science decreasing infant mortality, effectively raising the birth rate, while at the same time increasing longevity, and so lowering the death rate – trends that look set to continue.

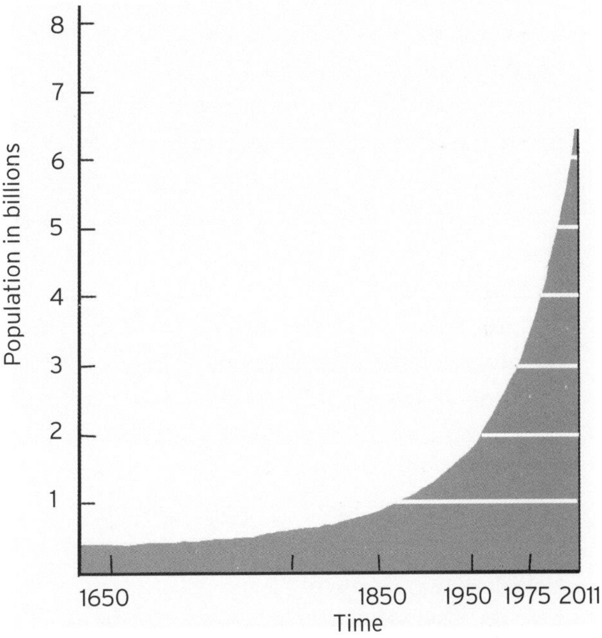

Left unchecked the growth of a population will proceed exponentially, because the rate of growth itself is proportional to the number of individuals in the population. This trend is sometimes known as Malthus's law, after the English clergyman Thomas Malthus who carried out some of the first research into population dynamics in the early 19th century. He was the first to spot the potential dangers of overpopulation – shortages of food and other resources as the population grows to exceed the Earth's means.

In the rest of the natural world, scientists use complex mathematical models to try to predict the population balance between different **species**, taking into account factors such as births, deaths, natural resources, and the **biological interactions** between different species – for example, prey and predator.

Flynn effect

It's all too easy to argue that school and university exams are getting easier and that if students today had to sit exam papers from 30 years ago, few of them would do well. Flying in the face of such statements, however, is a phenomenon known as the Flynn effect named after James Flynn, the scientist who first noticed the effect in the 1980s. Put simply it says that IQs – the measure of general intelligence – are increasing with time, and doing so at a rate of about 3 points per decade.

IQ is normalized so that the average score for each age group of the population at any time is 100. But the Flynn effect means that if children from the 1930s took an IQ test today, their average would be more like 75. Flynn believes it is caused by the increasing presence of technology and other manifestations of complexity in everyday life that humans have to get to grips with.

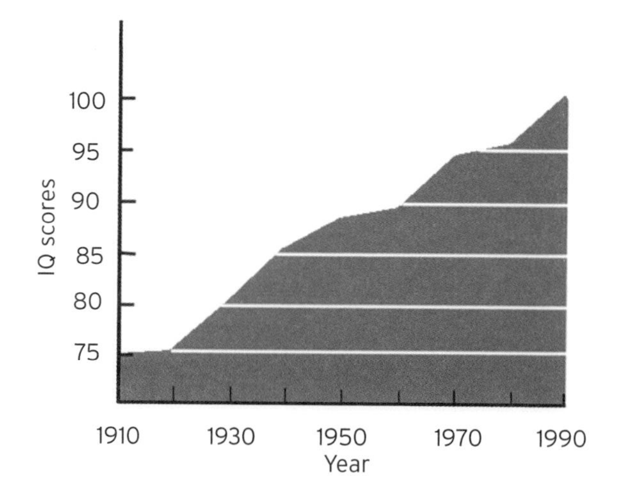

Height

Most people could be forgiven for thinking that human beings are, generation by generation, growing inexorably taller. After all, that certainly seems to be the case with our waistlines (see **Obesity**). But measurements of human skeletal remains from across the ages made by Richard Steckel, at Ohio State University, tell a different story. Far from gradual evolution from short to tall, Steckel finds that Northern European men living in the Early Middle Ages (9th to 11th centuries AD) were several centimetres taller than their counterparts at the start of the Industrial Revolution around 1750 – and almost as tall as modern humans today.

Steckel believes that the height of a population is an indicator of the overall prosperity of the region. He speculates that the differences may be due to a period of warm climate during the Early Middle Ages, which made food plentiful, followed by the Little Ice Age between the 16th and 19th centuries when it would have been scarce.

Luck

Why is that some people seem to have an endless supply of luck while others are sadly lacking? The answer has less to do with lucky charms and more to do with psychology. Richard Wiseman, a psychologist at the University of Hertfordshire in England, has carried out a study finding that people who rate themselves as lucky are often highly outgoing extroverts who are comfortable talking to strangers and mixing in new circles, and so are more likely to expose themselves to chance opportunities. On the other hand, those who believe themselves to be unlucky are more introverted and have lower expectations of themselves. Not only do they make fewer new acquaintances, leading to fewer lucky breaks, but even when they are offered an opportunity their self-doubt often holds them back. The personality traits of introversion and extroversion are to some degree inherited from our parents, meaning that some people really are born lucky.

Globalization

For better or for worse, we are all part of the global village now. Globalization is a movement that began in the 1980s, driven by banks and financial institutions who wanted to make it easier to do business overseas. Improved communication technology brought the means for these markets to operate seamlessly. Hot on its heels came the Internet, which brought globalization to everyone. Suddenly you could meet and chat in real time with a stranger on the other side of the planet. It has also meant that, someone in one country could legally post material online that might be outlawed in another, transcending national boundaries.

Today, globalization is about more than just communication – it is the integration of the countries of the world into a whole. With cheap air travel more people can actually *be* on the other side of the world – rather than just chatting with it online. Ultimately a product of technology, globalization has brought both benefits (greater awareness of world affairs, international disaster relief) and drawbacks (exploitation of poorer nations). But the safe money says it's here to stay.

Environmentalism

Growing awareness that human activities are compromising the environment in which we live has led to the birth of the environmental movement. It first began in the 1960s, when fears over atom bomb tests coincided with the realization of the impact on the environment of pollution and chemical pesticides, such as DDT. Suddenly, there was concern that human activity could have a significant negative impact on the Earth's **ecology**. At the same time, a small number of scientists were beginning to appreciate how the carbon content of the atmosphere seemed to correlate with rising global temperatures – science had discovered **climate change**. By the late 1970s, there was a chorus of scientific voices calling for curbs on human carbon emissions, lest we suffer the potentially catastrophic consequences of global warming during the 21st century. Even so, it was 20 years later before politicians started to listen.

Societal collapse

At its peak, Easter Island, one of the Polynesian islands in the Pacific Ocean, is thought to have been home to some 7,000 people. It was clearly once a cultured civilization, as evidenced by the hundreds of impressive stone statues, or 'moai', that remain there today. And yet when the first Europeans landed there in the early 18th century they found around 3,000 people living a savage, filthy life in primitive conditions. Cannibalism was rife as the warring factions on the island competed for scant food supplies.

Easter Island is thought to be an example of a society that collapsed because it exhausted its resources. A victim of its own success, this advanced people stripped the island of its trees and natural commodities to feed its flourishing skills in stone-working, architecture and transport. Today, the fate of the ancient Easter Islanders is a cautionary tale of the dangers of a civilization overexploiting its environment.

HUMANITIES

Anthropology

The humanities are areas of study through which human beings document, describe and explain human experience and the condition of being human – disciplines that include sociology, history and philosophy. But perhaps the most fundamental of them all is anthropology – the science of people, pulling together human evolution, history, culture and behaviour.

Anthropology breaks down into four main fields. 'Social anthropology' is concerned with people and their cultures, ideas and beliefs. 'Biological anthropologists' are concerned with the origins of the human species (see **Out of Africa**) and the biological factors that affect the human race today, such as disease. 'Linguistic anthropologists' investigate the influence of language and communication on human issues (see **Linguistics**), while 'archaeological anthropologists' study the lives and civilizations of people from ancient history (see **Archaeology**).

History

The study of humans in the past is known as history. A vast subject, it extends from recent times, for which there are living witnesses – known as 'living history' – back to the beginnings of humanity hundreds of thousands of years ago, which can only be investigated using remains recovered through **archaeology**. Historians are highly scientific in their approach, seeking strong evidence to back up their theories about the past. Evidence can take one of two forms – primary evidence, which consists of first-hand accounts, and secondary evidence, which are reports from parties not directly involved in events.

Historians break the past up into manageable eras. Generally speaking, anything before the fall of the Western Roman Empire in 476AD is classed as ancient history. The period from then until the end of the 15th century is known as the Middle Ages. And events happening between the end of the Middle Ages and the present day make up modern history. The first historians were the ancient Greek philosophers Herodotus and Thucydides, who lived in the 5th century BC.

Sociology

The study of human society is known as sociology. It is a subject that began to be looked at in earnest during the 18th and 19th centuries following important scientific advances, such as **Newtonian gravity** and Darwinian **natural selection**. Scientists and philosophers wondered whether not only the physical world but also humans, their culture and interactions could be unpicked through scientific inquiry. The 'father of sociology' was the French philosopher Auguste Comte, who believed that the answer to social problems lay in the philosophy of **positivism**.

Modern sociology is based upon the relationships formed between individuals and how these relationships build hierarchically to form the network upon which society rests. It is closely linked to social psychology. The effects of globalization now extend this network of relationships to form a single society encompassing the entire planet.

Philosophy

Philosophers spend their days contemplating the world. While not strictly a science in itself, philosophy is concerned with uncovering the deepest truths about reality. The pursuit underpins knowledge and logical thought – and thus makes up an essential component of the **scientific method**.

There are four major strands of philosophy: 'epistemology' is concerned with the growth of knowledge – what knowledge really is, and how it is acquired and processed; 'ethics' is about resolving moral dilemmas – scientific research dealing with the creation of life, such as **synthetic biology**, is increasingly being guided by ethical considerations; 'logic' determines our processes of reasoning and deduction; while finally 'metaphysics' examines questions of reality that lie outside the realm of what is testable by science – for example, is there a God? The word 'philosophy' itself comes from the Greek language and means 'love of wisdom'.

Humanism

Humanism is a kind of life philosophy that places at its core the value of the human individual. It rejects religion and the supernatural in favour of reason – humanists base their decisions on evidence rather than superstition. They believe that humans are able to choose between good and evil without the promise of reward – or the fear of suffering – in another life. But it's a philosophy also tempered by a humane code of ethics and **democracy** so that the views of all humans are represented. Human rights, as well as racial and sexual equality, all figure highly on the humanist agenda. Humanists have adopted the 'Happy Human' symbol (shown here) as their emblem.

ECONOMICS

Economic theory

Economics is the science of trade, governing the flow of money, goods, services and other resources – usually when there are insufficient quantities of these commodities to satisfy everyone. Scientists in other fields refer to it pejoratively as the 'dismal science'.

Economists often split their discipline into 'microeconomics', concerned with the dealings of small economic units such as businesses and households, and 'macroeconomics' – which is the big picture, dealing with economic growth and the wealth of nations (see **Economic growth**).

Capitalism is the dominant form of macroeconomy in the world today, in which the constituent businesses generate goods and services – the profits from which are returned to their owners. This increases the owners' 'capital', their means to purchase assets and services (usually workforce labour) through which they can generate further wealth. The opposite of capitalism is communism, in which profits are shared equally among the workers.

Supply and demand

At its most basic level economics is driven by supply and demand. When the supply of goods and services outweighs demand their monetary cost falls, and conversely when demand outweighs supply the price goes up.

Economists often study supply and demand graphically, plotting them as curves of price against quantity. The demand curve shows what people are willing to pay according to the quantity of the product available – it's high when quantity is low and decreases as the quantity available increases. Likewise, the supply curve shows how many units manufacturers are willing to make with a price. This is an upward sloping curve – the more a supplier can get for a product, the more of it they're willing to make so as to maximize revenue. Where they cross is the 'equilibrium point' towards which the price will converge. Supply and demand explains why raising interest rates (the cost of borrowing) is a way to keep inflation (rising prices – see **economic measures**) in check. When interest is high, consumers are less likely to borrow money and more likely to save it – driving down spending and so reducing demand, in turn lowering prices and inflation.

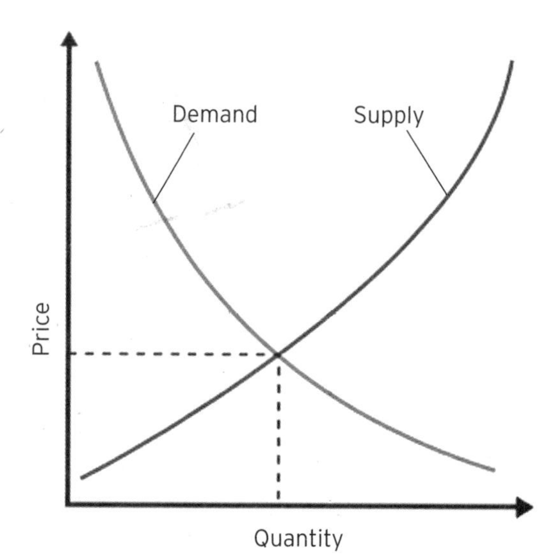

Marginalism

Marginalism is a concept in microeconomics that boils down to weighing up pros and cons when making decisions about how much of a product or service to buy or sell. Here's an example – you want to buy advertising to sell a product. Each advert costs $100 and past experience tells you that the first advert will bring in $400 of extra sales. The next advert will bring an additional $200, the next $100, and one after that just an extra $50. Clearly the first advert is worth placing as you will be $300 better off – referred to as your 'marginal gain'. The second is worth it too as your marginal gain from that is $100. But for the third your marginal gain is zero, and beyond that it is actually negative. Marginalism therefore dictates that you should place two ads, and no more.

Similarly, in manufacturing, marginalism reveals the optimum amount of a product to produce by weighing up the marginal cost (the cost of producing each extra unit) against the marginal benefit (the revenue selling that extra unit will bring in). As soon as the net marginal gain falls to zero, it is no longer cost effective to produce extra units.

Number of adverts	unit cost	benefit	marginal gain
0	$0	$0	$0
1	$100	$400	$300
2	$100	$200	$100
3	$100	$100	$0
4+	$100	$50	$-50

Diminishing returns

The law of diminishing returns is an economic principle which says that increasing a single resource required in the manufacture of a product (say, the size of the workforce), while keeping all other resources (such as tools and equipment) constant, leads to decreasing marginal gains (see **Marginalism**). An easy example is the size of the workforce in a small factory. Let's say you have one worker making ornamental flowerpots at the rate of 16 pots per day. You can sell each pot for $20 which, after paying your worker's wages of $150 per day, gives you a daily marginal gain of $170. Business is going well so you decide to employ two extra staff, expecting to triple your profits, but when you employ the first new worker you notice that rather than making 32 pots per day, your factory is only turning out 28. Your marginal gain from the new worker is only $90 a day. Worse still, when the second new worker starts, output only rises to 34 pots per day, not the 48 you were originally hoping for – and you're making a daily marginal loss from him of $-30.

Number of workers	daily wages per person	extra pots produced	revenue from extra pots	marginal gain
0	$0	0	$0	$0
1	$150	16	$320	$170
2	$150	12	$240	$90
3	$150	6	$120	$-30

Diminishing returns arises here because the workers all have to share tools and other equipment. It is a common phenomenon in businesses with 'fixed' capital (in this case the tools and equipment) and 'variable' labour. The possible solution is to increase the capital expenditure as the number of staff increases.

Economic measures

Macroeconomists use a host of economic measures to quantify numerically the state of a nation's economy – so that governments can keep tabs on their economic status and so the predictions of economic theories can be tested experimentally.

The principle measure of a country's productivity is its gross domestic product (GDP), given by adding up the monetary value of all the goods and services the country produces over the course of a year. The total GDP of the world in 2008 was about $60 trillion. Of a single nation the United States has the biggest GDP, at $14 trillion, followed by Japan and then China. A nation's GDP is often quoted 'per capita' – that is divided by its population.

Other common economic measures include inflation, which is the rise in the price of goods and services over time normally quoted as a percentage per year; and unemployment, the number of the population out of work.

Economic growth

A positive increase of a country's GDP per capita (see **Economic measures**) is referred to as economic growth. It is usually measured as a percentage, so if the GDP per capita increases from $10,000 to $11,000, then that's economic growth of 10 per cent. To help show trends in economic growth it is usually assessed every quarter throughout the year.

Economic growth is a sign of a nation's prosperity. An increase in GDP means that jobs are being created, which means that more people are earning money – and then spending that money, pushing up consumption. And as consumption goes up so more jobs are required to meet the increased demand, increasing consumption further – and so on. It seems like a self-perpetuating cycle, but chance events – such as the 2007 subprime mortgage crisis – can knock it out of kilter. When this happens growth can be reduced, and even made negative. Two consecutive quarters of negative economic growth are called a 'recession' and a deep recession – with many negative quarters – is known as a 'depression'.

Stock exchange

A stock exchange is a facility where traders can buy and sell stock, such as shares in companies, derivatives and other securities. Stock exchange traders speculate on the future performance of stocks. Shares in companies expected to do well will become popular and the increased demand will drive up the price (see **Supply and demand**), making them a good investment.

There are numerous stock exchanges around the world – such as the New York Stock Exchange (NYSE), the London Stock Exchange (LSE) and NASDAQ (National Association of

Securities Dealers Automated Quotations). A company's shares can usually only be listed on one exchange at a time. Market performance is gauged by 'stock market indices', which are averages of a representative sample of share prices. Commonly encountered indices are the NASDAQ Composite and the Dow Jones Industrial Average. These indices are used to identify market trends – such as a 'bull market', where share prices are climbing, and a 'bear market', where they are falling. A sudden, precipitous fall of market indices is known as a 'stock market crash'.

Derivatives

Derivatives – also known as 'options' or 'futures' – are a volatile kind of market commodity where traders don't deal in actual stock but rather in contracts that give the option to buy a particular stock for a particular price at a later time. If at this time, the stock is worth more than the pre-agreed price then the contract holder can buy it and resell it for an immediate profit. If the price is less, however, then the holder can decline to buy, but forfeits the cost of buying the derivative. Derivatives trading is governed by a mathematical formula known as the Black–Scholes equation, first put forward in 1973 by the American economists Fischer Black and Myron Scholes.

$$\frac{\partial V}{\partial t} + \frac{1}{2}\sigma^2 S^2 \frac{\partial^2 V}{\partial S^2} + rS\frac{\partial V}{\partial S} - rV = 0.$$

The fiendishly complex Black-Scholes equation that no derivatives trader should leave home without

Hedge funds

A hedge fund is an investment fund that attempts to deliver positive returns under all market conditions by hedging – buying a spread of stock and derivatives priced so as to ensure a net gain regardless of how the market performs. This means that when markets take a downturn, and the returns from traditional investment strategies based on stocks and shares diminish or even become negative, those from a hedge fund will continue to grow apace.

A common hedging strategy used by a fund manager is 'short-selling' – selling off stock and then buying it back at a later date. The manager will short-sell stock he thinks will perform worse than the market as a whole and buy stock he thinks will perform better. If the overall market rises his losses on short-sold assets are covered by gains on his stronger purchased stock – and vice versa.

Econometrics

Econometrics is an interdisciplinary field combining economics with mathematical statistics. It is used for spotting trends and correlations in economic data to form new economic theories, or to test out existing ones. Econometrics relies on being able to isolate the particular variable that the economist is interested in. For example, say an econometrician is trawling through statistics looking for the pattern connecting people's income with the amount they spend buying a car. He would have to account for regional trends in the data, gender differences, age differences, and a host of other confounding factors that could mask any effects due purely to income.

Econophysics

Econophysics is a field of mathematical economics based on theories originally developed for physics. Physicists at the University of Birmingham developed a theory of economics based on quantum electrodynamics (QED), a **quantum field theory** originally developed by Nobel prizewinning physicist Richard Feynman to describe the behaviour of electricity and magnetism. Whereas QED deals with positive and negative electric charge, mediated by an electromagnetic field, the economic version of the theory dealt with positive and negative amounts of money – credit and debt – mediated by what's called an 'arbitrage field', representing the actions of traders.

Just like QED, the theory gives rise to **virtual particles** of the arbitrage field which rapidly vanish again, and which the researchers interpret as random opportunities that are quickly snapped up by 'speculators', traders who act quickly to take advantage of such opportunities. In the absence of speculators, the researchers were able to use their theory to derive the standard form of the Black–Scholes equation for **derivatives**. Putting the speculators back in then gave them a new, modified form of the equation that enables derivatives traders to hedge their deals (see **Hedge fund**) against the action of speculators.

The long tail

The 'long tail' is the name given to a business model where retailers offer a large range of stock, but sell each item in small quantities – as opposed to the traditional model of trying to sell large volumes of a few 'best-seller' items. The name 'long tail' refers to the graph obtained by plotting the sales of products in order of decreasing popularity, which has a tail trailing off to the right (see diagram, below). Crucially, the volume of sales under the tail is the same as the volume under the best-seller portion of the graph. Increasingly, retailers have become able to apply the long-tail model thanks to the rise of online shopping – allowing a store to keep a vast range of stock in a way that's easily searched and browsed through by the customer. Indeed, online stores such as Amazon probably owe their success to it.

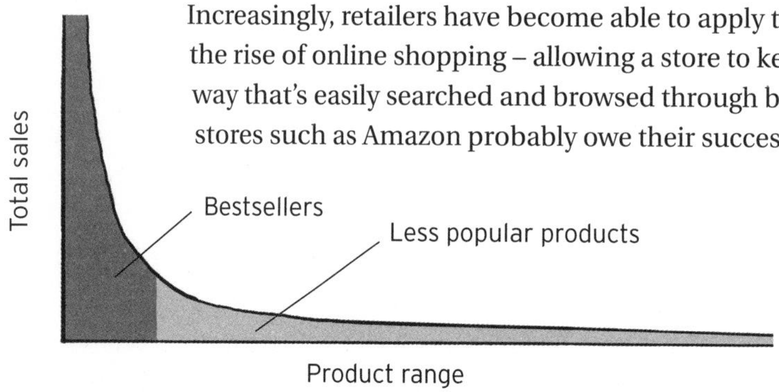

Total sales
Bestsellers
Less popular products
Product range

Prediction markets

In 1906, English scientist Francis Galton visited a country fair where a contest was being held to guess the weight of an ox. Each guess cost sixpence, and there were prizes for the closest. None of the 787 individual guesses came close to the animal's weight of 1,198 pounds. But their average, 1,197 pounds, was almost dead on. Galton published his findings in 1907 in the science journal *Nature*.

In 1945, Austrian economist Friedrich Hayek used Galton's result to argue that price is an

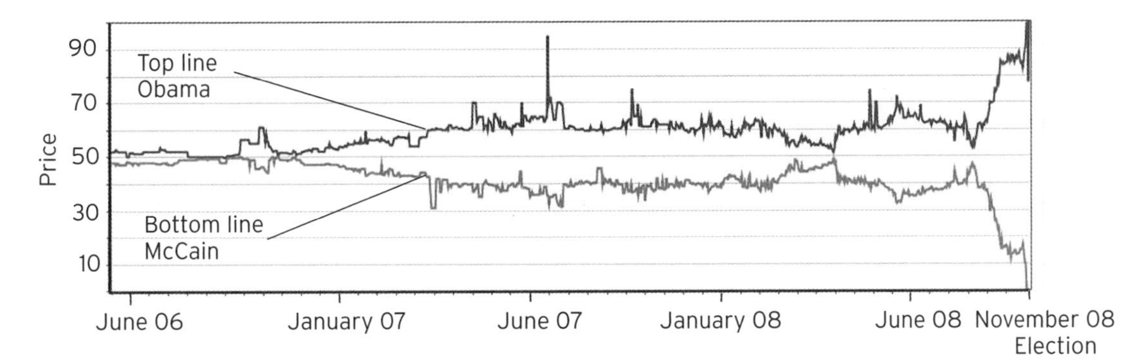

effective way to gauge public opinion on all sorts of issues – from the value of commodities to who will win elections. Later, the University of Iowa decided to find out if Hayek was right and set up a market on the US Presidential Election. The idea is that as the presidential hopefuls hit the campaign trail, traders can buy and sell shares in them. Market forces then drive the share prices to reflect each candidate's chances of being elected. The 'Iowa Electronic Market' turned out to be more accurate than opinion polls 74 per cent of the time, and correctly predicted Barack Obama's victory in 2008. Now there are online prediction markets in everything from box office takings for films to football scores.

POLITICS

Political science

Political science is the study of cause and effect in politics. Its status as a true science is questionable, given the difficulty of quantifying these causes and effects – and for this reason most academics would refer to it as a 'soft' science. Nevertheless, it can offer politicians useful strategies for picking their way through the mire that is modern politics.

One of the first political scientists was the Italian philosopher Niccolò Machiavelli, who summed up his musings on the political process in *The Prince*, a book published in the early 16th century. The book sets out strategies for being successful in politics. Machiavelli's key theme is a familiar one today – presenting the appearance of trustworthiness to the public, while behind the scenes using whatever means are necessary to achieve political aims. His duplicitous tactics are the root of the adjective 'Machiavellian'.

Liberalism and conservatism

The two opposing pillars of Western politics are liberalism and conservatism. Liberalism focuses on the rights and freedoms of the individual – such as the right to freedom of speech; conservatism, on the other hand, promotes traditional values, and often puts the needs of the country ahead of those of the people. Liberals and conservatives are sometimes referred to as 'left wing' and 'right wing', respectively, which are terms that date from the French Revolution at the end of the 18th century. The seating in the French parliament was such that those sympathetic to the Revolution and its movement for change sat on the left, while those wanting to retain order and tradition sat to the right.

In recent times, the distinctions between these two camps have become somewhat blurred, with the emergence of the 'neo-conservative'. This is an ideology that combines facets of liberalism and conservatism, and supports overseas interventionism – either through use of force or economic sanctions – something neither liberalism nor conservatism in their purer forms endorse.

Social justice

One aspect of government that politicians strive for, or at least like to be seen to be striving for, is a just and fair social system that offers equal opportunities for all. In his 1971 book *A Theory of Justice* the American political philosopher John Rawls put forward a complete system of social justice championing a thought experiment known as the 'veil of ignorance' for determining what is truly fair. The idea is that the fairest decisions would be made by policy-makers acting in complete ignorance of their own circumstances – because they cannot influence policy for personal gain without risking penalizing themselves. For example, no politician would ever have endorsed slavery without first knowing whether he will be the master or the slave.

Democracy

A democracy is a system of government that is determined by the people via a majority vote. In fact, democratic governments around the world today are more accurately described as 'representative democracies' – where policy decisions are made by elected representatives rather than taken directly by the populace. In most democracies everyone is permitted to vote, regardless of race, gender or social status. The only exceptions are prison inmates, who are disallowed from voting for the duration of their sentence – their democratic rights effectively suspended.

In England, the first democratic parliament sat in 1265, following the issue of Magna Carta by King John in 1215, which devolved some of the king's power down to the people. The first United States Congress gathered in 1789 in New York City.

Dictatorship

In contrast to a democracy, a dictatorship is a form of government where the people of a nation have no say in policy-making. A dictatorship is usually run by a single person – the dictator – and often accompanied by totalitarianism, where the state has absolute power over its citizens, and uses it to control every aspect of their lives. Dictatorships are sometimes also known as autocracies. Virtually all dictatorships throughout history – such as Hitler's Germany and Mussolini's Italy – have been malevolent. Yet some people insist that the most effective form of government might be a 'benevolent dictatorship'. Here, the ruling party acts in the best interests of the country and the people, but doesn't have to make concessions to an electorate every four years in order to secure its next term in office – especially when those concessions may be damaging in the long run (for example, excessive tax cuts). Most political commentators, however, regard benevolent dictatorship as an unrealistic ideal.

Anarchism

The lack of any government whatsoever is a state known as anarchy. While generally perceived as a negative situation, in which chaos prevails, some political philosophers have argued that a kind of self-organized anarchy might be the most efficient mode of operation for a country – and that people living in such a state might actually be happier. American philosopher and linguist Noam Chomsky has made clear his support for anarchist forms of self-government believing that power over people by authority should not be the default, but that authorities should be made to justify their existence – or else step down.

The principles of an anarchist society place great faith in the altruism of the individual, requiring citizens to care and provide for the disabled, make voluntary contributions to the provision of basic services and to put the needs and welfare of the community above their own. There seems little evidence that the human race in its present incarnation is up to meeting these demands.

Meritocracy

Whereas democracy hands the reins of a country over to whoever proves themselves to be the most popular on election day, meritocracy is a system of government that appoints leaders purely on the basis of their talents and ability, i.e. on merit. It means that those politicians who win office do so because they have demonstrated their suitability, either through experience or even by taking exams, rather than because of their political rhetoric or superficial likeability.

Some argue that meritocracy even encourages respect for a country's leaders that democracy and other forms of government cannot; and that it fosters an ethic of achievement in the population by demonstrating the rewards of talent and hard work. Modern-day Singapore is a living example of a meritocracy.

Social and political engineering

The United States government reportedly became interested in the findings of **social psychology** – in order to gauge how the people will respond in certain situations, and find out how best they can be influenced. This is an area of endeavour known as social engineering, where politicians use findings of sociology and other social sciences to indirectly control its citizens. It is usually carried off through propaganda; for example, announcing policies in the news expected to be unpopular on days when the public are likely more interested in other stories.

Social engineering is closely allied to political engineering, where governments influence public behaviour through the more traditional channels of passing laws and making policies – for instance, lowering the tax on environmentally friendly cars to encourage citizens to be greener drivers.

KNOWLEDGE, INFORMATION & COMPUTING

THE GREAT LIBRARY OF ALEXANDRIA, founded in the 3rd century BC in the city that was at the time the capital of Egypt, was a repository of books and manuscripts on everything from astronomy and mathematics to, some believe, even advanced technologies such as steam power. It is estimated to have contained between 600,000 and 1 million documents. But the library was burned to the ground in antiquity, taking its mass of accumulated knowledge with it. This single act may have set the advancement of human science and technology back by as much as 1,500 years.

In the modern world, as English philosopher Francis Bacon pointed out in 1597, knowledge is power. Never before

ERENDIPITOUS DISCOVERIES • OCCAM'S
POSITIVISM • INDUCTIVE REASONING •
ETHICS • PSEUDOSCIENCE • SCIENTIFIC
UMBERS • GOLDEN RATIO • INFINITY •
R THEOREM • POINCARÉ CONJECTURE
LETENESS THEOREM • CHAOS THEORY
TREME VALUE THEORY • SMALL WORLD
EM • GAME THEORY • KELLY CRITERION
• DATA COMPRESSION • INFORMATION
ETICS • CRYPTOGRAPHY • QUANTUM
ON • ALGORITHMS • TURING MACHINE
COMPUTERS • MICROCHIPS • MOORE'S
RALLEL COMPUTING • EVOLUTIONARY

has what we know about the world through science been such a powerful commodity, and it's going to become more so in the future. Information is the language in which our knowledge is encoded, and through information technology – computers and the Internet – the rate at which we can access, process and discover new knowledge is accelerating exponentially.

It is fitting then that now, on the approximate site of the old library in Alexandria harbour, stands a new great library – the Bibliotheca Alexandrina. It will bear witness to some of the great accomplishments of human knowledge that were held in abeyance by the destruction of its namesake so many centuries ago.

SCIENTIFIC METHOD

Scientific theories

Science is a method for discovering knowledge about the workings of nature – from subatomic particles to biological life forms to the Universe at large. It works by putting forward theories and then comparing their predictions with the results of observations and experiments. Scientific theories aren't just inklings or best guesses but are thoughtful, logical constructions – usually supported by rigorous mathematics and often the culmination of many years of research. Theories that are successful are the ones that predict a great deal more than they were originally formulated to explain.

Falsification

A scientific theory can only reflect the best state of our knowledge so far. One thing that sets scientific theories apart from all the others is that they can only be proven false. As scientists build better instruments and make more detailed observations of the world and the Universe, so new experimental data comes to light against which our theories can be tested and tested again.

If the theory passes each new test, that doesn't mean it is definitely correct – only that it's good enough until the next batch of observations comes in. But if the theory fails the test then we know for sure it's wrong. A good example is our view of the solar system. The early Greeks believed that the Earth was at the centre of the solar system with the Sun and other planets revolving around it. But observations in the 17th century by the Italian astronomer Galileo showed this theory was wrong, paving the way for the Sun-centred picture that prevails today. Falsification as the basis of science was first championed by the philosopher Karl Popper.

Serendipitous discoveries

Sometimes important scientific breakthroughs happen entirely by accident. Perhaps the most famous case was the discovery of penicillin – the wonder antibiotic that has saved countless millions of lives, including up to 15 per cent of wounded Allied soldiers during the Second World War, who would otherwise have died from infections such as gangrene. The antibiotic properties of penicillin were first realized in 1928 by Scottish biologist Alexander Fleming, who noticed the reluctance of bacteria to grow in a Petri dish that had been accidentally contaminated with penicillin mould. Other important discoveries have been made serendipitously – including **X-rays**, the anticoagulant drug warfarin, the 'vulcanized' rubber used in car tyres, and the planet Uranus.

Occam's razor

When constructing a scientific theory, should we go for one that's complicated or one that's simple and makes the fewest supporting assumptions?

Common sense would probably encourage the latter, but the 14th-century English logician William of Occam elevated this notion to the status of a principle underpinning the whole of science. It's become known as Occam's razor, and many since have imagined it as a metaphoric razor paring back candidate theories to make them as simple and straightforward as possible. Only if the theory fails an observational test should we consider making it more complex. Nowadays most scientists would be reluctant to call Occam's razor a 'principle', preferring to think of it as a **heuristic**.

Heuristics

Heuristics are non-rigorous methods or 'rules of thumb' that everyone – not just scientists – use to guide their decisions. An example would be 'double it and add 32' to roughly convert a temperature from Celsius to Fahrenheit.

The most common heuristic used by scientists is the 'trial and error' approach to mathematical or engineering problems, which works by trying a range of possible solutions, and then adjusting the best match until it fits exactly. Other scientific heuristics include Occam's razor, and the so-called 'gaze' heuristic – whereby watching a ball in flight normally gives you a better chance of catching it than trying to solve the complex equations governing its trajectory.

Reductionism

According to reductionism, every concept in science is just the sum of smaller parts. So, for example, in **kinetic theory** the temperature of a gas can be reduced to the effect of individual atoms and molecules bouncing around and colliding with one another. **Quantum theory** can then explain the behaviour of the atoms and molecules and, ultimately, **string theory** and **M-theory** explain quantum effects.

The idea of reductionism was advocated by the 17th-century French philosopher René Descartes, who believed that the world was like a machine that could be broken down into the action of many individual components. Taking reductionism to its logical extreme would seem to support the existence of grand unification theories, which underpin the theories we understand today. In opposition to reductionism is holism, which says that systems are much more than just the sum of their parts.

Positivism

The philosophy of positivism has grown from the belief that the only knowledge of any value is that which can be 'positively' perceived by the senses. In its modern form it says essentially that science is the only true path to knowledge. Some sceptics argue that certain areas of science, such as **cosmology** and some branches of quantum theory – for example, the **many worlds** interpretation – are at odds with positivism, because it's debatable whether these ideas could ever be tested observationally.

The opposite of positivism is metaphysics, the study of what lies outside or beyond physics. By definition metaphysics deals with concepts that are unanswerable by scientific enquiry and so most regard it as **pseudoscience**.

Inductive reasoning

The rain feels wet. Therefore all rain is wet. That's an example of inductive reasoning – it is an attempt to make a generalization from a specific observation. Another example is an exit poll during an election – a certain proportion of the people surveyed voting Democrat is used to imply that the same proportion of the entire population will vote that way. Scientists use inductive reasoning to construct theories; for example apples fall downwards, therefore gravity is an attractive force everywhere. It sounds obvious, but to infer that the same law applies in the depths of the solar system is quite a leap. Yet astronomical observations have shown it to be correct.

In contrast is 'deductive' reasoning – which makes inferences that follow with logical necessity (e.g. A=B and B=C, therefore we 'deduce' A=C). Inductive reasoning leads to more speculative notions, which must then be tested by experiment.

Scientific computing

Formulating a scientific theory – constructing a set of mathematical equations that describes the physical process you're trying to model – is hard enough, but solving these complex equations to get numerical answers is often almost impossible. Which is why many scientists turn to computers to do the dirty work for them. Giving a computer a set of initial conditions and then leaving it to crunch out the numbers is much easier than trying to work out a solution on paper.

Astronomers were among the early adopters of scientific computing – modelling complex processes such as **galaxy formation** and the evolution of the Universe. Nowadays most fields of study are sufficiently advanced to benefit from this approach. There is even computer software now that will attempt to solve mathematical equations for you algebraically – working out those elusive pen-and-paper solutions rather than crunching numbers. In 1977, the proof of the **four-colour theorem** became the first major non-numerical mathematical problem to be solved in this way.

Datamining

From supermarkets picking over your buying habits to online music providers capitalizing on download trends, datamining – scouring databases for statistical information – can be big business. Now scientists are catching on, realizing that it has the power to help make new discoveries. Set a datamining algorithm loose on archived information and who knows what hidden connections it might uncover that would otherwise be lost on human observers? For example, maybe it could perhaps reveal a correlation between meteor showers and volcanic eruptions, or solar eclipses and air quality – or something else equally strange that would betray new science at work, and provide the cue for researchers to investigate further. Already, at the Uppsala Monitoring Centre, in Sweden, scientists are using datamining as an early warning system to discover side-effects of clinical drugs. The centre combs medical databases and flags cases of adverse reactions to the same drug – a field called 'pharmacovigilance'.

Ethics

The consideration of ethical issues is becoming increasingly important in the advancement of science, especially biology and health and medicine. Is it right, for example, to carry out tests on animals, to clone humans for stem cells, or to engineer new life forms to do our bidding – as with the new field of **synthetic biology**? These are all questions for bioethicists to answer, adding a moral dimension to scientific research where previously there was nothing more than the drive to discover.

In the future, ethical issues may be more important in other scientific disciplines, as we develop artificial intelligence that approaches human levels, perhaps make alien contact, build nanorobots through **molecular engineering** that give us mastery over the microscopic world, or perhaps even discover the means to alter the past through time travel.

Pseudoscience

Pseudoscience is any pursuit portrayed in a scientific light, but which is actually far from scientific in its approach. Claims of the paranormal, UFO sightings, or methods of divination – such as palm reading and astrology – are often branded as pseudoscience. While some of these topics could be – and in some cases are – investigated scientifically, most practitioners in these fields make their claims so vague that they cannot be tested objectively. For example, a clairvoyant tells an audience of 400 people she has a message for someone called John (statistically, there will be 13 people present with this name). Many scientists aggressively oppose pseudoscience, believing it does harm to public understanding of true scientific pursuit and damages the capability for rational decision-making that's essential for a democratic society to function.

PURE MATHEMATICS

Scientific notation

Everyday experience teaches us how to deal with everyday numbers, such as 2, 37 and 875. But sometimes the numbers that crop up in science can be off the map; for example, astronomy deals with inordinately large length scales, spanning billions upon billions of kilometres. At the other end of the scale, particle physics and quantum theory chart the behaviour of matter on infinitesimally small scales – fractions of a billionth of a metre. Scientists handle these large numbers using a branch of mathematics called scientific notation, which works by writing numbers as powers of 10. For example, 100 is just 10 squared (10×10), or 10^2; similarly 100,000 is written 10^5; and 500,000,000 (500 million) would just be 5×10^8.

Small numbers can be handled in the same way. One-hundredth (1/100, or 0.01) is written 10^{-2}; likewise, a millionth (1/1,000,000 or 0.000001) is 10^{-6}, and 5 ten-billionths (5/10,000,000,000) becomes the slightly more manageable 5×10^{-10}.

Scale prefixes

Scientific notation is a neat way to condense numbers with long strings of zeros down into something more manageable, but it can still appear somewhat harsh on the eye. For this reason, commonly occurring powers of ten are condensed into a verbal prefix. Perhaps the best known is 'kilo' for 1,000, or 10^3 – a kilometre is 1,000 metres and a kilogram is 1,000 grams.

Other large-scale prefixes are mega for a million, or 10^6; giga for a billion, or 10^9; and tera for 10^{12}, also known as a trillion. Terabyte hard-disk drives, for example, hold a trillion bytes of data. A few years from now, we'll all be buying petabyte hard drives – each holding 10^{15}, or a quadrillion bytes of information.

Scale prefixes exist for small numbers too. Milli is a thousandth, or 10^{-3} – a millimetre, for instance, is a thousandth of a metre; micro is 10^{-6}, or a millionth; nano is a billionth, or 10^{-9}. So nanotechnology is engineering on scales of a billionth of a metre. Smaller still are pico, 10^{-12}, or a trillionth, and femto, 10^{-15}, a quadrillionth.

Prime numbers

Any whole number that cannot be divided exactly by any whole numbers other than itself and 1 is called a prime. The first few are: 2, 3, 5, 7, 11 and so on. At present, there is no neat mathematical formula to predict prime numbers; instead, primes are discovered by brute force, using computers. Finding new ones is harder than you might think: the largest recorded to date is nearly 13 million digits long. Every non-prime whole number can be written as a product of two prime factors. Like discovering new primes, splitting a very large number into its prime factors is extremely difficult and this fact has become the basis for codes and **cryptography**. Mathematicians hope to gain greater understanding of prime numbers through the **Riemann hypothesis**.

Golden ratio

Take a line and divide it so that the ratio of lengths of the smaller part to the larger part is equal to the ratio of the larger part to the whole. Numerically equal to 1.618, this is known as the 'golden ratio', and its proportions are especially pleasing to the eye – so much so, they appear in architecture and many great works of art, including the *Mona Lisa*. Artists were incorporating the golden ratio into their work as far back as the 5th century BC; while the first mathematical study of this number was carried out by Greek philosopher Euclid in 300 BC.

Mathematically, the ratio arises from the Fibonacci sequence of numbers, obtained by starting with the numbers 0 and 1, and generating each new number by adding together the previous two, so: 0, 1, 1, 2, 3, 5, 8 . . . The ratio between successive numbers in the sequence tends towards the golden ratio as the numbers get bigger. The Fibonacci sequence also crops up in biology, for example, describing the number of petals on certain flowering plants.

Infinity

When a number is so big that it is boundless, mathematicians refer to it as infinite. Any number multiplied by infinity gives infinity, and any number divided by infinity – chopped up into infinite small parts – is zero. Infinity is denoted using a symbol resembling a figure '8' turned on its side: ∞. Technically speaking, infinity is not a number and its overpowering effect on finite numbers means that inserting infinity into an equation can make little sense. Mathematicians must proceed with extreme caution when trying to determine how an equation behaves as its variables become large. Their usual tack is to make approximations that simplify a formula as the variables becomes very large – known as 'asymptotic analysis'.

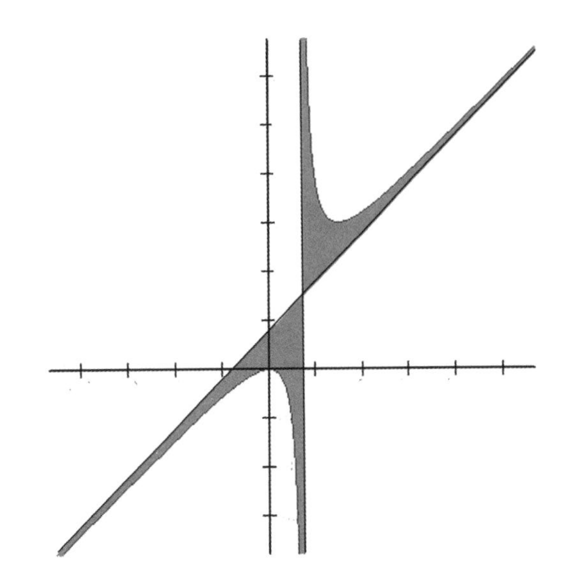

Asymptotic analysis of the graph $y = x^2 / (x - 3)$. Letting x go to infinity makes the top and bottom both infinite. But approximating the formula for large values of x shows it tends towards the straight line $y = x + 3$.

Fermat's last theorem

It looks innocuous enough. Choose three positive whole numbers – call them a, b and c – and now choose a whole number bigger than 2 – call that n. Fermat's last theorem says that there are no combinations of a, b, c and n that satisfy the equation $a^n + b^n = c^n$. The theorem was put forward by mathematician Pierre de Fermat in 1637. While it can clearly be verified for a handful of small numbers, a general proof of Fermat's last theorem that's valid for all values of a, b, c and n did not appear until 1994, when it was finally written down by British mathematician Andrew Wiles.

Four-colour theorem

Fermat's last theorem isn't the only seemingly simple puzzle that has occupied mathematicians for years. Another, known as the four-colour conjecture, says that given any map, only four different colours are required to shade all the different nations on the map distinctly, so that no two adjacent nations have the same colour. The theorem doesn't just apply to maps, but to any 2D plane that's divided up into a number of interlocking regions.

The four-colour theorem first appeared in

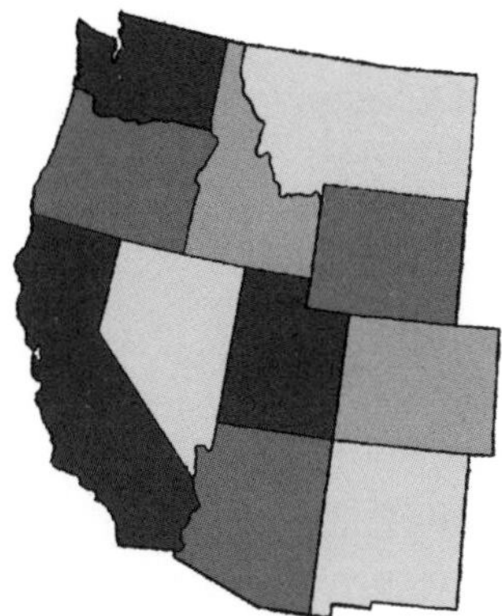

According to the four-colour theorem, four colours are enough to distinguish all the different states on a map of the western US.

mathematical literature in 1853, but a proof didn't emerge until 1977 – by mathematicians Wolfgang Haken and Kenneth Appel. Even then, doubt remained over the validity of their proof, because much of it had been performed by scientific computing. The most recent study, carried out in 2004, however, suggests that Appel and Haken's proof holds good.

Poincaré conjecture

Stretch an elastic band around the outside of a tennis ball. It's easy to shrink the band continuously down from a circle to a point by sliding it over the ball's surface. Compare that to threading a length of elastic through a coffee cup handle and then tying it up into a loop. Now, no amount of sliding the band can shrink it to a point.

In mathematical terminology, the tennis ball and all other spheres in three dimensions are 'simply connected' – coffee cups, on the other hand, are more complicated. French mathematician Henri Poincaré suspected that spheres in four dimensions are also simply connected but was unable to prove it – and this is the Poincaré conjecture. He threw the problem open to the mathematical community in 1904 and it was finally solved by Russian mathematician Grigori Perelman, in a series of papers published in 2002 and 2003.

Riemann hypothesis

Put forward by German mathematician Bernhard Riemann in 1859, the Riemann hypothesis offers an intriguing insight into the nature of **prime numbers**. It hinges around a complex mathematical formula, known as the Riemann zeta function. For any number, s, the zeta function is given by the sum over all positive whole numbers, n, of $(1/n)^s$, i.e. $1 + (1/2)^s + (1/3)^s$, and so on. Riemann was able to show that the quantity of prime numbers less than s is given by the number of times the zeta function of s passes through zero.

He went on to hypothesize that all the zeros of the zeta function lie on a single well-defined line, which has since become known as the Riemann hypothesis. While it can be verified for small numbers, Riemann himself was unable to arrive at a general proof. If he was right it's a massive step towards understanding how prime numbers arise. And that's why the Clay Mathematics Institute in Cambridge, Massachusetts, has offered a $1 million prize to anyone who can come up with a valid proof.

Gödel's incompleteness theorem

Published by Austrian mathematician Kurt Gödel in 1931, the incompleteness theorem says that there are questions within mathematics that mathematics itself cannot answer. Gödel proved his theorem by constructing what amounts to a numerical version of the Epimenides paradox. Named after the Greek philosopher Epimenides in the 5th-century BC, the paradox hinges on the sentence, 'This statement is false.' If the statement is genuinely false, then it must mean the opposite of what it says, i.e. it must be true. But if it's true, then it must mean exactly what it says, i.e. it's false. The paradox has no resolution.

Gödel constructed a similarly self-referential mathematical formula which stated that the formula itself was unprovable. If the formula was provable, then the formula itself must be false; but if that's the case then the laws of mathematics have proven true something which is actually false. Assuming our self-consistent laws of mathematics cannot allow this to happen, then there must exist mathematical truths which are unprovable using mathematics. The theorem remains a central result in logic theory and the philosophy of mathematics.

APPLIED MATHEMATICS

Chaos theory

Chaos is the emergence of wildly unpredictable behaviour from solid and seemingly innocent laws of physics. It's an immediate feature of the world around us, cropping up in esoterica such as quantum theory, space, and economics, as well as the weather and even the timing of a dripping tap.

Despite its apparent randomness, chaos is actually quite a well-ordered phenomenon, and is caused by extreme sensitivity of a physical system to its initial state, meaning that tiny differences in the initial state become magnified as the system evolves. Unpredictability arises because we cannot measure the initial state of the system accurately enough and this is partly why long-term weather forecasting is so difficult.

Fractals

Mathematicians sift chaos from randomness using what's called a 'phase portrait' – a diagram showing how the system evolves in time from its starting state. They look for regions in the phase portrait where the evolutionary paths from many initial states all converge, areas known as 'attractors'. For a simple pendulum, the phase portrait is just a graph plotting the pendulum's position against its speed, and the attractor is a circle. But as the complexity of a system increases, the shape of the attractor becomes more convoluted.

Chaotic systems have fractal attractors. Fractals are disjointed shapes which have the same appearance when viewed on many different length scales; their complex structure creates the illusion of randomness in systems. The simplest fractal is made by removing the middle third of a straight line and then repeating the process, ad infinitum, on each remaining segment (see diagram above). Fractal attractors are classified by a number, the so-called 'fractal dimension', which reveals the level of chaos present in the system.

Catastrophe theory

Closely related to chaos theory, catastrophe theory deals with abrupt discontinuous changes that can arrive from small, smoothly varying causes. The simplest example is gradually increasing the weight hanging on the end of a length of rope. To begin with, the rope stretches continuously in response to the weight increase but eventually comes the straw that breaks the camel's back – a small increment of extra weight that abruptly snaps the rope. A discontinuous change has arisen from a small, continuous cause. The 'point of no return', where the cause has just become sufficient to trigger the catastrophe, is sometimes referred to as the 'tipping point'.

Catastrophe theory was developed by the French mathematician René Thom in the 1960s and applies to many real-world phenomena such as landslides, the spread of disease and climate change.

Extreme value theory

Extreme events can throw an otherwise well-behaved system into turmoil – whether it's a freak tidal wave inundating flood defences, or a banking calamity that sends the stock market into freefall. Extreme value theory (EVT) is the branch of mathematics that deals with the likelihood of such extreme events. Central to EVT is the so-called 'extreme value distribution'; for any sequence of numbers the distribution gives the most likely values for the maxima and minima, in other words the extremities, in that sequence, based on past evidence. So, for example, given data on past tide heights, an extreme value distribution can predict the likely maximum tide height in any given period and this enables planners to construct cost-effective sea defences – high enough to cope with the biggest storm surges, but not so high as to waste resources.

The foundations of EVT were laid down by the German mathematician Emil Julius Gumbel in 1958. Most recently it has been used to predict the fastest time that's humanly possible in the 100-metres sprint – giving the answer of 9.51 seconds.

Small world theory

You may well have played the Kevin Bacon game – where you're given the name of an actor who you have to connect to Kevin Bacon via the movies they've been in, doing so in the least number of steps. So, for example, Michelle Pfeiffer is two steps away: she was in *Wolf* with Jack Nicholson, who in turn was in *A Few Good Men* with Kevin Bacon. This connectedness is a facet of what has become known as small world theory. Mathematicians believe a six-step small world network operates between all the 6.8 billion people on Earth – in other words, anyone on the planet can be linked to anyone else via the names in their contact books, in just six steps. This is the origin of the term 'six degrees of separation'.

Small world networks rely on a small number of long-distance connections between members. So, for instance, if you have a circle of friends who all live locally plus one friend in Kathmandu, that one long-distance connection links him and all his Nepalese friends to you and all of your local contacts.

Travelling salesman problem

Here's a conundrum. A salesman has to travel from his home to a number of cities, and in each city he will stop to sell products, before moving on to the next. After he's visited every city once, he's free to head back home again. What is the shortest route that gets him around all the cities (in no particular order) and then home? For a small number of cities the problem is relatively easy; indeed, if need be, it can solved by brute force, simply adding up the distance travelled along every route possible and then picking the shortest one. But as the number of cities increases so the problem becomes disproportionately harder. In general, for N cities plus the salesman's home, the number of possible routes is given by $N \times (N\text{-}1) \times (N\text{-}2) \ldots$ all the way down to 1. So if $N = 10$ there are over 3.6 million routes to choose from.

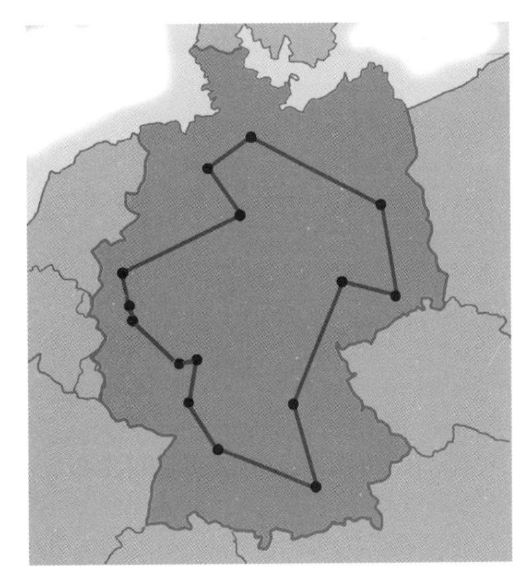

The shortest route around Germany's 15 largest cities.

There is still no general solution to the travelling salesman problem – solutions exist for small N, but they all break down as N becomes larger. The travelling salesman problem isn't restricted to transport; it crops up in areas as diverse as microchip design and **gene sequencing**.

Game theory

From card games to military campaigns, any competitive pursuit with a clear benefit for the winner can be analyzed using a branch of mathematics known as game theory. It works by comparing all the strategies available to a player against all the possible strategies that might be adopted by his or her opponent, and then assigning a numerical value known as the 'pay-off' to each one. The mathematically savvy gamer is then best advised to follow the strategy with the highest pay-off.

Game theory was first developed in the 1920s, dealing exclusively with what are known as 'zero-sum' games – where the winner benefits at the expense of the loser. Later, the theory was expanded to include non-zero-sum games, which mirror real life situations more accurately. It is now used by economists, political strategists and even biologists studying **evolution** – to explain the behaviour of competing organisms.

Kelly criterion

The Kelly criterion is a formula used by gamblers to calculate the optimal fraction of their money to bet when they have an edge. Let's say a gambler is offered money odds of 2/1 on a bet – meaning that for every \$1 they bet, they receive \$2 back if the bet wins and let's also say the gambler knows she actually has a 50 per cent chance of winning. So half the time she will lose \$1, but the other half of the time she gains \$2. She has an edge. But how much should she bet? Clearly, betting all her money would be foolish as she still has a 50 per cent chance of losing, in which case she'd go broke, and betting a tiny amount of money is also incorrect when the odds are stacked in her favour.

The Kelly criterion, formulated by US mathematician John Kelly in 1956, says that if the money odds on a bet are $b/1$ and the probability of winning the bet is p, then the gambler should stake a fraction of her wealth equal to $(b \times p) + p - 1$ all divided by b. For our example above, $b = 2$ and $p = 0.5$. So the gambler should bet a quarter of her money.

INFORMATION

Binary data

The data on your computer hard drive is stored in the form of a stream of binary 1s and 0s, which is the way electronic systems encode information – using microscopic switches in one of two positions, off (0) or on (1). Each switch inside a computer can record one such binary digit, or 'bit' of information; several bits can be put together to make a byte, capable of storing numbers bigger than 1. For example, two bits make a byte that can store any number from 0 to 3. Here, 3 is given by both bits switched on, 2 has the first switched on and the second off, 1 has the second on and the first off, and 0 has them both off. In an eight-bit computer each byte is made of eight individual bits and can store any number from 0 to 255.

The binary number system was first mentioned in the writings of Indian scholar Pingala in the 5th-century BC. It didn't appear in the West until the work of German mathematician Gottfried Leibniz, in the 17th century.

Information theory

We all rely on information, whether it's the stream of data arriving at your TV set, the flow of binary data on your computer or the signal beamed to your mobile phone.

The mathematical study of how information is stored, transmitted and manipulated is called information theory, a discipline pioneered in the late 1940s by American electronic engineer Claude Shannon. It encompasses crucial disciplines such as error correction, allowing signals to be extracted amid noise and interference, and data compression, enabling large data files to be squeezed down for efficient storage and transmission. Information theory has been key to the development of fields such as computing, cryptography, neurobiology and even the physics of black holes – leading to the idea of a **holographic universe**.

Data compression

Compressing data is a way to send and store large quantities of information efficiently. A simple example might be a picture file on a computer; if a large area of the picture is all the same colour, then it takes up much less storage space to specify the dimensions of the area and the colour than to specify the colour for every single pixel individually.

Data compression comes in two forms – lossless and lossy. Lossless data compression makes no compromise in the accuracy of the data being compressed – if you make a zip file of a document to send by email, then what the recipient gets out is exactly what you put in. This is lossless compression. On the other hand, lossy compression works by removing some parts of a data file. MP3 music files use a form of lossy compression; they typically take up 1/11 of the space occupied by a CD-quality audio track, and work by dumping sounds that the human ear is not sensitive to.

Information entropy

Entropy in information theory is related to the notion of thermodynamic **entropy**. However, whereas the thermodynamic quantity specifies the degree of disorder in a system, information entropy refers to the uncertainty in a random variable. The higher the uncertainty, the greater the entropy. For example, a coin that's weighted to give heads every toss has entropy zero, whereas a totally random coin has an entropy of 1 bit – it can either come up heads or tails, equivalent to a 0 or a 1 in binary data.

The concept was put forward by the father of information theory, Claude Shannon, who noticed a quantity arising in his theory that looked mathematically very similar to the definition of thermodynamic entropy in **statistical mechanics**. Indeed, thermodynamic entropy can be thought of as the amount of information needed to completely specify the state of every particle of matter in the system. Heating up the system enables it to occupy higher-energy states, increasing the number of possible states that each particle can be in – which is why entropy goes up with temperature. Information entropy places limits on lossless data compression.

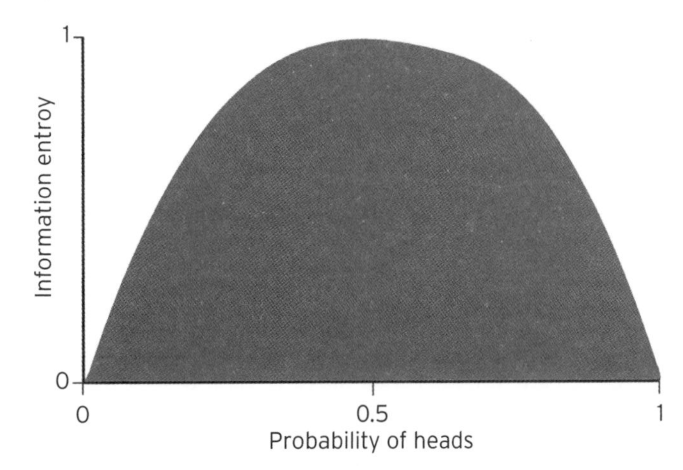

Information entropy of a coin, expressed as a function of its probability of coming up heads.

Error correction

Have you ever forgotten to put a CD back in its case after playing it, and it's become badly scratched as a result? Chances are the CD will still play fine despite the scratches and this is all down to error correction codes. They are essentially software **algorithms** able to check for the presence of noise or interference in a stream of data, and then correct for it. They can work in a number of ways. One scheme is to splice a sequence of known information into the data stream; measuring the errors that are introduced into this sequence tells the receiver how to correct all the other bits in the stream. Error correction is also used in supermarket barcode readers, enabling them to read the barcode – even on crumbled bags of potato crisps (chips).

Memetics

Popular YouTube videos, funny jokes and pictures of celebrities in compromising positions all have something in common – they are examples of 'memes'. A meme is a snippet of information that we feel the urge to pass on and share. In this sense, memes take on a life of their own – spreading across the Internet or by word of mouth. The term is a derivative of the word 'gene', the name given to the bundles of genetic information encoded in the DNA of living organisms. It's an apt link – where Darwinian **natural selection** ensures that only the fittest genes survive, a similar memetic process gives longevity to the funniest jokes and the best videos. Such parallels between evolution and information theory are often cited as examples of 'information ecology'.

Memes were first mentioned by biologist Richard Dawkins in his book *The Selfish Gene*, and the idea was developed further by British psychologist Susan Blackmore. They form the basis of viral marketing campaigns – where companies embed advertising within a rapidly spreading meme.

Cryptography

Cryptography is the science of encoding information, so that eavesdroppers intercepting a transmitted message won't be able to interpret it. That's not to say they'll never be able to decipher it; few codes are absolutely unbreakable. Perhaps the most famous codebreakers were the British team led by brilliant mathematician Alan Turing, who cracked the German Enigma code during the Second World War.

A common form of modern code is called public-key cryptography, the security of which relies on the difficulty in factorizing large numbers into their prime number factors. The sender chooses two large prime numbers and multiplies them together; this number, the 'key', is all that's needed to encrypt a message. The key is broadcast publicly; however, decrypting the code requires both the prime factors. Factorizing huge prime numbers is beyond the realm of today's computers, but may be possible in the forthcoming era of **quantum computing**.

Quantum communication

In 1984, researchers Charles Bennett, at IBM, and Gilles Brassard, of the University of Montreal, showed how the weirdness of quantum theory could be used to build a communication system absolutely secure against eavesdroppers. What Bennett and Brassard imagined was a kind of cryptography where the sender, Alice, transmits a private encryption key to the receiver, Bob, by encoding bits of binary data – 1s and 0s – using the polarization of photons of light.

Alice has two different sets of polarizing filters to encode her 1s and 0s, and she notes down which has been used for each photon. Each filter can put the photon into a polarization state corresponding to a 1 or a 0. Bob has the same filters and randomly chooses which one he uses to measure the photons; using the wrong filter to make a measure yields a nonsense result. Next Bob phones Alice and tells her the sequence of filters he used, but not the actual bit values he measured. Alice, in turn, tells Bob her filter sequence. Alice and Bob then use as their key the bit values corresponding to the photons for which they used the same filter. An eavesdropper trying to measure the photons en route would inevitably interfere with them, through the **uncertainty principle**. Bob and Alice would notice this when they compared their filter selections, in which case they would discard that key and start again.

Quantum cryptography is part of the broader field of quantum information, which also includes **quantum computing** and **quantum games**.

Information addiction

In February 2007, a 26-year-old Chinese man died after spending seven continuous days at his computer playing online games – only leaving for toilet breaks and occasional naps. Nick Bostrom of the Future of Humanity Institute, at Oxford University, calls this 'information addiction'. For information addicts all that matters is their next fix of data from online environments such as 'World of Warcraft' or 'Second Life'.

Bostrom likens this to our evolutionary craving for foods rich in fat or sugar, nutrients that were once scarce, but now plentiful – a fact that's been exploited by the food industry to create highly addictive snacks that many consume with abandon rather than eating a balanced diet. Likewise, says Bostrom, we now have online environments that are more intense and stimulating than the real world and for many this becomes an addiction.

COMPUTING

Algorithms

An algorithm is a finite sequence of instructions that can be used to solve a problem. It forms a step-by-step decision-making tree that tells a human, or a computer, what to do next in order to get a problem solved. A simple algorithm for solving the problem of making sure a room is lit might go something like:

1. Is the room lit (yes/no)? If 'no' go to 2. If 'yes' then go to 3.
2. Flick light switch. Go to 3.
3. Stop.

But algorithms exist for vastly more complex tasks and every piece of computer software has at its core an algorithm telling the computer what action it must take in the event of every possible contingency. Before they are translated into computer code, algorithms are often expressed using a 'flow chart' – a visual depiction of the decision-making process that shows each step as a box linked to the other steps by arrows. In some ways, algorithms can be thought of as the opposite of **heuristics**.

Turing machine

A Turing machine is a hypothetical mechanical computing machine put forward by British mathematician, and the father of modern computing, Alan Turing, in 1936. The machine uses a long strip of tape as its memory. Along its length the tape is divided into cells and each cell can hold a single symbol, which encodes a piece of information. A Turing machine can be in any one of a number of 'states', and each state consists of a set of instructions on what to do next depending what symbol is in the cell that the machine is currently acting on. Instructions take the form 'change symbol X to symbol Y', 'move tape three cells to the left', 'change machine from state 1 to state 2', and so on. Turing showed that his machine could solve any mathematical problem for which there exists a well-defined algorithm – and, conversely, that certain problems are uncomputable.

Analogue computers

Before the days of digital electronics, computers were devices that worked using mechanical, optical and non-digital electronic components to solve mathematical problems; these were called analogue computers. For example, a mechanical computer might use a complex arrangement of gears and linkages. If you take a cog with 45 teeth and mesh it with another that has 9 teeth, for each turn of the large cog, the smaller one goes round five times – that's the machine's way of saying $45 \div 9 = 5$. Mechanical computers were used by artillery gunners and aircraft bombardiers to calculate trajectories during the Second World War, and until the 1960s students used slide rules instead of electronic calculators.

The earliest-known analogue computer is the Antikythera mechanism dating from the 2nd-century BC. Discovered in a wreck off the coast of Greece, the mechanism is thought to have been used for astronomical calculations. Analogue computers gave way to faster and more efficient devices with the birth of digital computers in the 1940s.

Logic

Deductive and **inductive reasoning** are both branches of what's called logic – the science of reasoning. George Boole, the 19th-century English philosopher, developed a mathematical theory of logic that would prove essential in the development of computer science nearly 100 years later. Boolean logic enabled a mathematical description of operations such as AND and OR. These 'logic gates' take two binary input signals and give a single binary output. For example, an AND gate gives an output of 1 when the two inputs are both 1, and an output of 0 the rest of the time: while an OR gate gives 0 when the two inputs are both 0, and an output of 1 the rest of the time. Other logic gates exist, and they are the building blocks of information processing, enabling digital computers to make decisions; for instance, a computer could be programmed to switch on a heating system in an office only when the temperature falls below 13°C AND during working hours.

Digital computers

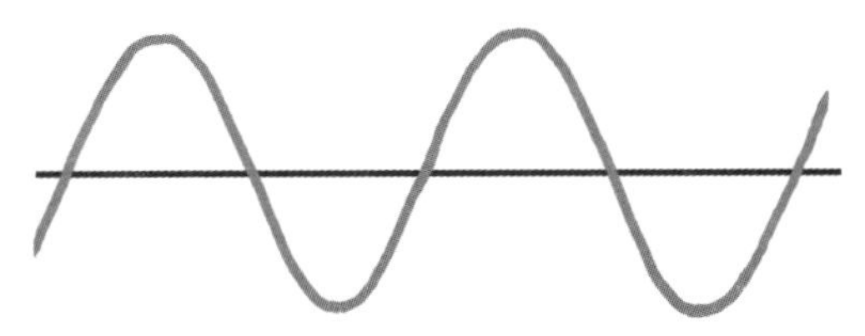

Analogue computers transfer information as continuous waves.

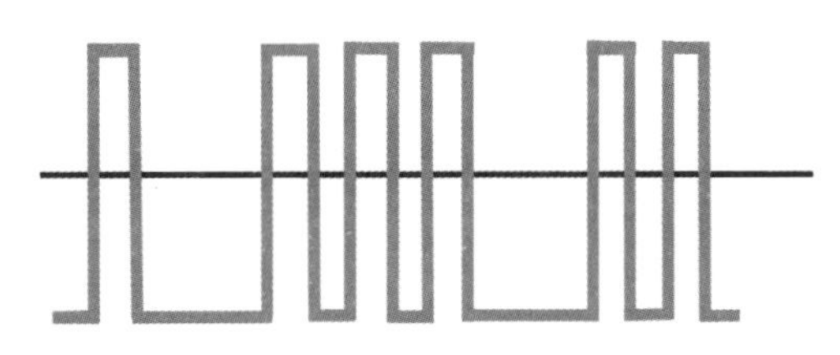

Digital computers transfer information as discrete pulses, on or off, representing the 1s and 0s of binary data.

Colossus, the world's first digital electronic computer, was built at the Post Office Research Station in London, England, in 1943. Whereas analogue electronics deals with electrical signals that are continuous in nature, digital devices produce and store discontinuous jumps in electrical voltage – perfect for recording the 1s and 0s of binary data. Colossus, and other early digital electronic devices, worked using electrical components called thermionic valves to switch electric current in such a way that simple logic operations could be carried out.

The development of **semiconductor** materials led, in 1947, to the invention of transistors – electronic switches that rapidly replaced the bulky, inefficient and fragile valves. The first transistor computer was switched on at the University of Manchester, England, in 1953. But an even bigger computer revolution was set to follow, with the invention of the microchip.

Microchips

In 1958, Jack Kilby, of American technology firm Texas Instruments, invented something that would revolutionize digital computers, and change the world for ever: the microchip. It's a wafer of **semiconductor** material, such as silicon or germanium, etched with corrosive chemicals to make a miniature, self-contained electronic circuit – hence the microchip's alternative name, the 'integrated circuit'. Small microchips can each hold many transistors, enabling computers – that used to be the size of a room – to be shrunk down to something near the size of a large typewriter, which could sit on a desk. The first 'microcomputers' were available to buy in the early 1970s. Microchips kick-started the miniaturization of electronics, and are now in virtually everything from cellphones to spacecraft.

Moore's law

In 1965, Gordon Moore, co-founder of American microchip manufacturer Intel, noticed an interesting trend – microchips were getting ever more compact. Specifically, the number of transistors that could be squeezed onto a single chip of fixed size was doubling roughly every two years (see diagram). Taking into account not just the number of transistors but also the gradual increase in their quality, it translated into a doubling in chip performance once every 18 months and this has become known as Moore's law.

And it's still happening; as of 2007, a low-cost microchip could hold billions of transistors, each so small it could fit on the surface of a blood cell. But this continual miniaturization can't go on for ever. Eventually, the transistors on a chip will start to approach the size of atoms, which is where nature says 'thus far and no further', and Moore's law will break down. Futurologists predict it will happen around 2020.

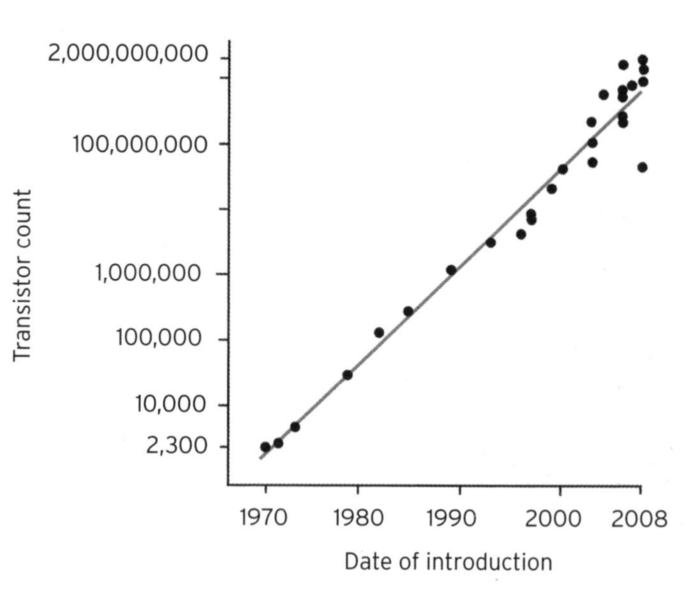

Software

Computer programmers translate the algorithm describing a problem-solving process into a series of instructions a computer can understand, known as software, a term coined by American mathematician John Tukey in 1958.

Central to writing software is the programming language. Early languages were known as low-level – that is, the programmer had to learn to speak in the numerical code used by the computer's processor. Later came high-level languages, which are more intuitive; BASIC (Beginner's All-Purpose Symbolic Instruction Code) is a high-level language developed in 1964 that uses plain English phrases – such as IF, THEN and PRINT – to convey instructions to the computer's processor. High-level languages require software of their own, called a 'compiler' program, which interprets the user's instructions into numerical code – effectively a low-level language – that the processor can then read. Most popular programming languages today – such as C++, Ruby and Python – are high-level.

Open source

Whereas most software organizations exist to make money, open source computer code is freely available for anyone to download, use, and even modify – providing those doing the modifying make the new versions freely available too. 'Source' code is the name programmers give to their high-level software scripts before they are run through a compiler and turned into a data stream for the processor to read.

The open source initiative was begun in 1998 by programmers Bruce Perens and Eric Raymond. It is not as altogether altruistic as it might sound – software development is an expensive business. Throwing programming tasks open to the masses saves money, and gives an organization a host of different versions of its software that programmers in the public domain have optimized for specific purposes. Recently, the term open source has been hijacked by other initiatives that seek to involve the public at a creative level – such as 'open source government'. Open source can be thought of as an example of **crowdsourcing**.

Parallel computing

The first microcomputers had just a single core processor to plod through a programmer's instructions. Then came parallel processors, and everything changed; a parallel processor has two or more core information processors, allowing the computer to execute different tasks simultaneously. And with extra pairs of hands to do the work, the job gets done much quicker.

Parallel processors first appeared in the 1970s for complex scientific computing applications, such as simulations of weather systems or the formation of vast swirling galaxies. Today, many desktop microcomputers feature dual- or even quad-core processors capable of a level of performance thousands of times better than the most advanced supercomputers of the 1970s. By late 2009, the fastest supercomputer was a modified Cray XT5, known as Jaguar, at the National Centre for Computational Science, at Oak Ridge National Laboratory, in Tennessee. The system had over 150,000 processors, making it over 10,000 times faster than a desktop computer and 10 million times quicker than its forebears in the 1970s.

Evolutionary algorithms

Evolutionary algorithms are a novel way of optimizing the solutions to computational problems and work by mimicking the process of Darwinian **natural selection** – in particular, its notion of survival of the fittest – to evolve a population of solutions and find one that's fittest to solve the problem at hand.

The basic idea is to start with a population of candidate solutions. Next, these solutions are allowed to 'breed' – aspects from pairs of solutions are spliced together, along with random mutations – to create a new generation of candidate solutions. The performance of each new offspring in solving the problem is then measured. Those which offer an improvement over their parents are allowed to produce another generation; the rest are killed off. And so the process repeats itself, gradually honing the solutions in the population. Evolutionary algorithms consume a great deal of computing power; however, with the advent of parallel processors they are becoming effective problem-solving tools.

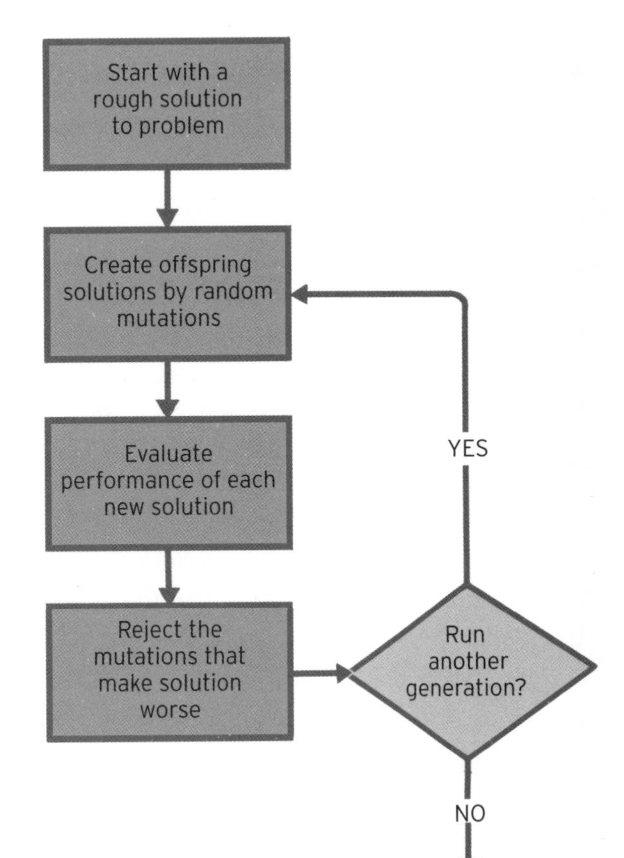

DNA computers

DNA molecules store information just like an electronic computer and just as electronic computers process information using tiny electrical switches stored on microchips, the information stored on DNA can be manipulated and processed by biological **enzymes**. These enzymes are able to mimic the action of logic gates, such as AND and OR, by altering the information encoded on DNA strands according to strict mathematical rules.

A DNA computer that's been built to play tic-tac-toe, for example, uses nine wells, each corresponding to a grid cell in the game, into which a culture of **microorganisms** has been placed. An enzyme acts on the DNA in each well, causing one to fluoresce – and this is the well where the computer has chosen to place its 0 or X. DNA processors naturally implement parallel computing, by applying different possible solutions to identical starting states – encoded as the information stored on hundreds of DNA molecules – and then picking the best final solution.

Quantum computing

The ability of particles to be in more than one place at the same time – a property of quantum theory known as the **uncertainty principle** – makes it possible to build staggeringly powerful computers. An ordinary computer encodes bits of binary data – and each bit can take the value 1 or 0. But in a quantum computer, a bit can be both 1 and 0 simultaneously, meaning that a byte, a chunk of data made from eight bits, can simultaneously represent 2^8 – or 256 – numbers. Performing a calculation on one such quantum byte simultaneously carries out the same calculation on each of the 256 numbers that it stores.

Quantum bits are known as 'qubits'; a quantum processor with a large number of qubits can carry out massive numbers of calculations in parallel – they are the ultimate parallel computers. Very basic quantum processors exist only in laboratories today; desktop versions are still many years away.

Quantum games

Quantum computers enable powerful information processing by exploiting the bizarre laws of the quantum world. Scientists have found that games played on quantum computers will offer new strategies and the study of these strategies is a branch of science derived from applying quantum principles to **game theory** – a new discipline known as quantum game theory.

A simple quantum game might be tossing a quantum coin, which could be represented in a quantum computer by the **quantum spin** states of a particle such as an electron with, say, 'spin up' as heads and 'spin down' as tails. An ordinary coin can be either heads or tails, but a quantum coin could be both heads and tails at the same time. Understanding the new strategies that quantum game theory permits will give an edge to those players wily enough to exploit them.

Components analogous to quantum coins turn up in the design of quantum computers. And programmers will need to bear quantum game theory in mind when they're developing the software that will ultimately run on these futuristic machines.

DATA STORAGE

Punched tape

The first digital computers, such as the British-built Colossus of the 1940s, didn't even use electronic media to record data. Very early computers used punched cards recording information through a pattern of holes in a piece of rigid cardboard. A human operator would need to feed the cards manually into the computer and remove them again. Not surprisingly, computer scientists soon lost patience with this and punched cards gave way to punched tape – which amounted to a long strip of punched cards stuck together, and which the computer itself could shuttle back and forth through. Punched cards and tape were read using systems of spring-loaded pins that could fit through the holes. Later readers were based on optical technology – using light beams shone through the holes. These data storage systems have a long history – the first were used to store 'programs' for textile weaving looms in the early 18th century.

Volatile memory

Early home computers were only able to store data in what's known as volatile memory – that is, memory that requires a constant source of power. Unplug your computer without saving the contents of its memory to a non-volatile storage device and you lose everything. The most common form of volatile memory is RAM (Random Access Memory), which records bits of binary data using the charges stored in an array of transistors on a microchip. The first desktop computers only had RAM; today, however, RAM is a kind of temporary, or 'working' memory, used by a computer to carry out the various subsidiary tasks involved in running a program, before the results are written to a non-volatile storage device.

Back in the early days of home computers, in the 1970s and 1980s, non-volatile memory might have consisted of a cassette tape or maybe a floppy disk if you were lucky. Nowadays, hard drives, **optical storage** and **flash memory** offer more convenient solutions.

Hard drives

Computer hard drives are the main variety of non-volatile memory used by most computer users today. Hard drives are a form of magnetic media, like cassette tapes and floppy disks that went before them; indeed, their heritage is the origin of the term 'hard' disk. Whereas floppy disks were flexible – made of thin plastic film with metallic particles embedded within them – hard-drive disks are rigid, made of a solid material such as glass coated with a layer of magnetic material to carry the data.

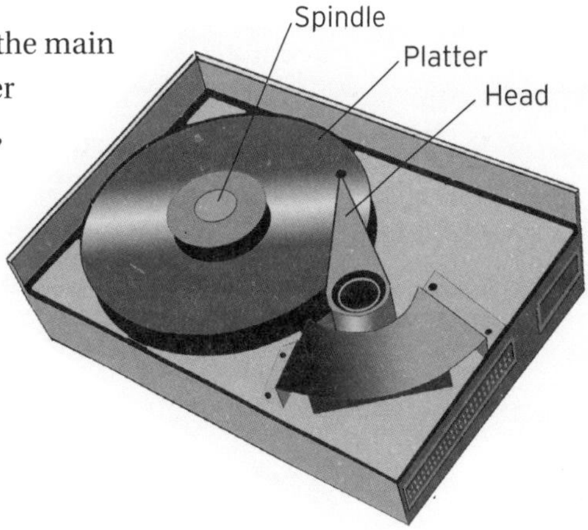

The surface of the disk is divided into many small regions. Each region can store one bit of data, recorded by magnetizing the region. This works because magnetization has a direction associated with it (think of the north–south direction of Earth's magnetic field) – and so shifting the direction of the magnetization can switch the value of a recorded bit from 0 to 1, or vice versa. Modern desktop hard drives spin at up to 10,000rpm and use **error correction** codes to ensure the accuracy of the data they store.

Flash memory

Whereas floppy disks and cassettes were once the portable data storage media of choice, flash memory has now taken over. SD cards, memory sticks and USB flash cards are all forms of flash memory. Flash is non-volatile memory, so it holds data even when its power source is removed by using semiconductor microchips containing a kind of transistor able to remain switched on or off even when unplugged from a computer. A switched-on transistor stores a 1 of binary data; a switched-off transistor stores a 0.

Flash memory chips are now used in iPods, digital cameras, cellphones and video camcorders. They have also begun to replace hard drives in some laptop and portable computers, where they are favoured for their resilience to impacts.

Optical storage

The first optical data storage medium was the compact disc – used by music publishers to market high-quality digital audio. This non-volatile memory works in a fundamentally different way to hard drives, which use magnetism to etch data onto a magnetic disk. A manufactured computer CD-ROM (CD Read Only Memory – meaning it can only be read and not written to) holds data as tiny 'pits' that have been pressed into the surface of a plastic disk. The depth of each pit encodes a bit of binary data, and bouncing a laser beam off the pit allows the value of the bit to be read out. Re-writable CDs can be written by the

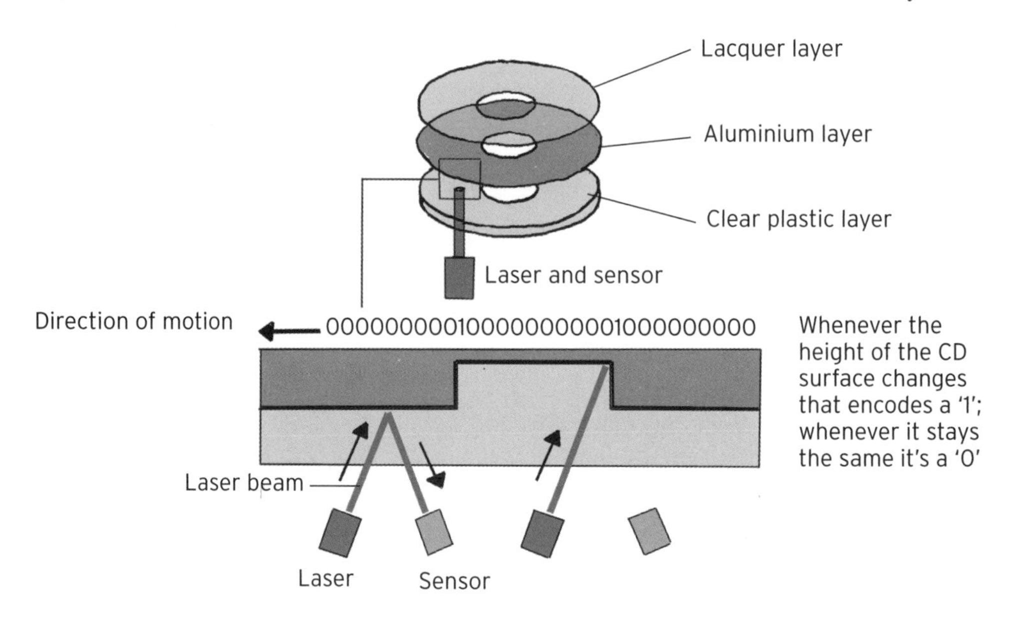

ccomputer user and work a little differently, using lasers to change the reflectivity of the disk in a way that mimics the presence of a pit.

In 1983 the first music CD was sold, closely followed in 1985 by the first CD-ROM. CD-ROMs have a limited capacity of 700 megabytes (MB). Newer optical media have improved on this considerably; for example, a DVD (Digital Versatile Disk) can store up to 9.4 gigabytes (GB) – 9.4 billion bytes of data.

Holographic memory

Optical storage media such as CDs and DVDs work by encoding data on the surface of the storage medium; holographic memory records data as a hologram within. It works using a laser beam to burn the hologram to the inside of a light-sensitive material. The beam is split into two – the first is shone through a filter that imprints on it the binary data to be stored; the other is then bounced off a mirror so that it intersects with the first. They meet inside the storage medium, where an image of the data is recorded. Colossal amounts of data can be stored this way; current prototype HVDs (Holographic Versatile Disks) can hold 500 gigabytes – with the theoretical limit anticipated to be around several terabytes per cubic centimetre of material.

Unlike hard drives and conventional optical storage, which degrade and lose data over time, holographic memory is an extremely robust form of data storage that can be used for the long-term archiving of information. Technically speaking, it's an example of WORM memory (Write Once Read Many).

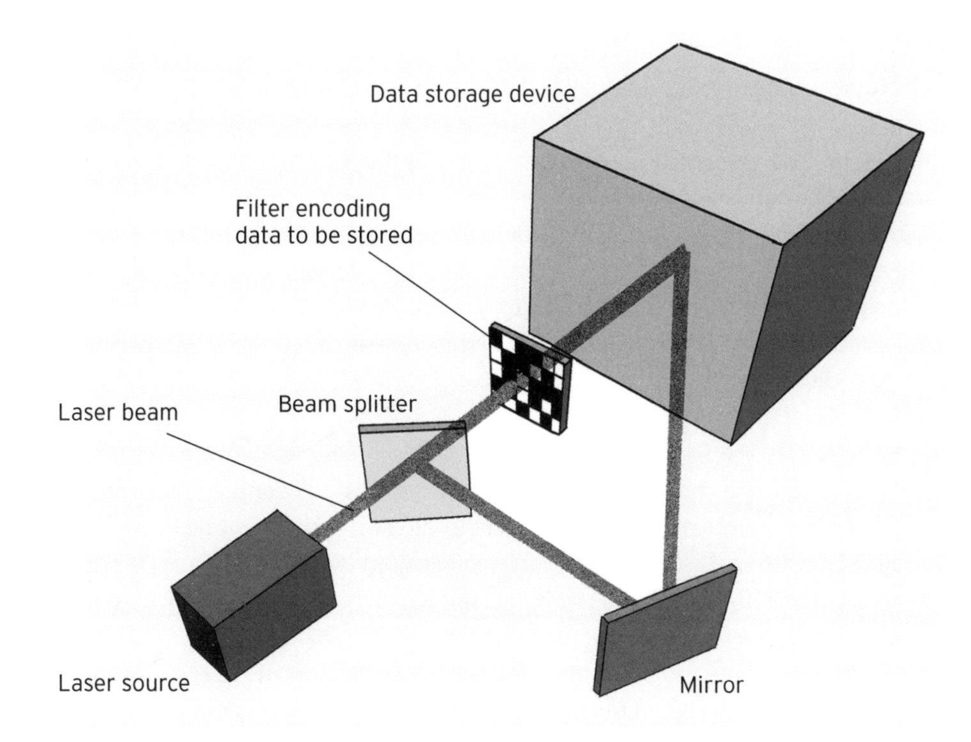

Spintronics

Traditional electronic storage media, such as volatile RAM and flash memory, work by encoding information using the electrical charge of subatomic particles called electrons. However, the emerging field of spintronics makes use of the electron's **quantum spin** as well. Because a particle's quantum spin can be in one of two states – up or down – the spin state of an electron can be used to represent the 1s and 0s of binary data.

The first spintronic devices were built in laboratories in the 1980s. Aspects of the technology have already been applied to improve the performance of hard drives but the major benefits are yet to come. Spintronic memory promises to be smaller, faster and more durable than existing options and it also offers improved memory density, while power consumption will be greatly reduced. In addition, spintronics may find applications in **quantum computing**, where it offers one possible way to store and transmit qubits of quantum information.

ONLINE TECHNOLOGY

The Internet

In 1958, the United States founded the Advance Research Projects Agency (ARPA) – a think tank and research unit intended to keep American innovation ahead of its rivals. A little over ten years later, ARPA came up with a way of linking up computers so they could exchange data over distance. In 1969, the scheme, known as ARPANET, was put into practice and two computers – one at UCLA and the other at Stanford University – were hooked up.

Another two nodes were soon connected and by 1971 there were 15. By the 1980s, similar networks were being established by academic communities in other countries. Commercial implementation soon followed, and enabled home users to connect to public networks using an early form of Internet access: dial-up modems. This allowed users to send email, view information posted on electronic 'bulletin boards' and play primitive multiplayer games. But the key development that turned the 'Internet' into the household word it is today would follow later: the World Wide Web.

Email

Most of the data traffic on the early Internet consisted of email – electronic mail. And most experts credit American programmer Ray Tomlinson with inventing it. He sent the first ever email in 1971; the message was transmitted between two computers at US company BBN Technologies, where Tomlinson was working, and which had been tasked with developing software for ARPANET – the forerunner of the Internet.

He had taken a messaging program that operated between users of the same computer and added elements of software designed to transfer files between different computers linked over a network, enabling all ARPANET users to send messages to each other. He also instigated the use of the @ symbol to separate the recipient's name from their location. Tomlinson can't remember what he wrote in that first message. Just seven years later the first piece of 'spam' – junk email – was delivered.

World Wide Web

Tim Berners-Lee, an English physicist working at the CERN **particle accelerator** lab on the Swiss–French border, wondered what it might be like if the Internet could be used to access documents hosted on remote computers. Clicking on highlighted words could then take the user to other documents – not necessarily hosted on the same computer. Better still, what if anyone could post documents of their own and link them to existing material. So the World Wide Web was born.

That was 1989. In 1990, Berners-Lee wrote the software language in which web pages are encoded – the Hyper Text Markup Language, or HTML. By 1992, there were a grand total of 26 web servers – computers storing the pages of websites – dotted around the world. Indeed, the web was still small enough that it had a single index page. With the release of the Mosaic graphical web interface – the first 'web browser' – in 1993, suddenly 'web surfing' became a pastime accessible to the masses. That set the stage for an exponential growth in popularity, leading ultimately to the sprawling electronic metropolis that is the World Wide Web of today.

Internet access

The most basic way to connect a computer to the Internet is via a dial-up modem, a device that links a computer to a telephone line. It can then use the phone line to call an 'Internet service provider', which acts as a relay station for the computer to communicate with the Internet – sending and receiving data, such as email or World Wide Web pages. Dial-up is now considered a painfully slow method of accessing the Internet, allowing speeds of just 56 kilobits of **binary data** per second.

In the late 1990s, the first broadband services became available. DSL, standing for Digital Subscriber Line, is an early example, still in use today, and uses digital technology to pack vast amounts of information down a standard telephone line. The fastest modern DSL lines offer download rates of up to 24 megabits per second; cable modems – delivering data via the same route as cable TV – offer speeds comparable to DSL. Nowadays, most people receive their household broadband to a router, which then beams it wirelessly around the house as a radio signal. This is known as Wi-Fi.

Mobile web

Social networking and **blogging** provide users with a continual feed of information, and this has stimulated demand for Internet access not just in the home and office, but on the move too. Many bars and cafes now offer free Wi-Fi Internet access to customers. Some cities have even established so-called citywide Wi-Fi, where transmitter and receiver stations create blanket coverage across the heart of an urban area. Anyone in the area with a wireless laptop computer or Wi-Fi cellphone can access the Internet.

Sunnyvale, California, became the first US city to do this, in 2005. Citywide Wi-Fi networks are now operating in other cities, including Philadelphia and Minneapolis.

Outside of the cities, many mobile devices are able to transfer smaller amounts of data over third generation – 3G – cellphone networks, offering download speeds of a few megabits per second, wherever there is a cellphone signal.

Bluetooth

Bluetooth is a short-range wireless network system for transferring files and data between hardware devices – like your cellphone and your computer. It's particularly useful for mobile devices, and for cutting down the number of cables in offices – thanks to Bluetooth-enabled keyboards, mice, and printers. Bluetooth can send binary data at a rate of up to three megabits of data per second. The range, officially, is limited to 100 metres, though some enthusiasts have succeeded in extending this significantly, to nearly 2km (just over a mile), using extra antennas. This practice, known as 'Bluesniping', is sometimes used to connect to another person's Bluetooth-enabled devices in order to gain unauthorized access. It is a form of electronic breaking and entering referred to as 'Bluesnarfing'.

Internet security

In a network spanning the entire world, not everyone will be trustworthy. And so connecting your computer to the Internet is inevitably going to bring security risks – especially as more and more financial transactions are conducted online. Dangers come in the form of hackers directly targeting computers, as well as attacks by hostile software applications. For example, viruses are programs that copy themselves onto your computer, from infected emails and data storage media, and then disrupt the computer's normal functions. Spyware, on the other hand, installs itself on a computer, monitors the user's

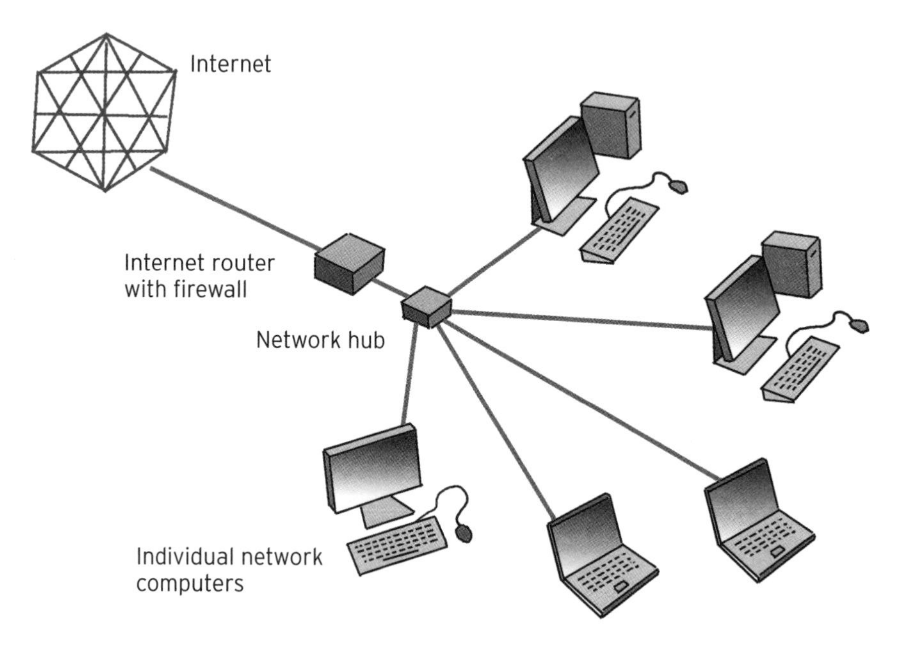

keystrokes and then passes them on to third parties – to gather information such as passwords. Bots, meanwhile, are programs that commandeer a computer's resources and use it to launch fresh attacks against other computers.

These malicious software agents are collectively known as 'malware'. Computer users can guard against it by installing software that regularly scans their hard drives searching for known malware programs and uninstalling them. Antivirus programs are one example of this. Most computers and Internet routers today are also equipped with 'firewalls' that block unauthorized attempts to access any computer placed behind them; this is the principal way of blocking attacks by hackers.

Grid computing

Grid computing is parallel computing gone mad; the idea is that the running of a program is divided up not between individual processors within one computer, but between different computers connected to one another via the Internet. In particular, grid computing emphasizes the use of computer resources that might otherwise be sitting idle – such as out-of-hours time on workstations normally used by human researchers. One example of grid computing is a project called SETI@Home, in which idle time on computers owned by the general public is used to analyze chunks of data from radio telescopes involved in the **SETI** search for alien life. A number of universities around the world have formed grid computing collaborations for scientific research.

Most recently, a new twist has emerged on the grid approach, known as 'cloud computing'. Here, companies offer their computing resources over the Internet to anyone who requires them, on a paid-for utility basis – just like gas and electricity.

Internet2

Academic networks involved in grid computing projects transfer vast amounts of data. Internet2 is a project to cater for this demand by providing super-high-speed Internet connections between over 200 US universities. The fastest ordinary Internet backbones today transfer data at a few gigabits (billion bits of binary data) per second (Gbps). With Internet2 technology, this could reach as high as 100Gbps. But it's not all about hardware; the Internet2 infrastructure also makes use of innovative data-handling software, known as 'middleware', to speed the flow of information across the network.

Semantic web

Anyone who's grappled with a web search engine to find a specific piece of information will know what a frustrating task it can be. The root of the problem is that computers don't really understand what you've asked them to look for – they just try to match the individual words in your search query to the key words on websites. For example, type in the search query 'What was the music used in the Levi's ad a few years ago, that featured people running through walls?' All you get are websites that feature the individual words in the query, but which capture none of the underlying meaning. You're very unlikely to get the correct answer: Handel's 'Sarabande'. That's where the semantic web comes in. Developed in part by father of the World Wide Web, Tim Berners-Lee, it's a project to index webpages according to the meaning of the information they store, rather than just on the basis of simple key words.

Berners-Lee instigated the semantic web project in 1999. Since then new computer languages for writing web pages – such as the Web Ontology Language (WOL) – have been developed, as well as semantic web search engines, such as semantic web Swoogle. However, despite the long gestation, use of the semantic web still seems to be very much in its infancy.

WEB TRENDS

RSS

No one has the time to continually skip from one news website or blog to another to keep checking for updates, which is where RSS comes in. To most people who use it, it stands for 'really simple syndication' (though others call it by its original acronym of 'rich site summary'). It works using software that monitors sites for updates. Users subscribe to so-called RSS 'feeds' from the sites they're interested in. Whenever new content is posted, the software automatically downloads the text and notifies the user of an update.

RSS has a long development history stretching back to the 1990s, but it only started to become widespread around 2005; it has proved to be a boon for professionals requiring up-to-the minute information feeds – such as journalists. RSS led to podcasting, whereby subscribers sign up not to text feeds but to audio content that can be listened to on an iPod or other MP3 player.

Filesharing

Filesharing does what it says on the tin – it allows anyone using the Internet to access files placed in a 'public' folder on your computer, and to do so for free. Doing this with music MP3s or, worse still, movie MPEG files is a serious breach of copyright that has landed users and networks – such as Napster and Kazaa – in deep trouble. The user searches a filesharing network by typing what they're after – say, the title of a music MP3 – into the network's software, which then displays a list of all the computers on the network where that file can be downloaded.

Napster was one network that was easy for authorities to suppress because it relied upon a central database stored on a central server – shut this down and the network dies. Today's filesharing networks, such as Gnutella, are much harder to take down because there is no central database – computers in the network communicate directly with one another instead. Filesharing networks are sometimes also known as peer-to-peer networks.

Blogging

Short for 'weblog', a blog is a public diary or commentary maintained on a World Wide Web site. Since the first ones were created in the late 1990s they have grown hugely in popularity. Early blogs were maintained by web enthusiasts, but now it's fashionable for politicians and prominent figures from all walks of life to have them. Usually conversational in tone, blogs are dashed out more quickly and with less polish than normally given to published writing – which is why journalists often use their blogs to break important stories quickly. At the end of 2007 there were an estimated 112 million blogs in existence.

Recently, short-form 'microblogging' has become popular, possibly due to the information overload created by the wealth of existing blogs. Entries on microblogs are often limited to a set number of characters. Probably the most popular microblogging site is Twitter, where each post is limited to 140 characters – the length of an SMS text message.

Social networking

For those who are bad at keeping in contact with their friends, social networking is a godsend, offering applications to make staying in touch easier. Most social networking websites provide users with email and live chat, and with a home 'profile' page where they can post information about themselves. They also have a facility to add other users as 'friends' who can see each other's profile pages and exchange messages. As with microblogging, users can post messages that are visible to not just one, but all of their friends – such as 'party at my house Saturday'. The biggest social networking sites today are MySpace and Facebook.

Social bookmarking

Whereas social networking systems enable users to stay connected to their friends, social bookmarking is a way of sharing bookmarks – that is, pointers to the web pages they use – either with your network of friends, or with the world at large. Unlike filesharing, there is nothing illegal about social bookmarking because all that gets shared and downloaded are web addresses – not their actual content. They work by users bookmarking the sites they like and indexing them with subject tags to enable others to search for popular sites on a given topic. In many ways this is more effective than using conventional search engines to look for websites, as the search tags are assigned by humans who understand the content of a web page – rather than by a computer, which doesn't. Popular social bookmarking systems include Digg and del.icio.us.

Lifelogging

Imagine what it would be like to have a comprehensive electronic record of everything that's happened in your life. Everything you've seen, heard, read and been told – stored in a searchable database that you could access with just a few keystrokes; this is the goal of so-called lifeloggers. They use wearable cameras, microphones and GPS devices to build up a time-indexed catalogue of their lives.

The field was pioneered by Steve Mann, a computer scientist at the University of Toronto. He began in 1994, wearing cameras and other lifelogging equipment to build up a 24/7 record of his life. Storing such large volumes of information needs large-capacity hard drives, which have been developed recently. However, many people are concerned about the privacy implications of lifelogging. Would the police and authorities have the right to seize your stored memories if they believed it might help with an investigation?

Virtual worlds

Multi-player online role-playing games have been around for years, first appearing on early precursors to the World Wide Web in the 1970s and 1980s. So-called 'multi-user dungeons' (MUDs) were games that ran on networks such as ARPANET and the UK's JANET academic network. Now these primitive text-based games have been superseded by massive multi-player online role-playing games (MMORPGs). Games such as EverQuest, World of Warcraft and Eve Online now offer sophisticated graphical interfaces, instead of text, and allow huge numbers of players and considerable interaction between them. MMORPGs are a subset of online environments known as virtual worlds, where users can effectively live out alternative lives through the actions of their online personas, or 'avatars'.

Other virtual worlds include Second Life and SmallWorlds, which differ from MMORPGs in that they do not necessarily have to have a fantasy or SF setting; some simply mirror everyday life. Chat rooms, where users can talk by exchanging lines of typed text in real time, were an early kind of virtual world.

Crowdsourcing

The most effective lifeline in the TV quiz show *Who Wants to Be a Millionaire?* is 'phone a friend'. Now companies are taking advantage of the power of the masses too, by outsourcing tasks to the general public – which they might otherwise have paid a contractor for. A cellphone manufacturer might want to find out which of their handsets works best with different networks and different operating-system software. Rather than exhaustively testing all the options themselves or hiring someone to do it, the company gives phones, with different combinations of operating systems and networks, to members of the public. Those who provide feedback on their experiences receive a cash reward. Of course this costs the company considerably less than paying for the research to be done by a contractor.

Crowdsourcing has even been adopted by the scientific community. For example, Galaxy Zoo is an online project that enlists the help of members of the public to classify pictures of galaxies taken by powerful astronomical telescopes, according to the **tuning fork diagram**.

Cyberslacking

Most of us have done it – used the Internet at work or school for purposes that are not altogether connected to work or school. In October 2009, it was estimated by the IT services group Morse that social networking sites such as Twitter and Facebook are now costing the British economy alone £1.4 billion per year in lost productivity. Meanwhile, a survey of over 10,000 office workers by America Online and Salary.com revealed that almost half consider personal Internet use to be their primary distraction at work. Many companies now prevent staff from accessing websites considered 'inappropriate', such as gambling, online chat and filesharing sites, and even make use of surveillance software to monitor employees' web-browsing habits. Even so, some employees have perfected the art of cyberslacking and are able to carry out several non-work activities out the same time – so-called 'multishirking'.

Wikis

A great application of crowdsourcing is the class of websites known as wikis, sites that throw open their content to the masses. The most famous wiki is the Wikipedia online encyclopedia, for which anyone can write or edit entries. Those knowledgeable in particular areas share their expertise, and the desire for kudos – or rather the reluctance to lose it – helps keep contributors loyal. In 2005, the British science journal *Nature* carried out a comparative test of encyclopedias, finding Wikipedia to be at least as good as the *Encyclopaedia Britannica*. In fact, one of the few groups to abuse Wikipedia are politicians – with members of US political parties vandalizing the opposition's pages.

ARTIFICIAL INTELLIGENCE

Turing test

In 1950, the British computing pioneer Alan Turing set out a way of gauging a machine's intelligence. The basic idea couldn't be simpler – you have a chat with it. Well, actually you have a chat with the machine and another human being. You aren't told which is which (the human talks to you through the same interface as the machine), and if you can't work it out from the conversation then the machine is considered to have demonstrated human intelligence. This scheme has since become known as the Turing test.

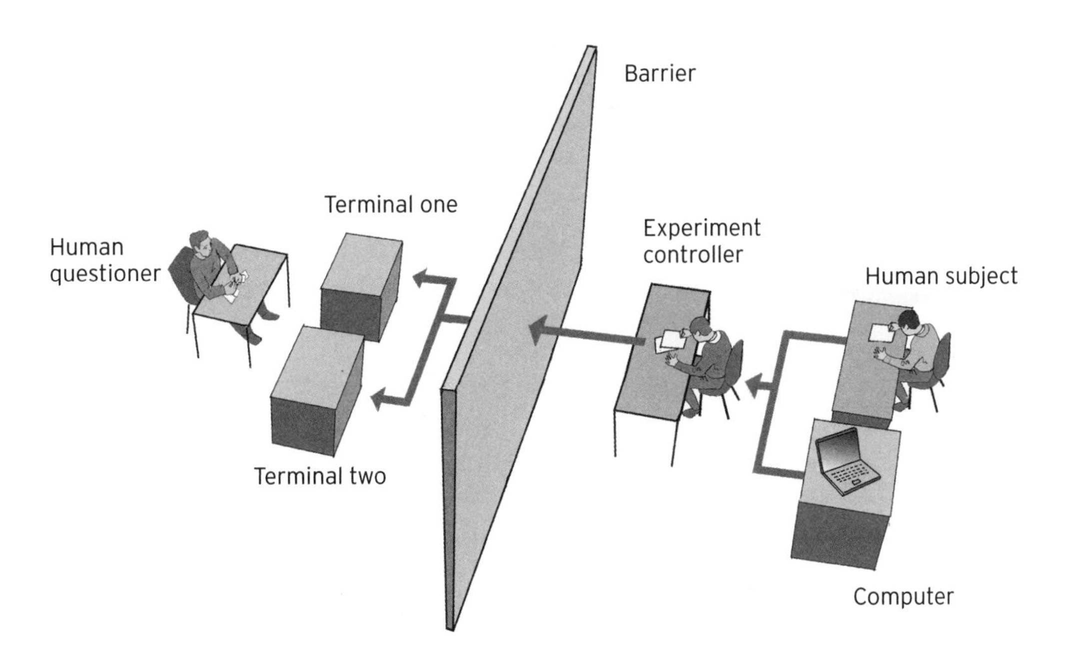

Computer scientists have since used the Turing test to assess whether their software creations exhibit any of the qualities of 'artificial intelligence' (AI) – the ability of a computer to demonstrate the human-like capability for decision-making and even **consciousness**.

Chatbots

Some programmers have produced what are known as chatbots – pieces of software that can hold a conversation with a human in an effort to pass the Turing test. While some of these do attempt true artificial intelligence, many work without any understanding whatsoever of what the conversation is about. They operate instead by making responses that are vague, or which take the conversation in a new direction – or by being programmed to come back with particular responses to certain cue phrases from the human. For example, if asked a question, the chatbot might be programmed to reply simply, 'Why?' or 'I don't know.'

In 1990, scientists began the annual Loebner prize contest where they pit their chatbots against one another by trying to pass the Turing test. The best chatbot each year receives a small cash prize, with $100,000 set aside for the first machine that can fool four of the contest's 12 judges. Winner in 2008 was Elbot (www.elbot.com), which managed to convince three of them.

Neural networks

Neural networks are pieces of computer software that seek to replicate the action of neurons in the human brain – in particular, the brain's ability to process and learn from information. They are extremely good at spotting patterns in large sets of data, and then using that knowledge to make predictions. For example, the UK company Epagogix uses neural net software to predict the box-office revenues of movies. The software is 'trained' using data on past movies, such as information on genre, budget, lead actors, and so on – as well as the bottom-line box–office takings. Entering the pre-release data for a new movie then yields a prediction for how much money it will make. It's quite accurate – Epagogix predicted that the 2007 movie *Lucky You*, made with a budget of $50 million, would tank earning $12.5 million in the US and Canada. And tank it did – in fact, making just $5.7 million.

Neural net software is also used in pattern recognition – for example, for spotting a human face in a photograph – as well as medical diagnosis and even lip-reading.

Natural language processing

There was a time when scientists would communicate with their computers in strings of numbers. We've come on from there somewhat, with 'high-level' software languages enabling programmers to give instructions using a predefined set of commands the computer can understand. But the ultimate goal is to develop an interface by which humans and computers can communicate using plain English – or, indeed, any other 'natural' human language. Accordingly, this area of research is known as natural language processing.

It's not as simple as you might think; chief among the problems to be overcome is the ambiguity of our natural languages. For instance, is a 'large, bearded pig farmer' a large, bearded farmer who keeps pigs, or someone who farms large, bearded pigs? Or, does 'feeding tigers can be dangerous' mean that it is dangerous to give food to tigers, or that tigers can be dangerous when they're eating? We can usually tell which is correct from the context and other cues, but it's far from clear how to teach these instincts to a computer. Natural language processing is an active area of research.

Autonomic computing

Autonomic computing refers to intelligent computer systems that manage themselves. The name is taken from the human body's autonomic nervous system which regulates involuntary processes such as respiration and heart beat, as well as healing. An autonomic computer system is able to manage system resources to complete tasks and optimize performance, not just on one computer but on an interconnected network. It also responds to unexpected problems such as power outages and viruses – doing so without any human intervention. Most of the research in this area has been carried out by IBM.

Autonomic technologies are now being applied in areas outside pure computing – for example, in anti-lock braking systems (ABS) for cars, which detect when a wheel has locked and then release the braking force momentarily.

Swarm intelligence

Swarm intelligence is a kind of artificial intellect that arises through the collective behaviour of many small units. The coherent behaviour of fish shoals or large flocks of birds are examples of swarm intelligence in the natural world. Swarm intelligence in computing is an attempt to mirror this behaviour, in order to solve computational problems.

A primary application is in optimization – solving problems as efficiently as possible. For example, scientists have found algorithms that closely model the foraging behaviour of large ant colonies. A colony of 'artificial ants', within a computer, and which obeys this algorithm, can be programmed to forage not for food, but for solutions to a problem. An example where this has been used effectively is in seeking solutions to the **travelling salesman problem**. Swarm intelligence has also been applied more literally by robotics researchers to control herds of small robots that collaborate to perform tasks – for instance, cleaning and even exploring the surfaces of other planets.

Cellular automata

Cellular automata are computer models that demonstrate how the complexity of living systems can arise from simple rules. The most famous cellular automaton is Conway's Game of Life, created by the Cambridge University mathematician John Conway in 1970. The idea is that you start with a large sheet of squared graph paper on which a small number of the squares are shaded and the rest are blank. The shaded population then evolves in a series of discrete steps, or 'generations', according to a set of rules. For Conway's Game of Life the rules are that a shaded square with less than two shaded neighbours becomes a white square; a shaded square with four or more shaded neighbours becomes a white square; a shaded square with two or three neighbours remains shaded; and a white square with three shaded neighbours becomes shaded. All sorts of complex behaviours result from these simple instructions – dependent on the starting conditions.

Cellular automata are sometimes cited as an example of 'artificial life' because they mimic the behaviour of biological cell colonies, with new cells growing into empty spaces, and cells that are surrounded by other cells – and thus have no food – dying out.

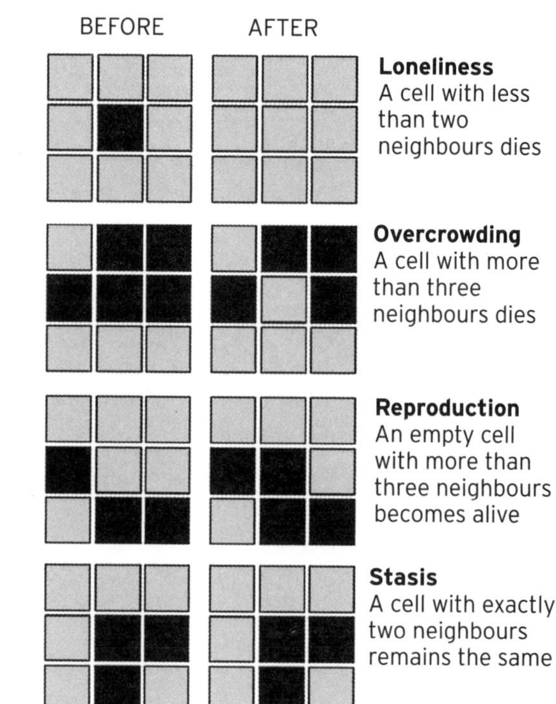

Strong AI

The ultimate goal of artificial intelligence is to create a computer that can equal or even exceed the capabilities of the human brain. Known as 'strong AI', the term was introduced by the philosopher John Searle, at the University of California, Berkeley. Existing artificial intelligence projects fall short of this – either feigning intelligence (like most **chatbots**) or implementing just a subset of the brain's functions (**neural networks**). A strong AI system would need to incorporate **natural language processing**, be able to reason and learn, and perhaps even exhibit human qualities of consciousness and self-awareness.

These challenges define the software that artificial intelligence researchers will need to develop. Even so, there are hardware requirements to be met too. American futurologist Ray Kurzweil estimates that implementing strong AI will require computers capable of 10 million billion (10^{16}) calculations per second, but he believes technology could achieve this as soon as 2013.

Artificial brains

One promising avenue for realizing strong AI is the construction of artificial brains – electronic systems that replicate the exact functionality of biological neurons, and their arrangement and interaction within the brain. Researchers at the Blue Brain Project, in Switzerland, are doing just this – picking apart the structure of the human brain in a bid to reverse engineer it inside a computer. Already they are close to building a complete working model of a rat brain and believe that a complete artificial human brain is achievable by 2020. As well as helping in the development of artificial intelligence, artificial brains may also have applications as tools for studying brain diseases such as **dementia**.

Androids

Ever-improving artificial intelligence brings the idea of intelligent robots closer to reality. And yet building a real-life C3PO is likely still years away. To date, one of the most impressive attempts to create a humanoid robot is the Honda Asimo, a bipedal robot capable of walking and running – even up and down stairs. Asimo's metal joints enable it to move with 34 'degrees of freedom' – the ability to move one body joint up/down or left/right. Humans, by comparison, have over 200 degrees of freedom; mimicking this flexibility and controlling it is a difficult problem.

Asimo is also able to recognize objects and faces, and can react to hearing its name called. Other humanoid robots have even been given lifelike human faces, though sometimes these additions are counterproductive. Robotics experts talk about the 'uncanny valley' effect, whereby a robot that appears nearly human but not quite appears slightly gruesome. Uncanny valley is a dip in the graph you get when you plot the positive response people experience to a humanoid robot against how realistic the robot appears. Nevertheless, research continues. Computer scientist Noel Sharkey, at the University of Sheffield in England, believes the first applications of humanoid robots may be as carers or companions for the elderly.

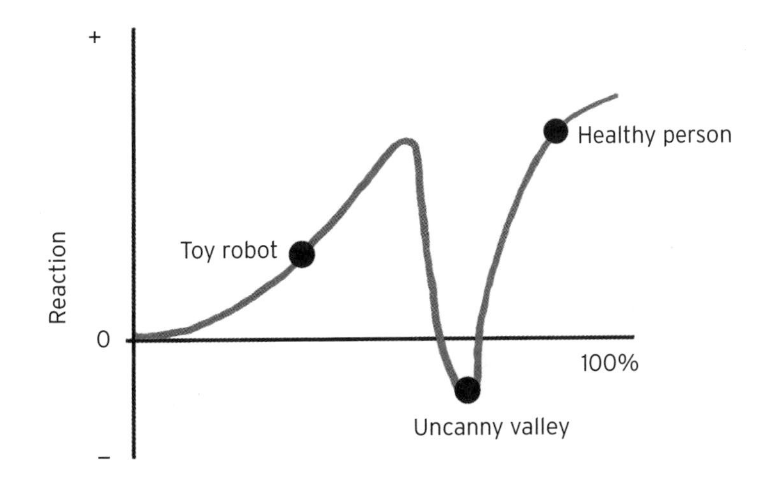

APPLIED SCIENCE

MOST PEOPLE ENCOUNTER SCIENCE purely through the impact on their daily experience of technology – also known as applied science, a discipline that's shaped more or less every aspect of modern life. It's given us transport, buildings, energy, communications, electronics, home entertainment, labour-saving appliances, heat, light, and ensures we have enough food to eat.

Historically, applications were always the driving force of science. Longbows and triremes gave a military advantage to nations. Thermodynamics yielded an efficient way of getting from A to B – via steam power. And working out how to measure longitude at sea was done for one reason: so humans could travel the world.

Today, dedicated applied science is still a thriving field. But it's supplemented by 'blue sky' science – research

EERING • ELECTRICAL ENGINEERING •
AEROSPACE ENGINEERING • CHEMICAL
• CERAMICS • COMPOSITES • AEROGEL •
ICTORS • SHAPE-MEMORY ALLOYS • D30
COAL • PETROL • NUCLEAR ELECTRICITY
PANELS • SOLAR THERMAL ENERGY •
T PUMP • MICROPOWER GENERATION •
• BIOFUEL • SYNTHETIC HYDROCARBON
COIL • ALPHABET • TELEGRAPHY •
EPHONE • FIBRE OPTICS • VOIP • INSTANT
• RADAR • SONAR • STEALTH • NUCLEAR
AL WARFARE • SPY SATELLITES • SPACE
NSGENICS • GENETICALLY MODIFIED

carried out that has no clear practical purpose. Blue sky research is important in its own right because it deepens our understanding of science. But occasionally it also leads to practical applications that no one could have foreseen at the outset.

Quantum computers and stem cell therapy are two technologies that emerged in just this manner. The former look set to play a defining role in information technology in the 21st century, while the latter could one day offer a cure for dementia and other serious illnesses. It's a healthy reminder that esoteric branches of pure science can lead to innovations of enormous practical significance – sometimes in the most unlikely ways imaginable.

ENGINEERING

Mechanical engineering

Combining the physical laws of mechanics, heat and fluids together with **materials chemistry** leads to a branch of innovation known as mechanical engineering. It's all about building structures that rely on the interaction of mechanical forces. The construction of cars, ships and aircraft are all sub-disciplines of mechanical engineering – as are the relatively high-tech areas of robotics and spacecraft design.

Early mechanical engineers of antiquity designed pulleys, gear wheels and other mechanisms that facilitated the building of the pyramids and other great monuments. Greek polymath Archimedes, in the 3rd century BC, invented the mechanical screw for lifting water. Most recently, mechanical engineers have sought to miniaturize their creations – building systems of gears and linkages that operate on scales of thousandths and even millionths of a millimetre. These are called microelectromechanical systems (MEMS) and nanotechnology (also known as **molecular engineering**), respectively.

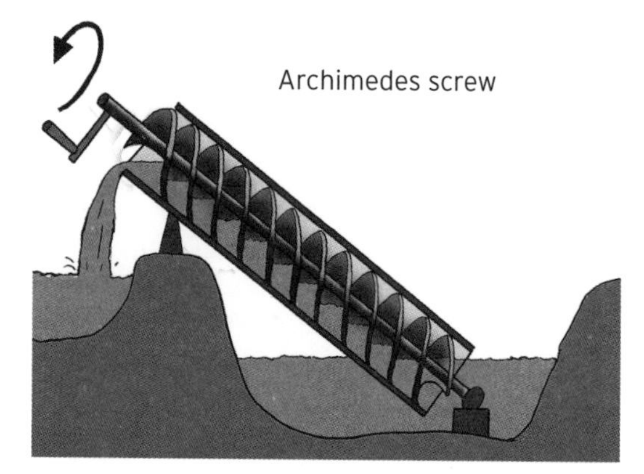
Archimedes screw

Civil engineering

Design and construction on a superscale – dealing with the building of roads, bridges, skyscrapers and dams – all falls under the umbrella term of civil engineering. It amounts to the science of building the structures which underpin civilization itself. Not surprisingly then, it is one of the oldest fields of engineering, pioneered by the Egyptians, the Mayans and later the Romans, who filled their early empire with all manner of roads, viaducts and fortresses.

One of the greatest civil engineers of all time was Britain's Isambard Kingdom Brunel, who built railways as well as bridges and the first transatlantic steamship to be driven by propellers rather than paddles.

Electrical engineering

It began with light bulbs and electric motors and led to computer engineering, TV and radio, as well as satellite communications. Electrical engineering is the application of the science of electricity and magnetism to design and build new devices.

The first electrical engineer could have lived in Iraq in the 3rd century BC, and built what is known as the 'Baghdad battery' – way before the discovery of electricity. The battery consisted of a terracotta jar with iron and copper terminals inside; when the jar was filled with an acid, like lemon juice, an electrochemical reaction created an electrical voltage across the terminals, in turn producing an **electric current**. Historians speculate that this device was used to coat cheap metals with gold, by a technique called electroplating. And, indeed, artefacts plated with thin layers of gold have been found in Iraq dating from the same period.

Electrical engineering developed greatly following the discovery of **Maxwell's equations** of electromagnetism in the 19th century. Many key inventions followed, such as the telephone, radio and electrical generator. It was quickly followed by electronics – using devices with precisely defined electrical properties such as resistance, capacitance and inductance to control the flow of electricity.

Computer engineering

Electrical engineering, or rather its subcategory of electronics, ultimately gave rise to perhaps the most widespread form of engineering today: computer engineering. Modern computers are electronic systems that are able to store and process **binary data** using the flow of subatomic particles called electrons. Whereas early computers used to fill entire rooms, sophisticated computers are now incorporated into tiny handheld devices such as cellphones and MP3 players. And they've become an integral part of TVs, cars and aircraft.

Computer engineering covers a diverse array of disciplines, from building **hard drives**, processors and interfaces (which connect computers to different devices), as well as the broad array of skills needed to write software – the instructions by which computers operate.

The field has an exciting future ahead of it, with radical new kinds of computer now becoming possible – including **quantum computers** and **DNA computers**. As information begins to dominate science, business and home life, some futurologists regard computer engineering as the ultimate branch of engineering – one that will survive when all the others have become obsolete.

Cybernetics

Cybernetics is an area of engineering that pulls together disciplines including mechanical engineering, electrical engineering, as well as computer engineering and even biology, applied mathematics and psychology. Strictly speaking, it deals with the control of systems, in particular regulating the system's response through control inputs. This may involve a mechanical system, such as a robot, controlled by electronics and computers, but also cybernetics describes how different social groups of people respond to the control inputs of their government.

In popular parlance, the term has become synonymous with robotics and **artificial intelligence** and has given rise to the term 'cyborg', short for 'cybernetic organism', to describe humans or other biological lifeforms that have been enhanced using robotic and electronic implants. Recently, however, there has been a resurgence of interest in cybernetics in its broader sense.

Aerospace engineering

Constructing craft that can fly both within and beyond the confines of the atmosphere is the province of aerospace engineering. The first designs for powered aircraft were produced by Leonardo da Vinci in the 15th century, though none of his primitive helicopters were ever built. In 1903, in the United States, Orville and Wilbur Wright conducted the first powered flight of a heavier-than-air flying machine. Since then flying has become considerably more sophisticated. The Wright brothers' aircraft used just very basic concepts in **fluid dynamics** and mechanical engineering.

Today, aircraft design and construction relies on electrical engineering, computer engineering, **applied mathematics**, modern materials, as well as jet and rocket propulsion. Extending the field to spacecraft, which travel beyond the atmosphere, means this list now also includes the mechanics of flying in Earth orbit, atmospheric re-entry, and space physics.

Chemical engineering

Designing machinery to mass-produce chemical compounds with interesting or useful properties is known as chemical engineering. Industrial demand for particular chemicals means that chemical engineering plants normally operate on a vast scale. Oil refining is a classic example, whereby a base material – in this case, crude oil – is processed to create useful substances such as **petrol**, lubricants and the chemicals from which plastics are manufactured. Oil refining makes use of a process called fractional distillation where heated crude oil vapour rises up inside a giant cooling tower. The temperature drops the further up the tower the vapour climbs; different chemicals condense out of the vapour at these progressively lower temperatures – which are then siphoned off. Other substances that are mass-produced in chemical engineering plants include fertilizers, pharmaceutical drugs and foodstuffs.

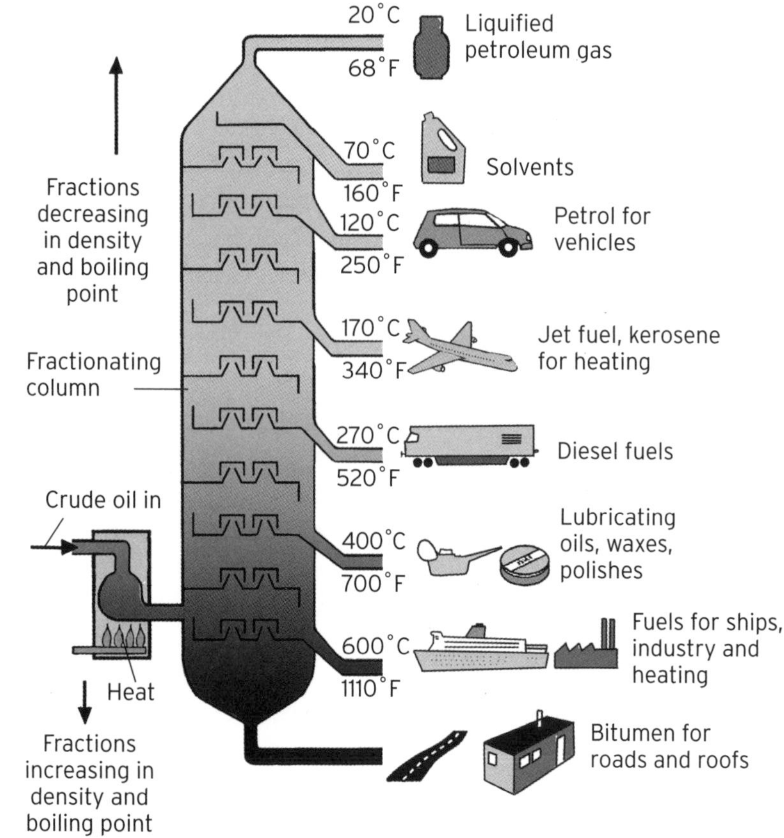

Molecular engineering

Nanotechnology is engineering on the scale of nanometres – billionths of a metre. Complete nanotech structures can measure as small as 100 nanometres – smaller than many bacteria and only slightly bigger than some molecules. Nanotechnology is quite literally engineering on the scale of the molecular world – hence its alternative name, molecular engineering. It has a number of important applications, including development of new materials, such as **carbon nanotubes**, and, in the future, the construction of tiny robots that can carry out maintenance on the nanoscale.

Most of the groundwork for molecular engineering was laid by American engineer Eric Drexler in the 1980s, while the field's practical development was facilitated by the invention of methods to scrutinize the molecular world, such as the **scanning electron microscope**. There are ethical concerns over nanotechnology, with some scientists and environmental groups worrying that, if left unchecked, intelligent nanorobots could wreak havoc by self-replicating and turning all the matter in the world into one big mass of nanorobots – the dreaded '**grey goo**' scenario.

MODERN MATERIALS

Ceramics

Any material made by firing non-metallic, inorganic substances at high temperature is called a ceramic. Ceramics have diverse applications in industry and perhaps the oldest examples of their use are in the making of pottery, and later fine china. The electrical properties of ceramics mean they are sometimes used as the dielectric material in capacitors – electronic components with **capacitance**. Their heat tolerance makes them ideal for use in home heating systems, cooking hobs, components for engines in cars and aircraft – and even heat shields for spacecraft re-entering the atmosphere from Earth orbit.

Many ceramic materials are harder than metals, making them ideal for mechanical parts, bullet-proof body armour, battle tank armour, and even knife blades that stay sharper for longer than steel. Although renowned for their strength, hardness and durability, ceramics are often brittle and need careful handling.

Composites

Putting together two materials with different properties can lead to a new material with properties that exceed either on their own. For example, hard setting **polymers** (plastics) and resins have rigidity but lack strength. Glass fibres have strength but lack rigidity. But combine the two and you get a material which is both strong and rigid: 'fibre-reinforced plastic' (FRP), also known as fibreglass. Lightweight, strong 'composites' such as this can outperform conventional materials, such as metals, and are revolutionizing engineering.

A prime use of composites is in aerospace engineering, where plastic-based composites that are strong as metal, yet weighing much less, allow planes to carry greater payloads and use less fuel. The Airbus A380, which began flights in 2007, is the biggest passenger airliner in the skies today – and is made from 25 per cent composite materials.

Aerogel

One of the weirdest materials known to man is called aerogel – it's the lightest-known solid with a lower density than air. Aerogel is also an incredible thermal insulator. A demonstration by NASA involved laying matches on a thin sheet of aerogel heated from below with a blowtorch. The matches remained unburnt. Aerogel was invented in 1931 by Samuel Stephens Kistler, an American chemical engineer; it is made by drying out the liquid component of a silica gel, to leave a desiccated, porous solid behind. It is translucent and slightly blue in appearance – materials scientists have given it the nickname 'frozen smoke'.

Because of its light weight aerogel is an excellent insulator on spacecraft, and inside astronauts' spacesuits. Meanwhile, the Stardust spaceprobe, which passed through the tail of a comet to gather dust particles, used blocks of aerogel to snag the microscopic particles.

Biomimetics

Shaped by millions of years of biological evolution, it's no wonder plants and animals have developed amazing solutions to deal with whatever the environment has to throw at them. Now engineers are cottoning on. Rather than toiling over their drawing boards they're taking inspiration for new products and materials from the natural world – a field known as biomimetics. Biomimicry is everywhere. Velcro – the combination of rigid hooks and fuzzy material used to stick objects together and pull them apart again, conveniently and over and over again – was invented by Swiss engineer George de Mestral in 1941. He was inspired by burdock seeds, which kept sticking to his clothes and the fur of his dog while he was out walking. On closer inspection he saw that the seeds were covered in tiny hooks.

Other examples of biomimetics include catseyes on roads; echolocation devices for the visually impaired, based on bat sonar; swimsuits based on the fluid dynamic texture of shark skin; and even pixels in electronic display screens that are inspired by the colour-shifting mechanism of scales in butterfly wings.

Biomaterials

Any material that can be implanted in a living organism to augment its capabilities or repair damage is called a biomaterial. Common examples in humans are the composite materials used in tooth fillings, the **polymers** used in contact lenses and the silicone in breast implants.

But some rather more technologically advanced biomaterials are used in medicine. Hip replacements use joints made of titanium with a ceramic and polyethylene ball joint. The implants are often coated with a mineral called hydroxyapatite to encourage new bone growth. Meanwhile, stents, mesh tubes used for holding arteries open, are made from stainless steel and even shape-memory alloys. Some biomaterials themselves are of biological origin; for example, heart valves from pigs are used to replace damaged valves in humans.

Semiconductors

Semiconductors are materials that aren't perfect conductors of electrical current and aren't perfect insulators either. Electrons can flow through them to a limited degree. But equally, the 'holes' left by the movement of an electron in a semiconductor creates a relative positive electric charge, and the flow of these positive holes, scientists realized, could be used to create electrical devices with interesting new properties.

One such device was the transistor. Invented in 1947, by Bell Labs researcher William Shockley, it's a sandwich of alternating layers of so-called P-type and N-type semiconductors. Positively charged semiconductors (with an excess of holes) are known as P-type; those with a negative electron excess are called N-type.

In an NPN transistor, for example, increasing the voltage of the middle layer increases the number of electrons, making the middle layer better at conducting electrons between the sandwich's two outer layers – and so the current through the transistor increases. Transistors made excellent electrical switches for digital computers in the 1950s. Soon after, they were miniaturized onto microchips. Semiconductors are the materials from which the modern computer age is built.

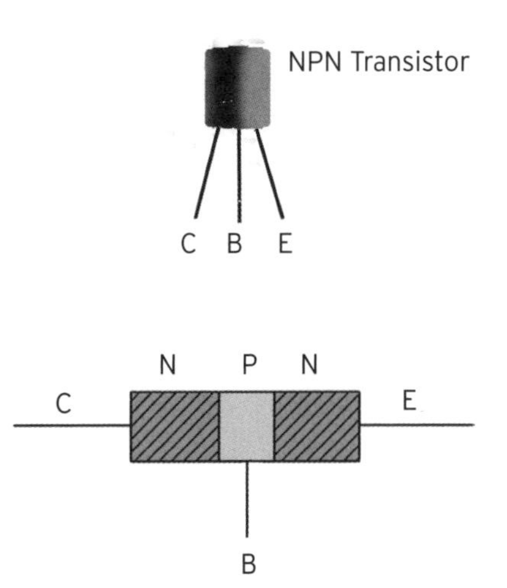

Current applied to B means more current flows across the transistor from C to E.

Shape-memory alloys

Thanks to these materials the days of having to buy a new pair of glasses when you accidentally sit on the old ones are gone. Shape-memory alloys are materials that can be bent and squashed and then instantly spring back into their original shape. Called the shape-memory effect, it is an extreme form of **elasticity** (see **Hooke's law**). It occurs in certain metals and **polymers** within a certain range of temperatures. The shape they return to after stretching can sometimes be controlled by varying the ambient temperature and magnetic field.

As well as helping those prone to misplacing their glasses, shape-memory materials have a host of applications in medicine – for example in filters to catch blood clots, and in wire tooth braces to correct misaligned teeth. And shape-memory foams are now used to make cushions and mattresses that mould to the user's body.

d3o

It's the snowboarder's dilemma – finding sportswear that balances safety with style. A chunky crash helmet, elbow guards and kneepads give greatest protection, but they restrict movement. Now there's a solution in the shape of a protective material called 'd3o' (pronounced 'dee-three-oh'), developed by UK inventor Richard Palmer. Under normal conditions the material is soft like rubber. But in the event of an impact it stiffens up to become an excellent shock absorber – allowing for flexible garments that still give protection when it's needed.

Palmer says the secret of d3o is its 'intelligent' molecules. Its behaviour is similar to that of a dilatant colloid, a type of non-**Newtonian fluid**. There are now numerous of d3o-based protective garments available for sportspeople and motorcyclists. Most recently Palmer's company, d3olab, collaborated with Puma on a d3o football – it's easier to control at low speed, yet flies further when kicked hard.

Carbon nanotubes

Superstrong networks of carbon atoms, arranged into sheet-like lattices and rolled into nano-scale tubes, look set to form the strongest and most versatile materials available to engineers in the years to come. Known as carbon nanotubes, they have over 300 times the tensile strength of steel. What's more, they are amazing conductors, offering 15 times the thermal conductivity of copper, and 1000 times the electrical conductivity.

The first carbon nanotubes were made in 1991 by Japanese scientist Sumio Iijima. By passing an electric current through two electrodes made of graphite and suspended in helium, some of the graphite evaporated and condensed on the walls of the container as nanotubes. It's hoped that once the cost of production drops and their properties can be better controlled during manufacture, carbon nanotubes will have applications in construction, energy storage, semiconductors and molecular engineering.

Metamaterials

Sometimes the needs of a material scientist go beyond what is readily available in nature and this is where metamaterials come into their own – materials that have been carefully constructed by human engineers to have a specific set of properties. An amazing example are metamaterials that have a negative index of **refraction**, which allows an object placed within a cloak made of such material to become invisible. Developed by scientists at Imperial College, London in 2006 they work by deflecting light around the object to be hidden in such a way that the light rays emerge from the other side on the same paths they would be on were the object, and the cloak, not there at all. The path of the rays looks rather like water flowing around an object in a stream. In the future, metamaterials may allow scientists to build superpowerful microscopes and 'optical computers', that operate using photons of light rather than electrons.

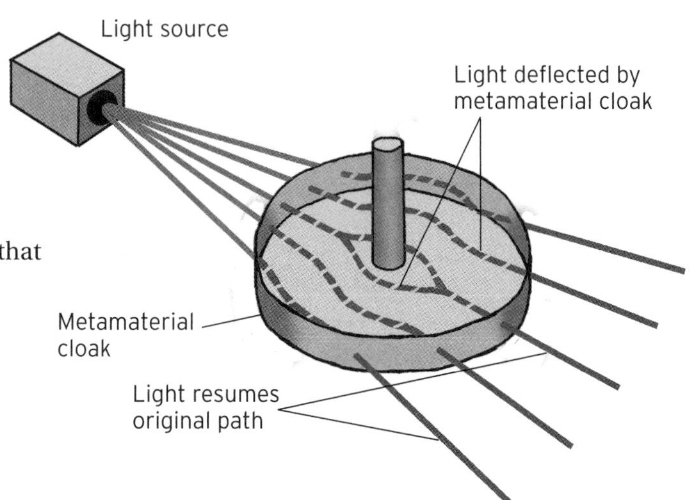

ENERGY GENERATION

Coal

Perhaps the oldest form of energy generation is the combustion of flammable material to release heat, given off as chemical energy during burning. Use of coal as a fuel is thought to date back to the year 3000 BC. Coal supplies half of America's electricity-generation needs. Power plants normally use coal that has been crushed to a powder to fire burners that heat water into steam, which then spins turbines to drive electrical generators.

Most coal in the United States is extracted from open-cast 'surface mines', which contrasts with the rest of the world, where much mining is sub-surface – requiring the digging of mine shafts down to coal seams. Along with oil, coal-fired energy production is a huge source of the greenhouse gases that are thought to be driving global warming today – accounting for a third of energy-related CO_2 emissions in the United States. For this reason, many see elimination of coal power as a key step in avoiding catastrophic **climate change**.

Petrol

The principle power source for transport is petrol, a fuel derived from oil in a chemical engineering process called fractional distillation. Like coal, oil is a fossil fuel extracted from underground. Formed from the remains of animals and plants that have been crushed by great temperature and pressure, the resulting liquid seeps into the pores of sedimentary rocks. Oil companies get at these deposits by drilling down into the ground and then piping the oil to the surface.

There is concern that oil deposits may one day, perhaps as little as ten years away, run out – a scenario known as '**peak oil**'. In addition, fumes from burning petrol in cars are a major source of a pollution, including the CO_2 responsible for global warming and climate change. For these reasons, there is much research into alternative fuels for motorcars. These include electricity and **biofuel**.

Nuclear electricity

Despite a number of high-profile accidents, nuclear power is regarded by many as the cleanest practical energy generation method at our disposal – arguing that renewables, such as solar panels and wind turbines, cannot meet demand. Nuclear power plants have also been used in naval ships, and in the future may even drive spacecraft. Nuclear power comes in two forms: **fission and fusion**. Existing nuclear generation systems operate on the principle of fission. Much of the energy given off is released as heat, which is used to turn water into steam and drive turbines.

Fusion, on the other hand, is the power source that drives the Sun but it is not yet established as a practical technology. This is mainly because it requires an extremely high temperature – 1 million degrees Celsius (1,800,000°F) – and no one has yet figured out how to confine the resulting plasma without putting in more energy than comes out. Magnetic confinement is perhaps the most promising approach, and the International Thermonuclear Experimental Reactor (ITER), currently under construction at the Cadarache research centre in France, will investigate this further.

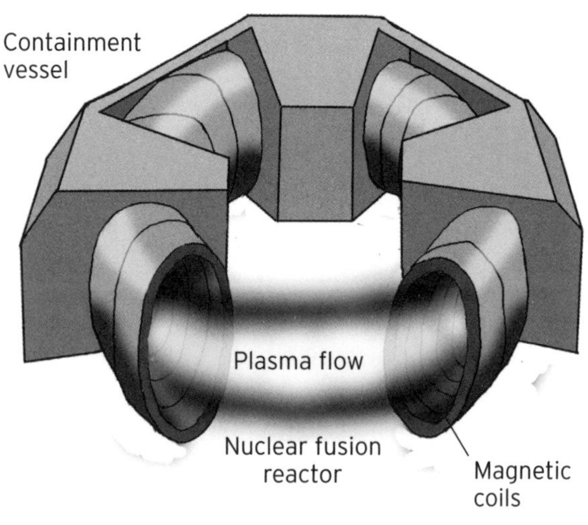

Wind turbines

Go for a walk by the sea and you'll get a first-hand demonstration of the power that's locked up in the wind. Wind turbines are a way of harnessing this to generate electricity. Looking a bit like a modern-day windmill, they consist of a set of rotary blades set spinning by the wind which turn a generator to produce **electric current**.

Wind power isn't without its critics, who argue that it cannot provide enough energy to meet demand, and that wind turbines are a blot on the natural landscape. New research seems to address these concerns, however – suggesting that the most effective place for turbines is offshore. Here, out of sight and with strong sea winds to turn their blades, it's estimated that a global network of wind turbines could crank out over five times the world's energy needs. Wind power is an indirect form of solar energy. Earth's weather systems are stirred up by the energy from the Sun; by comparison, distant planets have relatively quiescent atmospheres with little wind.

Sea power

The gravities of the Moon and Sun raise tides on the surface of the Earth; the powerful flows of water that these create can be harnessed to produce electricity. Undersea turbines resembling windmills turn in response to tidal currents and generate power in the same way as a wind turbine. But other kinds of sea-power generators exist: crossflow turbines resemble waterwheels, with paddles that catch on the flow of water; and helical turbines that use two wing-like hydrofoils entwined around one another to drive motion from the water's flow.

Stranger designs include serpent-like floats that extract power from the surface of the water as it undulates with the passage of waves. There are plans for tidal power projects under development at a number of locations around the world – including San Francisco Bay and England's River Severn.

Solar panels

The Sun gives out a colossal amount of energy per second, 4×10^{26} watts (in **scientific notation**). If we could harness even the small fraction of this falling on the Earth it would feed our energy needs 10,000 times over. With increasing emphasis on renewable energy, interest in solar power is growing.

It can take one of two forms: one is solar thermal energy; the other – photovoltaic cells, or solar panels – are devices that produce an electric current in response to photons of sunlight falling on them. These are based loosely on the photoelectric effect, discovered by Albert Einstein, and are a convenient way to harness the Sun's energy, which is why some enterprising people have installed them on the roofs of their homes as a form of **micropower generation**. The ultimate solar panel would be a **Dyson sphere**, allowing humans to harness every watt of power our star gives out.

Solar thermal energy

As well as solar panels there's another class of solar power generators, known as solar thermal energy. Solar concentrators use arrangements of mirrors to focus the Sun's heat to boil water, so generating steam that is then used to drive turbines. Some solar concentrators use molten salts to store the heat energy they gather. A number of solar concentrators are now in action around the world, including a large experimental set-up in the Mojave Desert, California.

Other devices use flat panels through which water is pumped via a network of pipes – rather like a central heating radiator. As the panels are exposed to sunlight, the water is heated. These systems are suitable for home installation – for example, to heat water.

Geothermal energy

Geothermal energy works by pumping water down to hot rocks beneath the surface of the Earth, turning the water into steam that then blast back up to drive turbines. Most geothermal wells are less than 3km (1.8 miles) deep, but this is enough for the temperature to rise by nearly 100°C (212°F). The internal temperature of the Earth is mainly due to heat left over from the violent process of **planet formation** and the **radioactive decay** of unstable elements.

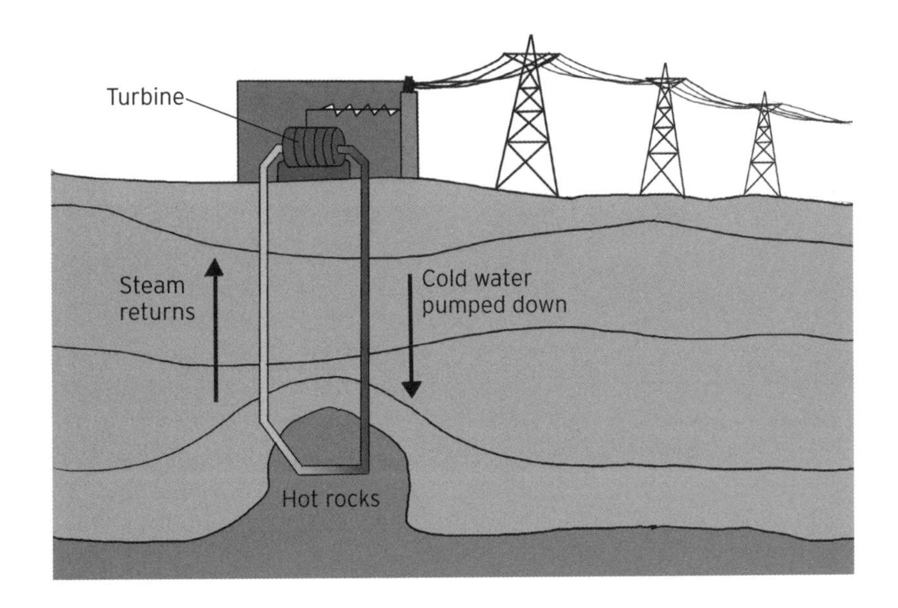

Geothermal heat pumps

Individual homes can take advantage of **geothermal energy** – for both heating and cooling. It's done using a heat pump, a device that can transfer heat from one place to another, and does so against the temperature gradient – for example, transferring heat from a cold place to a hotter one, like a refrigerator.

You can get an idea how this works by imagining a piston full of room-temperature gas. Push in the piston rod and the gas is compressed, making it hotter. Now take the piston outside to cool down – so it dumps its heat into the atmosphere. Now bring it back inside and pull the piston rod back – the gas will expand and cool, to much lower than the temperature of the room. Homes implementing this idea use a network of pipes in the ground underneath the house – called a 'heat exchanger'. Water is pumped through the pipes to transfer heat either to or from the ground – depending on whether it is summer or winter, respectively.

Micropower generation

With the availability of small-scale, affordable solar panels, solar thermal generators, wind turbines and geothermal heat pumps, many home owners are making their own electricity on-site. The proces is known as micropower generation. Most people who do it aren't entirely dependent on their home-grown electricity but simply use it to supplement their supply from traditional utility companies. Some utility providers even allow homeowners to feed their power back into the grid, allowing them to sell surplus electricity – and thus lower their bills still further, which is known as 'net metering'.

Another trend that is catching on is 'cogeneration', whereby rather than using a gas boiler simply to heat water, the hot exhaust gas – normally vented to the environment – is also harnessed and used to drive a turbine for electricity. Such devices are known as combined heat and power (CHP) boilers.

ENERGY STORAGE

Flywheels

Perhaps the simplest way to store energy is using a heavy, rotating mass known as a flywheel. Energy is put into the flywheel by spinning it – turning the shaft on which it revolves. Because of its large mass the wheel keeps spinning and loses very little energy through **friction** or **viscosity**. Now connect the flywheel to a mechanical system or generator and its energy can be used to do work, or generate electricity.

A 100kg flywheel spinning at 3,000rpm holds enough energy to light a 100-watt light bulb for about half an hour. Trains use giant flywheels weighing thousands of kilos in their 'regenerative' braking systems. As the train slows down, its energy of motion is transferred to the flywheel, which can then be harnessed to get the train moving again. Rotating planets act as superscale flywheels which, with virtually no resistive forces in the vacuum of space, spin for billions of years.

Water reservoirs

Reservoirs are another way of storing energy mechanically – by lifting a mass of water to a greater height in the Earth's gravitational field. Raising water up a hill takes energy – you either have to pump it or carry it in order to get it to the top. If you let the water flow back down then you get the energy back in the form of a fast-flowing body of water at the bottom of the hill which you can harness – using, say, turbines in a dam.

Wily utility companies use electricity to pump water to high elevation during off-peak hours when it's cheap; they can then tap into these reservoirs to generate extra power at times of high-demand, when the electricity they produce is more expensive – earning them greater profit.

Batteries

Ipods, cellphones and laptops would be nothing without a compact, lightweight source of electricity – the battery, which works by converting chemical energy into electricity. Usually two electrodes, made of different metals, are separated by a chemical solution known as an **electrolyte**. A chemical reaction takes place between one of the electrode metals and the electrolyte that liberates electrons. Meanwhile, at the other electrode the opposite reaction happens, adding electrons. The net effect is to make one of the electrodes negatively charged and the other positive, so that when the battery is connected to an electrical circuit an electric current flows.

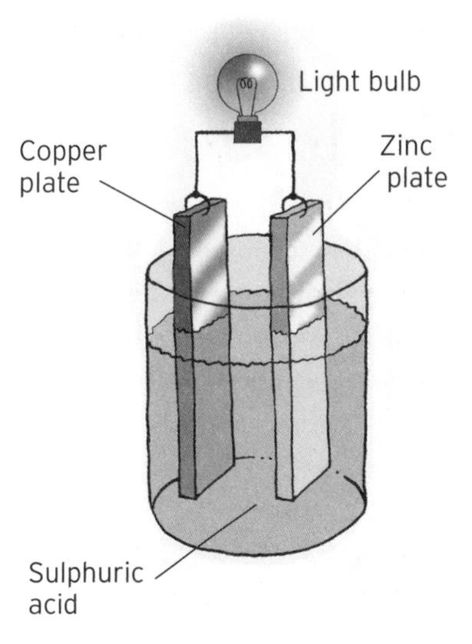

Some batteries have liquid electrolytes, such as sulphuric acid, but more modern batteries use dry electrolytes such as lithium. Rechargeable batteries use a combination of electrode metals and electrolytes so that applying an electric current across the electrodes reverses the chemical reaction and 'recharges' the battery.

Biofuels

Plant-based combustible fuel to power motor cars and generators is known as biofuel. Biofuels come in various forms. Alcohol-based fuels are made by fermenting plant matter and then distilling off the ethanol – a strong spirit-based fuel. Another type are the biodiesels, which are made from oil-rich crops, such as palm and rape, and can be burned in modified diesel engines – hence their name.

Burning biofuel does not contribute to climate change because the carbon emitted was already part of the natural carbon cycle – the plant simply absorbed it from the atmosphere while it was growing. On the other hand, burning fossil fuels such as petrol is harmful because it releases extra carbon that was previously locked away underground. There is, however, concern that the expansion of agriculture needed to grow the required quantities of biofuel is destroying some natural habitats.

Synthetic hydrocarbon fuel

In addition to biofuel, another way to produce fuel for cars that doesn't add extra carbon to the environment is to simply make it in **chemical engineering** plants. For example, passing an electric current through water releases hydrogen, which can then be combined with CO_2 in the atmosphere to create flammable methane gas. If the electricity used in the first step comes from renewable sources this is an environmentally friendly process.

It could even help in space exploration. A big problem in getting humans to Mars is the weight of the huge quantity of fuel the astronauts would need to take with them for the journey home. One solution is a scheme whereby humans could manufacture a large amount of methane rocket fuel in situ on Mars, by taking just a small tank of hydrogen with them and combining it with the abundant CO_2 in the Red Planet's atmosphere.

Hydrogen

Hydrogen fuel cells are a potential power source for motor cars that work by combining hydrogen with oxygen from the atmosphere to produce electricity and a clean by-product: water. Hydrogen must first be manufactured via the reverse process. In this way, hydrogen effectively becomes a way of storing electricity. If this electricity comes from clean, renewable sources, such as wind turbines, then hydrogen power is an environmentally clean technology.

Its success relies on car manufacturers developing vehicles that can give acceptable performance from an electric motor. And they're achieving this goal. In 2008, the Tesla Roadster, a sports car that runs on lithium batteries, went into production. A full charge of the battery gives the car a range of nearly 400km and costs just $5. Meanwhile its performance is blistering, accelerating from 0 to 100km/h (60mph) in just 3.7 seconds. The only drawback is that a full charge takes 3.5 hours. Replacing the batteries with hydrogen tanks would make recharging as convenient as refilling with petrol.

Superconducting coils

Superconductivity is a property observed in some metals, whereby their electrical resistance drops to zero when they are cooled to low temperatures, around -270°C (-454°F). This means that a current introduced into a loop of superconducting wire will continue to circulate forever – it's stored.

In practice, the cost of keeping superconductors cool eats into the efficiency of this technique, making it only really viable for storage over short periods – such as buffering excess electricity generated at night to use as a top-up to meet daytime demand. However, research continues into room-temperature superconductors. These would remove the need for refrigeration and make superconducting coils a viable form of long-term energy storage.

COMMUNICATION TECHNOLOGY

Alphabet

The most simple invention that we use to help us communicate is the alphabet – the set of 26 letters that enable us to write words and sentences, which can then be communicated over distance by post or email. The first written script was probably the hieroglyphs of the ancient Egyptians, their use beginning nearly 5,000 years ago.

Written English evolved from runic Old English in the 5th century AD. One of the most significant works of literature penned in Old English was the epic poem Beowulf, written in the 8th century.

Beginning in the late 11th century, Old English evolved into Middle English, following the Norman conquest of Britain. Modern English, in a form that most people would recognize, has been written since around the mid-1500s. The first English dictionary was published in 1603.

Telegraphy

With an alphabet in place, the increasing understanding of **electricity and magnetism** in the 19th century brought a new way to convey the written word quicker than the mail. Telegraphy was born. The basic idea was that information can be encoded as pulses of current sent down a wire stretching between, say, two towns, enabling virtually instantaneous communication between them.

A •–	J •–––	S •••	1 •––––
B –•••	K –•–	T –	2 ••–––
C –•–•	L •–••	U ••–	3 •••––
D –••	M ––	V •••–	4 ••••–
E •	N –•	W •––	5 •••••
F ••–•	O –––	X –••–	6 –••••
G ––•	P •––•	Y –•––	7 ––•••
H ••••	Q ––•–	Z ––••	8 –––••
I ••	R •–•		9 ––––•
			0 –––––

Early telegraph designs used multiple wires to send signals, each corresponding to different letters of the alphabet. However, in 1837, Samuel Morse, a painter and part-time inventor, worked out how to send messages down a single wire. He came up with a code where each letter of the alphabet was represented by a unique sequence of dots and dashes (short pulses and long pulses, respectively). The first message using this 'Morse code' was sent in 1838, by tapping out dots and dashes with a hand-operated electrical contact. It soon caught on and the first transatlantic telegraph cables were laid in 1866.

Telephones

The next step on from telegraphy was the telephone – a device that enabled spoken words to be transmitted using electricity. The patent for the invention of the telephone was granted to Scottish scientist Alexander Graham Bell in 1876.

A telephone's mouthpiece has a microphone inside – a device that uses vibrations of air caused by sound to vibrate a magnet inside a coil of wire, which then generates a current in the coil by **induction**. The current varies in accordance with the vibrations. The varying current is then transmitted down a wire to the telephone at the other end, where it's fed to a speaker in the earpiece – effectively the reverse of a microphone, which translates the electrical vibrations back into sound so they can be heard. Calls are routed via a telephone exchange, which uses the number dialled by a caller (transmitted as a series of tones) to identify the receiver and send a current to their phone, causing it to ring.

Radio

Telegraphy and telephones were one thing, but wouldn't it be great if you could lose the wires? Scientists achieved just that in the late 19th century. The work of a number of brilliant minds including those of Guglielmo Marconi, Nikola Tesla and Thomas Edison led to the first demonstration of so-called wireless telegraphy in the 1890s. Quite who got there first is controversial, but a number of groups around the world all succeeded in beaming messages over short distances.

Radio works by piggybacking information on **electromagnetic radiation** broadcast from a transmitter. The transmitter has a large metal structure – the aerial – which, when an electrical current is passed through it, emits radio waves. The current varies in time according to the signal being transmitted – say, a piece of music – and the variation is imprinted on the radio waves given off. A receiver then uses a similar aerial to pick up the waves, and convert them back into an electric current, which is then fed to a speaker – from which the music can be heard.

Cellphones

Put the radio and the telephone together and you get the cellphone. In 1947, two engineers at AT&T's Bell Labs, New Jersey – Douglas H. Ring and W. Rae Young – put forward the idea of portable phones that were small and light enough for personal use. Their key innovation was to divide the signal coverage area into hexagonal 'cells' served by a network of low-power transmitter/receiver base stations to relay calls between each 'cellphone' and the main landline network.

NTT launched the world's first commercial cellular network in 1979, serving customers in Japan. The first cellphone networks sent and received analogue signals – messy streams of data that were plagued by background noise. In the early 1990s, cellphone technology made the jump to digital, encoding signals as a stream of **binary data**. The switch to digital also meant that phones could now send and receive any kind of data, not just sound. So text messaging was born, and picture and video messaging quickly followed. As much a fashion statement as a functional technology, the cellphone has now evolved into perhaps the most pervasive and iconic gadget of the 21st century.

Satellite phones

The discovery of **Clarke orbits** made it possible for satellites to hover over the surface of the Earth, bouncing radio signals from one continent to another, and making possible the ultimate cellphone.

Satellite phones are similar in size to a cellphone of the late 1990s, with a prominent antenna protruding from the handset. Using a satellite phone is slightly less straightforward than operating a cellphone, as they rely on the user being able to obtain a clear line of sight to the satellite in the sky. Tall trees and buildings can hinder this, but in remote areas it's not a problem – making sat phones especially suited for exploration or disaster relief in regions where there is no cellphone signal.

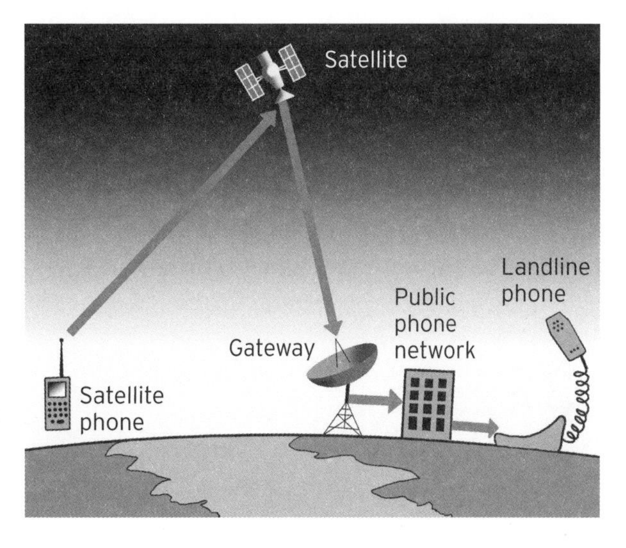

Fibre optics

Telegraphy works by sending signals on wires; radio works using electromagnetic waves. Fibre optics are yet another method, which sends information by encoding it on pulses of light beamed down the middle of long, thin lengths of plastic or glass, known as optical fibres.

Binary data can be transmitted at speeds of up to 40 gigabits per second in these communication systems, along a single fibre sometimes spanning thousands of kilometres. As well as long-distance data transfer, fibre optics also has a role to play in piping data over small-scale networks such as in offices – with the large data rate per fibre keeping bulky cables to a minimum.

VOIP

Broadband **Internet access** makes it possible for people to exchange electronic sound files as easily as they can exchange voice signals over the traditional telephone networks. Websites such as Skype take advantage of this by offering users a service whereby they can place calls both to other Skype users, and to main network telephone numbers, via the Internet. Calls between Skype users are free of charge, and those to telephone numbers are typically cheaper than using your telecom provider – especially when calling internationally.

This whole practice is known as 'voice over Internet protocol', or VOIP. Internet protocol is just the set of formal rules by which information is exchanged over the Internet. VOIP just means using these rules to send voice data. Initially, VOIP users were forced to plug headsets into their computers in order to make calls. But now many companies offer dedicated VOIP phones – telephones that look identical to an ordinary handset, but which place calls across cyberspace.

Instant messaging

Instant messaging enables Internet users to type messages to friends and reply to them instantly – it's essentially about using the Internet to hold a text conversation in real-time. Internet chat rooms work in a similar way to instant messaging. Though rather than appearing as a single message, as in an email, in a chat room each new line of dialogue appears beneath the previous one in a list that scrolls up the screen as the conversation proceeds.

The ease of exchanging a basic stream of text correspondence means that chat is a simple application to run online – which is why enthusiasts have been chatting online since as far back as the 1970s. Today, chatters don't even need to access remote websites; they run self-contained chat software applications on their computers, which then link up over the Internet with similar software being run by their friends.

Video chat

In the late 1990s high-speed Internet access allowed chat rooms to begin including video feeds. Two people, chatting on other sides of the world, could have a virtual conversation and see each other at the same time – this is video chat. It works using a webcam, which plugs into your computer and captures frames of video which are then relayed via the Internet. Systems like this haven't just been used socially; businesses now use souped-up versions of this set-up to conduct videoconferences – holding virtual meetings across cyberspace.

Videoconferencing cameras are typically higher resolution than ordinary webcams, allowing high-definition pictures to be transmitted. Top-spec videoconferencing systems occupy a whole room and feature a normal conference table with seats on one side and large screens on the other – making it seem like your absent colleagues really are in the room with you. Internet video is even used by some TV news organizations to transmit pictures from locations where other methods are not possible – for example, in war zones.

Smartphones

It wasn't long after the widespread uptake of cellphones that the devices became able to do more than just make phone calls. Texting became available, soon followed by picture messaging and later video messaging. New handsets can do even more – and these are the smartphones. There is no strictly accepted industry definition of what a smartphone is, but most function as miniature computers. They are able to run a cut-down 'operating system' – the software that manages a computer's activity, like Windows or Mac OS. And this means that smartphones can host all manner of applications.

Modern smartphones offer email and Internet access – in other words, **mobile web** – but also mini software applications, or 'apps', that users can download from the phone manufacturer's website. There are apps for microblogging, currency converters and even software that lets you use your phone as an impromptu spirit level. Some phones also come with **sat nav (GPS)**. The first smartphones were released in the late 1990s, though it wasn't until the first years of the 21st century that they really came into their own.

MILITARY TECH

Radar

Not all radio waves are the sort that you want to receive – if you're a military pilot flying over enemy terrain, you'll be trying your best to dodge enemy radar. It's an acronym for 'radio detection and ranging', and consists of radio waves beamed across the sky from a broadcasting station. An aircraft in the path of the radar beam reflects the waves back to their source, revealing the aircraft's presence as a blip on the radar system's display screen. Once a plane's position is known, fighter interceptors or surface-to-air missiles can be despatched to bring it down.

Radar was first implemented successfully during the Second World War, when the British used it against German air raids – as an early warning system and to direct searchlights and thus guide anti-aircraft fire. It is now widely used in both military and civilian aviation, shipping and even in space exploration. A spacecraft orbiting a planet can use the timing of radar pulses bouncing off the ground below to map out its topography.

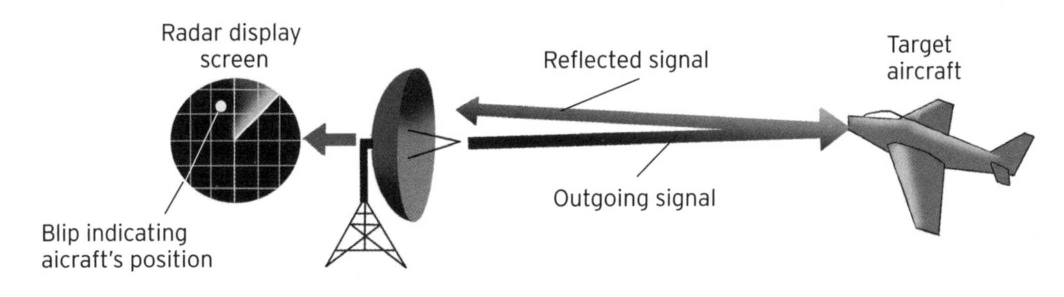

Sonar

Sonar is the underwater equivalent of radar and works using **sound waves** instead of radio. Its primary use is in submarine warfare. Ships use sonar to listen for nearby submerged contacts, while subs themselves use it to detect approaching vessels both above and below the waves. Early sonar development was carried out by British scientists to detect icebergs in response to the sinking of the *Titanic* in 1912. But in 1916, the technology was commandeered for the war effort to help British destroyers combat the threat posed by German U-boats.

American scientists took the concept further during the Second World War and coined the modern name of sonar (short for 'sound navigation and ranging'). Sonar can be used in either passive or active forms. Passive sonar involves simply listening for the sounds given off by a moving vessel – modern passive sonar can even tell different types of ship from their acoustic signature. Active sonar involves the listener sending out 'pings' of sound and then listening for the echo as they come back. Active sonar is the more effective of the two, but the drawback is that you risk disclosing your location to a potentially better-armed foe.

Stealth

It looks like some kind of alien spacecraft, an ominous black triangle powering skywards with its engines screeching. But the B2 stealth bomber is an example of one of the most sophisticated military technologies developed by humans: stealth. The basic idea is to make a plane hard to detect – by sight, sound, radar or heat. To this end, stealth aircraft incorporate camouflage technologies to make them hard to spot. Cool air is added to the engine exhaust to mask its heat signature, and the plane is jet black, making it virtually impossible to see in the dead of night.

But most important is radar stealth. The plane's shape is carefully designed so as not to **reflect** any radar beams back to the ground – there are no sharp edges and all weapons are mounted internally. Meanwhile, radar-absorbent paint soaks up as much of the incident radar as possible. All things considered, the B2 has about the same radar signature as an aluminium marble. Researchers are now experimenting with fitting stealth technology to other vehicles, such as ships and tanks.

Nuclear weapons

Nuclear weapons work by letting the controlled **nuclear reactions** that take place in nuclear power plants – nuclear **fission and fusion** – to proceed unchecked, leading to a colossal release of energy: a nuclear explosion. The first nuclear bomb, a fission device that released the equivalent energy of detonating 15,000 tons of TNT, was exploded in 1945 over the Japanese city of Hiroshima.

Yet this was nothing compared with what was to follow. During the years following the war, scientists in both the East and the West worked on harnessing the far more potent force of nuclear fusion. The first fusion device, or 'hydrogen bomb', was exploded by the United States in 1952; it had an output equivalent to 10.4 megatons (millions of tons of TNT) – nearly 700 times the power of the Hiroshima blast. The biggest device ever detonated was the Russian Tsar Bomba in 1961, which created a 57 megaton explosion.

Nuclear weapons deliver damage in three principle ways: first, a flash of intense thermal radiation lights fires; next a shock wave from the epicentre of the blast destroys structures; and finally, fallout – radioactive ash from the explosion – rains down over a wide area, poisoning all life.

Biological warfare

Any use of dangerous diseases or biological organisms as a weapon is an example of biological warfare. It ranges from dropping venomous snakes on your enemies to releasing anthrax in their cities.

During medieval sieges in Europe, bodies infected with bubonic plague were flung into enemy strongholds using war catapults. There are reports of biological agents being used during the Second World War, and in the Korean War where North Korea accused the United States of seizing on the war as an opportunity to road test its biological arsenal.

Many nation states have now signed the Biological and Toxic Weapons Convention of 1972 banning the development of biological warfare agents. Today, the biggest concern is over their use by terrorists. In 2001, packages containing anthrax spores were sent to a number of American newspapers and political offices, killing five people.

Chemical warfare

Chemical warfare is the art of poisoning the enemy using toxic substances. History is replete with stories of warring factions having their supplies of water and food poisoned by the enemy. Its first high-profile use was in the trenches of the First World War, where toxic chemicals such as chlorine and mustard gas were exchanged by both sides. The effects were horrific; mustard gas causes hideous and painful burns to the skin – and to the respiratory system if inhaled – and can be fatal.

Use of poison gas during the Second World War was more restrained, although the Allies kept stockpiles for fear the Axis powers would use them. But there was a more deadly chemical threat brewing: nerve gas. Invented by the Germans in 1936, nerve agents disrupt the human nervous system, by inhibiting the chemical processes upon which it operates, leading to death by suffocation as the lungs cease to function.

Although it has never been used in war, terrorists used sarin nerve gas to attack the Tokyo subway in 1995.

Spy satellites

Imagery intelligence (or IMINT) is the military name for using cameras to gather intelligence on enemy activities. And spy satellites, **artificial satellites** fitted with high-resolution cameras, are an extremely effective way of doing this. During the first Gulf War, satellite imagery gave allied forces a week's warning of the Iraqi invasion of Kuwait, clearly showing troops massing on the Kuwaiti border. Satellites are able to image the ground below in stunning detail – showing up details as fine as 10cm (4in) and even smaller.

The United States has been developing spy satellites since the 1950s, and has been fielding them in space since the 1960s. Most of these craft employed visual imaging capabilities – but this is no use when the target is shrouded by cloud. A solution in this case is to use **radar**. So-called synthetic aperture radar (SAR) works by capturing a number of radar images as the satellite flies over, and then combining the images gathered in a computer to make a single radar image that rivals the resolution of visual images – and which can be see through cloud.

The Lacrosse satellites were the first American spy sats to incorporate SAR, entering service in 1998. They occupy low Earth orbits and can be seen from the ground as they fly overhead.

Space weapons

In February 2008, a missile fired from an American Navy cruiser shot down USA 193, a defunct satellite. The satellite was already out of control and falling back to Earth, so destroying it removed a potential threat to those beneath its path. But the downing of USA 193 was also a key test of a relatively new addition to America's offensive arsenal: space weapons. Western nations have long relied on space-based systems, such as spy satellites, to give them the edge in conflict situations. But now the days of satellites simply as observers are over, as we gain the capability to involve space-based systems in the actual shooting war.

Over the coming years, a frightening array of weapons could be stationed in Earth orbit. These include lasers, particle beam cannons (which project streams of high-energy subatomic particles), and so-called 'rods from God' – metal rods the size of telegraph poles that can be thrown javelin-like from satellites towards the ground, striking with the force of a small nuclear weapon.

UCAVs

UCAVs, or unmanned combat air vehicles to use their full title, are fighter/bomber aircraft that have no onboard pilot – instead, they are flown remotely from the ground. This practice not only saves risking the lives of pilots – it's also cheaper. UCAVs are typically much lighter than a full-scale crewed aircraft and can stay in the air for longer intervals. Not only are they an alternative to long-duration patrols, but they can also be used on single-strike missions – taking off from a carrier to deliver a large munitions payload to a target. In this sense a UCAV can be thought of as rather like a reusable cruise missile, again reducing costs.

The first UCAVs were helicopter drones flown from destroyers to drop torpedoes on enemy submarines in the 1960s. The idea fell from favour for several decades before returning in the 1990s with the benefit of new technologies – in particular, stealth. Some military researchers are even examining the idea of removing humans from the loop entirely to create autonomous robot warplanes. However, many experts in artificial intelligence have voiced their opposition.

Cyber warfare

With the infrastructure of every nation increasingly reliant on information and the Internet, military planners have realized the value of conducting attacks, not with bombs and bullets, but across cyberspace. Cyber warfare can mean using the Internet to spread propaganda, intercepting electronic communications, or slipping 'malware' applications past enemy **Internet security** to disable key infrastructure targets such as energy generation systems, communications, and banking.

One particularly effective form of cyber warfare is known as the denial-of-service attack (DoS), where a massive number of computers all attempt to access the same website at once. The webserver is unable to cope with the volume of traffic, making it virtually impossible for anyone who genuinely needs to use the website to do so. Many of the computers involved are owned by civilians but get hijacked by software agents known as bots, that roam the Internet attacking vulnerable computers. In 2007, a DoS attack against Estonia brought government websites, banks and online news sites to a standstill.

How a denial-of-service attack works

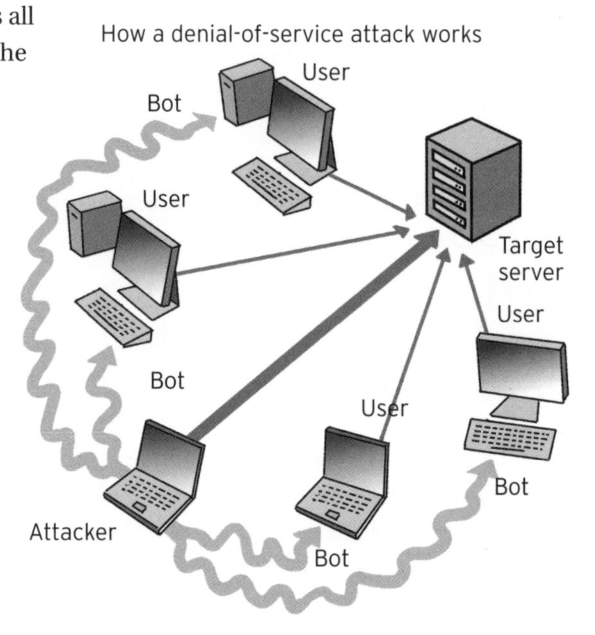

GENETIC MODIFICATION

Transgenics

The practice of taking **genes** from one organism and inserting them into the **DNA** of another is known as transgenics. The new genes are cut and pasted into place using recombinant DNA techniques. In plants, a basic kind of transgenics is the traditional method of cross-pollination, where pollen from one species is used to fertilize the flowers of another, which then produces seeds that are a cross between the two.

This is a rather haphazard process though. Advanced techniques make more deliberate changes possible. In plants this is done by soaking slivers of plant material in a bacterium that contains the new DNA. This new DNA works its way into the plant cells, from which transgenic plants can then be cultured.

In animals, scientists splice the new DNA into an embryo which is then implanted into a mother animal and brought to full term – yielding an animal that has the new genetic make-up. Transgenics is the empowering technology for genetically modified organisms, for instance transferring human DNA into cows so they can produce human milk for babies.

Genetically modified organisms

Genetically modified organisms (GMOs) are life forms that have had their genetic make-up altered by transgenic techniques. The field began in the 1970s, when bacteria were engineered to express genes spliced in from other strains. In 1978, the first application of this technology emerged when scientists created a strain of E. coli bacteria that was tweaked to manufacture insulin, the drug used to treat **diabetes**. GMOs made headlines in the 1990s, when biotechnology companies tried to market GM food. Today, genetic modification has been overtaken by synthetic biology.

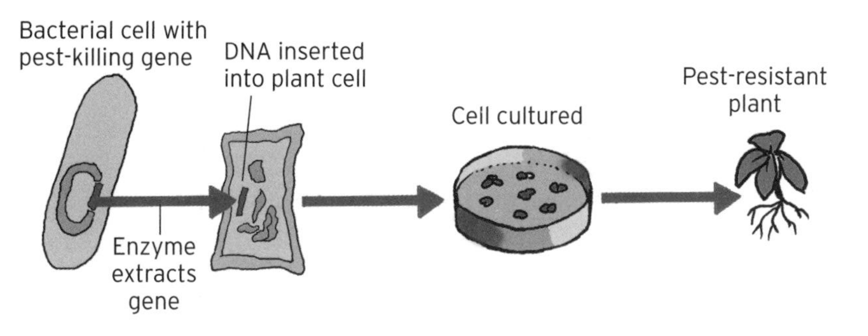

Synthetic biology

Genetic modification in its most elementary form deals with tweaking the odd gene here and there to introduce new properties into organisms. The field of synthetic biology has much grander ambitions – literally engineering entire lifeforms from the ground up. It's transgenics taken to the extreme.

Synthetic biology scientists take the building blocks of life – chunks of genetic material – and bolt them together to produce something never before seen in nature. The process is a little like building models from Lego. One popular way of doing this is using freely available genetic blocks known as BioBricks. These are put together to form organisms that perform useful functions – like growing clean biofuels, or seeking out and destroying cancers. Completed BioBrick creations are then made available to other scientists to incorporate into their own work. The whole process can be thought of as a kind of **open source** biology.

Reporter genes

Reporter genes are used in genetic modification, in particular synthetic biology, as a marker to reveal when a transgenic gene has made its way into an organism. A common choice is the gene which codes for the green fluorescent protein found in some jellyfish. The reporter gene is inserted next to the gene that's being modified so that when that gene is activated, the reporter gene is activated too – causing the organism to glow.

Scientists use reporter genes to check that a new piece of transgenic DNA has been inserted into an organism correctly. They're also used as genetic indicators in synthetic biology; for example, certain bacteria cells only become active above a certain temperature. Add a fluorescent reporter gene next to the DNA coding for this heat sensitivity yields cells that will glow, but only if their temperature has ever been raised above the minimum threshold. This can serve as a safety marker, for example, on the packaging of chilled, perishable foods.

Genetic pollution

One of the big objections to the development of genetic modification is the possibility that artificial organisms may 'escape the lab' – or even escape the field where they are being grown, in the case of genetically engineered crops, introducing their new DNA into the wild with potentially devastating results. Proponents of genetic modification argue that their experimental organisms are usually too feeble to survive in the wild. However, some researchers have fitted their creations with terminator genes to ensure they cannot breed.

Other critics worry that the tools of synthetic biology could be abused if the genetic sequence for a dangerous disease fell into the wrong hands. In 2005, there was controversy over the decision by the journal Science to publish the genome of the Spanish influenza virus that killed 50 million people worldwide during a pandemic in 1918. Critics argued that terrorists with inexpensive equipment could use this information to recreate the virus.

Terminator genes

DNA inserted into the genome of a genetically modified organism to render it sterile, and so stop it spreading in the wild, is called a terminator gene. In the 1990s, biotech companies expressed interest in introducing terminator genes into GM food crop seeds. The idea was to prevent farmers from harvesting seeds for free from the plants they'd grown – forcing them to buy new ones each season.

Strong opposition to terminator genes has prevented the technology being brought to market. As well as objections from farmers on financial grounds, many experts worried that if terminator genes jumped to natural plant species as a result of genetic pollution, it could inhibit growth or even lead to extinction of these species.

Pharming

Genetically modifying organisms to make them produce pharmaceutical drugs is a practice known as pharming. Common lifeforms to be pharmed are microorganisms, such as yeast, which do their work in vats – in a process rather like brewing wine or beer. This isn't the only approach though. Mammals are sometimes pharmed, their genetic make-up being reprogrammed to make them produce useful chemicals in their milk. For example, Massachusetts-based company Genzyme is using a herd of GM goats to produce anticoagulant drugs that treat the dangerous blood condition known as thrombosis.

Plants are used too, with genes added to make certain therapeutic **proteins** within them, which can then be extracted by chemical processes – in just the same way that, say, sugarcane is processed to extract the sugar. Canadian company Medicago is using genetically modified alfalfa to produce influenza vaccines.

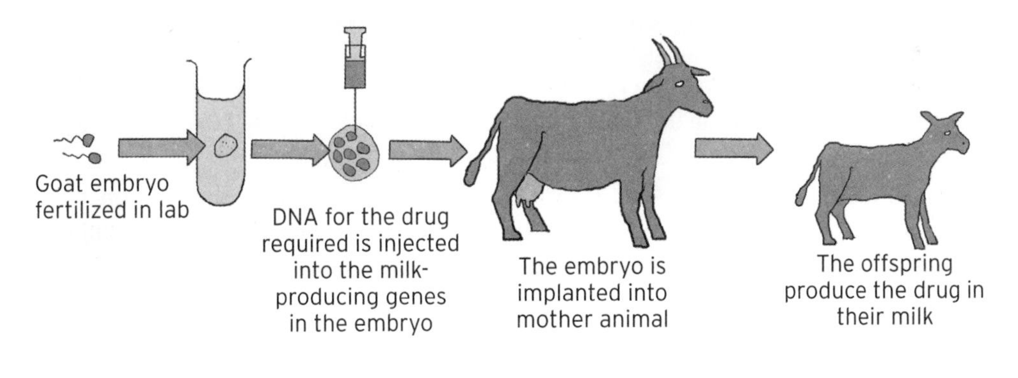

Gene doping

The use of drugs to enhance the performance of athletes generates headlines every time the Olympics come round – with athletes accused of using chemicals to boost their strength and stamina, so-called 'doping'. Gene doping takes this a step further, by using **gene therapy** techniques to enhance athletic performance. For example, a gene therapy called Repoxygen coaxes the body to produce a hormone known as EPO which increases the body's production of red blood cells. These are the cells that carry oxygen around the body, improving an athlete's aerobic capacity and giving them an edge in endurance sports such as cycling and long-distance running. There were allegations of Repoxygen use at the 2006 Winter Olympics in Turin. Athletics governing bodies are now trying to regulate gene doping in sport; however, scientists are only just beginning to develop reliable tests for it.

Conservation genetics

Using genetic techniques to save endangered species is a discipline known as conservation genetics. Scientists in this field study the gene make-up of organisms to gain a better understanding of the genetic diversity of life in the natural world – and so direct their conservation efforts toward saving species that are truly endangered. Genetic diversity is measured by the looking at the **alleles** of different genes. For example, the colour of a flower's petals are determined by different alleles of the same gene. Alleles in other genes, however, can be less obvious to spot. If a population all have the same alleles of a gene, then it has low genetic diversity, and the greater the number of genes for which the diversity is low, the greater the risk posed to the species.

Genetic diversity is what allows a species to survive upheaval. For example, if temperatures suddenly drop, a few individuals with genes that give them resistance to the cold will survive to keep the species going. But when diversity is low the existence of such individuals is less likely.

Eugenics

Eugenics is the idea of using science to guide the course of human evolution, with the aim of making the species stronger. Put that way, it might almost sound like a good idea; yet in reality that's far from the truth. It works by preventing the genes responsible for human characteristics that are perceived as weak from being passed on to future generations. At its worst, it led to Nazi Germany's attempt during the Second World War to eradicate the Jewish race from the human gene pool. This culminated in the mass extermination of millions of Jews in Nazi gas chambers.

But other nations, including the United States and Canada, conducted their own eugenics programmes during the early 20th century. While none of these were as extreme as Nazi Germany's, they were far from pleasant. They centred around compulsory sterilization, whereby those suffering from genetic diseases, mental illness – and, in some cases even those failing to score sufficiently high in IQ tests – were forced to undergo surgery or chemical treatment that prevented them from having children. Today, compulsory sterilization, and the eugenics movement as a whole, are considered abhorrent.

FOOD

Agriculture

The development of agriculture – farming – allowed early humans to relinquish their nomadic hunter-gatherer lifestyle and establish permanent settlements. The first farmers figured out how to cultivate the land and developed techniques for irrigation, allowing them to grow food crops. In so doing, they laid the foundations of modern civilization.

These pioneering people lived in the Far East during the 9th millennium BC. With the surplus crops not required for human consumption, they soon began keeping livestock – first as work animals, and later to be bred and raised for food. Long-distance sea travel in the 15th century enabled transfer of livestock and crop species from abroad – agriculture became global. The Industrial Revolution brought the benefits of mechanization in the 19th century. This was followed shortly after by the development of the first fertilizers, which massively increased food production. In the modern world, agriculture provides more than just food – **biofuels** could offer an environmentally friendly alternative to petrol and diesel oil.

Hydroponics

Hydroponic crops are grown in nutrient-rich water rather than soil. The nutrient levels in the growing liquid can be precisely controlled and made higher than the levels found in soil – meaning hydroponics can produce crop yields per acre many times that obtainable through conventional land farming. There are records of the ancient Egyptians growing plants in water, yet the breakthroughs that enable modern hydroponics came in the mid-19th century when scientists discovered the nutrients needed to keep hydroponically grown plants healthy.

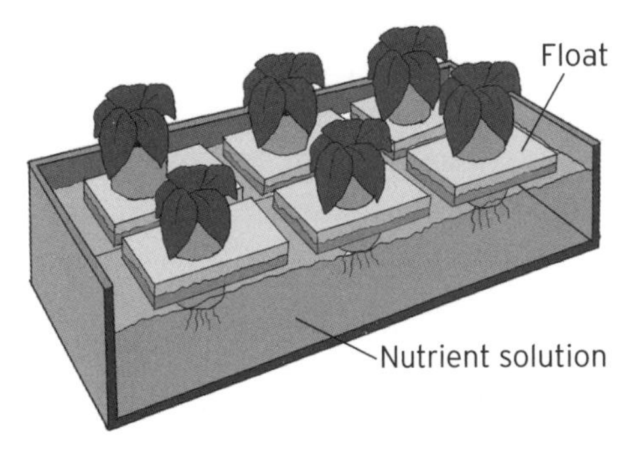

As well as increasing yields, hydroponics reduces the risk from pests and disease – because the plants are often grown in a sealed environment. For the same reason, the water in the system can be continuously recycled, which makes hydroponics a promising method for feeding astronauts on long-duration space missions, such as future crewed flights to Mars. A number of working hydroponic farms are in operation around the world today; the largest is Eurofresh Farms, in Willcox, Arizona, which spans 1.3 square kilometres. Home hydroponics kits are also available.

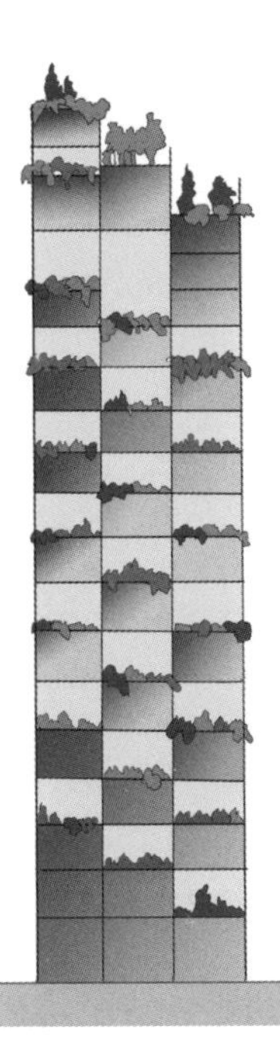

Farmscrapers

Feeding the world's population today requires farming a land area equal in size to the entire continent of South America. And as the number of people on the planet grows over the next 50 years that figure is set to rise by 40 per cent. One way the amount of land required could be reduced is by putting farms inside high-rise buildings, called farmscrapers. The idea is the brainchild of Professor Dickson Despommier, a microbiologist at Columbia University, New York. He says that a 30-storey farmscraper could provide fruit, vegetables, meat, fish, poultry – and clean water – for 50,000 people.

Crops are grown using **hydroponics**, fed by waste water from nearby urban areas. Plants give off water vapour from their leaves by a process called **transpiration** – and this clean, naturally filtered water can be collected for drinking. The buildings are powered by renewable sources, such as solar panels and wind turbines, and by burning waste plant material from finished crops in high-efficiency incinerators.

As of 2009, the world's first working farmscraper has yet to be built. Despommier believes he could have a prototype working within a few years, though it would require funding to the tune of $60 million.

Food science

The overarching field of science dealing with the production, safety and preparation of foodstuffs is known as food science. It includes areas such as **microbiology**, which is used to check for the presence of potentially harmful microorganisms in food, and so-called 'food engineering' which pulls together knowledge from agriculture and **chemical engineering** to develop processes needed for the manufacture of food.

It also deals with day-to-day tasks such as the processing, preservation, packaging and delivery of foodstuffs to consumers. Specific techniques include pasteurization, whereby milk is heated to reduce the number of harmful bacteria, and freeze drying, where foodstuffs such as coffee are dehydrated for preservation and ease of transport. In the United States, the main body for regulating food science procedures is the Institute of Food Technologists, founded in 1939.

GM food

The application of **transgenic** techniques to improve the quality of foodstuffs is an area known as genetically modified food, sometimes shortened to GM food. For example, by taking the genes from an Antarctic fish that stop it freezing in the icy polar waters, and inserting them into potatoes you get frost-resistant potatoes. The same technology yields pest-resistant crops, tomatoes with longer shelf lives, and foods with richer nutritional content. However, public reaction to GM foods has been extremely hostile, leading supermarkets to remove them from shelves.

Some scientists consider the failure of GM food to gain traction to be almost as dangerous as the threats perceived by its opponents. They say that in an increasingly populous world, where feeding every mouth is becoming even harder, we should think long and hard before rejecting technologies that promise to increase the nutritional content, shelf life and volume of food we produce.

Molecular gastronomy

Science not only comes into the production and delivery of food, but also its cooking and preparation – turning the traditional kitchen into a scientific laboratory. This area of research is referred to by scientists and chefs alike as molecular gastronomy. The field examines how all of the body's senses react to food and uses this information to design dishes that have not only a pleasing taste, aroma and appearance, but also an attractive texture – the so-called 'mouth-feel' of the food – and even **psychology** to enhance the diner's experience.

Much of molecular gastronomy was pioneered by Hungarian-born physicist Nicholas Kurti. Among his many food experiments, Kurti demonstrated how to make a perfect meringue in a vacuum chamber and used a microwave oven to create a dish known as 'Frozen Florida' – a reverse Baked Alaska, that's hot inside and cold outside.

One of the leading proponents of molecular gastronomy today is the British chef Heston Blumenthal. At his restaurant The Fat Duck, in Berkshire, England, diners can sample the delights of snail porridge, mustard ice cream, and green tea and lime mousse poached in liquid nitrogen.

Flavour chemistry

Flavour chemistry is a subset of food science and molecular gastronomy, and deals with the chemical compounds responsible for different flavours. For example, the characteristic taste and smell of almonds is due to a molecule called benzaldehyde, while the aroma of vanilla is down to 4-hydroxy-3-methoxy-benzaldehyde, and boiled beef is essentially 3,5-dimethyl-1,2,4-trithiolane. These chemicals can be engineered artificially, which is why beef-flavoured potato chips are often suitable for vegetarians.

Flavour chemistry also deals with undesirable flavour compounds and how they are produced, to ensure that none of these unwanted chemicals are created during food production processes. The mouth can detect five distinct flavours: sweet, salty, bitter, sour, and a savoury fifth taste called umami. Chemical taste enhancers are used in food to boost these components of flavour. They include salt, sugar, and mild acids, as well as monosodium glutamate (MSG), which enhances umami.

In-vitro meat

What if we could grow slabs of meat in the laboratory, removing the suffering of livestock reared for food? This process, known as in-vitro meat, works in roughly the same way as growing yoghurt and yeast cultures. A small sample of animal meat would be taken from a living creature and then grown in a liquid solution containing all the nutrients it needs. Crucially, the cultures grown in this way have no nervous system (see **Neurobiology**), and so feel no pain. In addition, they would generate no harmful greenhouse gases, such as the methane produced by cattle, take up less land area and require less food than living, breathing animals. Like **hydroponics**, in-vitro meat is a technology that has been examined by space agencies, including NASA, with a view to using it for food production on long-haul space missions.

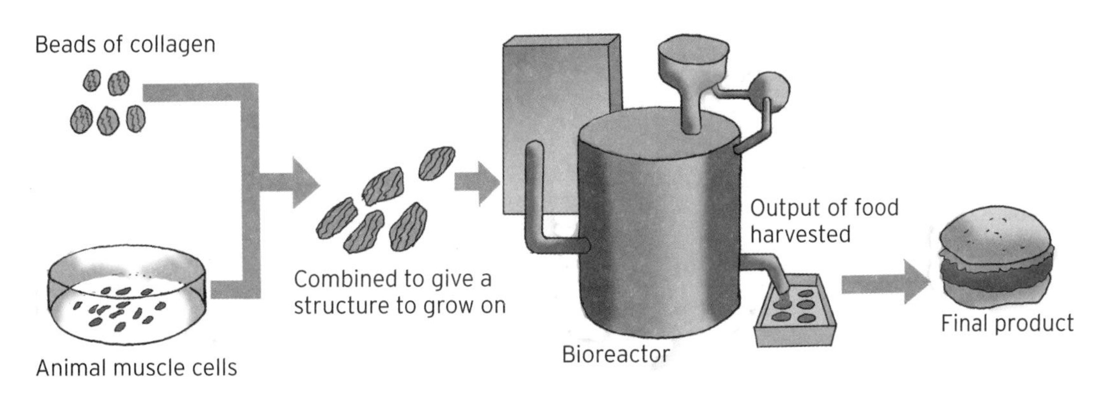

Nanofood

Applying the techniques of nanotechnology, or **molecular engineering**, to the production of foodstuffs is a discipline scientists call nanofood. Experts in the field claim that nanotechnology can be used to change the texture and flavour of food, and improve safety by using nano-scale sensors to monitor for dangerous bacteria.

Food company Kraft is even looking into using nanotechnology to design so-called smart food – interactive foodstuffs, allowing consumers to change the colour and flavour and even release nutrients in response to body deficiencies detected by sensors in the food. Kraft envisages a colourless, flavourless liquid containing programmable nanoparticles. Consumers do the programming by selecting their choices on a keypad, which are then encoded as a beam of microwaves that's fired into the liquid, activating the necessary particles. Food scientists at Wageningen Biotechnology Centre for Food and Health Innovation, in the Netherlands, believe that one day nanotechnology could be engineering complete meals from the ground up, molecules by molecule, while we wait.

FORENSICS

Fingerprints

The patterns of ridges on the tips of our fingers are unique – no two individuals have the same pattern. When we touch anything, the natural oils on the skin leave an imprint of this pattern behind. Forensic officers can search for prints left at the scene of a crime and use these to identify suspects. Each person's unique set of fingerprints are formed in the womb by the crumples and folds of the developing baby's skin and that imprint remains with us for life The ridges are also thought to amplify our sense of touch. Forensic teams check crime scenes by dusting with fine powder that sticks to the greasy residue of a fingerprint. Detected prints are then scanned into a computer where they can be cross-referenced against a database.

Some criminals have resorted to drastic measures to remove their fingerprints – such as burning them off with acid, or even taking a razor to them. Others simply wear gloves. Concerns have arisen over fingerprint-based security systems for homes and cars that use fingerprints to verify the owner's identity – for fear that robbers will amputate victims' fingers in order to gain access.

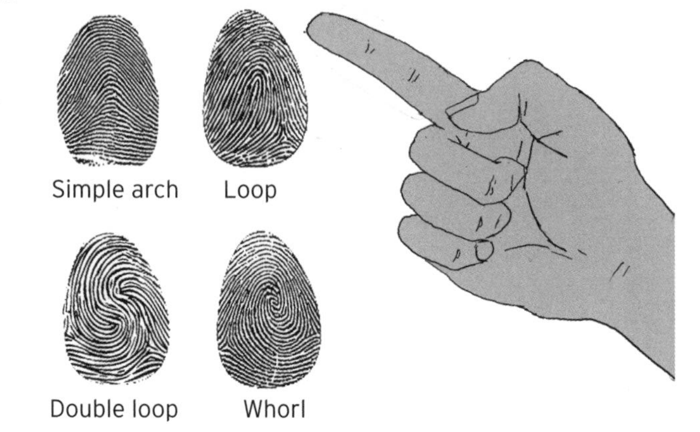

Trace evidence

Searching for the tiniest traces of material left at the scene of a crime – such as hair, blood and flakes of skin – can sometimes provide the biggest clues in the search for the culprit. Gathering such trace evidence is a painstaking process, requiring officers to work on their hands and knees combing a crime scene in search of clues. Samples are returned to forensic laboratories for examination using a gamut of techniques including optical microscopes, scanning electron microscopes and tests to determine chemical composition. Sometimes trace evidence from the victim can be used to implicate a suspect in a crime – for example, if spatters of blood matching that of a murder victim are found on the suspect's property.

DNA profiling

Just as we all have a unique set of fingerprints, so the genetic make-up of our DNA is unique too, which means that trace evidence, such as blood specks, saliva, skin or hair, can be used to place a suspect at a crime scene. It works because there are millions of base pairs of DNA in the human genetic code, making the odds of any two people having an identical genetic make-up vanishingly small. DNA profiling doesn't work by sequencing all of these pairs – that would take far too long. Instead, forensic scientists look for repeating patterns in the DNA. This isn't quite the same as taking a complete genetic fingerprint but it's usually sufficient to tell whether a sample gathered at a crime scene and that of a suspect have come from the same person.

Toxicology

Toxicology is a branch of science concerned with analyzing and identifying poisons. In forensics, it's used to ascertain cause of death – as well as testing for drug use, for example if someone is suspected of driving under the influence of narcotics or alcohol. The forensic toxicologist can carry out chemical tests on any number of bodily fluids, such as urine and blood, and even hair samples to spot the presence of poisonous compounds. This is often combined with a physical examination – which may, for instance, reveal needle marks. Other supporting evidence can be important too – for example, was the deceased on any prescription drugs, on which they may have overdosed? Forensic toxicology was pioneered by Spanish chemist Mathieu Orfila in 1813. It pulls together the fields of **human physiology**, **pathology**, **biochemistry** and **epidemiology**.

Forensic entomology

It might seem strange to think that entomology (see **Zoology**), the study of insects, could be of any use in a criminal investigation. And yet studies of insects have proved extremely useful in cases ranging from murder to drug smuggling.

A principle application is in determining the time of death in a murder inquiry. Flies lay their eggs on a corpse within seconds of death. Once laid, the eggs hatch into maggots, which then turn into pupae and finally new adult flies. Other creatures, such as ants, beetles and dung flies, gradually join in the parasitic activities of the flies. The whole process happens with such clockwork timing that by examining the state of insect infestation of a corpse it's possible to tell the time of death, often to within the hour.

The crimes that forensic entomology can tackle needn't involve bodies. For example, when presented with a shipment of drugs, forensic entomologists can tell what part of the world it has come from simply by examining the insects it contains.

Ballistics

Ballistics, in a forensic context, is the study of firearms and bullets. A gun's rifling – the spiralling grooves down the inside of the barrel that make the bullet spin, improving its accuracy – leaves a unique set of scratches on each bullet it despatches. Forensic scientists use this to match bullets with the weapons that fired them. Ballistics experts also look at the passage of a bullet, both through the victim and any nearby objects, giving clues to the direction from which it was it was fired. They can also chemically identify burnt powder, which clings to clothes and skin – revealing whether a suspect has fired a gun recently.

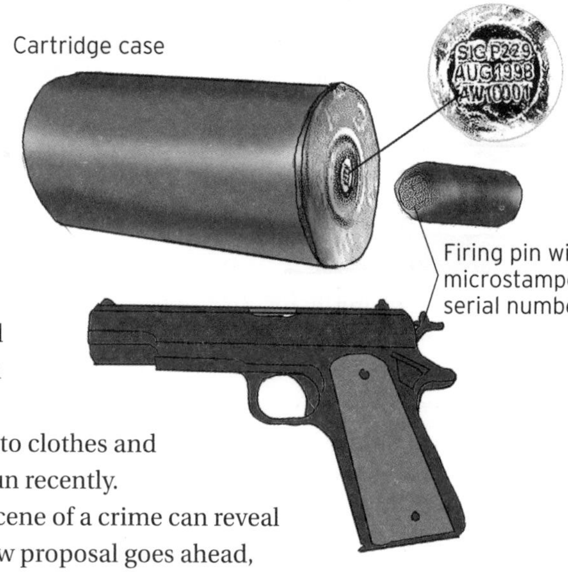

Recovering used cartridge cases from the scene of a crime can reveal what type of gun was used in a shooting. If a new proposal goes ahead, firing pins of guns may even be fitted with a microstamp which imprints a gun's serial number on the ejected cartridges.

Criminal profiling

It's a grim job: getting under the skin of killers, rapists and arsonists. But the work of the criminal profiler is invaluable. Profilers are experts at reading the thoughts of the dangerous and the violent. They use evidence from crime scenes, and statements from victims and witnesses, to deduce the personality and behaviour patterns of criminals. Profiles give police crucial information with which to winnow down their pool of suspects – and in some cases identify new ones.

The organization best known for its use of criminal profiling is the US Federal Bureau of Investigation (the FBI). Its Behavioral Analysis Unit (formerly the Behavioral Science Unit) was established in 1972. But the field of profiling was born earlier still when George Metesky, the New York bomber, was tracked down in 1957 from a profile drawn up by psychiatrist Dr James Brussel. Detectives working on the case had scoffed at Brussel's analysis but after the arrest many law enforcement officers began to take profiling seriously.

The technique isn't infallible; profilers investigating the 2002 Beltway Sniper attacks, in Washington DC, suggested the killer was a middle-aged, white male with right-wing political views. The murders were actually carried out by two black men, the youngest of whom was only seventeen. Nevertheless, profiling has helped catch some of the world's most notorious villains and remains a vital resource for serious crime investigation the world over.

Forensic odontology

Sometimes the body of a murder victim won't be discovered until long after the crime has been committed. In this case, decomposition means there may be few remaining features by which the body can be identified. This is where forensic odontology comes in – the science of identifying human remains using dental records. A straight examination of teeth can yield an accurate estimate of the victim's age at the time of death. Taking photographs and impressions of the victim's teeth and scanning them into a computer enables scientists to cross-match the teeth with the dental records of known missing persons.

But odontology has another use in criminal investigations – it can be used to match bite marks on victims to the teeth of suspects. In the trial of serial killer Ted Bundy in 1979, bite marks he had left on victim Lisa Levy were key to his conviction.

Forensic anthropology

Like practitioners of forensic odontology, forensic anthropologists assist investigators by identifying decayed human remains. These scientists are able to deduce details such as the age and sex of the deceased, their stature, ancestry, and in many cases cause of death. As well as working on individual murder cases, forensic anthropologists are often called in to help identify remains in mass graves – such as those discovered in the course of war crimes investigations.

Forensic anthropologists occasionally carry out facial reconstructions from skulls – building up layers of muscle and then skin to reveal what the owner of the skull may have looked like. In 2005 scientists used computed **tomography** scans of the skull of Tutankhamun to create a striking likeness of the boy pharaoh's face.

Computer forensics

The majority of people own computers. Because of the increasing role these devices play in our lives – through email, Internet access and social networking – police seize the computers of those implicated in a criminal investigation to scan the hard drive for clues. This is known as computer forensics.

Nowadays, the scope of computer forensics is expanding to include the multitude of smaller computers that we carry around in our pockets – cellphones, digital cameras, and portable **flash memory.**

ARCHAEOLOGY

Archaeometry

Archaeology is the science of studying the ancient past through the analysis of artefacts and architecture unearthed by careful excavation. Many disciplines of science contribute to archaeology and these collectively form what is known as archaeometry.

It includes fields such as **metal detecting**, **aerial archaeology** and archaeological **geophysics**, but it also aims to apply the broader principles of science to archaeology such as rigorous measurement and the development of theories that can be scientifically tested through **falsification**.

Metal detecting

It's amazing what you can find wandering around in a field with a metal detector. In September 2009, an amateur metal detector enthusiast in Staffordshire, England, unearthed a hoard of Anglo-Saxon gold dating from the 7th century AD, thought to be worth millions of pounds.

Metal detectors work using an alternating current (see **AC/DC**) to produce an alternating magnetic field in a coil by the principle of **induction**. Any metallic objects in the ground, cause the reverse effect to kick in – the field from the metal detector creates a small current in them. And this induces its own magnetic field that can be picked up by the detector. Metal detectors are also used for prospecting, where they can pick up natural metallic deposits in the ground, sweeping for land mines in war zones, and for security scanning at airports.

Aerial archaeology

Standing back and looking at a landscape can offer deep insights into the secrets it holds. It doesn't necessarily involve aircraft – any view from above, be that from a plane, a satellite, or simply any high point of land, can yield clues to the ancient history of a site.

A classic example is the Chicxulub meteor crater in Mexico. Scientists had suspected since 1980 that the dinosaurs were wiped out by a colossal meteor impact with the Earth 65 million years ago. But in that case where was the impact crater? The definitive answer came in 1996 when satellite images of the Yucatán Peninsula revealed a faint ring in the ground – the remnant of an ancient crater 180km (110 miles) across.

Surveying the lay of the land from above can reveal lumps and bumps in the ground that betray the presence of buried ancient settlements. Differences in soil colour and images taken at other electromagnetic wavelengths – such as infrared – can also identify hiddden structures. The work of aerial archaeologists tells archaeologists on the ground where they should dig.

Geophysics

Archaeologists use geophysics to probe the ground beneath their feet using a range of remote sensing techniques based on elementary principles of physics. By connecting electrodes to the ground archaeologists can measure the electrical resistance of the soil. Subsurface structures create an increase or decrease in the resistance measured; for example, stone has a higher electrical resistance than damp soil. Another common technique is ground-penetrating radar, which works by firing radar beams into the soil. Any large structures present will bounce the beams back up to the surface, alerting archaeologists to their presence.

Finally, magnetometry measures the **magnetism** properties of the soil and objects in it; anything made of iron has a strong magnetic signature, but other materials produce a measurable response too – including brickwork and stone.

Computational archaeology

Computers, and other forms of digital technology, are powerful tools in all areas of science, and archaeology is no exception. An important computational resource that archaeologists use are the so-called geographic information systems. These are computerized databases of land **topography**; Google Earth is an example of one such database. They give scientists an initial overview of an area of suspected archaeological interest, allowing them to gather cheaply the information needed to make the decision whether or not to dig there. Another digital tool is GPS, or Global Positioning System, otherwise known as **sat nav (GPS)**, which helps excavation teams take accurate position readings.

The power of computers to crunch numbers also lets archaeologists carry out detailed statistical analysis of archaeological data to spot patterns. Some sophisticated software packages even allow researchers to reconstruct dig sites as walk-through virtual reality simulations.

Dating

Being able to establish the date of an artefact is a key skill in archaeology; and scientists have developed a number of reliable methods to tell archaeologists just what era of history they're looking at. Radiocarbon dating revolves around the radioactive decay of an **isotope** of carbon, known as carbon-14. Plants absorb this carbon during their lives, but once they die, it isn't replaced and begins to decay – with its radioactivity diminishing by half every 5,730 years. Measuring how much carbon-14 there is in a sample then gives an idea how old it is.

Another method is dendrochronology, which is used for dating ancient trees by the pattern of rings inside them. Another still works using thermoluminescence – a phenomenon whereby pottery and other ceramic materials give off **photons** of light when they are heated. The amount of light emitted is proportional to the amount of **electromagnetic radiation** the sample has absorbed since it was heated last. We are bathed in a constant sea of weak electromagnetic radiation which all objects, including archaeological artefacts, soak up. The intensity of the thermoluminescent light from an archaeological sample thus reveals how long it's been absorbing this radiation for, i.e. how old it is.

Experimental archaeology

How did people living in the Iron Age smelt iron from ore, the unprocessed form of the metal dug out of the ground? The best way to find out is by building an Iron-Age smelter; that would be an example of experimental archaeology. It's the art and science of reconstructing ancient methods, buildings and technologies, based on the examination of artefacts, to find out how they would really have worked.

Another example of experimental archaeology – one that made international news – was the construction of *Kon-Tiki*, a balsa wood raft built in the 1940s and sailed from Peru to the

Polynesian Islands in the Pacific. The idea was to demonstrate how early inhabitants of South America possessed the technology to colonize Polynesia. Experimental archaeology is an important discipline because it allows archaeologists to test their **scientific theories** about how ancient peoples lived and worked.

Underwater archaeology

Coastlines change as the millennia pass, and this means that some important archaeological sites are not just underground but underwater too. Underwater archaeologists investigate these sites by combining archaeological techniques with diving skills. Coastal areas become inundated as sea levels rise for various reasons, from earthquakes to subsidence to climate change. In northern Egypt, much of the ancient port of Alexandria is now beneath the waves as a result of seismic activity 1,700 years ago. But this hasn't stopped a team of divers led by French archaeologist Franck Goddio recovering ornate statues and other relics from the site, which is thought to encompass the former quarters of Egyptian queen Cleopatra. But it's not just about submerged settlements; shipwrecks and the remains of downed aircraft are also viable targets.

Underwater archaeology is plagued with difficulties; visibility is poor, and the water greatly hampers the delicate work of excavation. Some sites are so deep underwater, that remotely piloted submersibles have to be used. And then, even when artefacts are recovered their sudden exposure to the air can cause rapid deterioration – demanding special preservation techniques.

Archaeoacoustics

Ancient peoples knew a thing or two about the science of **sound waves**. The ancient Mayan pyramid El Castillo, at the Chichen Itza site in Mexico, has an unusual sonic property. Knock two stones together nearby and the echo that comes back sounds like a bird's 'chirp' – a high-pitched sound followed by a descending tone of lower pitch. Some people believe it was an attempt by the Mayans to capture bird song – the cry of the Central American quetzal bird is remarkably similar.

Meanwhile, there are numerous examples around the world of 'ringing rocks' – carved standing stones which, when struck, produce a pure note like a bell. It has even been suggested that some artefacts may encode primitive sound recordings from the time of their creation. The reasoning goes that as an ancient craftsman worked with wet clay the vibrations in his tools, caused by nearby sounds, became imprinted on the clay. This was first suggested in the 1960s, although so far there are no convincing examples of sounds extracted from ancient pottery.

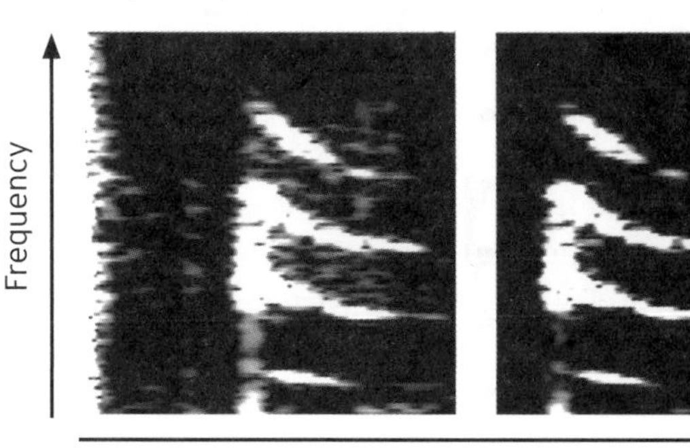

Archaeoastronomy

Even though the astronomical telescope wasn't invented until the 17th century, that didn't stop ancient astronomers from studying and mapping out the heavens. The first astronomical **constellations** were named by the ancient Greeks, demonstrating an interest in the night sky dating back thousands of years. Meanwhile, all of the planets out to Saturn were discovered and named in prehistoric times by early astronomers using nothing but sharp eyes to spot planetary movement across the sky.

A number of ancient astronomical observatories have been discovered. At Stonehenge, a 4,500-year-old stone circle in England, it's believed that the alignment of certain stones with astronomical objects was used to gauge the time of the year. For example, the rising Sun only lines up exactly with a large rock called the Heel Stone at the time of the Summer Solstice. Other ancient observatories have been found at Abu Simbel, in Egypt, and at Chichen Itza in Mexico.

Palaeontology

Archaeology and biology collide in a field known as palaeontology. It deals with the study of prehistoric lifeforms, including the dinosaurs, much of the evidence for which must be gathered using archaeological techniques. Palaeontologists unearth the fossil remains of ancient creatures to try to reconstruct the evolutionary tree of life on our planet – formed as new species were created by the process of **natural selection**, while others became extinct. Fossils are formed when organic material is compressed for millions of years by layers of dirt and sediment, turning it to stone.

Broadly speaking, palaeontologists are split between two camps – palaeobotany, concerned with ancient plant life, and palaeozoology, the study ancient animals. They have succeeded in tracing life on Earth back across 3 billion years of time. Studying fossils naturally involves the layers of rock in which they form – and for this reason palaeontology research occasionally makes contributions to the field of geology.

Tomb raiding

Some archaeologists, in fields such as Egyptology and the study of other ancient civilizations that were capable of feats of architecture, spend their time trying to discover the grand tombs of deceased leaders and kings. The grandest tomb discovered to date is that of the Egyptian boy pharaoh Tutankhamun, who died in 1,323 BC – a period of Egypt's history called the New Kingdom, during which the nation was at its most prosperous. A team led by British archaeologist Howard Carter discovered the tomb in 1922. Tutankhamun's tomb had already been robbed several times through antiquity. This practice is now greatly frowned upon by serious archaeologists – and is illegal in many countries.

LANDMARK INVENTIONS

Bicycles

One of the most widespread devices ever invented, the bicycle is everywhere. Far cleaner than a car and quicker than walking – bicycles have become a popular form of transport for commuters. They are also used by mail delivery services, police, medics and, of course, for recreation.

Bicycles in various forms were built by a host of inventors during the early 19th century. They were clunky, uncomfortable and difficult to ride, but a quick succession of innovations towards the end of the century – such as chain drives, air-filled tyres, cable brakes and gears – delivered bicycles that more closely resemble the machines we know and love today. Modern bicycles are extremely complex, often incorporating high-performance disc brakes, and suspension to improve handling over rugged terrain.

Flush toilets

Flush toilets, water closets – call them what you will – few of us would want to be without them. Yet many countries in the Third World, in Asia and Africa, still make do with the most basic toilet facilities – a primitive hole in the floor, or worse. The flushing toilet is believed to have been invented in the 16th century by English writer John Harington, a member in the court of Queen Elizabeth I. A number of developments since have improved the design, including the addition of a cistern – removing the need to manually flush from a bucket – and the U-bend, to keep smelly sewage gases at bay.

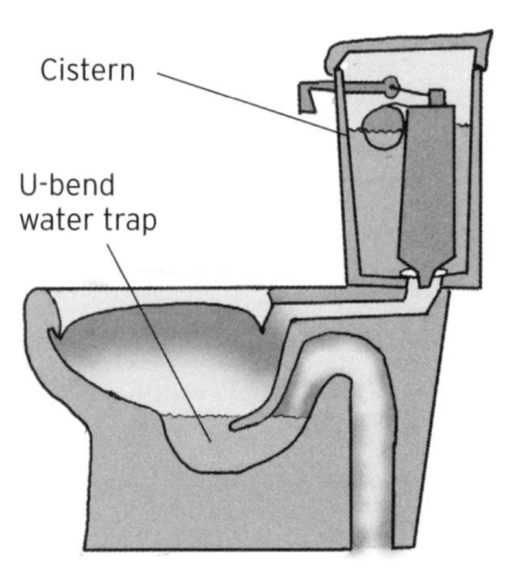

Flush toilets greatly improved sanitation, allowing excreta to be swept away in a stream of water and kept separate from drinking water supplies. One early toilet pioneer was an English plumber called Thomas Crapper, whose name will be forever linked with toilets and their use.

Money

Xi-Lin has a chicken that she doesn't want, yet she needs some bread. Hasok has a loaf of bread to spare but is in need of a chicken for his evening meal. Clearly, they can help each other, but Xi-Lin thinks her chicken is worth more than Hasok's bread. What if they could represent the price of both items with cowrie shells – say, one loaf of bread is worth 10 shells and the chicken is worth 30. Now Xi-Lin would be happy to hand over her chicken in return for Hasok's bread plus 20 shells that she can spend tomorrow.

This encounter is fictional, but it illustrates how the concept of money could have begun. People in Asia started using the shells of the cowrie – a type of marine mollusc – as a unit of currency around the year 1,200 BC. The first actual coins were minted in Lydia – part of modern-day Turkey – in the 7th century BC. Money allowed people to accumulate wealth beyond the value of their immediate possessions, produce or services. Today, computers and the Internet have made money rather more intangible – with banks able to transfer vast sums internationally as a burst of electrons down a wire.

Contraception

At the beginning of the 1960s something was invented that would change the world – birth control in the form of a pill. Invented by Dr Gregory Pincus and first put on sale in 1960, it was not only more reliable than other methods, but it empowered women to make the decision to use contraception. It was the beginning of the sexual revolution.

Of course, other methods of contraception had been around for longer and the oldest device is probably the condom, which is thought to date back thousands of years, to the ancient Egyptians. The first condoms were made of linen, and later sections of animal intestines. One of their principle uses in Medieval times was to guard against the deadly sexually transmitted disease of syphilis. It was in this capacity for preventing infection that the condom made a comeback in the 1980s – as the **AIDS** epidemic, spread by the HIV virus, began to take hold.

Motor cars

For anyone bothered by the fact that there are over 500 million cars trundling over the planet, blame the man who started it all: Karl Benz. German engineer Benz built his first car in 1885, a 1000cc petrol-engined vehicle that looked, quite literally, like a carriage without horses. The engine in this 'Motorwagen' was woefully inefficient, giving a top speed of just 10 miles per hour.

But engine quality rapidly improved and in 1903 Henry Ford founded the Ford Motor Company. In 1908, it began manufacturing the Model T – one of the most successful production cars of all time. Not only did it look less like a carriage and more like a motor car, but Ford did away with the frills so that ordinary people could afford it – no wonder more than 15 million of them were sold.

Today, cars are as much about electronics and computers as mechanical engineering – with all of a car's functions run by a finely tuned management computer. This trend towards electronics looks set to continue, as manufacturers turn to electric cars powered by **hydrogen** as a clean alternative to petrol.

Microwave ovens

Microwave cooking was stumbled upon during the development of radar, and heats food in a fraction of the time, and at half the cost of traditional methods. It was discovered in 1945, by Percy Spencer, an electronics engineer working for Raytheon, in Waltham, Massachusetts. Spencer was testing a magnetron – a type of radio signal generator used in military radar sets – when he noticed that a candy bar in his pocket was melting. Suspecting that the microwaves from the magnetron were responsible, he tried an experiment, placing some popcorn kernels in the beam. Sure enough, as he switched on the power, the kernels began popping into puffed snacks.

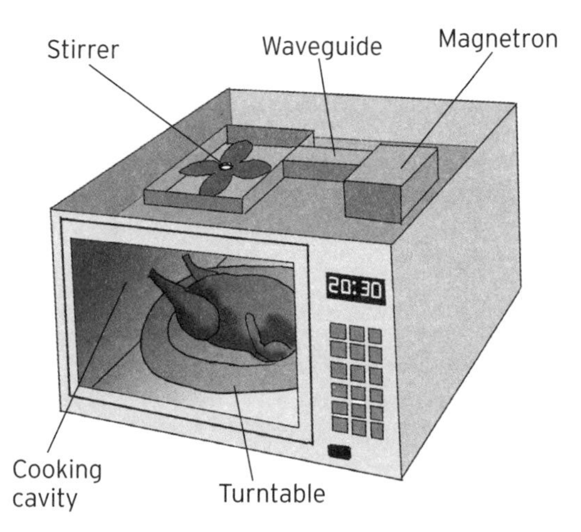

Microwaves are a form of electromagnetic radiation, with a wavelength of around 12cm. In a microwave oven, microwaves cook food through a process called dielectric heating. Molecules of water in the food align themselves with the field. They do this because electrical charges are distributed unevenly within the molecules and the jostling that results makes the molecules bump into one another and vibrate. In accordance with **kinetic theory**, we feel this vibration as heat.

Lasers

The laser was invented in 1957 and has since changed the world, underpinning everything from DVD players to fibre-optic communications. The word is an acronym for 'light amplification by the stimulated emission of radiation'. Ordinary light sources – such as filament bulbs – emit photons randomly and at all wavelengths. But lasers offered a new method whereby all the photons have the same wavelength and are emitted in lockstep, creating light that is 'coherent'. In addition, laser light doesn't spread out, but stays in a narrow beam.

Lasers work on the principle of stimulated emission, the theoretical foundations of which were laid by Albert Einstein. Atoms are raised to a predetermined **energy level** and then stimulated to drop back down by bombarding them with photons of the same energy. Some of these in turn can crash into other excited atoms creating a cascade of coherent light. DVDs, and other forms of **optical storage**, use lasers to read information from a reflective disk; the smaller the wavelength of the laser's light, the more information the disk can store. In the 1990s, scientists developed lasers that produce short-wavelength blue light, and this has brought about a new generation of high-capacity DVDs – called Blu-ray.

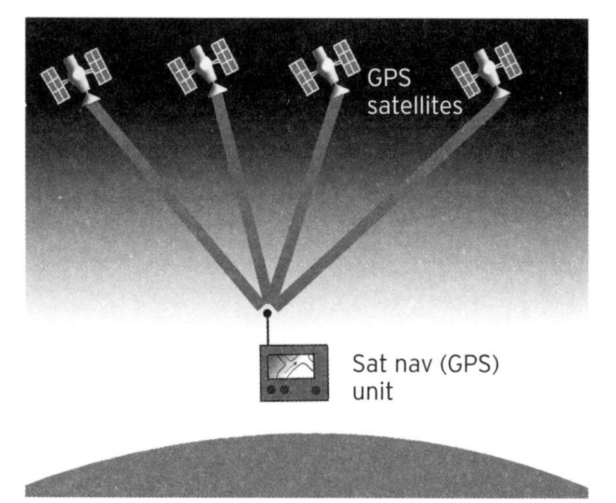

Sat nav (GPS)

Some of us would be horribly lost were it not for that helpful voice from the dashboard. Sat nav, or Global Positioning System (GPS) to use its full name, uses signals from **artificial satellites** in Earth orbit to fix the user's location on the planet's surface. As of 2009, there were 29 healthy GPS satellites circling the Earth in a cloud of criss-crossing orbits, like electron particles buzzing around an atomic nucleus. Each satellite emits a regular time pulse. This is compared to a clock in the receiver so that the travel time, and hence the distance, from the satellite can be calculated. Taking such a fix from four or more satellites is enough to pinpoint your exact position.

The GPS system has been launched gradually, starting in 1989. It was first intended solely for military use, until President Ronald Reagan decreed that it should be available for civilians after a Korean airliner was shot down having accidentally strayed into Soviet airspace. GPS is now used in commercial aviation and shipping, personal navigation devices and even cellphones.

Scanning electron microscopes

Scanning electron microscopes (SEM) are incredible. Like other microscopes, they are machines for producing images of small objects, but for an SEM the definition of small is extremely small indeed. Image resolutions can reach scales of 1 nanometre (nm) – allowing crisp images of blood cells, microorganisms, and crystal structures in materials. Microscopes work by illuminating the target material with a beam of light; the level of detail that can be revealed increases as the light's wavelength gets smaller. SEMs work by replacing the light with a beam of electron particles. Because of **wave-particle duality**, electrons have a wavelength too – which turns out to be 250,000 times smaller than that of light, allowing them to reveal far greater detail.

SEMs were developed in the 1930s. In 1981 the scanning tunnelling microscope (STM) was introduced; this is a variant that works by dragging a pencil-like probe back and forth across the target's surface. Lumps and bumps on the surface cause electrons to 'quantum tunnel' (see **Fission and fusion**) between the probe and the surface. This sets up electrical currents which can be measured, enabling scientists to map the surface in ultrafine detail – down to 0.1nm, in some cases even revealing individual atoms.

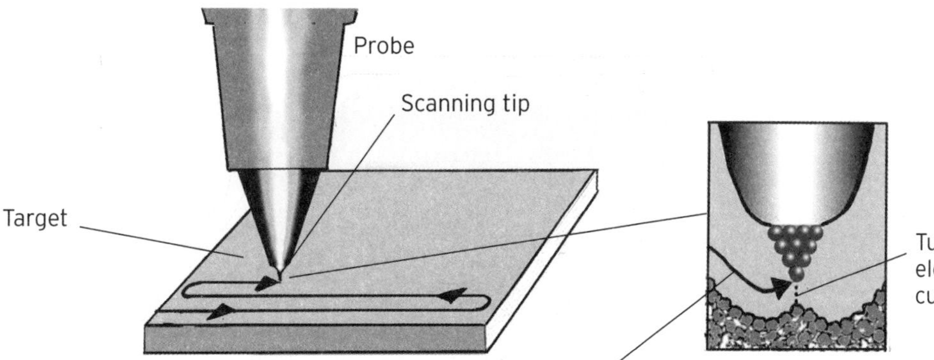

FUTURE

WHAT WILL LIFE BE LIKE IN THE YEAR 2100, 2500 or even AD 10,000? How will society be changed by contact with extraterrestrial life? Will we ever unlock the secret to travelling through time? And will deadly and debilitating medical conditions such as AIDS and spinal injury ever be curable? Speculation about the breakthroughs, discoveries and developments which are yet to come is one of the most exciting areas of science.

Once the province of soothsayers and charlatans, predicting the future is a now a solid area of science – called futurology. Futurologists work by extrapolating from

RAVITY • FASTER-THAN-LIGHT TRAVEL
TROPICS • SOBER-UP PILLS • CLONING
RAS • DIY GENETICS • HUMAN CLONING •
ERVOLCANOES • JOURNEY TO THE CORE
TION • MAGSAILS • SPACE ELEVATOR •
NG • VON NEUMANN MACHINES • OMEGA
NTROL • NANOMEDICINE • AIDS VACCINE
FUTARCHY • GALACTIC EMPIRES
ELEPATHY • ONLINE CONSCIOUSNESS
THE SINGULARITY • AUTOMATIC
PROSTHETICS • FREE ENERGY •
TRAVEL • TELEPORTATION
NEWTONIAN DYNAMICS

technological and societal trends, by looking for historical parallels to current events and by canvassing the opinions of experts. Their findings are invaluable for governments, economists, businesses and other scientists.

Lately, however, futurologists have cast doubt on the very future of their own profession. They've predicted the occurrence of the so-called 'singularity', a point in the future (the relatively near-term future, according to some) where technological advancement undergoes a sudden, exponential rise. So great will be the changes introduced in the singularity that it is impossible to forecast the future beyond it.

FUTURE PHYSICS

Time travel

When Albert Einstein put forward his theory of **special relativity** in 1905, some of the weirdest consequences were its predictions of **length contraction and time dilation**. In particular, time dilation said that a person flying off on a spaceship at close to the speed of light would return home a short number of years later to find that many years had passed back on Earth. The traveller had jumped into the future.

When Einstein published the theory of **general relativity** ten years later, it brought the possibility of travel into the past. Taking the mouths of two **wormholes** – tunnels through space made possible by general relativity – and putting one on a spaceship and flying it off at close to light speed would introduce a time difference between them, again by time dilation. Someone then jumping into the 'future' mouth of the wormhole would emerge from the other mouth in the past. The technical difficulties of putting such a scheme into practice are considerable, though some physicists have speculated that it may become possible in the centuries to come.

Teleportation

It's long been a science fiction dream, to be able to measure the state of all the atoms in an object and 'beam' that information to a receiver where the object can be reassembled. But could such teleportation ever become possible? For years it was thought not, because of the Heisenberg **uncertainty principle**, which says you can never know all of the information about an atom at the same time – making it impossible to measure. However, in 1993, an international team of researchers worked out how it is possible to sidestep uncertainty using **quantum entanglement** – in a similar process to **quantum communication**.

Entangling each atom with another particle and sending this particle to the receiver, it carries exactly the information about the atom which is hidden by uncertainty. An exact copy of the atom can be assembled at the receiver, while the original is destroyed in the process. In 2004, researchers in the United States and Austria verified the scheme, teleporting atoms across their lab. However, the teleportation of large objects is still many years away.

Antigravity

In 1996, Eugene Podkletnov, a physicist at Tampere University in Finland, came forward claiming to have built the world's first antigravity device – a machine that literally shields the force of gravity from other objects. He claimed that his 30cm disc made from the **superconductor** yttrium-barium-copper oxide, cooled to -230°C (-382°F) and spun round at 5,000rpm, made objects placed above it get lighter – by about 2 per cent.

If correct, it would have massive implications for launching spacecraft. However, teams around the world have tried to verify the claim – including scientists from NASA, Boeing and the British aerospace firm BAE Systems – so far without success.

Faster-than-light travel

One of the predictions of Einstein's relativity was that nothing can travel faster than light. However, in 1994 Miguel Alcubierre, a physicist at the University of Wales, came up with a theoretical framework by which it could be done. He named it the 'warp drive'.

The central thrust of Einstein's theory of general relativity is that space can be curved, and that the curvature is determined by the matter it contains. Alcubierre worked out that by using matter with a very special set of properties and arranging it around a spacecraft in just the right way could make the space in front of it contract quickly while the space behind expanded at the same rate – the net effect being to sweep the piece of space in the middle, carrying the spacecraft along to its destination quicker than a light ray. The trouble is that Alcubierre's special kind of matter – called exotic matter – has negative mass (see **Wormholes**). While small amounts of this stuff have been produced by the **Casimir effect**, building a working warp drive would demand a supply of exotc matter equal to a third of the mass of the Sun.

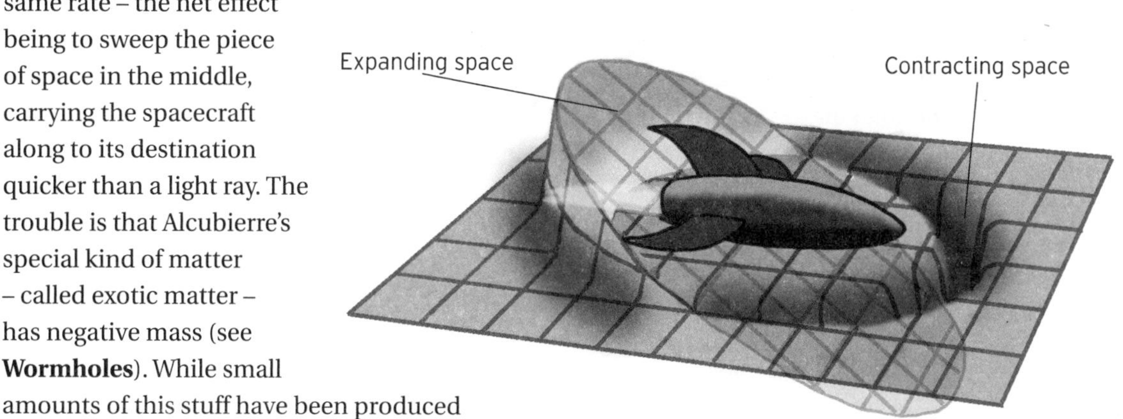

Modified Newtonian dynamics

Some scientists think **dark matter** doesn't exist and say the effects normally attributed to it can be explained by applying a small correction to **Newtonian gravity**. In 1983, physicist Mordehai Milgrom at the Weizmann Institute in Tel Aviv, Israel, put forward a theory known as modified Newtonian dynamics, often abbreviated to just MOND. Essentially it says that the law of gravity at long distance deviates from Newton's theory. The theory is able to explain the flat rotation curves of galaxies – the graph you get when you plot the galaxy's rotation speed against distance from the centre – without the need for dark matter. The shapes of these rotation curves were one of the foundations upon which dark matter theory was built.

More recently, Jacob Bekenstein, at the Hebrew University of Jerusalem, has developed a theory called TeVeS (tensor-vector-scalar gravity) – a version of MOND that is consistent with Einstein's principles of relativity. TeVeS is able to explain **gravitational lensing** observations – a test that MOND on its own had previously failed.

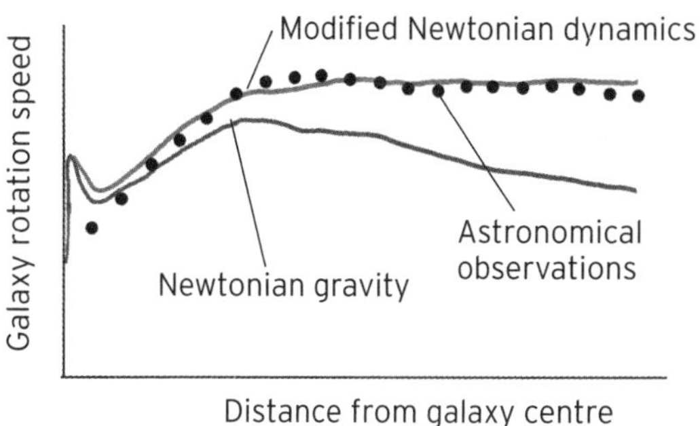

FUTURE CHEMISTRY

Nootropics

Nootropics are chemicals that enhance cognitive powers such as intelligence, memory and concentration. Already, stories abound in the press of students turning to drugs such as modafinil (normally used to treat the sleep disorder narcolepsy) and Ritalin (used to treat attention-deficit hyperactivity disorder, ADHD) to boost their attention, memory and energy levels while cramming before exams. But a raft of other pills are also available. They work by topping up levels of crucial neurotransmitters, as well as the **enzymes** and **hormones** essential for a sharp mind. For example, acetylcholine is a neurotransmitter involved in the formation of memories; while amino acids, such as tyrosine and phenylalanine, trigger the body to produce the natural stimulants epinephrine and dopamine.

Few doctors would prescribe nootropic drugs to boost brain power – most are obtained instead through online pharmacies. Meanwhile, many more nootropics are available as **dietary supplements**, where regulation is even slacker. A report by the UK government in 2005 concluded that before long there could be a pill to boost practically every aspect of brain function – even drugs that can block painful memories.

Sober-up pills

Imagine having a pill you could pop at the end of an evening out that would counteract the effects of alcohol, leaving you safe to drive home and free from the pain of a hangover the next morning. A team at the University of California, Los Angeles, has researched a potential drug to do this called Ro15-4513. First discovered in 1984, it was found to block the brain pathways through which alcohol acts. It does this by binding to receptors on cells in the brain to stop them absorbing alcohol.

The researchers suggest that as well as sobering up partygoers, such an alcohol antidote could be used to treat alcoholics – and could enable doctors to use the mood-enhancing properties of alcohol as a treatment, without getting patients drunk.

FUTURE BIOLOGY

Cloning extinct species

Scientists have used cloning to create genetically identical copies of sheep, mice, horses and other animals. But could it ever be used *Jurassic Park*-style to bring vanished species back from the dead? The idea is to extract genetic material from preserved remains of the extinct species and then insert this into an egg cell from a female of a related species still living in the world today. The hope is that implanting the egg back into the female animal would then allow it to be brought to term and delivered.

In 2007, the body of a 10,000-year-old woolly mammoth was found preserved in ice in Siberia. Some biologists have suggested that DNA from cells in this animal could be inserted into the egg of an elephant. However, there are a host of practical and ethical concerns to overcome first. Cloning is notoriously hard – it took 277 attempts to produce Dolly the sheep, the world's first cloned mammal. Also is it right to bring an animal to life without its habitat or parents to show it the way in the world?

Xenobiology

If life exists on planets elsewhere in the Universe what exactly might it look like? One thing is certain: it won't be human. Even if we could rewind the evolutionary clock of life on Earth and run the process all over again the chances of ending up with people a second time are next to nil – evolution is just too random.

Aliens on other planets would be shaped by **natural selection**, which drives organisms to evolve to match their environment. So, for example, on a world where the gravity is high, creatures would grow to be short and squat, with thick chunky bones to support their extra weight. On a low-gravity planet they would be much taller. The atmosphere would be thinner on such a world, so they would probably also have large chest cavities to accommodate the large lungs they would need to breathe. Other worlds, with their bizarre environments, might give rise to forms of life that we can barely imagine. American astronomer Carl Sagan once speculated about giant balloon-like life forms that drifted like jellyfish in the atmosphere of the gas giant planet Jupiter.

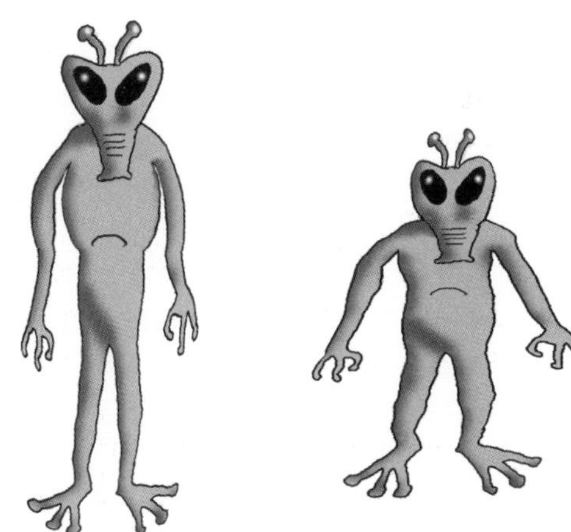

Low-gravity inhabitant High-gravity inhabitant

Chimeras

A chimera is an animal made up of cells from two different species. Unlike a hybrid – where the fusion between the two species takes place at the genetic or cellular level – chimeras contain distinct cells from each species. Any recipient of a **xenotransplant** is a mild chimera – they have cells inside them from their animal organ donor.

Chimera research became a hot topic in April 2005, following publication of a report into **stem cell therapy**. A whole section was dedicated to 'interspecies mixing' since, as the authors pointed out, all human stem cell procedures must be validated in animals first, forcing the creation of chimeras. Some researchers speculated that these animal tests could create a creature with human levels of intelligence and self-awareness trapped inside an animal's body. In response, a strict code of practice for chimera experiments was drawn up, in which experiments must be halted the moment any human-like traits in the subject are spotted.

DIY genetics

Just as computers were once the preserve of rich corporations and labs carrying out research at the cutting edge – yet now there's one in every home – so one day it may be that all of us will be tinkering with the genetic building blocks of life on our kitchen tables. That's the prediction of Freeman Dyson, a physicist and futurologist at the Institute for Advanced Study, at Princeton. He believes the creativity which has driven the computer industry to its current heights ultimately came from play – when computers reached the point at which they became children's toys.

And he thinks the same must happen with biotechnology if we are to unlock its true potential. If he's right, we might all one day be designing genetically modified plants (see **Genetic modification**) for our gardens and using **synthetic biology** to build our own pets.

Human cloning

Therapeutic human **cloning**, for the growth of cells to be used in **stem cell therapy**, has been approved in some parts of the world. And yet the difficulties in creating a clone – and the ensuing health problems that sometimes afflict them during their development – make it unlikely that that full reproductive cloning of humans will be sanctioned any time soon.

Why would anyone want to? The principle reason is as a method of fertility treatment, for childless couples who have exhausted all other methods. Genetic material from the father would be implanted into an egg from the mother. The resulting child would be genetically identical to the father. In 2003, maverick Italian fertility doctor Severino Antinori claimed to have used a cloning technique to help three women get pregnant. However, his claims were greeted with scepticism by other experts.

Transhumanism

Transhumanism is a term used to describe the evolution of human beings beyond the reach of natural selection, using deliberate alterations introduced by scientists. This includes **genetic modifications**, **cybernetic implants**, and other technologies designed to improve the length and quality of human life.

A future transhuman might have **nanomedicine** robots swimming in their blood to repair damage, they might take regular injections of anti-ageing drugs (see **Ismmortality**), and they might have **gene Therapy** to guard them against serious illnesses like diabetes, cancer and heart disease. Critics, however, worry that transhumanism is akin to **eugenics** and could breed a technologically superior master race.

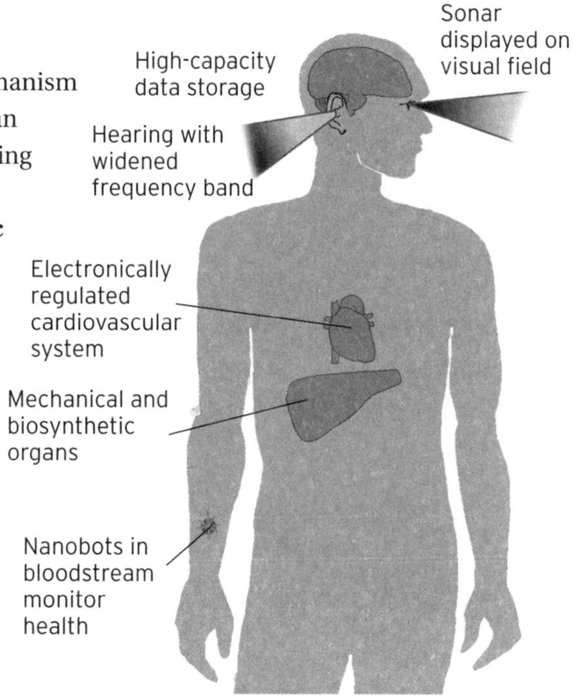

FUTURE EARTH

Pangaea Ultima

Continental drift means that the layout of the continents changes over many millions of years. In the past, this movement has aggregated all the land masses on the planet's surface together into so-called **supercontinents**. Scientists believe the continents will collide again in the future too – around 250 million years from now. The precise form of the resulting land mass is still up for debate but the leading possibilities have already been named, Pangaea Ultima and Amasia. It is thought Pangaea Ultima will form if the widening of the Atlantic stops and reverses – as some geologists believe it could – bringing Africa, Eurasia and the Americas back together. Subduction, however, will shift Africa northwards, between Eurasia and America, with its southern tip approaching the equator.

Amasia is an alternative scenario where the Atlantic continues to widen and Eurasia and North America merge instead at Siberia. In this case, the Pacific Ocean will shrink to become smaller than the Atlantic, before disappearing altogether.

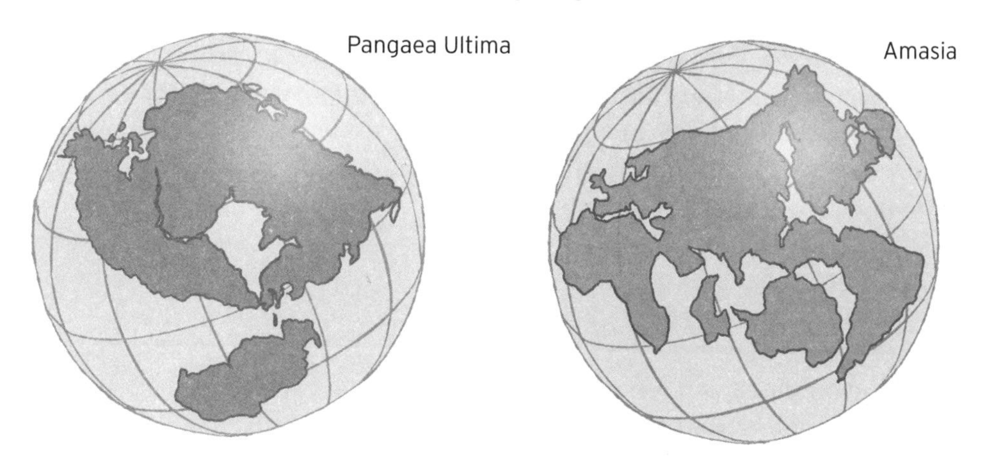

Supervolcanoes

Earth's history is punctuated by calamitous **volcanic eruptions** known as supervolcanoes. A supervolcano can hurl up a volume of magma and other material a million times greater than the biggest ordinary volcanoes such as Vesuvius or Krakatoa – and explodes with 30 times the power. Supervolcanoes don't even resemble the familiar form of other volcanoes, where lava spouts from a mountain-like cone. Instead, magma wells up in a chamber beneath seemingly level ground, tens of kilometres across; the mass of rock holds the magma down until the pressure beneath becomes so great the whole area explodes.

Supervolcano eruptions have the power to smother entire continents with volcanic debris and fling ash into the atmosphere that can affect the climate for hundreds of thousands of years. There is currently concern over a supervolcano in Yellowstone National Park, which erupts roughly once every 600,000 years; the last eruption was 640,000 years ago. Measurements of the caldera height showed it to rise by 20cm between 2004 and 2008 – the biggest increase since measurements began in 1923; however, the rate of uplift slowed in 2009.

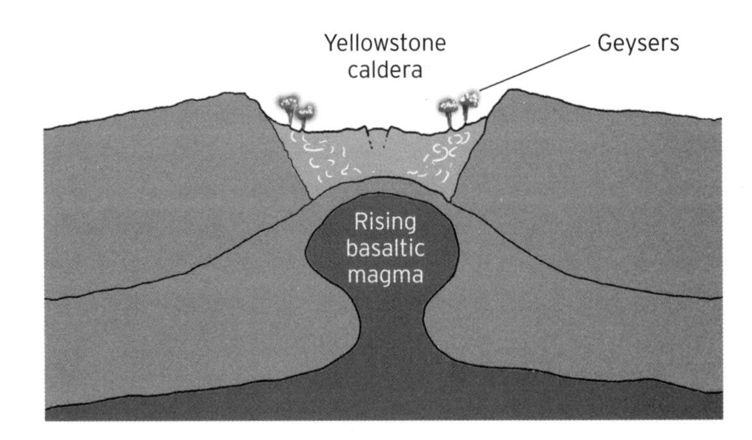

Journey to the core

In 2003, David Stevenson, a scientist at the California Institute of Technology, published a theoretical scheme by which a small unmanned scientific probe could be sent all the way down to the Earth's core. His plan was to open up a giant crack in the Earth's crust, perhaps using a nuclear bomb. Into that would be dropped the grapefruit-sized probe, along with hundreds of thousands of tonnes of molten iron – about the amount turned out by all the world's foundries every week. Because the iron is twice as heavy as the surrounding rock, it sinks to the core in a few days, carrying the probe down with it. Stevenson proposed to communicate with the device using seismic waves – sound vibrations transmitted through the Earth. He thinks the project could be done for around $10 billion – a fraction of what has been spent so far on unmanned space exploration.

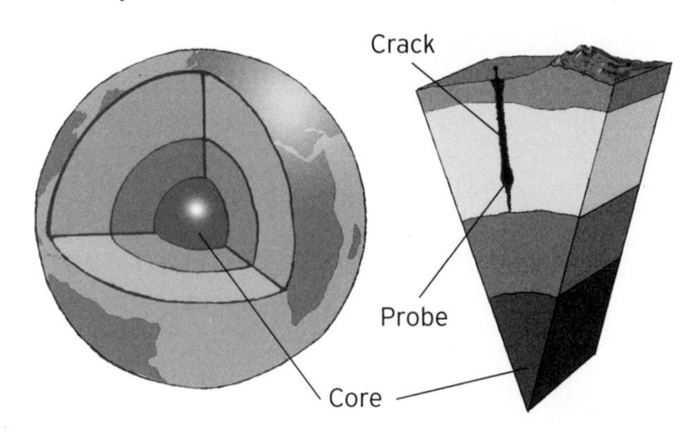

Weather control

Altering the weather isn't as difficult as might be thought. In 1946, American chemist Vincent Schaefer discovered the principle of cloud seeding – scattering chemical particles (such as silver iodide) into the air that act as condensation nuclei around which water droplets can form, creating clouds. The idea is to increase rainfall over dry areas.

The opposite process is possible too. During the 2008 Olympic Games, Chinese authorities deployed chemicals using aircraft, rocket launchers and anti-aircraft guns to disperse rain clouds before they could reach the stadium. Meanwhile scientist Ross Hoffman, at NASA's

Institute for Advanced Concepts, has even grander plans. He has run computer simulations that reveal how relatively small changes in temperature, of just 2–3°C, can drastically alter the path of a **cyclone**. He proposes a network of satellites in Earth orbit that could bombard incipient cyclones with microwaves. Because the clouds in the cyclone are made of water, they would absorb the radiation and heat up. It's a plan that could potentially avert disasters such as that caused by Hurricane Katrina in 2005.

Asteroid deflection

Cosmic impacts – comets and asteroids from space slamming into the planet – present a great danger to the Earth. The good news is that scientists have come up with various strategies to deflect a hazardous object in space before it strikes. Nuclear weapons are an option explored many times by Hollywood, although for them to be effective we would need to be sure that the explosion didn't just blow the asteroid apart into a number of smaller pieces – effectively turning a rifle bullet into a shotgun blast.

'Kinetic impacts' are another option – colliding a massive object into the comet or asteroid to alter its path. A more left-field plan is to paint the asteroid. Different colours absorb and reflect heat from the Sun at different rates – and radiating heat exerts a force that, given sufficient time, can change an asteroid's course.

FUTURE SPACE

Magsails

Whereas solar sails work by harnessing the momentum (see **Inertia and momentum**) carried in the light from the Sun, some scientists believe it might be possible to hitch a ride on the **solar wind** of electrically charged particles that constantly billow out from the surface of our nearest star. But rather than using a solid surface as a sail to catch the particles, these so-called magnetic sail, or 'magsail', spacecraft generate a vast magnetic field around themselves. The field deflects each incoming solar wind particle and the craft in turn receives a kick that boosts its speed.

Standard designs for magsails involve a gigantic loop of superconducting wire (see **Superconductivity**), roughly 50km (30 miles) in diameter. However, this would weigh somewhere in the region of 40 tonnes – a colossal mass to blast into orbit. One way around this that's been investigated by NASA is to use a plasma (see **Plasma physics**) to inflate the magnetic field from a smaller loop of wire – in much the same way as the plasma surrounding the Sun extends the reach of its magnetic field to encompass the whole Solar System.

Space elevator

Launching satellites and spacecraft beyond Earth's atmosphere is extremely expensive, costing thousands of dollars per kilogram. One scheme that space scientists think could reduce this outlay is the so-called space elevator, an orbiting platform from which payloads could be winched into space on a long cable. It's estimated that a space elevator could cut costs to as low as $50 per kg – once the amount spent on construction is recovered (currently estimated at around $40 billion).

The rough principle was first hinted at in 1895 by Russian space pioneer Konstantin Tsiolkovsky, who suggested a tower leading into space. In 1959, Russian Yuri Artsutanov, was the first to suggest a cable suspended from an orbiting counterweight. Modern versions use a 'climber' that travels up and down a fixed cable between Earth's surface and orbit.

A stumbling block in the design of space elevators has been finding a material for the cable strong enough to take its own weight. But the development of **carbon nanotubes** may now have solved this problem.

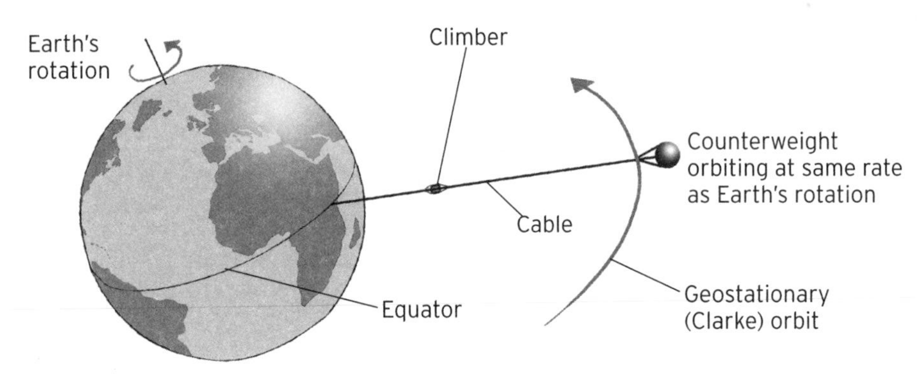

Moonbase

Uses for a base on the Moon would be manifold; the Moon is a natural staging post from which to launch craft outwards to other destinations in the Solar System such as Mars, and a base there would serve as the perfect training ground for future Mars colonists. Others have suggested locating an astronomical observatory on the Moon's far side, shielded from the Earth's glow. And there has been the inevitable military interest. But the enormous cost of building and operating a project on this scale, and the lack of international collaboration, have so far always led to cancellation. That may change in the future; the Moon is thought to be rich in an **isotope** known as helium-3, which is a fuel source for proposed fusion-based **nuclear electricity** systems. Perhaps the first lunar colonists will be miners.

Alien contact

In 1960, an astronomer named Frank Drake began sweeping the skies with a 26-metre (85-foot) radio dish based in Green Bank, West Virginia. But he wasn't looking at stars or galaxies – Drake was searching for radio signals from extraterrestrial life. It was the first in a series of experiments to scan the cosmic airwaves for messages from other civilizations, that go under the umbrella term of SETI (search for extraterrestrial intelligence). However, so far all of these searches have drawn a blank.

SETI scientists believe, given the number of stars in the Milky Way, that it is almost impossible for Earth to be the only world with intelligent life. So much so, they are now working on an array of 350 six-metre radio dishes, that will combine to give the power of a single fully-steerable radio telescope 100 metres (330 feet)across. If there is intelligent life in our corner of the galaxy then this new telescope – called the Allen Telescope Array (after Microsoft co-founder Paul Allen, who provided much of the funding) – may well find it.

Terraforming

Terraforming is a hypothetical procedure by which a planet's atmosphere is engineered to make it like the Earth's. The primary candidate for terraforming is Mars, a world similar in size to our own and known to have water-ice locked away beneath its surface that could be melted. The process would require technologies to first generate significant quantities of oxygen – say by using genetically modified plants (see **Genetically modified organisms**) – and then use known greenhouse gases to increase the surface temperature to melt Mars's ice reserves. Some scientists have suggested this could be done by steering asteroids rich in the greenhouse gas ammonia onto a collision course with Mars. Proponents of terraforming argue that it is a technology we must develop for the very survival of our species – the human population will soon become too great for Earth to sustain on its own.

Von Neumann machines

Our first alien contact might not be with little green men, but rather little shiny robots. That was the prediction of Hungarian-born mathematician John von Neumann, who in the late 1940s suggested that the most efficient way for a civilization to explore the galaxy is to build a fleet of self-replicating robots that could travel from star to star, making scientific observations and building copies of themselves. These robots have since become known as von Neumann machines.

The idea is that the robot would arrive at a distant star system and use the raw materials there, such as minerals from asteroids, to assemble many copies of itself – perhaps using **molecular engineering**. Each of these new craft would then be sent off on journeys to other worlds, where the process repeats. Von Neumann robots could colonize the entire galaxy this way in around 300 million years.

Omega point

The Omega Point is a theory for the ultimate fate of the Universe, put forward by physicist Frank Tipler, at Tulane University, Louisiana. In essence it says that the Universe will ultimately recollapse on itself – a scenario known as the Big Crunch (see **End of the Universe**). By this time he says life will have grown to encompass the entire cosmos and be able to harness all of its resources as a computer. Tipler claims to have used established laws of physics to calculate that the power of this computer will increase exponentially – faster than the Universe dies in the Big Crunch – meaning that the available computing time inside the device will become infinite, even though the time left in the outside Universe is finite. He suggests life could continue for ever as a simulation running within this computer. The theory is controversial, though it does have its supporters. Most scientists, however, have criticized Tipler's final assertion – that the Omega Point is God.

FUTURE HEALTH AND MEDICINE

Cryonics

Cryonics is a technique for freezing the bodies of the dead in supercold (-130°C/-202°F) liquid nitrogen in the hope that one day the technology will exist to revive them. The viability of cryonics relies on the fact that cells in the body take several hours to die after the heart has stopped beating. The idea is that immediately after death the body is frozen and kept in that state until, theoretically, medical science has advanced to the point that, firstly, the freezing process can be reversed and, secondly, that the condition from which the patient 'died' can be treated.

At the end of 2009, the Arizona-based Alcor Life Extension Foundation reportedly had 89 patients preserved in liquid nitrogen, some in full-body cryopreservation and some in 'neuropreservation' – where just the severed head is frozen. Alcor charges $150,000 plus a $500 annual membership fee during life for full-body preservation and $80,000 for neuropreservation.

Immortality

Some scientists are coming to the conclusion that death, at least by natural causes, may soon be a thing of the past. One of them is the American futurologist Ray Kurzweil, who in 2005 told *New Scientist* magazine: 'I'm not planning on dying.' He calls his plan for achieving immortality 'a bridge to a bridge', which means taking advantage of the rapidly accelerating development of medical science in a number of steps, or bridges. His 'bridge one' means taking some 150 **dietary supplements** per day, ten glasses of alkaline water and ten cups of green tea to fight **free radicals**. This he hopes will hold back the years until science reaches 'bridge two' – **genetic medicine** techniques such as **personal genomics** and **gene therapy** by which his likelihood of developing serious illness can be assessed and, if necessary, treated. This, he hopes, will hold the fort until 'bridge three' will be possible – nanomedicine, using a swarm of tiny robots in his bloodstream to constantly repair the damage done to his cells by the ageing process.

Mind control

Cybernetic implants in the brain are enabling disabled people to control their computers by the power of thought. In 2004, 24-year-old Matthew Nagle was fitted with an experimental device called a BrainGate. Nagle had been rendered tetraplegic following a knife attack in which he was stabbed through the vertebrae in his neck. The BrainGate consists of electrodes resting on the surface of the brain's motor cortex – the region controlling movement. The device was programmed to recognize the patterns of electrical activity in the cortex produced when Nagle thought about moving left, right, up and down. Soon he was able to use its output to control his television and his computer – with sufficient dexterity to play the game Tetris.

In 2009, a monkey was able to use the device to control a robotic arm – offering hope to amputees that this technology may one day enable them to control prosthetic limbs, simply by thinking.

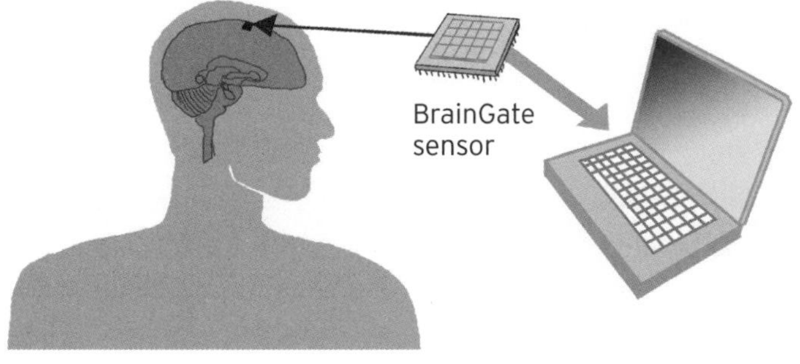

Nanomedicine

Molecular engineering, or nanotechnology, is the science of creating machines measuring just a few hundred nanometres (billionths of a metre) in size. This rapidly developing field promises a plethora of amazing new technologies – one is the idea that medical 'nanorobots' could swim through your bloodstream, repairing cuts and other injuries, fighting infection and even repairing the DNA damage that can lead to cancer. Robert Freitas, at the Institute of Molecular Manufacturing in Palo Alto, California, believes that medical nanorobots, probably constructed from durable forms of carbon such as diamond or graphite, could become a reality by 2030.

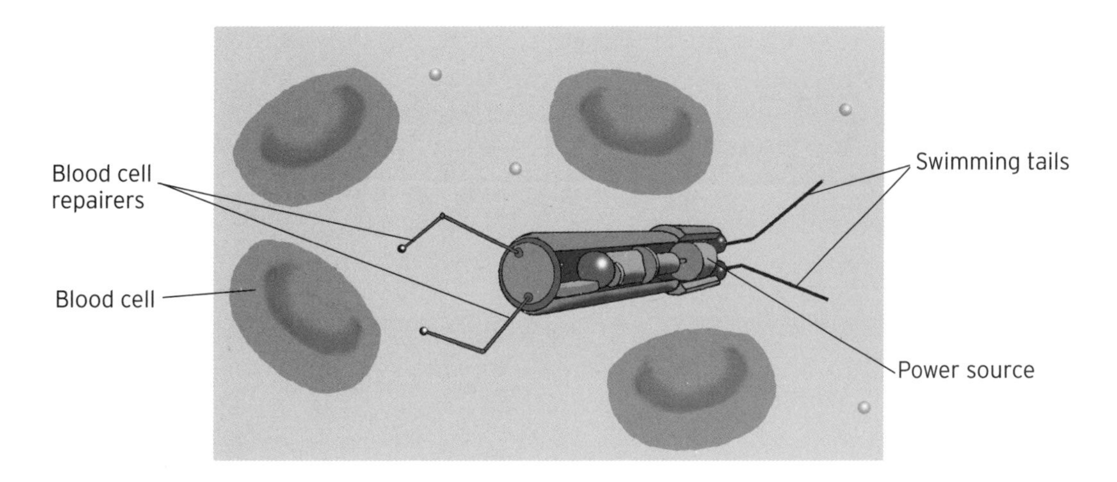

AIDS vaccine

When the HIV virus was confirmed as the source of the disease **AIDS** in 1984, a vaccine (see **Vaccines and antivirals**) was expected within two years. That was over 25 years ago and it still hasn't been found. By the end of 2008, AIDS had killed over 33 million people worldwide. One of the problems is that HIV mutates rapidly – adapting to new environments through **natural selection**, which means that a vaccine developed to fight one strain of the virus may well be ineffective a few generations down the line when that strain has evolved. Research is further confounded because most vaccines work by training the patient's immune system with an inactivated virus – but inactivated HIV has no effect.

There is one glimmer of hope though. In 2008, an HIV-positive patient underwent a bone marrow transplant for leukaemia; twenty months after the procedure his HIV was undetectable. The reason seems to be that the marrow donor had a particular mutation in the gene of a **receptor** called CCR5 – and that's the receptor by which HIV enters cells. Bone marrow transplants are hazardous, and so can't be regarded as a cure in themselves. However, there is hope that **genetic medicine** techniques such as **RNAi** may now be able to target the CCR5 gene.

Spinal reconnection

Spinal injuries are particularly devastating because, while damage to the peripheral nervous system can sometimes be repaired, damage to the central nervous system – which includes the spinal cord – cannot. A promising area of treatment is **stem cell therapy**, in which stem cells are introduced to the injury site and coaxed to grow into new nerve fibres. As of early 2010, research was well underway into a promising stem cell treatment for spinal trauma, known as GRNOPC1. Perfecting spinal reconnection surgery would enable patients experiencing multiple organ failure or severe injury to have whole-body transplants – literally having their head transplanted onto a complete new donor body. This sort of medical technology will also be necessary before patients in 'neuropreservation' (see **Cryonics**) can be revived.

FUTURE SOCIAL SCIENCE

Future of money

The late, great science fiction writer Arthur C. Clarke once predicted that by the year 2016 all the world's currencies would be abolished in favour of a universal unit of money called the megawatt-hour – a unit of energy, used today for measuring electricity generation and consumption. Clarke's reasoning was that conventional money based on a nation's material wealth, gauged for example in terms of its gold reserves, would become irrelevant in a future where the application of technology is set to become the defining measure of national success. The most powerful nations would be the ones with greatest capacity to drive technology – those with the greatest energy reserves. In the UK in 2009 the average cost of a megawatt-hour of electricity was £139 (in the United States it was $120) – enough energy to run a 100W light bulb for 10,000 hours.

Futarchy

Prediction markets are a powerful tool for prognosticating about everything from the weather to who will win political elections. But if Robin Hansen, an economist at George Mason University, Virginia, gets his way then whoever wins the elections might be using markets too – to set their policies. Hansen has come up with an entire system of government based on the predictions of markets. His idea, called a futarchy, is that market investors get to put their money behind key policy decisions and only those policies with the biggest share of the market get to become law. Elected representatives would still handle the day-to-day running of the country, but their policy decisions are made by market speculators.

Hansen suggests that a good measure by which bets on the market could later be settled might the country's GDP (see **Economic measures**), which is a general measure of how successful a nation is. So investors would bet on the policies they believe will increase GDP, and only those that end up increasing GDP will yield a payout to their backers.

Galactic empires

How long would it take for a spacefaring civilization to establish an empire stretching across the galaxy? Seth Shostak, a senior astronomer at the Search for Extraterrestrial Intelligence Institute, in Mountain View, California, draws a parallel with empires that have arisen on Earth throughout history. The Roman Empire was able to span most of Europe, and was able to respond to threats – i.e. march its troops from one side of the empire to the other – on timescales of a few months. Similarly the British Empire extended its influence around the world, but, with better transport technology, was still able to respond to threats in a matter of months.

Shostak thinks the same constraints will apply to empires in space – that the biggest empire possible is one which can be crossed in no more than a couple of months. Even travelling at the speed of light, this puts the nearest stars – which are a few light-years away – beyond reach, and makes galactic empires unlikely, though an 'Empire of the Sun', spanning the planets and moons of the Solar System, could perhaps be sustained.

Exopolitics

How would the arrival of extraterrestrial life affect our society? In 1938, Orson Welles's radio adaptation of the H.G. Wells novel *War of the Worlds* caused mass panic across America when listeners believed they were hearing genuine news reports of a Martian invasion. But what if contact was peaceful? Albert Harrison, a psychologist at the University of California Davis, thinks this is the most likely scenario, and that the detection of an alien signal could be beneficial to society, having a unifying effect that could bind together our fractured world.

FUTURE KNOWLEDGE, INFORMATION AND COMPUTING

Future VR

Virtual reality (VR) creates an artificial environment for the user by stimulating the senses, such as sight and hearing, with images and sounds. New technologies are now emerging that bypass the body's sensory network so computers can interface with the brain directly – by, say, feeding images directly into the visual cortex. Home entertainment firm Sony has patented a technique where this can be done using **ultrasound**. Combined with **mind control** technology, it would allow the user's brain to roam **the Internet** and **virtual worlds** with no need for a mouse or keyboard.

Synthetic telepathy

Just as brain implants have given disabled people **mind control** powers over their computers, some scientists have wondered whether similar technology might one day enable a kind of telepathy between humans.

Dr Robert Freitas, of the Institute for Molecular Manufacturing in California, imagines a swarm of nanoscale robots (see **Molecular engineering**) that would sit inside the human brain, monitoring the thoughts of the host. Any thoughts intended for transmission could be beamed from the nanorobots by ultrasound to a central hub implanted within the skull, which then converts the data to radio and beams it out. Like a cellphone call from inside your head, you would select someone to contact from your mental address book and the technology would then direct your 'call' to the similar hub inside the recipient's head. The nanorobots in the recipient's brain then stimulate the auditory nerves directly, allowing them to hear your thoughts – and then reply to you by reversing the whole process. Freitas speculates that synthetic telepathy could become reality by 2050.

Online consciousness

Could technology ever progress to the point where humans could upload their brains into a computer? This may not be as far away as it seems. Projects today to build **artificial brains** – electronic devices that emulate the function of human brain architecture – are highly advanced. Indeed, futurologist Ian Pearson, of the Futurizon consultancy in Switzerland, believes that by 2050 we could be dispensing with biology and uploading our conscious brains permanently into computers. Patients held in neuropreservation (see **Cryonics**) might one day be revived in this way. It could offer immortality – assuming the computer never crashes.

Galactic Internet

Originally proposed by the science writer Timothy Ferris in the late 1980s, the galactic Internet is a concept that could speed up contact between civilizations dotted around the galaxy. The basic idea is that every intelligent species in the galaxy uploads all the information about its civilization onto what amounts to a web server mounted inside a spacecraft. The craft are launched out into interstellar space and begin broadcasting their information towards other nearby star systems. As each probe receives the data from the others, it beams it home and also adds it to the data that it is transmitting outwards.

In some versions of the idea the server spacecraft are also self-replicating – like **von Neumann machines**. A network of galactic beacons would mean that information about civilizations hundreds of thousands of light-years away could be within just a few hundred light-years of Earth.

Bekenstein bound

Data storage devices are getting ever smaller, but what's the absolute smallest they can get in the future? Physicist Jacob Bekenstein, at the Hebrew University of Jerusalem, has worked out the smallest volume that a given quantity of information can be compressed into. Bekenstein drew upon an analogy with black hole physics, where information obeys rules analogous to the **laws of thermodynamics**. He was able to prove mathematically that if a storage medium contained a quantity of information greater than its so-called Bekenstein bound then dropping it into a black hole would violate the second law of thermodynamics – a sacrosanct pillar of physics. His calculations suggest that a future computer memory chip measuring just a thousandth of a millimetre across could store up to a whopping 10^{22} bits of data (see **Scientific notation**) – that's about a trillion or 10^{12} gigabytes.

The singularity

Strong AI brings the possibility that scientists will one day create a computer brain that far exceeds the power of the human intellect – a so-called superintelligence. When this happens, the superintelligence will be able to redesign its own architecture to make itself more intelligent still. It would be both smarter and quicker than humans. Whereas a biological brain cell can fire around 200 times per second (working at 200Hz), modern computers can already tick over at around 2GHz (2 billion Hz), meaning that all the thoughts possible by a human brain in a year could be processed by a computer in a little over 3 seconds. And with each new

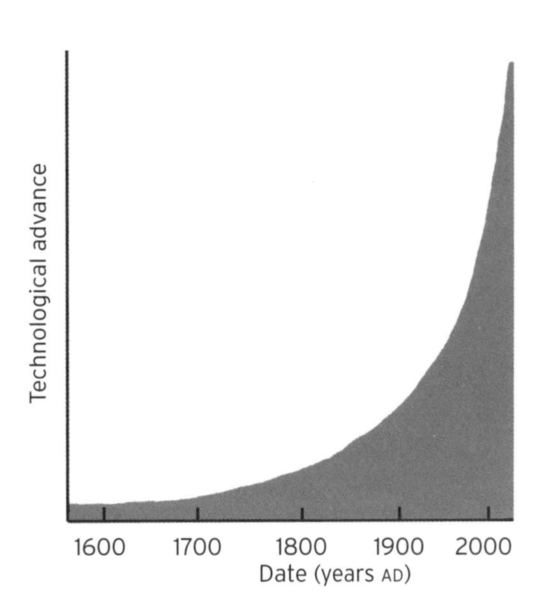

redesign the superintelligence becomes smarter and quicker still, leading to an ever increasing rate of advancement, not just in computing but in all technologies. The rate soon becomes exponential, and tends towards infinitely fast.

The point in time at which technological advancement actually becomes infinite is known as the singularity. The term was first coined by science fiction author Vernor Vinge, and has been popularized by the American inventor and futurologist Ray Kurzweil who predicts it will happen before the end of the 21st century.

FUTURE APPLIED SCIENCE

Automatic highways

How do you fancy driving at 110km/h (70mph) just a metre behind the car in front? Today, this would be near suicidal but in the future it might be the norm thanks to automatic highways. Once you joined the highway, you would transfer control of your vehicle to a centralized computer system that manages all the traffic on the road simultaneously, via radio links. It controls your accelerator, brake and steering using magnetic markers in the tarmac to gauge speed and road position. Cars would travel in 'platoons' of around a dozen vehicles. All the cars in each platoon brake and accelerate in lockstep – which is why it's OK to tailgate – and each platoon stays several stopping distances behind the one in front, limiting the damage if there is an accident. Proponents of the system say it's safer, and allows a greater volume of traffic to use the roads.

In 1997, California carried out trials of such a system on a section of interstate north of San Diego. And another test took place between 2007 and 2010 along California's interstate 805.

Flying cars

It's a science fiction vision that's been with us for years – personal flying transport that you can hop in and out of as easily as you can your own car. Sadly, though, it's yet to materialize in the real world. The closest we've come to date is the Moller Skycar, a four-seater vehicle, powered by eight fan engines, that its designer, Paul Moller, claims will give a top speed of 576km/h (360mph) and a range of 1207km (750 miles). According to Moller, owners would not require a full pilot's licence, owing to the largely automated nature of the controls.

However, after some 40 years of development the Skycar is still only at the prototype stage, and only capable of hovering a few metres above the ground. Moller has yet to gain FAA certification, and his anticipated date for obtaining it is continually slipping. It seems this particular science fiction dream may remain that way for some time yet.

Peak oil

Oil, and the fuels that we derive from it – such as **petrol** – power our transport, electricity generation, industry and agriculture. But in 1956, American geophysicist M. King Hubbert predicted that supplies of oil are not going to last for ever. His concept, known as the 'Hubbert peak theory', predicted that oil production in the continental United States would reach a maximum during the 1970s. And indeed it did – peaking at 10.2 million barrels per day, and it has been in steady decline ever since.

Now experts worry about peak oil as a phenomenon that might affect oil supplies the world over. A report in 2009 by the UK Energy Research Council even suggested that global oil production could start to go into decline by 2020. Peak oil is often cited as a reason to develop alternative energy sources to oil, such as hydrogen power and **biofuel**. Though the big concern is that the oil may run out before these new technologies can reach maturity.

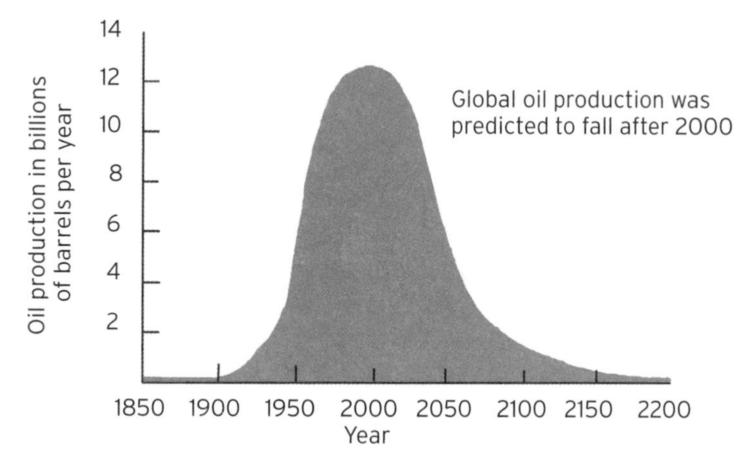

Memory prosthetics

Researchers at the University of Southampton are working on a kind of artificial body part with a difference: memory prosthetics. At present, their technology is an extension of **lifelogging**, using wearable aids to record your life experiences by video and audio, and keeping a time-stamped record of your email, your web-browser history, even what documents and files you've had open on your computer. Key to the project is clever search technology that is able to work across different media. Say, for example, you were trying to recall a piece of information that you read somewhere. You don't remember where or when, but you do remember that you were listening to Handel's 'Sarabande' at the time. The software finds when you have listened to that MP3 and then cross-matches those times with the documents and webpages you were viewing. The research promises memory aids for people working in business, education and research – and for the average person prone to misplacing their keys.

Free energy

Constructing a device that can conjure energy from thin air is a challenge that's baffled the great thinkers for many centuries. But in 2007 Dr Thorsten Emig, of

Cologne University, in Germany, came up with a free energy machine that could actually work. His idea is based on the **Casimir effect**, in which two metal plates a short distance apart in a vacuum feel a small attractive force towards each other. Emig has designed a 'Casimir ratchet' that can extract useable motion from this effect; it works by substituting smooth plates for corrugated ones to introduce a lateral force making the plates slip past one another. By making the corrugations asymmetric, Emig keeps the slipping motion in one direction so that it can be harnessed.

A team at the University of California, Riverside, has already demonstrated the lateral Casimir force experimentally. Emig now believes that his design could be used to power tiny nanorobots (see **Molecular engineering**), which could have umpteen applications – such as **nanomedicine**.

Geo-engineering

Some of the more pessimistic climatologists believe that climate change on Earth has already passed the 'tipping point' beyond which it is not possible to halt the slide into global meltdown as **sea-level rise** and scorching temperatures bring civilization to its knees. But other groups of scientists think that even if the tipping point

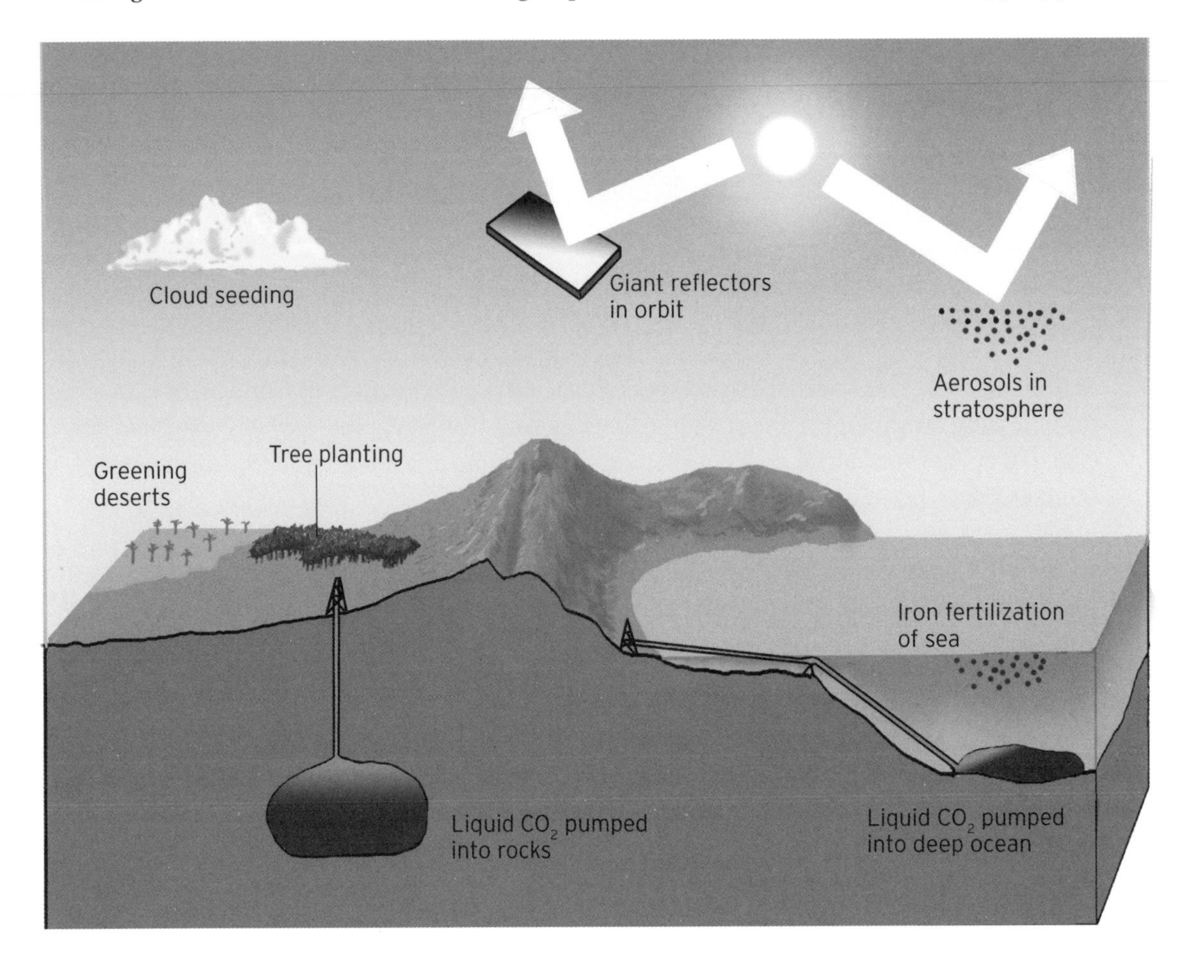

has been and gone there might still be hope for the planet in the form of geo-engineering, which is superscale design-and-build work that could, it is hoped, fix our broken planet and reverse the effects of climate change.

The ideas put forward include lofting giant sunshades into space to block out some of the Sun's light; building a network of industrial plants around the globe that would suck in carbon dioxide, compress it into liquid, and then pump it underground; and tipping gallons of fertilizer into the oceans to encourage the growth of marine plant life to absorb carbon dioxide from the air by photosynthesis. Critics, however, worry about the potential unintended consequences of such schemes.

Robot soldiers

What if wars around the world could be settled by robots, saving humans from having to risk their lives in front-line combat duties? It sounds rather like a scenario from the *Terminator* movies, but it's one that's edging increasingly closer to reality. In 2009, the US military was estimated to own some 7,000 unmanned combat air vehicles (**UCAV**s) and 12,000 robot ground vehicles, used for reconnaissance and for defusing explosive devices. Many of these have seen action. In 2007, the US Air Force began deploying 'Reapers' in Iraq and Afghanistan – these are armed UCAVs, each capable of carrying missiles and bombs, and are remote-controlled by pilots on the ground. But now defence contractors are building autonomy into their mechanical soldiers. The US Defense Advanced Research Projects Agency (DARPA) is reportedly working on robots that can destroy human opponents on the ground by hunting in packs, communicating with one another and making many of their own decisions. Experts believe robots such as these will bring a revolution to warfare comparable to the discovery of gunpowder or nuclear weapons.

Grey goo

Grey goo is the name given to a future disaster scenario, where nanotechnology (see **Molecular engineering**) runs amok, turning the entire planet into a mass of self-replicating nanorobots. The term 'grey goo' was coined by the father of nanotechnology Eric Drexler in his book *Engines of Creation* and was popularized by the late science fiction author Michael Crichton in his novel *Prey*.

Nanotechnology works by building miniature machines that can carry out tasks on the tiniest of scales – for example, **nanomedicine**. But if a nanorobot was programmed to build copies of itself then what would happen if the robot got out of control – is it possible that every molecule of matter on the planet could be consumed in its quest for building materials?

In fact, there is as yet no such thing as a self-replicating nanorobot, and even if we did know how to make one, *self*-replication would be inefficient compared with simply running off nanorobots in a dedicated – and stationary – factory. The only way the grey-goo scenario could actually happen is if someone built and released self-replicating nanorobots deliberately – as a form of weapon.

INDEX

Listings correspond to text entry headings; **bold** denotes sub-section headings

L

M

N

O

P

Q

Published by Firefly Books Ltd. 2014

First Paperback Printing, 2014

Publisher Cataloging-in-Publication Data (U.S.)

Parsons, Paul.
Science 1001 : absolutely everything that matters in science / Paul Parsons.
[400] pages : ill. ; cm.
Includes index.
Summary: A comprehensive guide, spanning all of the key scientific disciplines including physics, chemistry, biology, the earth, space, health and medicine, social science, information science, the applied sciences and futurology.
ISBN-13: 978-1-77085-501-4 (pbk.)
1. Science – Popular works. 2. Technology – Popular works. I. Title.
500 dc23 Q162.P37 2014

Library and Archives Canada Cataloguing in Publication

Parsons, Paul, 1971-
Science 1001 : absolutely everything that matters in science / Paul Parsons.
ISBN 978-1-55407-718-2 bound). – ISBN 978-1-77085-501-4 (pbk.)
1. Science–Popular works. I. Title.
Q162.P373 2010 500 C2010-900219-9

Published in the United States by
Firefly Books (U.S.) Inc.
P.O. Box 1338, Ellicott Station
Buffalo, New York 14205

Published in Canada by
Firefly Books Ltd.
50 Staples Avenue, Unit 1
Richmond Hill, Ontario L4B 0A7

Interior design: Patrick Nugent
All illustrations by Patrick Nugent except pp2–3 Martin Krzywinski/Science Photo Library, and p303 symbol of the Humanist Movement
All jacket images © Shutterstock

Printed in China